ELECTROMAGNETIC WAVES

David H. Staelin

Ann W. Morgenthaler

Jin Au Kong

Massachusetts Institute of Technology

An Alan R. Apt Book

PRENTICE HALL, Upper Saddle River, New Jersey 07458

Library of Congress Cataloging-in-Publication Data
Staelin, David H.
    Electromagnetic waves / David H. Staelin, Ann W. Morgenthaler, Jin
Au Kong.
        p. cm.
    Includes index.
    ISBN 0-13-225871-4
    1. Electromagnetic waves. 2. Electrodynamics. I. Morgenthaler,
Ann W. II. Kong, Jin Au. III. Title.
QC661.S74   1994
539.2—dc20   93-15705
                                                          CIP

Publisher: Alan Apt
Production Editor: Bayani Mendoza de Leon
Cover designer: Bruce Kenselaar
Cover concept: Katharine E. Staelin
Copy editor: Shirley Michaels
Prepress buyer: Linda Behrens
Manufacturing buyer: Dave Dickey
Supplements editor: Alice Dworkin
Editorial assistant: Shirley McGuire

©1998 by Prentice-Hall, Inc.
A Pearson Education Company
Upper Saddle River, NJ 07458

The author and publisher of this book have used their best efforts in preparing this book. These efforts include the development, research, and testing of the theories and formulas to determine their effectiveness. The author and publisher shall not be liable in any event for incidental or consequential damages in connection with, or arising out of, the furnishing, performance, or use of these formulas.

ISBN 0-13-225871-4

Printed in the United States of America

ISBN 0-13-225871-4

9 780132 258715

Prentice-Hall International (UK) Limited,London
Prentice-Hall of Australia Pty. Limited, Sydney
Prentice-Hall Canada Inc., Toronto
Prentice-Hall Hispanoamericana, S.A., Mexico
Prentice-Hall of India Private Limited, New Delhi
Prentice-Hall of Japan, Inc., Tokyo
Pearson Education Asia Pte. Ltd., Singapore
Editora Prentice-Hall do Brasil, Ltda., Rio de Janeiro

To our spouses:

Ellen,

Carey,

and

Wen-Yuan

CONTENTS

Contents

PREFACE

This text presents electromagnetic wave phenomena so as to develop the reader's competence to solve a broad range of practial problems. It is sufficient for students to have had prior exposure to Maxwell's equations (second-term freshman physics), integral theorems and differential operators (second-term freshman calculus), and introductory circuit theory. Although the material in this book was developed principally for a required course taken by undergraduate electrical engineering students at the Massachusetts Institute of Technology, the text can also be used for a more advanced undergraduate elective or a first-year graduate course. By using many of the physics, circuit, signal, and electromagnetic concepts taught in other courses, this text provides a "capstone" experience integrating a wide range of material students often regard as unrelated. Sufficient explanations and examples have been provided so that practicing engineers should also find it a useful reference and a tool for self-teaching.

The text has evolved continuously for more than thirty years since its predecessor, *Electromagnetic Energy Transmission and Radiation* by Adler, Chu, and Fano, presented similar material with a strong emphasis on *TEM* transmission lines, electromagnetic plane waves at boundaries, and dipole antennas. Continued student interest in gaining a deeper, more physical understanding of electromagnetic theory and of more modern topics, such as optical fibers, forces, more complex antenna, complex media, and acoustics, has gradually let us to the present formulation which emphasizes concepts, problem solving techniques, and examples having wide applicability. In meeting this challenge, the work has benefited greatly from the contributions and insight of many faculty members who have taught this subject with us over the past few decades.

As the text progresses from simple wave propagation in unbounded free space to antenna and resonator design, students are also exposed to a considerable array of mathematical concepts, raning from partial differential equations, vector calculus, complex vectors, and different equations to separation of variables, orthogonal function expansions, perturbation techniques, and duality. Basic electromagnetic theorems (Poynting, energy, uniqueness, and reciprocity) are also presented at appropraite places in the text and explained from a physical perspective. This text repeatedly stresses both the connection between wave phonomena and the quivalent circuit models and the use of insightful "back-of-the-envelope" approximations. In some cases, alternative problem-solving approaches are compared, and both mathematical and physically intuitive approaches are taken. Interesting practical examples are provided throughout and also underlie many of the problems provided at the end of each chapter.

The principal factors contributing to the novelty of the text are the improved breadth and depth of the subject achieved through efficient and clear presentation, the emphasis on generic problem-solving techniques, and the reliance upon only basic physics and mathematics prerequisites rather than electrostatics, magnetostatics, and quasistatics.

The text begins directly with Maxwell's equations and their solutions in unbounded free space. The fundamental concepts of plane waves, phasors, polarization, energy, power, and force are also presented in the first chapter and applied repeatedly throughout the text to problems with progressively more complex boundary conditions. Chapter Two focuses on radiation from Hertzian dipoles and simple two-dipole arrays. This early treatment of radiation motivates students, promotes intuition, gives practice with phasor manipulation, and makes more plausible phenomena such as polarizable media, Brewster's angle, and current sheets, which are all introduced before discussing antennas. In the third chapter, the discussion of plane waves in unbounded vacuum is extended to unbounded media characterized by polarizability, anisotropy, loss, inhomogeneity, dispersion, or gyrotropicity, such as crystals, plasmas, and ferrites. Chapter Four derives boundary conditions from Maxwell's equations, enabling a discussion of reflection and transmission at normal and oblique incidence from planar interfaces.

Discussion of parallel-plate, coaxial cable and arbitrary *TEM* lines from a physical perspective forms the basis of Chapter Five, which includes time-domain and discrete-element line analysis. Impedance, *ABCD* matrices, and Smith chart techniques are used to discuss sinusoidal steady state *TEM* line behavior in Chapter Six, and transient wave behavior on *TEM* lines is also presented. Increasingly restrictive boundary conditions are applied to media in Chapters Seven and Eight, which describe structures bounded in two dimensions (metallic and dielectric waveguides) and three dimensions (resonators). Chapter Nine discusses antennas, where the number of modes is usually so great that a different (approximate) approach is required. After discussing both elementary and complex antennas, analyzing them as transmitting and receiving circuits, and introducing diffraction, the text concludes in Chapter Ten with a review of wave behavior. Essentially all of the phenomena encountered in Chapters One through Nine are succinctly presented anew in the context of acoustic waves. This approach also clarifies which electromagnetic phenomena result from the nature of waves in general, and which are unique to electromagnetism.

Undergraduate subjects would typically omit those sections of the text marked by an asterisk. Additional sections might be omitted without penalty, such as 1.9, 3.3, 3.6, 4.4, 5.6, 6.2, 6.4, 6.5, 7.2, and any consecutive sections in each of Chapters 8–10 that do not precede a section of interest within that chapter. Although the sections treated can be chosen still more erratically, care should then be taken to ensure that critical material has not been omitted.

Of the many faculty who have aided in the development of this text, special recognition must be given to K. N. Stevens for his contributions to the chapter on acoustics, and to A. Bers, F. R. Morgenthaler, and L. D. Smullin for their contributions in the area of resonators. We are also indebted to R. Briggs, J. G. Fujimoto, H. A. Haus, E. P. Ippen, J. R. Melcher, R. R. Parker, W. T. Peake, C. M. Rappaport, and many other faculty who have helped teach and shape the course. Special mention should also be given to C. F. Smith and K. Lai, who helped type and retype many versions of the text as it evolved. Finally, the enthusiasm and feedback of the many students of this subject have been absolutely essential and will continue to help refine the text in its future editions.

David H. Staelin
Ann W. Morgenthaler
Jin Au Kong

Cambridge, Massachusetts

1

INTRODUCTION TO MAXWELL'S EQUATIONS AND WAVES

1.1 INTRODUCTION

Classical electromagnetic theory is an elegant, accurate description of electric and magnetic phenomena, dependent on only a few simple equations. The most common formulation of the laws of electricity and magnetism is credited to James Clerk Maxwell (1831–1879), who in 1864 unified observations of Michael Faraday (1791–1867), Karl Friedrich Gauss (1777–1855), and Andre-Marie Ampere (1775–1836). In his honor, the fundamental equations governing the behavior of electromagnetic fields are called Maxwell's equations. The fascinating historical development of electromagnetism has been described by many authors, but the basic justification for Maxwell's equations is their enormous success in explaining physical experiments.

Because Maxwell's equations are a set of coupled, partial differential equations, solving problems in electrodynamics is intimately connected with understanding partial differential equations. But mathematical understanding, though necessary, is not sufficient. Maxwell's equations are rich in physical meaning, and gaining an appreciation for these equations is closely linked with developing physical intuition. Therefore, this text explores both the physical and mathematical consequences of Maxwell's equations by using them to solve problems in a variety of media (e.g., dielectric materials, conductors) with differing boundary conditions.

In this chapter, a general discussion of Maxwell's equations is presented, along with a characterization of the simplest medium possible—unbounded vacuum or free space. In subsequent chapters, solutions to Maxwell's equations are described for increasingly restrictive boundary conditions.

1.2 MAXWELL'S EQUATIONS AND THE LORENTZ FORCE LAW

Electric charges have been found experimentally to move in response to forces that are linearly related to quantities called electric and magnetic fields existing at every point in space at every instant of time. Maxwell's equations describe the relation of these electric and magnetic fields to each other and to the position and motion of charged particles. Although these equations are simple, they are sufficient to understand the full range of electromagnetic phenomena discussed in this text.

In differential form,[1] *Maxwell's equations* are written as:

$$\nabla \times \overline{E}(\overline{r}, t) = -\frac{\partial \overline{B}(\overline{r}, t)}{\partial t} \qquad \text{(Faraday's law)} \qquad (1.2.1)$$

$$\nabla \times \overline{H}(\overline{r}, t) = \frac{\partial \overline{D}(\overline{r}, t)}{\partial t} + \overline{J}(\overline{r}, t) \qquad \text{(Ampere's law)} \qquad (1.2.2)$$

$$\nabla \cdot \overline{D}(\overline{r}, t) = \rho(\overline{r}, t) \qquad \text{(Gauss's law)} \qquad (1.2.3)$$

$$\nabla \cdot \overline{B}(\overline{r}, t) = 0 \qquad \text{(Gauss's law)} \qquad (1.2.4)$$

where the field variables are defined as:

$\overline{E}$: electric field   (volts/meter; V/m)
$\overline{H}$: magnetic field   (amperes/meter; A/m)
$\overline{B}$: magnetic flux density   (tesla; T)
$\overline{D}$: electric displacement   (coulombs/meter^2; C/m^2)
$\overline{J}$: electric current density   (amperes/meter^2; A/m^2)
ρ: electric charge density   (coulombs/meter^3; C/m^3)

Except for the scalar charge density $\rho(\overline{r}, t)$, these fields are all vectors; that is, $\overline{E}$, $\overline{H}$, $\overline{B}$, $\overline{D}$, and $\overline{J}$ assign not only a magnitude but also a direction to every point in space and time. This text uses mks (SI) units throughout, though it is noted that other systems of units (e.g., cgs) are possible, in which case Maxwell's equations appear in slightly altered form.[2]

The charge and current sources which give rise to electric and magnetic fields

---

1. See Appendix B for definitions of the differential operators.

2. The cgs form of Maxwell's equations is:

$$\nabla \cdot \overline{D} = 4\pi\rho$$

$$\nabla \cdot \overline{B} = 0$$

$$\nabla \times \overline{E} = -\frac{1}{c}\frac{\partial \overline{B}}{\partial t}$$

$$\nabla \times \overline{H} = \frac{4\pi\overline{J}}{c} + \frac{1}{c}\frac{\partial \overline{D}}{\partial t}$$

where c is the speed of light. The fields are measured in the following units: $\overline{E}$ in statvolts/cm, $\overline{D}$ in statcoulombs/cm^2, $\overline{H}$ in oersteds, $\overline{B}$ in gauss, ρ in statcoulombs/cm^3, and $\overline{J}$ in statamperes/cm^2. One coulomb equals 3×10^9 statcoulombs, one ampere equals 3×10^9 statamperes, one volt equals 1/300 statvolt, one tesla equals 10^4 gauss, and one ampere/m equals $4\pi \times 10^{-3}$ oersteds.

are $\rho(\bar{r}, t)$ and $\bar{J}(\bar{r}, t)$, respectively. Often ρ and $\bar{J}$ are completely specified in a problem, and the field quantities are then determined by Maxwell's equations subject to boundary conditions. A more difficult problem arises when ρ and $\bar{J}$ are **not** uniquely specified, but are able to change in response to the fields themselves. Maxwell's equations must then be solved self-consistently so as to satisfy whatever constraints are imposed. This text is primarily concerned with situations in which ρ and $\bar{J}$ are known a priori (or can be accurately estimated, as in the case of the current distribution on certain antennas).

Maxwell's equations also contain a very important relationship between the charge and the current. Using the vector identity $\nabla \cdot (\nabla \times \bar{A}) = 0$ for any vector $\bar{A}$, the divergence of Ampere's law is taken:

$$\nabla \cdot (\nabla \times \bar{H}) = \nabla \cdot \frac{\partial \bar{D}}{\partial t} + \nabla \cdot \bar{J} = 0$$

The order of the time and space derivatives operating on $\bar{D}$ may be interchanged, and use of Gauss's law (1.2.3) then yields

$$\frac{\partial \rho}{\partial t} + \nabla \cdot \bar{J} = 0 \tag{1.2.5}$$

which is the *continuity equation* for the conservation of charge and current. This equation has the simple physical interpretation that charge distributions which change in time give rise to currents, and has the important consequence that $\rho(\bar{r}, t)$ and $\bar{J}(\bar{r}, t)$ may not be specified independently.

The *Lorentz force law* relates the electromagnetic fields to measurable forces. The force $\bar{f}$ on a positive charge q moving at velocity $\bar{v}$ through the fields $\bar{E}$ and $\bar{B}$ is given by

$$\bar{f} = q(\bar{E} + \bar{v} \times \bar{B}) \tag{1.2.6}$$

In mks units, $\bar{f}$ is measured in newtons, q in coulombs, and $\bar{v}$ in meters/second (m/s). If the charge is not discrete, but is instead given by a charge density ρ, then the Lorentz force law relates the *force density* $\bar{F}$ (N/m^3) to ρ and $\bar{J}$:

$$\bar{F} = \rho \bar{E} + \bar{J} \times \bar{B} \tag{1.2.7}$$

where (1.2.6) is the volume integral of (1.2.7).

In order to solve electrodynamics problems, a further set of relationships between the field quantities $\bar{D}$, $\bar{E}$, $\bar{B}$, and $\bar{H}$ must be established as there are more variables than equations contained within Maxwell's equations (1.2.1)–(1.2.4) These additional equations are known as *constitutive relations*, and they arise from physical consideration of the media in the problem to be solved. Initially, we consider only fields in vacuum; other media are discussed beginning in Chapter 3. In vacuum, the constitutive relations are particularly simple:

$$\bar{D}(\bar{r}, t) = \epsilon_0 \bar{E}(\bar{r}, t) \tag{1.2.8}$$

$$\overline{B}(\overline{r}, t) = \mu_0 \overline{H}(\overline{r}, t) \qquad (1.2.9)$$

where ϵ_0 is called the *permittivity of free space* and μ_0 is the *permeability of free space*. In mks units,

$$\epsilon_0 = 8.8542 \times 10^{-12} \text{ (farads/m)}$$

$$\mu_0 = 4\pi \times 10^{-7} \text{ (henries/m)}$$

For *linear media*, the constitutive relations may be written

$$\overline{D}(\overline{r}, t) = \epsilon(\overline{r})\overline{E}(\overline{r}, t) \qquad (1.2.10)$$

$$\overline{B}(\overline{r}, t) = \mu(\overline{r})\overline{H}(\overline{r}, t) \qquad (1.2.11)$$

where ϵ and μ are independent of field strength but may be functions of position. In this type of media, Maxwell's equations are linear, and therefore the *superposition principle* applies. That is, if $\{\overline{E}_1(\overline{r}, t),\ \overline{H}_1(\overline{r}, t),\ \overline{B}_1(\overline{r}, t),\ \overline{D}_1(\overline{r}, t)\}$ is a solution of Maxwell's equations for the sources $\{\rho_1(\overline{r}, t),\ \overline{J}_1(\overline{r}, t)\}$, and $\{\overline{E}_2(\overline{r}, t),$ $\overline{H}_2(\overline{r}, t),\ \overline{B}_2(\overline{r}, t),\ \overline{D}_2(\overline{r}, t)\}$ is a solution for the sources $\{\rho_2(\overline{r}, t),\ \overline{J}_2(\overline{r}, t)\}$, then any linear combination of these two solutions is **also** a solution to Maxwell's equations for the corresponding linear combination of the sources.

Superposition is extremely useful in solving problems. In general, Maxwell's equations may be solved by first looking for a *particular solution* for the source distribution specified. The particular solution is **any** solution that satisfies Maxwell's equations, and is often constructed by breaking up the current and/or charge into simpler distributions for which Maxwell's equations are more readily solved. The particular solutions for the individual source distributions are then added together (superposed) to give the total particular solution, which has been established without regard to boundary conditions and is therefore not unique. But because Maxwell's equations are not fully solved until **all** the boundary conditions are satisfied, it is often necessary to add a *homogeneous solution* to the particular solution. The homogeneous solution is found by setting ρ and $\overline{J}$ to zero and again solving Maxwell's equations. Because there are adjustable coefficients in the homogeneous solution, the total solution (homogeneous plus particular) will be a unique solution satisfying both Maxwell's equations and the boundary conditions if the problem is well-posed.[3]

That superposition applies for linear media can be easily shown by adding Maxwell's equations (1.2.1)–(1.2.4) for the two separate solutions:

$$\nabla \times \overline{E}_1 = -\mu \frac{\partial \overline{H}_1}{\partial t} - \overline{H}_1 \frac{\partial \mu}{\partial t} \qquad \nabla \times \overline{E}_2 = -\mu \frac{\partial \overline{H}_2}{\partial t} - \overline{H}_2 \frac{\partial \mu}{\partial t}$$

$$\nabla \times \overline{H}_1 = \epsilon \frac{\partial \overline{E}_1}{\partial t} + \overline{E}_1 \frac{\partial \epsilon}{\partial t} + \overline{J}_1 \qquad \nabla \times \overline{H}_2 = \epsilon \frac{\partial \overline{E}_2}{\partial t} + \overline{E}_2 \frac{\partial \epsilon}{\partial t} + \overline{J}_2$$

---

3. See Section 1.8 for a further discussion of uniqueness.

$$\epsilon \nabla \cdot \overline{E}_1 + \overline{E}_1 \cdot \nabla \epsilon = \rho_1 \qquad \epsilon \nabla \cdot \overline{E}_2 + \overline{E}_2 \cdot \nabla \epsilon = \rho_2$$

$$\mu \nabla \cdot \overline{H}_1 + \overline{H}_1 \cdot \nabla \mu = 0 \qquad \mu \nabla \cdot \overline{H}_2 + \overline{H}_2 \cdot \nabla \mu = 0 \qquad (1.2.12)$$

Multiplying one set of Maxwell's equations by an arbitrary constant α and the other set by another constant β and adding the corresponding equations yields:

$$\nabla \times (\alpha \overline{E}_1 + \beta \overline{E}_2) = -\mu \frac{\partial(\alpha \overline{H}_1 + \beta \overline{H}_2)}{\partial t} - (\alpha \overline{H}_1 + \beta \overline{H}_2) \frac{\partial \mu}{\partial t}$$

$$\nabla \times (\alpha \overline{H}_1 + \beta \overline{H}_2) = \epsilon \frac{\partial(\alpha \overline{E}_1 + \beta \overline{E}_2)}{\partial t} + (\alpha \overline{E}_1 + \beta \overline{E}_2) \frac{\partial \epsilon}{\partial t} +$$

$$(\alpha \overline{J}_1 + \beta \overline{J}_2)$$

$$\epsilon \nabla \cdot (\alpha \overline{E}_1 + \beta \overline{E}_2) + (\alpha \overline{E}_1 + \beta \overline{E}_2) \cdot \nabla \epsilon = (\alpha \rho_1 + \beta \rho_2)$$

$$\mu \nabla \cdot (\alpha \overline{H}_1 + \beta \overline{H}_2) + (\alpha \overline{H}_1 + \beta \overline{H}_2) \cdot \nabla \mu = 0 \qquad (1.2.13)$$

Examination of (1.2.13) makes clear that the superposition of the two solutions (i.e., $\{\alpha \overline{E}_1 + \beta \overline{E}_2, \alpha \overline{H}_1 + \beta \overline{H}_2, \alpha \overline{B}_1 + \beta \overline{B}_2, \alpha \overline{D}_1 + \beta \overline{D}_2\}$) is also a solution, with charge and current sources given by $\alpha \rho_1 + \beta \rho_2$ and $\alpha \overline{J}_1 + \beta \overline{J}_2$ respectively, even when ϵ and μ are functions of position and time. For vacuum and most materials at field strengths normally encountered, ϵ and μ are not functions of the electric and/or magnetic fields, and therefore superposition holds. Exceptions include nonlinear media such as ferrites, certain dielectrics exposed to high field strengths generated by powerful lasers, and even vacuum itself when the field strengths approach those needed to generate fundamental particles like electrons and positrons. All physical media become nonlinear in sufficiently strong fields.

Example 1.2.1

If $\overline{E}(x, y, z, t) = \hat{y} E_0 \cos(\omega t - kz)$, what is the magnetic field $\overline{H}(x, y, z, t)$ in vacuum?

Solution: From Faraday's law (1.2.1)

$$\frac{\partial \overline{B}}{\partial t} = -\nabla \times \overline{E} = - \begin{vmatrix} \hat{x} & \hat{y} & \hat{z} \\ \partial/\partial x & \partial/\partial y & \partial/\partial z \\ E_x & E_y & E_z \end{vmatrix}$$

$$= \hat{x} \frac{\partial}{\partial z} E_y = \hat{x} k E_0 \sin(\omega t - kz)$$

Therefore,

$$\overline{H} = \frac{\overline{B}}{\mu_0} = -\hat{x} \frac{k}{\omega \mu_0} E_0 \cos(\omega t - kz) + \overline{H}_0$$

where $\overline{H}_0$ is a constant of integration.

1.3 UNIFORM PLANE WAVES

Perhaps the most important implication of Maxwell's equations is the prediction of waves that can carry energy and information across space, even in vacuum. The nature of these waves follows directly from Maxwell's equations (1.2.1)–(1.2.4) and the constitutive laws for vacuum (1.2.8)–(1.2.9) in the absence of sources ρ and $\overline{J}$:

$$\nabla \times \overline{E} = -\mu_0 \frac{\partial \overline{H}}{\partial t} \tag{1.3.1}$$

$$\nabla \times \overline{H} = \epsilon_0 \frac{\partial \overline{E}}{\partial t} \tag{1.3.2}$$

$$\nabla \cdot \overline{E} = 0 \tag{1.3.3}$$

$$\nabla \cdot \overline{H} = 0 \tag{1.3.4}$$

If we have set ρ and $\overline{J}$ to zero in Maxwell's equations, the question naturally arises as to what is producing these waves. In general, when solving electrodynamics problems, we confine our attention to a restricted region of interest in which we wish to know the solution to Maxwell's equations. It is in this region of interest (which may be infinitely large) that the charge and current densities vanish; obviously, we must have nonzero ρ and $\overline{J}$ **outside** the region generating the waves. We eventually account for all of the charges and currents outside the region of interest by specifying boundary conditions at the surface of this region. This need for boundary conditions is not surprising, since boundary conditions are an essential ingredient for uniquely solving any partial differential equation.

We can eliminate $\overline{H}$ by taking the curl of (1.3.1) and substituting (1.3.2):

$$\nabla \times (\nabla \times \overline{E}) = -\mu_0 \epsilon_0 \frac{\partial^2 \overline{E}}{\partial t^2} \tag{1.3.5}$$

Applying the vector identity

$$\nabla \times (\nabla \times \overline{A}) = \nabla(\nabla \cdot \overline{A}) - \nabla^2 \overline{A} \tag{1.3.6}$$

for an arbitrary vector $\overline{A}$ to (1.3.5) and noting that $\nabla \cdot \overline{E} = 0$ in a source-free region, we obtain a *wave equation* for $\overline{E}$:

$$\nabla^2 \overline{E} - \mu_0 \epsilon_0 \frac{\partial^2 \overline{E}}{\partial t^2} = 0 \tag{1.3.7}$$

The symmetry of Maxwell's equations (1.3.1)–(1.3.4) with respect to $\overline{E}$ and $\overline{H}$ permits derivation of an identical wave equation for $\overline{H}$.

In the wave equation (1.3.7), ∇^2 is the linear, three-dimensional *Laplacian* operator. In Cartesian coordinates,[4] the Laplacian is

$$\nabla^2 = \frac{\partial^2}{\partial x^2} + \frac{\partial^2}{\partial y^2} + \frac{\partial^2}{\partial z^2}$$

The most general solution to the wave equation is

$$\overline{E}(\overline{r}, t) = \overline{E}_0 f(\omega t - \overline{k} \cdot \overline{r}) \tag{1.3.8}$$

where $\overline{E}_0$ is a constant vector, f is **any** function of the argument $\omega t - \overline{k} \cdot \overline{r}$ and $\overline{k} \cdot \overline{r} = k_x x + k_y y + k_z z$ in Cartesian coordinates. The dependence of the solution (1.3.8) on the time t is the same as its dependence on the spatial coordinate $\overline{r}$ except for differing values of the constants $\overline{k}$ and ω.

The general wave solution (1.3.8) must satisfy the wave equation, and its substitution into (1.3.7) yields

$$(-k_x^2 - k_y^2 - k_z^2 + \omega^2 \mu_0 \epsilon_0) \overline{E}_0 f'' = 0 \tag{1.3.9}$$

where f'' is the second derivative of f with respect to its entire argument $\omega t - \overline{k} \cdot \overline{r}$. For nontrivial solutions, f'' is nonzero, and (1.3.9) leads to the *dispersion relation* for vacuum:

$$k^2 = \omega^2 \mu_0 \epsilon_0 \tag{1.3.10}$$

In Cartesian coordinates, $\overline{k} = \hat{x} k_x + \hat{y} k_y + \hat{z} k_z$ so $k^2 = \overline{k} \cdot \overline{k} = k_x^2 + k_y^2 + k_z^2$.

Note that not every solution to the wave equation is also a solution to every one of Maxwell's equations. The general solution (1.3.8) must **also** satisfy Gauss's law in free space (1.3.3), or $\nabla \cdot \overline{E}(\overline{r}, t) = \overline{k} \cdot \overline{E}_0 = 0$, to be accepted as a solution of Maxwell's equations. This means that the direction of $\overline{k}$ must be perpendicular to the direction in which the solution points. For example, $\overline{E} = \hat{z} E_0 \cos(\omega t - kz)$ satisfies the wave equation (1.3.7) but violates $\nabla \cdot \overline{E} = 0$ and must therefore be discarded as a solution.

How does the electric field $\overline{E}$ appear in space and time? We choose to examine a solution to the wave equation that has a particularly simple form:

$$\overline{E} = \hat{x} E_0 \cos(\omega t - kz) \tag{1.3.11}$$

We note first that the electric field points in the $\hat{x}$ direction for all time, so the wave is said to be $\hat{x}$-polarized. If we look at $\overline{E}$ at a particular point in space (say $z = 0$), then

$$\overline{E}(z = 0, t) = \hat{x} E_0 \cos \omega t$$

which is plotted in Figure 1.1(a). Therefore, the electric field oscillates at every point in space with *angular frequency* ω measured in radians/ second. The angular frequency ω is equal to $2\pi f$, where f is the *frequency* of the wave measured in

---

4. See Appendix B for the form of the differential operators in other coordinate systems.

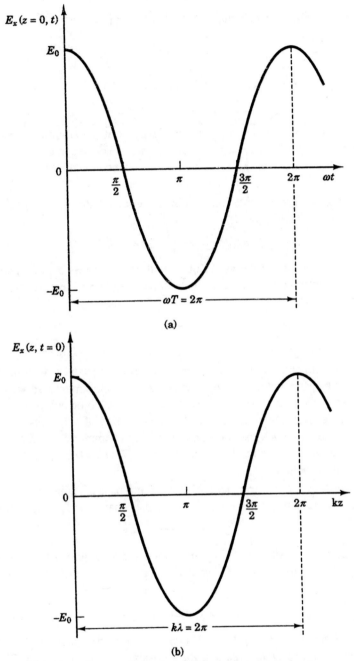

Figure 1.1 Possible electric field solution to the wave equation.

hertz (s^{-1}), and the temporal period of the wave T is equal to $1/f$ seconds; the positive scalar quantity E_0 is the *amplitude* of the wave.

We can also examine the spatial variation of $\overline{E}$ for a fixed time (say $t = 0$)

$$\overline{E}(z, t = 0) = \hat{x} E_0 \cos kz$$

which is shown in Figure 1.1(b). This wave is clearly periodic in space as well as in time. The spatial period of the wave is called the *wavelength* λ where $\lambda = 2\pi/k$ meters, and k (m^{-1}) is called the *wave number* or *propagation constant*. If k is positive, we note that the wave travels in the $+\hat{z}$ direction as time increases, and we can calculate the velocity at which the wave moves by focusing on one point on the waveform. If we are "riding" the waveform at the point z such that the intensity of the field at our location does not change, then

$$\omega t - kz = \text{constant} \qquad (1.3.12)$$

This means that as time progresses, we have to move in space to keep up with the wave. The velocity v at which we must move to stay on the waveform is given by

$$v = \frac{dz}{dt} = \frac{\omega}{k} \qquad (1.3.13)$$

which follows by taking the derivative of (1.3.12). This velocity is called the *phase velocity* because it gives the speed of the phase fronts of the wave. But the dispersion relation for waves in vacuum (1.3.10) implies that the ratio of the angular frequency to the wave number is

$$\frac{\omega}{k} = \frac{1}{\sqrt{\mu_0 \epsilon_0}} \equiv c \qquad (1.3.14)$$

where the velocity of electromagnetic radiation (e.g., light) in free space is given the special symbol $c = 2.998 \times 10^8$ meters/second. The correct prediction of the speed of light was one of the early triumphs of Maxwell's equations!

The magnetic field associated with the electric field given by (1.3.11) may be calculated from either Faraday's or Ampere's law. Applying Faraday's law (1.3.1) to (1.3.11) yields

$$\frac{\partial \overline{H}}{\partial t} = -\frac{1}{\mu_0} \nabla \times \overline{E} = -\hat{y} \frac{k E_0}{\mu_0} \sin(\omega t - kz)$$

or

$$\overline{H} = \hat{y} \sqrt{\frac{\epsilon_0}{\mu_0}} E_0 \cos(\omega t - kz) \qquad (1.3.15)$$

after application of the dispersion relation (1.3.10): $k/\omega\mu_0 = \sqrt{\epsilon_0/\mu_0}$.

The electric and magnetic fields in free space can thus be seen to have identical functional dependencies and to point in orthogonal directions. The direction of propagation is normal to the plane formed by the electric and magnetic field vectors. We note that the **phase** of the fields is independent of x and y, so that there is no phase variation over the planar surfaces orthogonal to the direction of propagation. Waves that do not exhibit any phase variation in a plane are called *plane waves*. If the amplitude of the wave is also a constant in that plane (as it is in this example), the waves are called *uniform plane waves*. Figure 1.2 suggests the form of such a propagating plane wave. The ratio of the electric to magnetic field amplitudes is known as the *wave impedance* and is $\sqrt{\mu_0/\epsilon_0}$ for the uniform plane wave in free space discussed above. The wave impedance $\sqrt{\mu_0/\epsilon_0}$ is given the symbol η_0; for vacuum, $\eta_0 \simeq 120\pi \simeq 377$ ohms (Ω).

Example 1.3.1

Does the electric field $\overline{E}(x, y, z, t) = \hat{y}[f_+(z - ct) + f_-(z + ct)]$ satisfy Maxwell's equations in free space, where f_+ and f_- are arbitrary analytic functions of their arguments $(z \pm ct)$?

Solution: Faraday's and Ampere's laws (1.3.1) and (1.3.2) can be satisfied if their combination, the wave equation (1.3.7), is satisfied:

$$\left(\nabla^2 - \mu_0\epsilon_0\frac{\partial^2}{\partial t^2}\right)\hat{y}[f_+(z - ct) + f_-(z + ct)] \stackrel{?}{=} 0$$

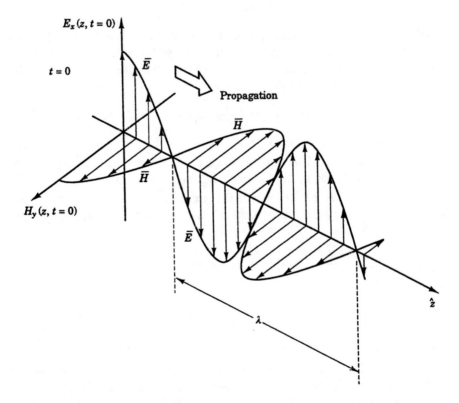

Figure 1.2   Electric and magnetic field vectors for a uniform plane wave.

where $\nabla^2 = \partial^2/\partial x^2 + \partial^2/\partial y^2 + \partial^2/\partial z^2$. We find

$$f_+''(z - ct) + f_-''(z + ct) - \mu_0\epsilon_0[(-c)^2 f_+''(z - ct) + c^2 f_-''(z + ct)] \overset{?}{=} 0$$

where $f_\pm''$ are the second derivatives of $f_\pm$ with respect to their arguments $(z \pm ct)$. Thus the wave equation for $\overline{E}$ is satisfied if $c = 1/\sqrt{\mu_0\epsilon_0}$, which is the velocity of wave propagation in vacuum.

Gauss's law (1.3.3) is satisfied because $\nabla \cdot \overline{E} = \dfrac{\partial}{\partial y} E_y = 0$. Use of Faraday's law (1.3.1) yields an expression for $\overline{H}$:

$$\overline{H} = -\int \frac{\nabla \times \overline{E}}{\mu_0}\, dt = -\hat{x}\sqrt{\frac{\epsilon_0}{\mu_0}}\, [f_+(z - ct) - f_-(z + ct)]$$

This expression also clearly satisfies Gauss's law (1.3.4). The given $\overline{E}$ and its associated $\overline{H}$ thus satisfy all of Maxwell's equations in free space.

1.4 TIME-HARMONIC FIELDS

Maxwell's equations are first-order differential equations in space and time, and in vacuum the constitutive relations are linear. Therefore, superpositions of solutions to Maxwell's equations are also valid solutions, as demonstrated by (1.2.13). For example, the single frequency or "monochromatic" solution $\overline{E} = \hat{x} E_0 \cos(\omega t - kz)$ could be added to solutions at different frequencies ω to produce arbitrary waveforms.[5] Thus, it is useful to understand first the wave behavior of monochromatic waves, which have only a single frequency. These basic sinusoidal solutions are called *time-harmonic fields*, and from them any other waveform may be constructed.

Consider a time-harmonic, $\hat{x}$-directed electric field with angular frequency $\omega = 2\pi f$:

$$\overline{E}(\overline{r}, t) = \hat{x} E_x(\overline{r}) \cos(\omega t - \phi_x(\overline{r})) \tag{1.4.1}$$

where $E_x(\overline{r})$ and $\phi_x(\overline{r})$ are real, scalar quantities which do not depend on time, but which may depend on coordinate. For ease of mathematical manipulation, we can write (1.4.1) as

$$E_x(\overline{r}, t) = \mathrm{Re}\left\{ E_x(\overline{r})\, e^{j(\omega t - \phi_x(\overline{r}))} \right\} = \mathrm{Re}\left\{ \underline{E}_x(\overline{r})\, e^{j\omega t} \right\}$$

---

5. Mathematically, this superposition can be expressed in terms of a Fourier series or integral. For example, the integral

$$\overline{E}(z, t) = \hat{x}\, \frac{1}{2\pi} \int_0^\infty \overline{E}(\omega) \cos(\omega t - k(\omega)z + \phi(\omega))\, d\omega$$

is the general solution for an $\hat{x}$-polarized uniform plane wave propagating along the $+\hat{z}$-axis. Furthermore, such a wave could be superimposed with other arbitrarily polarized waves propagating in any of an infinite number of other directions. We note that $\overline{E}(z, t)$ and $\overline{E}(\omega)$ are Fourier transform pairs, where $\overline{E}(\omega)$ is a Fourier amplitude and does not have the usual electric field dimensions of V/m.

where the operator Re{ } extracts the real part of its argument and

$$\underline{E}_x(\overline{r}) = E_x(\overline{r})\, e^{-j\phi_x(\overline{r})} \qquad (1.4.2)$$

The quantity $\underline{E}_x(\overline{r})$ is called a *phasor*, written with an underbar to remind us that it is a complex quantity with both magnitude and phase. In general, we may combine all three vector components of $\overline{E}(\overline{r}, t)$ in a phasor representation to yield

$$\overline{E}(\overline{r}, t) = \mathrm{Re}\left\{\underline{\overline{E}}(\overline{r})\, e^{j\omega t}\right\} \qquad (1.4.3)$$

where we note that $\underline{\overline{E}}(\overline{r})$ has **six** numbers associated with it (three vectors, each with magnitude and phase). Other field variables such as $\overline{H}$, $\overline{D}$, $\overline{B}$, $\overline{J}$, and ρ are also expressible as phasors. Essentially all we have done in constructing phasors is to isolate the spatial and temporal dependencies of the fields so that the separation-of-variables technique for solving partial differential equations may be applied to Maxwell's equations. Phasors and their properties are reviewed in Appendix A.

Substituting $\overline{E}(\overline{r}, t) = \mathrm{Re}\left\{\underline{\overline{E}}(\overline{r})\, e^{j\omega t}\right\}$ and $\overline{B}(\overline{r}, t) = \mathrm{Re}\left\{\underline{\overline{B}}(\overline{r})\, e^{j\omega t}\right\}$ into Faraday's law (1.2.1), we obtain

$$\mathrm{Re}\left\{(\nabla \times \underline{\overline{E}}(\overline{r}) + j\omega\underline{\overline{B}}(\overline{r}))\, e^{j\omega t}\right\} = 0$$

Because this equation is valid for all time, it must be valid at $\omega t = 0$:

$$\mathrm{Re}\left\{\nabla \times \underline{\overline{E}}(\overline{r}) + j\omega\underline{\overline{B}}(\overline{r})\right\} = 0$$

and at $\omega t = \pi/2$:

$$\mathrm{Im}\left\{\nabla \times \underline{\overline{E}}(\overline{r}) + j\omega\underline{\overline{B}}(\overline{r})\right\} = 0$$

It follows that

$$\nabla \times \underline{\overline{E}}(\overline{r}) = -j\omega\underline{\overline{B}}(\overline{r})$$

must be valid in general. We thus have the important result that the time-harmonic form of Faraday's law is **independent** of time, consistent with all phasor field quantities being complex functions of coordinate alone.

The time-harmonic form of the rest of Maxwell's equations may be determined using the same procedure, which basically amounts to changing real, time-dependent fields into time-independent phasors and replacing $\partial/\partial t$ with $j\omega$:

$$\nabla \times \underline{\overline{E}} = -j\omega\underline{\overline{B}} \qquad (1.4.4)$$

$$\nabla \times \underline{\overline{H}} = j\omega\underline{\overline{D}} + \underline{\overline{J}} \qquad (1.4.5)$$

$$\nabla \cdot \underline{\overline{D}} = \underline{\rho} \qquad (1.4.6)$$

$$\nabla \cdot \underline{\overline{B}} = 0 \qquad (1.4.7)$$

The charge continuity equation becomes

$$j\omega\underline{\rho} + \nabla \cdot \underline{\overline{J}} = 0 \qquad (1.4.8)$$

and the time-harmonic form of the wave equation in free space is given by

$$(\nabla^2 + \omega^2 \mu_0 \epsilon_0)\underline{E} = 0 \qquad (1.4.9)$$

which is also known as the *Helmholtz equation*.

To keep the notation in this text simple, the underbars denoting complex phasor fields will be dropped after this section, as it is almost always obvious from the context of a discussion whether the fields are explicitly time-dependent or are time-harmonic. In cases where the fields are time-dependent, t will be included explicitly in the argument of the field quantity; i.e., the electric field will be written as $\overline{E}(\bar{r}, t)$, because a time-dependent field can **never** be a phasor! In the few cases where ambiguities might arise (such as Section 1.6, which discusses both the real and complex Poynting theorem), the underbars will be retained.

We now study the time-harmonic solution to the wave equation, recalling that a possible time-dependent electric field solution discussed in Section 1.3 was

$$\overline{E}(z, t) = \hat{x} E_0 \cos(\omega t - kz)$$

where this solution represents a $+\hat{x}$-directed uniform plane wave. This time-dependent electric field can easily be converted to a time-harmonic form:

$$\overline{E}(z, t) = \text{Re}\left\{\underline{E}(z) e^{j\omega t}\right\}; \qquad \underline{E}(z) = \hat{x} E_0 e^{-jkz} \qquad (1.4.10)$$

We check first to see whether $\underline{E}(z)$ satisfies the Helmholtz equation by substituting $\underline{E}(z)$ into (1.4.9), which yields

$$(-k^2 + \omega^2 \mu_0 \epsilon_0)E_0 e^{-jkz} = 0$$

Thus for nonzero E_0, we obtain the same dispersion relation (1.3.10) given in Section 1.3 for the time-dependent fields. We also see that since $\underline{E}(z)$ points in the $\hat{x}$ direction but has only z dependence, Gauss's law in free space is satisfied ($\nabla \cdot \underline{E} = 0$). The corresponding time-harmonic magnetic field may be calculated from the time-harmonic form of Faraday's law (1.4.4)

$$\overline{H}(z) = -\frac{\nabla \times \underline{E}}{j\omega\mu_0} = \hat{y} \frac{kE_0}{\omega\mu_0} e^{-jkz} \qquad (1.4.11)$$

where the constitutive law in vacuum (1.2.9) has also been used.

If we want to see the real, time-dependent form of $\overline{H}$, we just multiply $\underline{H}$ by $e^{j\omega t}$ and take its real part

$$\overline{H}(z, t) = \text{Re}\left\{\underline{H}(z) e^{j\omega t}\right\} = \hat{y} \frac{kE_0}{\omega\mu_0} \cos(\omega t - kz)$$

which is the same solution as (1.3.15) because

$$\frac{k}{\omega\mu_0} = \sqrt{\frac{\epsilon_0}{\mu_0}} = \frac{1}{\eta_0}$$

in view of the dispersion relation (1.3.10).

Comparison of the expressions for $\overline{E}(z)$ (1.4.10) and $\overline{H}(z)$ (1.4.11) yields the useful result that for a $\hat{z}$-directed uniform plane wave,

$$\overline{H} = \hat{z} \times \overline{E}/\eta_0 \tag{1.4.12}$$

Example 1.4.1

Determine the equivalent electric field $\overline{E}(x, y, z, t)$ for each representation below and find the corresponding magnetic field in both phasor and time-dependent form.

(a) $\overline{E} = \hat{x}\, e^{-jkz}$

(b) $\overline{E} = -\hat{x}\, j\, e^{-jkz} = \hat{x}\, e^{-j(\pi/2+kz)}$

(c) $\overline{E} = (\hat{x} - j\hat{y})\, e^{jkz}$

Solution:

(a) $\overline{E}(x, y, z, t) = \text{Re}\left\{\hat{x}\, e^{-jkz}\, e^{j\omega t}\right\}$

$= \hat{x}\text{Re}\{\cos(\omega t - kz) + j\sin(\omega t - kz)\}$

$= \hat{x}\cos(\omega t - kz)$

$\overline{H}(x, y, z) = \hat{z}\,\dfrac{1}{\eta_0} \times (\hat{x}\, e^{-jkz}) = \hat{y}\,\dfrac{1}{\eta_0}\, e^{-jkz}$

$\overline{H}(x, y, z, t) = \hat{y}\,\dfrac{1}{\eta_0}\cos(\omega t - kz)$

(b) $\overline{E}(x, y, z, t) = \hat{x}\sin(\omega t - kz)$

$\overline{H}(x, y, z) = -\hat{y}\,\dfrac{j}{\eta_0}\, e^{-jkz}$

$\overline{H}(x, y, z, t) = \hat{y}\,\dfrac{1}{\eta_0}\sin(\omega t - kz)$

(c) $\overline{E}(x, y, z, t) = \hat{x}\cos(\omega t + kz) + \hat{y}\sin(\omega t + kz)$

$\overline{H}(x, y, z) = -\hat{z}\,\dfrac{1}{\eta_0} \times (\hat{x} - j\hat{y})\, e^{jkz}$

$= \dfrac{1}{\eta_0}(-\hat{y} - j\hat{x})\, e^{jkz}$

$\overline{H}(x, y, z, t) = \dfrac{1}{\eta_0}\left(-\hat{y}\cos(\omega t + kz) + \hat{x}\sin(\omega t + kz)\right)$

Note that the time-dependent electric and magnetic fields have the same spatial dependence and that $\overline{E} \times \overline{H}$ points in the direction of propagation ($+\hat{z}$ for the fields in (a) and (b), $-\hat{z}$ for the fields in (c)). The significance of this observation will be made clear in Section 1.6.

1.5 POLARIZATION

We would now like to consider how the direction of the electric field vector changes with time at a fixed point in space for a uniform plane wave. If the direction of propagation of an electromagnetic wave is defined to be in the $+\hat{z}$ direction, then the electric field may have only $\hat{x}$ and $\hat{y}$ components in order to satisfy $\nabla \cdot \overline{E} = 0$. Thus, the most general form of this electric field phasor is given by

$$\overline{E}(z) = (\hat{x} E_x + \hat{y} E_y) e^{-jkz} \tag{1.5.1}$$

where E_x and E_y are complex numbers possessing both magnitude and phase. If $E_y = 0$, then the electric field is said to be $\hat{x}$-*polarized* or linearly polarized in the $\hat{x}$ direction because the phasor always points in the $\hat{x}$ direction; if $E_x = 0$, the field is $\hat{y}$-*polarized*.

But what happens if the quantities E_x and E_y are arbitrary complex numbers? Because the overall magnitude and phase of the electric field does not affect the polarization, we shall only be concerned with the **relative** differences between the $\hat{x}$ and $\hat{y}$ components of $\overline{E}$. That is, we could multiply both E_x and E_y by the same arbitrary complex number and the polarization would remain unchanged. So it is convenient to rewrite the electric field as

$$\overline{E}(z) = E_x e^{-jkz} (\hat{x} + \underline{A} \hat{y}) \tag{1.5.2}$$

where $\underline{A} = A e^{j\phi}$, and the information about the polarization is contained in A and ϕ. It is useful to consider how the electric field vector varies in time in order to gain an understanding of what polarization means, so we convert the phasor $\overline{E}$ in (1.5.2) back to the real, time-dependent $\overline{E}(t)$, and we suppress the common factor $E_x e^{-jkz}$ because it does not affect the type of polarization. (Although, as we shall see, the "handedness" of the polarization depends on the direction of propagation of the wave, which is implicit in the e^{-jkz} dependence.)

$$\overline{E}(t) = \text{Re}\left\{(\hat{x} + A e^{j\phi}\hat{y}) e^{j\omega t}\right\} = \hat{x} \cos \omega t + \hat{y} A \cos(\omega t + \phi) \tag{1.5.3}$$

The quantities A and ϕ are arbitrary real numbers, but certain special cases give insight into the phenomenon of polarization and are discussed below. Figure 1.3 shows how the direction of the electric field vector changes with time for each special case.

Case 1: $A = 0$, Figure 1.3(a)

$$\overline{E} = \hat{x}, \quad \overline{E}(t) = \hat{x} \cos \omega t$$

The electric field is $\hat{x}$-polarized, which is a special case of linear polarization. The tip of the electric field vector oscillates back and forth along the $\hat{x}$-axis.

Case 2: $A \to \infty$, Figure 1.3(b)

$$\overline{E} = A\hat{y}, \quad \overline{E}(t) = A\hat{y} \cos \omega t$$

Only the $\hat{y}$ component of the electric field survives, and therefore the field is $\hat{y}$-polarized. (This is just another way of saying that $E_x = 0$.)

Case 3: $A = 1$, $\phi = 0$, Figure 1.3(c)

$$\overline{E} = \hat{x} + \hat{y}, \quad \overline{E}(t) = (\hat{x} + \hat{y}) \cos \omega t$$

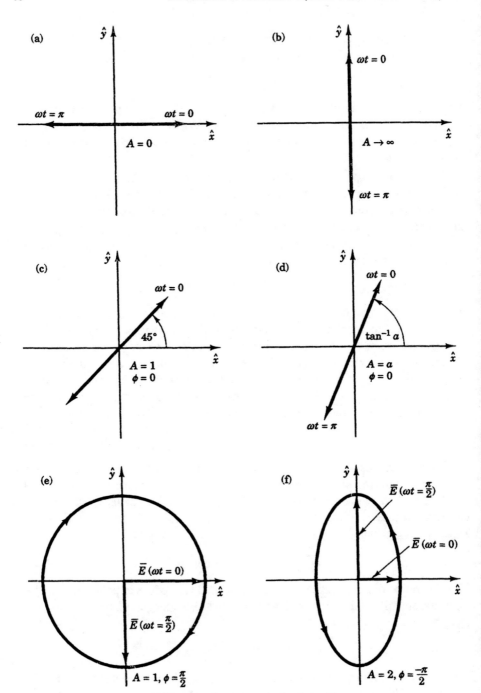

Figure 1.3 Special cases of polarization.

At $\omega t = 0$, the electric field tip is at the point $(1, 1)$ in the x–y plane, and moves through $(0, 0)$ at $\omega t = \pi/2$ to $(-1, -1)$ at $\omega t = \pi$, tracing out a line in space. Thus, the field is linearly polarized at an angle of $45°$ to the $\hat{x}$-axis because at any given instant in time, the $\hat{x}$ and $\hat{y}$ components of the electric field are equal and in phase.

Case 4: $A = a$, $\phi = 0$, Figure 1.3(d)

$$\overline{E} = \hat{x} + a\hat{y}, \quad \overline{E}(t) = (\hat{x} + a\hat{y})\cos\omega t$$

The field is again linearly polarized at an angle of $\tan^{-1} a$ radians from the $+\hat{x}$-axis. It appears that as long as the $\hat{x}$ and $\hat{y}$ components of the field are in phase ($\phi = 0$), the polarization is *linear*. If $\phi = \pi$, the field is also linearly polarized at $\pi - \tan^{-1} a$ radians from the $+\hat{x}$-axis.

Case 5: $A = 1$, $\phi = \pi/2$, Figure 1.3(e)

$$\overline{E} = \hat{x} + j\hat{y}, \quad \overline{E}(t) = \hat{x}\cos\omega t - \hat{y}\sin\omega t$$

The $\hat{y}$ component of the field lags the $\hat{x}$ component by $90°$ and the two components are of equal length. The magnitude of the electric field is seen to be $|\overline{E}(t)|^2 = \cos^2\omega t + \sin^2\omega t = 1$, so the tip of the electric field vector traces out a circle in time. The polarization is therefore *circular*. We see that at $\omega t = 0$, $\overline{E} = \hat{x}$ and at $\omega t = \pi/2$, $\overline{E} = -\hat{y}$, so the tip of the electric field appears to be rotating in a clockwise direction as time increases if the $\hat{x}$- and $\hat{y}$-axes are oriented as in Figure 1.3. The Institute of Electrical and Electronics Engineers (IEEE) defines the following handedness convention for polarization: Orient your **right** hand so that your fingers curl in the direction that the electric field is rotating as time increases. If your thumb is pointing in the direction of propagation, then the field is right-polarized, but if your thumb is pointing the other way, then the field is left-polarized. For Case 5, where propagation is in the $+\hat{z}$ direction, the wave is *left circularly polarized*. If ϕ were to switch to $\phi = -\pi/2$, then the wave would become *right circularly polarized*. For $A = 1$, $\phi = \pi/2$, and propagation in the $-\hat{z}$ direction, the field would also be right circularly polarized. So it is important not only to specify A and ϕ but also the propagation direction of the field. Outside the engineering community, the opposite direction of handedness is sometimes used, so caution in reading the literature is recommended.

Case 6: $A = 2$, $\phi = -\pi/2$, Figure 1.3(f)

$$\overline{E} = \hat{x} - 2j\hat{y}, \quad \overline{E}(t) = \hat{x}\cos\omega t + 2\hat{y}\sin\omega t$$

At $\omega t = 0$, $\overline{E} = \hat{x}$ and at $\omega t = \pi/2$, $\overline{E} = 2\hat{y}$. Therefore, we suspect that the tip of the electric field vector traces out a counterclockwise ellipse oriented in the $\hat{y}$ direction with a ratio of semi-major to semi-minor axes equal to 2. The field is thus *right elliptically polarized*.   The equation for the tip of the field vector may be established by letting $x = \cos\omega t$ and $y = 2\sin\omega t$. Noting that $\sin\omega t = \sqrt{1 - x^2} = y/2$ gives the resulting equation for an ellipse: $x^2 + y^2/4 = 1$. Elliptical polarization is the most general type of polarization possible; linear and circular polarization are simply special cases of ellipses with eccentricities of infinity and 1 respectively. In general, the ellipse is tilted with respect to the $\hat{x}$-axis.

We can easily derive equations for arbitrary polarization by letting

$$x = \cos \omega t$$

$$y = A \cos(\omega t + \phi) = A \cos \omega t \cos \phi - A \sin \omega t \sin \phi$$

in (1.5.3). Eliminating ωt from these two equations by replacing $\cos \omega t$ with x and $\sin \omega t$ with $\sqrt{1 - x^2}$ eventually yields

$$A^2 x^2 + y^2 - 2Axy \cos \phi = A^2 \sin^2 \phi \tag{1.5.4}$$

which is the general equation for an ellipse.

We can rotate the coordinate system (x, y) to (x', y') as shown in Figure 1.4, where the semi-major axis lies along the $\hat{x}'$-axis and is tilted by an angle ψ from the $\hat{x}$-axis. This coordinate transformation is given by

$$x = x' \cos \psi - y' \sin \psi$$
$$y = x' \sin \psi + y' \cos \psi \tag{1.5.5}$$

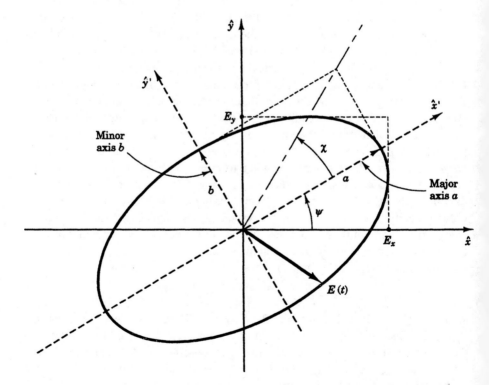

Figure 1.4   Polarization ellipse.

These two equations may be substituted into (1.5.4) and the $x'y'$ cross-term eliminated to give the following expression for ψ:

$$\tan 2\psi = \frac{2A}{1 - A^2} \cos \phi \qquad (1.5.6)$$

With the semi-major and semi-minor axes (a and b, respectively) defined as in Figure 1.4, and the angle $\chi = \tan^{-1}(b/a)$, the following relations may be shown to hold:

$$\sin 2\chi = \frac{2A}{1 + A^2} \sin \phi$$

$$. \qquad a^2 + b^2 = 1 + A^2 \qquad (1.5.7)$$

$$ab = A \sin \phi$$

Thus we can represent the polarization of monochromatic waves in several alternative ways, including (A, ϕ), (E_y/E_x), (ψ, χ), or $(\psi, a/b)$, remembering that we need to know the direction of propagation to determine the handedness of the wave. For waves propagating in the $+\hat{z}$ direction, and with A and ϕ defined by (1.5.2), we have right-handed polarization for $-\pi < \phi < 0$ and left-handed polarization for $0 < \phi < \pi$.

All possible polarizations for monochromatic (single-frequency) waves can be represented by proper choice of A and ϕ, and these are suggested in Figure 1.5 where A and ϕ are polar coordinates. Equivalent coordinates are $\text{Re}\{E_y/E_x\}$ and $\text{Im}\{E_y/E_x\}$, as illustrated. The figure shows that the most general type of polarization is elliptical except for the special cases of the real axis ($\phi = 0, \pi$) where the polarization is linear, and the two points ($A = 1$, $\phi = \pm\pi/2$) where it is circular.

Note that by (1.4.12), $\overline{H}(t)$ for a uniform plane wave propagating in the $\hat{z}$ direction in vacuum is

$$\overline{H}(t) = \frac{\hat{z} \times \overline{E}(t)}{\eta_0} \qquad (1.5.8)$$

The tip of the vector $\overline{H}(t)$ simultaneously executes the same figure that $\overline{E}(t)$ does, but rotated by 90° in the x–y plane. Thus, the electric and magnetic fields are proportional and orthogonal at all times. By convention we associate the type of polarization with the electric field vector, so an $\hat{x}$-polarized wave has an $\hat{x}$-directed electric field.

Many signals are not monochromatic, but instead occupy some narrow bandwidth B (hertz) which is much less than the frequency f. In this case, we can use the same expression (1.5.3) for the electric field but assume that the real amplitudes E_x and E_y and the differential phase ϕ are each independently and slowly varying on a time scale of $\sim 1/2\pi B$ seconds. The polarization of such waves can be char-

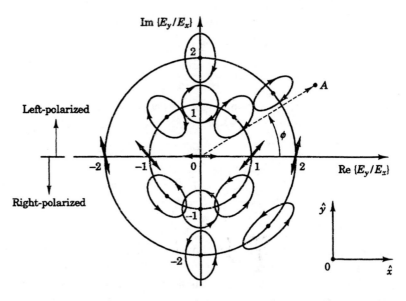

Figure 1.5   Polarization diagram for $+\hat{z}$-propagating waves.

acterized by any set of three parameters specifying the average polarization ellipse, in addition to a fourth parameter m, which is defined as the *degree of polarization*

$$m = \frac{S_0 - S_u}{S_0} \qquad (1.5.9)$$

representing the fraction of the total power S_0, which is carried by the polarized component of the wave. Here, the total wave is the superposition of an elliptically polarized wave and an independent, unpolarized wave carrying power S_u.[6]

The phenomenon of polarization must be taken into account in almost any

---

6. *Stokes's parameters*, which have units of watts/m^2, are defined as:

$$S_0 = \left[\langle E_x^2(t) \rangle + \langle E_y^2(t) \rangle \right] / \eta_0$$

$$S_1 = \left[\langle E_x^2(t) \rangle - \langle E_y^2(t) \rangle \right] / \eta_0$$

$$S_2 = \frac{2}{\eta_0} \langle E_x(t) E_y(t) \cos \phi(t) \rangle$$

$$S_3 = \frac{2}{\eta_0} \langle E_x(t) E_y(t) \sin \phi(t) \rangle$$

Sometimes S_0, S_1, S_2, and S_3 are represented by Stokes's original symbols, I, Q, U, and V, respectively.

For a completely polarized narrow-band wave $E_y(t) = d \cdot E_x(t - \phi(t)/\omega)$ where d and $\phi(t)$ are constants; in this case, the vector $\overline{E}(t)$ traces an ellipse of constant shape but slowly varying size. Examination of the definitions for Stokes's parameters then yields

$$S_0^2 = S_1^2 + S_2^2 + S_3^2 \qquad (1)$$

electromagnetics application. For example, communication satellites typically receive and transmit circularly polarized radiation because linearly polarized satellite signals are rotated an uncertain amount by the ionosphere. If the waves are rotated so as to be nearly orthogonal to the polarization of the receiving antenna, then very little of the wave is received and the satellite signal will occasionally fade. This phenomenon of Faraday rotation (discussed in Section 3.8) arises because the ionosphere is not an isotropic medium; use of circular polarization circumvents this problem. Because it is possible to decompose an arbitrarily polarized field into two orthogonal (independent) polarizations, an electromagnetic signal may be used to carry two independent channels of information simultaneously. When the signal is processed, it may be decomposed into its two component waves (e.g., $\hat{x}$- and $\hat{y}$-linearly polarized fields or right and left circularly polarized fields). Signal processing may then be carried out on each channel independently, as there is no interaction between orthogonal polarizations in simple media.

Example 1.5.1

(a) For each uniform plane wave traveling in vacuum characterized below, describe the polarization and direction of propagation.

(i) $\overline{E} = \hat{x}\, e^{+jkz}$

(ii) $\overline{H} = (j\hat{x} - \hat{y})\, e^{-jkz}$

(iii) $\overline{E} = (\hat{x} + \sqrt{3}\hat{y})\, e^{-jkz}$

(iv) $\overline{E} = j\left(\hat{x} + e^{j\pi/4}\hat{y}\right) e^{-jkz}$

(b) Show that a uniform plane wave with arbitrary polarization can be created from the superposition of two waves, one with right circular and one with left circular polarization.

Solution:

(a)   (i) $\hat{x}$-polarized, propagating in the $-\hat{z}$ direction.

(ii) $\overline{H}(x, y, z, t) = -\hat{x}\sin(\omega t - kz) - \hat{y}\cos(\omega t - kz)$;
$\overline{E}(x, y, z, t) = \eta_0 \overline{H}(x, y, z, t) \times \hat{z}$
$= \hat{y}\eta_0 \sin(\omega t - kz) - \hat{x}\eta_0 \cos(\omega t - kz)$;
left circular polarization, $+\hat{z}$ propagation.

(iii)   linear polarization at an angle $\tan^{-1}\sqrt{3} = 60°$ from the $\hat{x}$-axis;
$+\hat{z}$ propagation.

(iv) $\overline{E}(x, y, z, t) = -\hat{x}\sin(\omega t - kz) - \hat{y}\sin(\omega t - kz + \pi/4)$;
left elliptical polarization, $+\hat{z}$ propagation.

(b) The most general polarization of a $\hat{z}$-propagating wave is given by (1.5.2):

$$\overline{E} = E_0(\hat{x} + A\, e^{j\phi}\hat{y}) \qquad (1)$$

---

for a perfectly polarized wave. For a more general *partially polarized wave*, the equality (1) does not hold. Instead, the wave power S_0 may be divided into two parts, $S_0 - S_u$ which satisfies (1), and S_u, the unpolarized power in the wave. The Stokes's parameters for the completely polarized wave and the unpolarized wave are then $S_0 - S_u$, S_1, S_2, S_3, and S_u, 0, 0, 0, respectively.

A wave with right circular polarization is given by $a(\hat{x} - j\hat{y})$ while a left circularly polarized wave has the form $b(\hat{x} + j\hat{y})$; a and b are complex coefficients to be determined. The superposition of the two circularly polarized waves is set equal to the general wave (1):

$$E_0(\hat{x} + A\,e^{j\phi}\hat{y}) = a(\hat{x} - j\hat{y}) + b(\hat{x} + j\hat{y})$$

Equating the $\hat{x}$ and $\hat{y}$ components separately yields

$$a + b = E_0$$

and

$$a - b = jE_0A\,e^{j\phi}$$

Therefore, $a = (1 + jA\,e^{j\phi})E_0/2$ and $b = (1 - jA\,e^{j\phi})E_0/2$. Note that an arbitrary polarization may be represented as a linear combination of **any** two orthogonal polarizations.

1.6 THE POYNTING THEOREM

To study how the flow of power is related to energy storage in electromagnetic fields, we first go back to the time-dependent form of Faraday's and Ampere's laws (1.2.1) and (1.2.2):

$$\nabla \times \overline{E} = -\frac{\partial \overline{B}}{\partial t} \tag{1.6.1}$$

$$\nabla \times \overline{H} = \frac{\partial \overline{D}}{\partial t} + \overline{J} \tag{1.6.2}$$

We take the dot product of $\overline{H}$ and (1.6.1) and the dot product of $\overline{E}$ and (1.6.2). Subtracting one from the other yields

$$\overline{E} \cdot (\nabla \times \overline{H}) - \overline{H} \cdot (\nabla \times \overline{E}) = \overline{E} \cdot \frac{\partial \overline{D}}{\partial t} + \overline{H} \cdot \frac{\partial \overline{B}}{\partial t} + \overline{E} \cdot \overline{J} \tag{1.6.3}$$

Using the vector identity

$$\nabla \cdot (\overline{E} \times \overline{H}) = \overline{H} \cdot (\nabla \times \overline{E}) - \overline{E} \cdot (\nabla \times \overline{H}) \tag{1.6.4}$$

with (1.6.3) and rearranging terms leads to the differential form of the *Poynting theorem*:

$$-\overline{E} \cdot \overline{J} = \left(\overline{H} \cdot \frac{\partial \overline{B}}{\partial t}\right) + \left(\overline{E} \cdot \frac{\partial \overline{D}}{\partial t}\right) + \nabla \cdot (\overline{E} \times \overline{H}) \tag{1.6.5}$$

We note that $\overline{E}$ has dimensions of V/m while the dimensions of $\overline{J}$ are A/m^2, so $\overline{E} \cdot \overline{J}$ has dimensions of watts/m^3 (as do each of the other terms in (1.6.5)). The differential form of the Poynting theorem is therefore an equation in power density.

In order to interpret this equation in terms of power flow, we integrate (1.6.5) over a closed volume V. Since $\overline{E} \cdot \overline{J}$ has units of W/m^3, volume integration gives a power equation in watts. We then obtain a relation known as the integral form of the Poynting theorem

$$-\int_V \overline{E} \cdot \overline{J} \, dv = \int_V \left(\overline{H} \cdot \frac{\partial \overline{B}}{\partial t} + \overline{E} \cdot \frac{\partial \overline{D}}{\partial t} \right) dv + \oint_A (\overline{E} \times \overline{H}) \cdot \hat{n} \, da \qquad (1.6.6)$$

where we have made use of *Gauss's divergence theorem* to simplify the last term of (1.6.5):

$$\int_V (\nabla \cdot \overline{G}) \, dv = \oint_A \overline{G} \cdot \hat{n} da \qquad (1.6.7)$$

Gauss's divergence theorem is **not** to be confused with Gauss's law! Instead, Gauss's divergence theorem equates an integral over a closed volume to an integral over the surface enclosing that volume, where $\hat{n}$ is an outwardly directed unit vector normal to the surface A which encloses the volume V, $\overline{G}$ is an arbitrary vector field, and da and dv are differential areas and volumes, respectively.

Consider the case of isotropic media where $\overline{D} = \epsilon \overline{E}$ and $\overline{B} = \mu \overline{H}$. Then the Poynting theorem can be written as

$$-\int_V \overline{E} \cdot \overline{J} \, dv = \frac{d}{dt} \int_V \left(\frac{1}{2}\mu|\overline{H}|^2 + \frac{1}{2}\epsilon|\overline{E}|^2 \right) dv + \oint_A (\overline{E} \times \overline{H}) \cdot \hat{n} da \qquad (1.6.8)$$

where a total time derivative may be taken outside the integral as long as the volume V is not moving or deforming in time.

We see that the Poynting theorem is an expression of conservation of energy, and the individual terms in (1.6.8) can be interpreted as follows. The instantaneous stored *magnetic energy density* is given by $W_m = \frac{1}{2}\mu|\overline{H}|^2$ (joules/m^3), and $W_e = \frac{1}{2}\epsilon|\overline{E}|^2$ is the instantaneous stored electric energy density (joules/m^3). Therefore, it follows that

$$\frac{d}{dt} \int_V \left(\frac{1}{2}\mu|\overline{H}|^2 + \frac{1}{2}\epsilon|\overline{E}|^2 \right) dv$$

is the rate at which stored energy increases in the volume V. We interpret the term $-\int_V \overline{E} \cdot \overline{J} \, dv$ as the power generated by the source $\overline{J}$ in the volume V, and $\oint_A (\overline{E} \times \overline{H}) \cdot \hat{n} da$ as the power flowing **out** of the volume V. We therefore define

$$\overline{S} \equiv \overline{E} \times \overline{H} \qquad (1.6.9)$$

where $\overline{S}$ is called the *Poynting vector*, characterizing the direction and magnitude of power flow in the fields.

The Poynting theorem states that the power flowing out of the volume plus the rate of increase of energy storage inside the volume is equal to the total power supplied by the source, and is essentially a statement of conservation of energy. This formulation must be modified for use with moving media or boundaries, which will not concern us here.

As an example, consider the wave solution that we discussed in Section 1.3, where

$$\overline{E} = \hat{x} E_0 \cos(\omega t - kz)$$

and

$$\overline{H} = \hat{y} \sqrt{\epsilon_0/\mu_0} \, E_0 \cos(\omega t - kz)$$

The Poynting vector for these fields is

$$\overline{S} = \overline{E} \times \overline{H} = \hat{z} \sqrt{\frac{\epsilon_0}{\mu_0}} E_0^2 \cos^2(\omega t - kz) \tag{1.6.10}$$

which indicates that the power carried by the wave is flowing in the direction that the wave is propagating. The instantaneous stored energy density in the magnetic field is

$$W_m = \frac{1}{2}\mu_0 H_0^2 \cos^2(\omega t - kz) = \frac{1}{2}\epsilon_0 E_0^2 \cos^2(\omega t - kz) \tag{1.6.11}$$

where $E_0 = \eta_0 H_0$ has been used. The instantaneous stored energy density in the electric field is

$$W_e = \frac{1}{2}\epsilon_0 |\overline{E}|^2 = \frac{1}{2}\epsilon_0 E_0^2 \cos^2(\omega t - kz) \tag{1.6.12}$$

which is identical to W_m in this special case. In regions where there is no source $(\overline{J} = 0)$, the differential form of the Poynting theorem (1.6.5) states that

$$\nabla \cdot \overline{S} + \frac{\partial}{\partial t}(W_m + W_e) = 0$$

Evidently this is true for uniform plane waves, as seen from (1.6.10)–(1.6.12).

Before examining the equivalent power relations for time-harmonic fields, let us evaluate the scalar product of two sinusoids

$$A(t)B(t) = \text{Re}\left\{\underline{A}\,e^{j\omega t}\right\} \text{Re}\left\{\underline{B}\,e^{j\omega t}\right\}$$

where $\underline{A} = A\,e^{j\phi}$ and $\underline{B} = B\,e^{j\theta}$ (A and B are real numbers). The product of $A(t)$ and $B(t)$ thus becomes

$$A(t)B(t) = A\cos(\omega t + \phi)B\cos(\omega t + \theta)$$

or equivalently

$$A(t)B(t) = \frac{AB}{2}[\cos(\phi - \theta) + \cos(2\omega t + \phi + \theta)] \tag{1.6.13}$$

We define the time average of a function $f(t)$ over the time interval T to be

$$\langle f(t) \rangle = \frac{1}{T}\int_0^T f(t)dt \tag{1.6.14}$$

where often $f(t)$ is periodic with period T. If we take the time average of $A(t)B(t)$ over one period $T = 2\pi/\omega$, we find that

$$\langle A(t)B(t)\rangle = \frac{AB}{2}\cos(\phi - \theta) \tag{1.6.15}$$

The first term in (1.6.13) is independent of time, so it is just a constant within the time integral and thus survives the time-averaging, but the $\cos(2\omega t + \phi + \theta)$ term averages to zero over one period.

We can rewrite (1.6.15) in terms of phasors:

$$\langle A(t)B(t)\rangle = \mathrm{Re}\left\{\frac{AB}{2}e^{j\phi}e^{-j\theta}\right\} = \frac{1}{2}\mathrm{Re}\left\{\underline{A}\,\underline{B}^*\right\} \tag{1.6.16}$$

Therefore, we see that the time average of two scalar time-harmonic quantities is equal to half of the real part of the product of one scalar and the complex conjugate of the other. The above discussion of scalar products may be easily extended to vector products as well, and thus we can define the *complex Poynting vector* as

$$\overline{\underline{S}} = \overline{\underline{E}} \times \overline{\underline{H}}^* \tag{1.6.17}$$

The relationship between the complex Poynting vector and the real, time-dependent Poynting vector is given by

$$\langle \overline{S}(t)\rangle = \langle \overline{E}(t) \times \overline{H}(t)\rangle = \frac{1}{2}\mathrm{Re}\left\{\overline{\underline{E}} \times \overline{\underline{H}}^*\right\} = \frac{1}{2}\mathrm{Re}\{\overline{\underline{S}}\} \tag{1.6.18}$$

which may be shown by considering each of the scalar components of the vector $\overline{S}(t)$ separately. The quantity $\langle \overline{S}(t)\rangle$ is the time-averaged power flow per unit area in the direction perpendicular to $\overline{E}$ and $\overline{H}$. It cannot be overemphasized that the real Poynting vector $\overline{S}(t)$ and the complex Poynting vector $\overline{\underline{S}}$ do **not** have a phasor relationship with each other. That is, the real Poynting vector may **not** be found by multiplying $\overline{\underline{S}}$ by $e^{j\omega t}$ and then taking the real part. The two Poynting vectors are instead related by (1.6.18).

For uniform plane waves propagating in free space, we can write the Poynting vector in a particularly simple manner. If the time-harmonic fields have $e^{-j\overline{k}\cdot\overline{r}}$ spatial dependence, Maxwell's equations (1.4.1)–(1.4.4) become

$$\overline{k} \times \overline{E} = \omega\mu_0\overline{H} \tag{1.6.19}$$

$$\overline{k} \times \overline{H} = -\omega\epsilon_0\overline{E} \tag{1.6.20}$$

$$\overline{k} \cdot \overline{E} = 0 \tag{1.6.21}$$

$$\overline{k} \cdot \overline{H} = 0 \tag{1.6.22}$$

where the del operator ∇ is replaced by $-j\overline{k}$. Faraday's law for uniform plane waves (1.6.19) is a generalization of (1.4.12), which was applicable only to

$\hat{z}$-directed propagation. From Faraday's law, we can find the magnetic field of a uniform plane wave

$$\overline{H} = \frac{\overline{k} \times \overline{E}}{\omega \mu_0} \tag{1.6.23}$$

so the Poynting vector may be written in terms of the electric field alone:

$$\overline{S} = \frac{\overline{E} \times (\overline{k} \times \overline{E})^*}{\omega \mu_0}$$

Using the vector identity $\overline{A} \times (\overline{B} \times \overline{C}) = \overline{B}(\overline{A} \cdot \overline{C}) - \overline{C}(\overline{A} \cdot \overline{B})$, we find that

$$\overline{S} = \frac{1}{\omega \mu_0} \left[\overline{k}^*(\overline{E} \cdot \overline{E}^*) - \overline{E}^*(\overline{k}^* \cdot \overline{E}) \right] \tag{1.6.24}$$

But the direction of propagation for a uniform plane wave is perpendicular to the electric field vector from Gauss's law (1.6.21). Since a uniform plane wave propagating in space has a real propagation vector ($\overline{k} = \overline{k}^*$), we find a general expression for the Poynting vector:

$$\overline{S} = \frac{\overline{k}}{\omega \mu_0} |\overline{E}|^2 = \hat{k} \frac{|\overline{E}|^2}{\eta_0} \tag{1.6.25}$$

where $\hat{k}$ is the unit vector in the direction of propagation ($\overline{k} = |\overline{k}|\hat{k}$) and the dispersion relation for free space (1.3.10) has been used.

To obtain the *complex Poynting theorem*, we write the time-harmonic Maxwell's equations (1.4.4) and (1.4.5) but we take the **complex conjugate** of each term in Ampere's law:

$$\nabla \times \overline{E} = -j\omega \overline{B} \tag{1.6.26}$$

$$\nabla \times \overline{H}^* = -j\omega \overline{D}^* + \overline{J}^* \tag{1.6.27}$$

Using the identity $\nabla \cdot (\overline{E} \times \overline{H}^*) = \overline{H}^* \cdot (\nabla \times \overline{E}) - \overline{E} \cdot (\nabla \times \overline{H}^*)$ and dotting $\overline{H}^*$ into (1.6.26) and $\overline{E}$ into (1.6.27), we find

$$-\overline{E} \cdot \overline{J}^* = j\omega(\overline{H}^* \cdot \overline{B} - \overline{E} \cdot \overline{D}^*) + \nabla \cdot (\overline{E} \times \overline{H}^*) \tag{1.6.28}$$

which is the differential form of the complex Poynting theorem. We can write the integral form by using Gauss's divergence theorem (1.6.7) as before

$$-\int_V \overline{E} \cdot \overline{J}^* \, dv = j\omega \int_V (\overline{H}^* \cdot \overline{B} - \overline{E} \cdot \overline{D}^*) \, dv + \oint_A \overline{E} \times \overline{H}^* \cdot \hat{n} \, da \tag{1.6.29}$$

where $\hat{n}$ is an outwardly directed unit vector normal to the surface A enclosing the volume V.

If $\overline{B} = \mu \overline{H}$ and $\overline{D} = \epsilon \overline{E}$, then (1.6.29) can be rewritten as

$$-\frac{1}{2} \int_V \overline{E} \cdot \overline{J}^* \, dv = 2j\omega \left[\langle \mathrm{w}_m \rangle - \langle \mathrm{w}_e \rangle \right] + \frac{1}{2} \oint_A \overline{S} \cdot \hat{n} \, da \tag{1.6.30}$$

which is the complex Poynting theorem for linear media. The time-averaged stored magnetic $\langle w_m \rangle$ and electric $\langle w_e \rangle$ energies (joules) are defined as:

$$\langle W_m \rangle = \frac{1}{4}\mu_0|\overline{\underline{H}}|^2; \qquad \langle w_m \rangle = \int_V \langle W_m \rangle \, dv \qquad (1.6.31)$$

$$\langle W_e \rangle = \frac{1}{4}\epsilon_0|\overline{\underline{E}}|^2; \qquad \langle w_e \rangle = \int_V \langle W_e \rangle \, dv \qquad (1.6.32)$$

In dispersive media where ϵ and/or μ depend upon frequency, the time-averaged stored energies are generally different from (1.6.31) and (1.6.32).

Because the complex Poynting theorem involves complex numbers, it actually contains both a real and an imaginary equation. The real part of the complex Poynting theorem (1.6.30) is an equation relating time-averaged power flow and power dissipation, while the imaginary part describes energy storage in the electric and magnetic fields. Continuing the example of the uniform plane wave with electric field $\overline{E}(z,t) = \hat{x}E_0\cos(\omega t - kz)$ discussed in Sections 1.4 and 1.6, we recall that $\overline{\underline{E}} = \hat{x}E_0\,e^{-jkz}$ and $\overline{\underline{H}} = \hat{y}(E_0/\eta_0)\,e^{-jkz}$. Therefore, the complex Poynting vector is $\overline{\underline{S}} = \hat{z}|E_0|^2/\eta_0$ and the time-averaged power density is $\langle \overline{S} \rangle = \hat{z}|E_0|^2/2\eta_0$. The time-averaged electric and magnetic energy densities are both $\frac{1}{4}\epsilon_0|E_0|^2$. We thus can see by inspection that the complex Poynting theorem (1.6.28) is satisfied in this case, because the imaginary terms involving the energy densities cancel, and both the divergence of $\overline{\underline{S}}$ and the current density $\overline{\underline{J}}$ are zero for uniform plane waves in free space.

The negative sign in front of $\overline{\underline{E}} \cdot \overline{\underline{D}}^*$ in (1.6.29) is often a source of confusion, since the real Poynting theorem involves the **sum** of electric and magnetic energy densities. We can understand the complex Poynting theorem by making an analogy with a circuit description of the electric and magnetic fields. If we think of energy stored in an electric field as capacitive and energy stored in a magnetic field as inductive, we can write the circuit analogy to the Poynting vector as $P = \underline{V}\,\underline{I}^* = \underline{Z}|\underline{I}|^2$, where $\underline{V}$ and $\underline{I}$ are voltage and current corresponding to $\overline{\underline{E}}$ and $\overline{\underline{H}}$, respectively. The impedance $\underline{Z}$ is the ratio of voltage to current. Because the impedance of a capacitor is $Z_C = -j/\omega C$, the stored electric energy is associated with a **negative** imaginary quantity. The impedance of the inductor, on the other hand, is $Z_L = j\omega L$, and so the stored magnetic energy is a **positive** imaginary number, leading to the correct signs in (1.6.28) and (1.6.29). The analogies between lumped-element circuits and electromagnetic fields will be developed further in subsequent chapters.

Example 1.6.1

Two $\hat{x}$-polarized uniform plane waves of different strength are traveling in opposite directions and are superimposed:

$$\overline{\underline{E}} = \hat{x}\left(e^{-jkz} + a\,e^{jkz}\right)$$

What average power (W/m^2) is associated with each wave considered separately (in the absence of the other), and what average power is associated with the combination?

Solution: The wave traveling in the $+\hat{z}$ direction has an average power given by

$$\frac{1}{2} \operatorname{Re}\{\overline{S}\} = \hat{z} \frac{1}{2\eta_0} |\overline{E}|^2 = \hat{z} \frac{1}{2 \cdot 377} = \hat{z} \, 1.3 \times 10^{-3} \text{ (W/m}^2)$$

The wave traveling in the $-\hat{z}$ direction has an average power

$$\frac{1}{2} \operatorname{Re}\{\overline{S}\} = -\hat{z} \, 1.3 \times 10^{-3} a^2 \text{ (W/m}^2)$$

Both waves superimposed have a time-averaged power

$$\begin{aligned} \frac{1}{2} \operatorname{Re}\{\overline{S}\} &= \frac{1}{2} \operatorname{Re}\{\overline{E} \times \overline{H}^*\} \\ &= \frac{1}{2} \operatorname{Re}\left\{ \hat{x} \left(e^{-jkz} + a \, e^{jkz}\right) \times \frac{1}{\eta_0} \hat{y} \left(e^{-jkz} - a \, e^{jkz}\right)^* \right\} \\ &= \hat{z} \frac{1}{2\eta_0} \left(1 - |a|^2\right) \quad \text{(W/m}^2) \end{aligned}$$

Note that the total power flowing in the $+\hat{z}$ direction is the difference between the powers in the $+\hat{z}$ and $-\hat{z}$ directions. Power does not generally superimpose like this unless the two waves are in some sense orthogonal, as they are here.

Example 1.6.2

Repeat Example 1.6.1 for the wave:

$$\overline{E} = \hat{x} \left(e^{-jkz} + a \, e^{-jkz}\right)$$

Solution: The first plane wave has average power

$$\frac{1}{2} \operatorname{Re}\{\overline{S}\} = \hat{z} \frac{1}{2\eta_0} |\overline{E}|^2 = \hat{z} \frac{1}{2\eta_0}$$

while the second plane wave has average power

$$\frac{1}{2} \operatorname{Re}\{\overline{S}\} = \hat{z} \frac{|a|^2}{2\eta_0}$$

Both waves superimposed have average power

$$\frac{1}{2} \operatorname{Re}\{\overline{S}\} = \hat{z} \left(\frac{|1 + a|^2}{2\eta_0}\right) = \hat{z} \left(\frac{1 + |a|^2 + 2\operatorname{Re}\{a\}}{2\eta_0}\right)$$

which is not equal to the sum of the individual wave powers

$$\hat{z} \left(\frac{1}{2\eta_0} + \frac{|a|^2}{2\eta_0}\right)$$

unless a is purely imaginary. Therefore, powers do not superpose here because the two waves are clearly not orthogonal (unless the real part of a is zero).

1.7* THE ENERGY THEOREM

In Section 1.6 we studied the Poynting theorem, which describes conservation of energy for electromagnetic systems. An alternative way of combining Maxwell's equations leads instead to the energy theorem, which describes how the frequency derivatives of electric and magnetic fields are related to energy stored in these fields. Although the energy theorem is less widely used than the Poynting theorem, it will give physical insight into how fast information can be carried by an electromagnetic wave, as well as enable us to relate energy stored in a resonator to the resonator terminal impedance in Section 8.8.

We obtain the energy theorem from the following manipulations of the time-harmonic forms of Faraday's and Ampere's laws (1.4.4) and (1.4.5): (1) Dot $\partial \overline{H}/\partial\omega$ into both sides of the complex conjugate of Faraday's law, (2) dot $-\partial \overline{E}/\partial\omega$ into the complex conjugate of Ampere's law, (3) dot $\overline{H}^*$ into the derivative of Faraday's law with respect to frequency ω, and (4) dot $-\overline{E}^*$ into the derivative of Ampere's law with respect to ω

$$\frac{\partial \overline{H}}{\partial\omega} \cdot (\nabla \times \overline{E}^*) = \frac{\partial \overline{H}}{\partial\omega} \cdot (j\omega\overline{B}^*) \tag{1.7.1}$$

$$-\frac{\partial \overline{E}}{\partial\omega} \cdot (\nabla \times \overline{H}^*) = -\frac{\partial \overline{E}}{\partial\omega} \cdot (-j\omega\overline{D}^* + \overline{J}^*) \tag{1.7.2}$$

$$\overline{H}^* \cdot \left(\nabla \times \frac{\partial \overline{E}}{\partial\omega}\right) = \overline{H}^* \cdot \left(-j\overline{B} - j\omega\frac{\partial \overline{B}}{\partial\omega}\right) \tag{1.7.3}$$

$$-\overline{E}^* \cdot \left(\nabla \times \frac{\partial \overline{H}}{\partial\omega}\right) = -\overline{E}^* \cdot \left(j\overline{D} + j\omega\frac{\partial \overline{D}}{\partial\omega} + \frac{\partial \overline{J}}{\partial\omega}\right) \tag{1.7.4}$$

Equations (1.7.1)–(1.7.4) are the mathematical representations of these four steps, where we note that the time-dependent factor $e^{j\omega t}$ is **not** part of any of these phasors. If we sum (1.7.1)–(1.7.4) and use the identity $\nabla \cdot (\overline{A} \times \overline{B}) = \overline{B} \cdot (\nabla \times \overline{A}) - \overline{A} \cdot (\nabla \times \overline{B})$, we find

$$\nabla \cdot \left(\frac{\partial \overline{E}}{\partial\omega} \times \overline{H}^* + \overline{E}^* \times \frac{\partial \overline{H}}{\partial\omega}\right) = j\omega \left(\overline{B}^* \cdot \frac{\partial \overline{H}}{\partial\omega} - \overline{H}^* \cdot \frac{\partial \overline{B}}{\partial\omega}\right)$$

$$- j\omega \left(\overline{D}^* \cdot \frac{\partial \overline{E}}{\partial\omega} - \overline{E}^* \cdot \frac{\partial \overline{D}}{\partial\omega}\right) - j(\overline{H}^* \cdot \overline{B} + \overline{E}^* \cdot \overline{D}) \tag{1.7.5}$$

$$- \left(\frac{\partial \overline{E}}{\partial\omega} \cdot \overline{J}^* + \overline{E}^* \cdot \frac{\partial \overline{J}}{\partial\omega}\right)$$

If the medium is linear (but possibly dispersive), so that $\underline{\overline{D}} = \epsilon(\omega)\underline{\overline{E}}$ and $\underline{\overline{B}} = \mu(\omega)\underline{\overline{H}}$, we can rewrite (1.7.5) as

$$\frac{1}{4} \nabla \cdot \left(\frac{\partial \underline{\overline{E}}}{\partial \omega} \times \underline{\overline{H}}^* + \underline{\overline{E}}^* \times \frac{\partial \underline{\overline{H}}}{\partial \omega} \right) = -j \left[\langle W_e \rangle + \langle W_m \rangle \right]$$

$$- \frac{j\omega}{4} \left(\frac{\partial \mu}{\partial \omega} |\overline{H}|^2 + \frac{\partial \epsilon}{\partial \omega} |\overline{E}|^2 \right) - \frac{1}{4} \left(\frac{\partial \underline{\overline{E}}}{\partial \omega} \cdot \underline{\overline{J}}^* + \underline{\overline{E}}^* \cdot \frac{\partial \underline{\overline{J}}}{\partial \omega} \right) \tag{1.7.6}$$

where we have divided each term in (1.7.6) by four and the stored electric and magnetic energy densities $\langle W_e \rangle$ and $\langle W_m \rangle$ are given by (1.6.31) and (1.6.32). Equation (1.7.6) is the differential form of the *energy theorem*, relating the frequency derivatives of power-like terms to the total stored energy density $\langle W_T \rangle = \langle W_e \rangle + \langle W_m \rangle$. The integral form of the energy theorem is found by integrating (1.7.6) over an arbitrary volume V and then applying Gauss's divergence theorem (1.6.7):

$$\frac{1}{4} \oint_A \left(\frac{\partial \underline{\overline{E}}}{\partial \omega} \times \underline{\overline{H}}^* + \underline{\overline{E}}^* \times \frac{\partial \underline{\overline{H}}}{\partial \omega} \right) \cdot \hat{n} \, da = -j \left[\langle w_e \rangle + \langle w_m \rangle \right]$$

$$- \frac{j\omega}{4} \int_V \left(\frac{\partial \mu}{\partial \omega} |\overline{H}|^2 + \frac{\partial \epsilon}{\partial \omega} |\overline{E}|^2 \right) dv - \frac{1}{4} \int_V \left(\frac{\partial \underline{\overline{E}}}{\partial \omega} \cdot \underline{\overline{J}}^* + \underline{\overline{E}}^* \cdot \frac{\partial \underline{\overline{J}}}{\partial \omega} \right) dv \tag{1.7.7}$$

The quantities $\langle w_e \rangle$ and $\langle w_m \rangle$ are the total electric and magnetic energies stored in the volume V.

We shall interpret the energy theorem in later sections as it becomes relevant; here, we check that it applies to a free space uniform plane wave with electric fields given by (1.4.10) and (1.4.11):

$$\underline{\overline{E}} = \hat{x} \, E_0 \, e^{-jkz} \tag{1.7.8}$$

$$\underline{\overline{H}} = \hat{y} \, \frac{E_0}{\eta_0} \, e^{-jkz} \tag{1.7.9}$$

Here, E_0 is a complex constant, k is a function of frequency given by the dispersion relation (1.3.10), and free space is dispersionless so that $\partial \epsilon / \partial \omega = \partial \mu / \partial \omega = 0$. The left side of (1.7.6) is then

$$\frac{1}{4} \nabla \cdot \left\{ \hat{x} \left(-j \frac{\partial k}{\partial \omega} z \right) E_0 e^{-jkz} \times \hat{y} \frac{E_0^*}{\eta_0} e^{+jkz} + \right.$$

$$\hat{x} E_0^* e^{+jkz} \times \hat{y} \left(-j \frac{\partial k}{\partial \omega} z \right) \frac{E_0}{\eta_0} e^{-jkz} \right\}$$

$$= \frac{1}{4} \nabla \cdot \left\{ \hat{z} \left(-2j \frac{\partial k}{\partial \omega} z \frac{|E_0|^2}{\eta_0} \right) \right\}$$

$$= -j \frac{\partial k}{\partial \omega} \frac{|E_0|^2}{2\eta_0} \tag{1.7.10}$$

which is equal to $-j(\partial \overline{k}/\partial \omega) \cdot \langle \overline{S} \rangle$ for an arbitrary uniform plane wave in free space. The right side of (1.7.6) is just $-j$ times the total time-averaged energy density $\langle W_T \rangle$, since $\overline{J} = 0$ for a uniform plane wave in free space:

$$-j\langle W_T \rangle = -j \left(\frac{1}{4} \epsilon |\overline{E}|^2 + \frac{1}{4} \mu |\overline{H}|^2 \right) = -j \left(\frac{1}{4} \epsilon_0 |E_0|^2 + \frac{1}{4} \mu_0 \frac{|E_0|^2}{\eta_0^2} \right)$$

$$= -\frac{j}{2} \epsilon_0 |E_0|^2 \tag{1.7.11}$$

The last equality follows because $\mu_0/\eta_0^2 = \epsilon_0$. As expected, the electric and magnetic energy densities for the uniform plane wave are equal, a result we also showed using the complex Poynting theorem. We now equate (1.7.10) and (1.7.11), yielding

$$\frac{\partial \overline{k}}{\partial \omega} \cdot \langle \overline{S} \rangle = \langle W_T \rangle \tag{1.7.12}$$

for the general free space plane wave. Since $\overline{k}$ and $\overline{S}$ are parallel in this case,

$$\frac{\partial \omega}{\partial k} = \frac{|\langle \overline{S} \rangle|}{\langle W_T \rangle} = \frac{1}{\epsilon_0 \eta_0} = \frac{1}{\sqrt{\mu_0 \epsilon_0}} = c \tag{1.7.13}$$

which is confirmed by the dispersion relation $\omega = k/\sqrt{\mu_0\epsilon_0} = ck$ for a uniform plane wave in free space. The quantity $\partial \omega/\partial k$ is called the *group velocity*, and is the velocity at which energy (information) is propagated in the wave. That is, the velocity of energy propagation (m/s) is equal to the average power (J/s) divided by the average energy density (J/m). The differences between group and phase velocities will be the subject of Section 3.5. For now, we simply note that in an unbounded isotropic medium (such as free space), the group velocity $\partial \omega/\partial k$ and phase velocity ω/k are the same.

1.8* UNIQUENESS

So far, we have assumed that if charge and current are specified along with appropriate boundary and initial conditions, Maxwell's equations will produce a unique field solution. In this section, we prove this *uniqueness* property and describe under what conditions uniqueness fails. Our general strategy is to assume that a unique solution does **not** exist and then show that this assumption leads to a contradiction.

We shall begin by considering a volume V (which may be infinite in size) bounded by the closed surface S which contains charge density ρ and current density $\overline{J}$. The medium inside the volume possesses a constant permeability μ and constant permittivity ϵ so that the linear constitutive relations (1.2.10) and (1.2.11) apply. At some initial time, we specify all fields at each point within the volume V, and we also specify each field on the surface S for all subsequent times. We shall see that we have greatly overconstrained the problem; i.e., these conditions will guarantee uniqueness but not all of them are necessary. After we demonstrate uniqueness, we shall see how we can weaken the boundary conditions and the requirements on ϵ and μ while still maintaining a unique field solution.

If we assume that two distinct field solutions exist, then each set independently satisfies Maxwell's equations with source terms ρ and $\overline{J}$:

$$
\begin{aligned}
\nabla \cdot \overline{D}_1 &= \rho & \nabla \cdot \overline{D}_2 &= \rho \\
\nabla \cdot \overline{B}_1 &= 0 & \nabla \cdot \overline{B}_2 &= 0 \\
\nabla \times \overline{E}_1 &= -\partial \overline{B}_1/\partial t & \nabla \times \overline{E}_2 &= -\partial \overline{B}_2/\partial t \\
\nabla \times \overline{H}_1 &= \overline{J} + \partial \overline{D}_1/\partial t & \nabla \times \overline{H}_2 &= \overline{J} + \partial \overline{D}_2/\partial t
\end{aligned}
\tag{1.8.1}
$$

We now form the difference fields $\overline{E}_d = \overline{E}_1 - \overline{E}_2$, $\overline{H}_d = \overline{H}_1 - \overline{H}_2$, $\overline{B}_d = \overline{B}_1 - \overline{B}_2$, and $\overline{D}_d = \overline{D}_1 - \overline{D}_2$. If we can show that all of these difference fields vanish within V for all time, we shall have shown that only one field solution $\{\overline{E}_1, \overline{H}_1, \overline{D}_1, \overline{B}_1\}$ equal to $\{\overline{E}_2, \overline{H}_2, \overline{D}_2, \overline{B}_2\}$ exists and this solution is therefore unique.

By subtracting the fields on the right in (1.8.1) from those on the left, we easily see that the difference fields themselves satisfy Maxwell's equations with no source terms:

$$
\nabla \cdot \overline{D}_d = 0
$$

$$
\nabla \cdot \overline{B}_d = 0
$$

$$
\nabla \times \overline{E}_d = -\frac{\partial \overline{B}_d}{\partial t}
\tag{1.8.2}
$$

$$
\nabla \times \overline{H}_d = \frac{\partial \overline{D}_d}{\partial t}
$$

The difference fields obey the simple initial conditions:

$$
\overline{E}_d(\overline{r}, t = 0) = \overline{H}_d(\overline{r}, t = 0) = \overline{D}_d(\overline{r}, t = 0) = \overline{B}_d(\overline{r}, t = 0) = 0
\tag{1.8.3}
$$

since both fields 1 and 2 are equal to the field specified by the initial conditions. Similarly, fields 1 and 2 must also equal the field specified on the surface S for all time, so the difference fields must vanish at that boundary:

$$
\overline{E}_d(\overline{r} \in S, t) = \overline{H}_d(\overline{r} \in S, t) = \overline{D}_d(\overline{r} \in S, t) = \overline{B}_d(\overline{r} \in S, t) = 0
\tag{1.8.4}
$$

Equations (1.8.3) and (1.8.4) will be essential to the proof of uniqueness. In effect, they state that many fields will satisfy Maxwell's equations, but in order to uniquely specify one particular field solution, that field will not only have to obey Maxwell's equations with particular sources, but will also have to satisfy specific initial conditions and boundary conditions.

With the source-free Maxwell's equations given by (1.8.2), we can repeat the Poynting theorem derivation in Section 1.6, arriving at the integral form of the

Poynting theorem for the difference fields which may be compared to (1.6.6)

$$\int_V \left(\overline{H}_d \cdot \frac{\partial \overline{B}_d}{\partial t} + \overline{E}_d \cdot \frac{\partial \overline{D}_d}{\partial t} \right) dv + \oint_S (\overline{E}_d \times \overline{H}_d) \cdot \hat{n}\, da = 0 \qquad (1.8.5)$$

From (1.8.4), we know that the electric and magnetic difference fields must vanish on the surface S of the volume V, forcing the surface term in (1.8.5) to vanish.[7] We may also rewrite the first two terms in (1.8.5) using the constitutive relations $\overline{D}_d = \epsilon \overline{E}_d$ and $\overline{B}_d = \mu \overline{H}_d$, yielding the useful expression:

$$\frac{d}{dt} \int_V \left(\frac{1}{2}\mu |\overline{H}_d|^2 + \frac{1}{2}\epsilon |\overline{E}_d|^2 \right) dv = 0 \qquad (1.8.6)$$

where we recognize the integrand of (1.8.6) as the energy stored in the difference fields. We therefore can interpret (1.8.6) as stating that the total energy of the difference fields does not change with time and evaluate the constant energy by considering the one time ($t = 0$) at which we know $\overline{E}_d$ and $\overline{H}_d$ at every point within V:

$$\frac{1}{2}\mu |\overline{H}_d(t)|^2 + \frac{1}{2}\epsilon |\overline{E}_d(t)|^2 = \frac{1}{2}\mu |\overline{H}_d(t=0)|^2 + \frac{1}{2}\epsilon |\overline{E}_d(t=0)|^2 = 0 \qquad (1.8.7)$$

The last equality follows from the initial conditions in (1.8.3). If ϵ and μ are positive quantities, then the left side of (1.8.7) is always nonnegative, and can be zero only when $\overline{E}_d$ and $\overline{H}_d$ are both zero at every point in the volume V at all times. This is another way of saying that $\overline{E}_1$ and $\overline{E}_2$ are identical and therefore equal to a unique field, as are $\overline{H}_1$ and $\overline{H}_2$. The magnetic flux $\overline{B}$ and displacement field $\overline{D}$ are therefore also unique via the constitutive relations. We have thus shown that by specifying charge and current densities, media parameters ϵ and μ (which are both positive constants), and boundary and initial conditions, only a single, unique field solution exists.

We have been rather restrictive in how ϵ and μ must be specified and also in how we impose boundary conditions at the surface S. We shall now relax these conditions and see what is necessary to maintain uniqueness. We recall that boundary conditions were used to eliminate the term $\oint_S (\overline{E}_d \times \overline{H}_d) \cdot \hat{n}\, da$ from (1.8.5). Since the normal component of $\overline{E}_d \times \overline{H}_d$ alone is relevant to the surface integral, we see that only the tangential components of E_d or H_d need vanish to guarantee the disappearance of this surface term in (1.8.6). It is therefore possible to demonstrate uniqueness if **only** the tangential components of $\overline{E}$ are specified at every point on the surface (while $\overline{H}$ and the normal component of $\overline{E}$ are left unspecified), or if the tangential components of $\overline{H}$ alone are specified on the boundary.

---

7. Care must be taken when integrating the second term of (1.8.5) when the surface is infinitely large. For sources of finite extent, the electric and magnetic field amplitudes decrease by at least $1/r$ where r is the distance from the sources. The Poynting power in the difference fields $\overline{E}_d \times \overline{H}_d$ therefore goes as α/r^2 where $\alpha = 0$ on the boundary and the surface integral $\alpha/r^2 \cdot 4\pi r^2 = 4\pi\alpha$ does vanish, even though the surface is infinitely large.

It is also possible to impose a mixed boundary condition specifying the tangential electric field components at some places on the surface and the tangential magnetic field components at the rest, with the normal fields unconstrained insofar as they are consistent with Maxwell's equations.

Likewise, we used the constitutive relation $\overline{D}_d = \epsilon \overline{E}_d$ to replace $\overline{E}_d \cdot \partial \overline{D}_d / \partial t$ with $\frac{\partial}{\partial t}(\frac{1}{2}\epsilon |\overline{E}_d|^2)$. However, this did not require that ϵ be a constant; an inhomogeneous permittivity $\epsilon(\overline{r})$ would still enable this substitution.[8] The only problems in demonstrating uniqueness would occur if ϵ or μ were time-dependent, or if the medium were nonlinear, which would imply that ϵ was a function of the electric field or μ was a function of the magnetic field. In either of these cases, the volume integral in (1.8.5) could not be expressed as a total time derivative of a nonnegative quantity. It is also necessary that ϵ and μ have the same sign at all points within V. Since both permeability and permittivity are positive quantities, this does not restrict the validity of the uniqueness theorem.[9]

In summary, a medium with positive, time-independent, field-independent permeability and permittivity will give rise to a unique solution for the electric and magnetic fields as long as: (1) the initial conditions $\overline{E}(\overline{r}, t = 0)$ and $\overline{H}(\overline{r}, t = 0)$ are given, and (2) the tangential components of $\overline{E}$ or $\overline{H}$ are specified at every point on the boundary for $t > 0$.

1.9 NONELECTROMAGNETIC WAVES

We should also be aware that electromagnetic wave phenomena have counterparts in acoustics, mechanics, and other fields, and that the results presented in this text have applications beyond electromagnetics alone. In this section we briefly consider a few examples of nonelectromagnetic waves.

Many nonelectromagnetic systems obey a wave equation similar to (1.3.7) for small amplitude signals and can therefore propagate waves. It is usually necessary to consider only small amplitude waves so that higher order terms due to system nonlinearities may be neglected.

Consider, first, a straight string supported at its two ends and in constant tension T (kg $\cdot$ m/s^2). Let the lateral displacement of the string from the $\hat{z}$-axis be denoted by $y(z)$ and the mass density of the string by ρ (kg/m) as illustrated in

---

8. In fact, ϵ could even be a symmetric tensor with constitutive relation $\overline{D}_d = \overline{\overline{\epsilon}} \cdot \overline{E}_d$ without affecting the main demonstration of uniqueness. If $\overline{\overline{\epsilon}} = \overline{\overline{\epsilon}}^T$, then $\overline{E}_d \cdot \partial \overline{D}_d / \partial t$ would become $\frac{\partial}{\partial t}(\frac{1}{2}\overline{E}_d \cdot \overline{\overline{\epsilon}} \cdot \overline{E}_d)$ and the stored electric energy would be equal to the nonnegative quantity $\frac{1}{2}\overline{E}_d \cdot \overline{\overline{\epsilon}} \cdot \overline{E}_d$ if $\overline{\overline{\epsilon}}$ were a positive definite matrix.

9. We have also omitted the possibility that the medium might be conductive, and we shall postpone a discussion of conductivity until Section 3.4. It is not difficult to show, however, that a conductivity with the same attributes as either permeability or permittivity (i.e., positive, field-independent and time-independent) will only slightly modify the demonstration of uniqueness, and so uniqueness will still apply for fields with appropriate boundary and initial conditions.

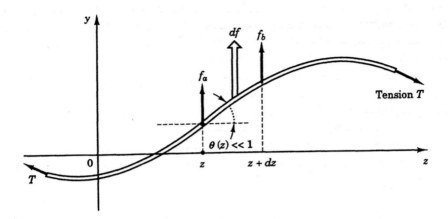

Figure 1.6   Incremental model of a string.

Figure 1.6. The equations of motion for the string can be derived by considering the dynamics of a single incremental element of length dz, as illustrated.

The net external force applied to the string element in the $+\hat{y}$ direction is $f_b - f_a$, where f_a and f_b are the $\hat{y}$-directed forces at the two ends of the incremental element:

$$f_a \cong T \left.\frac{dy}{dz}\right|_z$$

$$f_b \cong T \left.\frac{dy}{dz}\right|_{z+dz} \tag{1.9.1}$$

Therefore,

$$f_b - f_a \cong T \left[\left.\frac{dy}{dz}\right|_{z+dz} - \left.\frac{dy}{dz}\right|_z \right] = T \frac{d^2y}{dz^2} \cdot dz \tag{1.9.2}$$

We require a second equation relating forces and string shape, and Newton's force law (force = mass times acceleration) provides it; i.e., the net $\hat{y}$-directed force df on the segment dz is

$$df = (\rho \, dz) \frac{d^2y}{dt^2} \tag{1.9.3}$$

But since $df = f_b - f_a$, we find that the string displacement $y(z, t)$ obeys the wave equation

$$\frac{d^2y}{dz^2} - \frac{\rho}{T} \frac{d^2y}{dt^2} = 0 \tag{1.9.4}$$

One possible solution is

$$y(z, t) = A \cos(z - c_s t) \tag{1.9.5}$$

where the wave velocity c_s is $\sqrt{T/\rho}$. Note that increased tension and decreased mass density both result in higher wave velocities.

The key approximations that result in the linearized string wave equation (1.9.4) are: (1) the string is inelastic, and (2) dy/dz is sufficiently small that the tension T is approximately constant for all z. If these approximations are not valid, then the waves typically distort as they move and interact slightly when two waves are superimposed.

Three-dimensional waves occur in acoustics and can also be characterized by a wave equation. A rigorous derivation of the acoustic wave equation is deferred until Section 10.2 [see (10.2.15)], but the linearized small-signal acoustic wave equation is just

$$\left(\nabla^2 - \frac{\rho_0}{\gamma p_0}\frac{\partial^2}{\partial t^2}\right)p = 0 \tag{1.9.6}$$

The perturbation pressure of the wave p (N/m^2) thus obeys a simple wave equation that is valid only when p is much less than the average pressure p_0. The velocity of sound is easily found from the wave equation to be $c_s = \sqrt{\gamma p_0/\rho_0}$, where ρ_0 is the average gas density (kg/m^3) and γ is a constant near unity (~ 1.41); γ is the ratio of the specific heat at constant pressure to that at constant volume. As in the case of the string, the velocity is proportional to the square root of the ratio of pressure (tension) to mass density, or approximately 344 m/s in air at $20°$ C and one atmosphere pressure. Several approximations are necessary to linearize the equations of motion for sound, and we have neglected convection, thermal effects, the statistics of molecular collisions, and so forth.

The third example is even more removed from electromagnetic waves because the interacting elements are discrete, well-spaced, and governed by human decisions rather than physical forces. Consider an infinite chain of automobiles on a highway. The cars move with velocity $dz/dt = U = \langle U \rangle + u(z, t)$, and the space between two adjacent cars is $D(z, t)$. If each driver accelerates his car to equalize the space between his car and the cars in front and behind him, then his acceleration du/dt might be governed by

$$\frac{du}{dt} = J \cdot \frac{dD}{dz} \tag{1.9.7}$$

where J is the "jumpiness" factor of the driver. In addition, we know that the space D between two cars will approximately change as

$$\frac{dD}{dt} \cong \langle D \rangle \frac{du}{dz} \tag{1.9.8}$$

where $\langle D \rangle$ is the average value of D, and $|D - \langle D \rangle| \ll D$.

Combining (1.9.7) and (1.9.8) we find

$$\frac{d^2u}{dz^2} - \frac{1}{\langle D \rangle J}\frac{d^2u}{dt^2} = 0 \tag{1.9.9}$$

which is the standard wave equation, but now applied to density waves of automobiles in a line of traffic. The velocity of these waves relative to the moving line of cars is $\pm\sqrt{J\langle D\rangle}$, where the waves are perturbations to the car density $1/D$ (cars per meter) and to the average velocity $\langle U\rangle$. The average automobile velocity $\langle U\rangle$ must be added to yield the total wave velocity because the wave equation is in terms of the incremental velocity u of a moving automobile. Note that $\sqrt{J\langle D\rangle}$ may sometimes equal $\langle U\rangle$, and then backward-moving waves can even be stationary relative to the ground. Traffic jams can thus sometimes be long lasting in one spot on a straight road without any apparent reason.

This last example required even more approximations than before. For example, waves moving in the same direction as the traffic would require drivers constantly to look behind them, which is not realistic. Furthermore, the notion that there is any rational plan to driving, let alone that everyone adopts the same plan, is also a gross approximation, particularly in Boston. Nonetheless, the observation of density waves moving through traffic is a common occurrence on crowded highways.

The range of situations that exhibit wave phenomena is limited largely by one's imagination. The principal requirement is that many similar independent units be coupled to their nearest neighbors in some simple way and that the motion of each unit be characterized by two coupled variables such as $\{dy/dz, dy/dt\}$, $\{p, u\}$, or $\{u, D\}$ for the three examples illustrated here, i.e., strings, acoustics, and automobile traffic, respectively. An extended discussion of acoustic counterparts to most electromagnetic phenomena is found in Chapter 10.

1.10 SUMMARY

The classical behavior of charged particles is explained by hypothesizing the existence of electric and magnetic fields through which the charges interact. A charge distribution generates an *electric field*; if the charges are moving (so that a current exists), then a *magnetic field* is also generated. The *Lorentz force law* describes how electric and magnetic fields exert forces on charges, and *Maxwell's equations* relate these fields to each other and to the positions and motions of the charges.

$$\nabla \times \overline{E} = -\frac{\partial \overline{B}}{\partial t}$$

$$\nabla \times \overline{H} = \overline{J} + \frac{\partial \overline{D}}{\partial t}$$

$$\nabla \cdot \overline{D} = \rho$$

$$\nabla \cdot \overline{B} = 0$$

Although an infinite number of solutions to Maxwell's equations exist, we considered only a few special cases in Chapter 1, namely *uniform plane waves* in vacuum, where the *constitutive relations* for free space are given by

$$\overline{D} = \epsilon_0 \overline{E}$$

$$\overline{B} = \mu_0 \overline{H}$$

If boundary and initial conditions are properly applied to an electromagnetics problem, then Maxwell's equations are shown to produce a *unique* solution provided that the constitutive relations are linear.

Both the electric and magnetic fields obey the homogeneous *wave equation* in vacuum

$$\left(\nabla^2 - \mu_0 \epsilon_0 \frac{\partial^2}{\partial t^2} \right) \left\{ \begin{matrix} \overline{E} \\ \overline{H} \end{matrix} \right\} = 0$$

where the electric field that satisfies the wave equation can be written quite generally as

$$\overline{E}(\overline{r}, t) = \overline{E}_0 f(\omega t - \overline{k} \cdot \overline{r})$$

and f can be an arbitrary function of its argument $(\omega t - \overline{k} \cdot \overline{r})$. This wave propagates in the $\hat{k}$ direction with velocity $c = \omega/|k| = 1/\sqrt{\mu_0 \epsilon_0}$ in vacuum. The relation between the *angular frequency* ω and *wave number* $\overline{k}$ of a wave is called the *dispersion relation*. The *wavelength* λ is $2\pi/|\overline{k}|$, and the *frequency* f is $\omega/2\pi$. In order for **all** of Maxwell's equations to be satisfied in free space, the direction in which the electric field points ($\overline{E}_0$) must be perpendicular to $\overline{k}$.

An alternative representation for a wave at a single frequency ω is

$$\overline{E}(\overline{r}, t) = \text{Re} \left\{ \underline{\overline{E}}(\overline{r}) e^{j\omega t} \right\}$$

Substitution of this *phasor representation* into Maxwell's equations yields the *time-harmonic Maxwell's equations*:

$$\nabla \times \underline{\overline{E}} = -j\omega \underline{\overline{B}}$$

$$\nabla \times \underline{\overline{H}} = \underline{\overline{J}} + j\omega \underline{\overline{D}}$$

$$\nabla \cdot \underline{\overline{D}} = \underline{\rho}$$

$$\nabla \cdot \underline{\overline{B}} = 0$$

In complex notation, a sinusoidal wave propagating in the $+\hat{z}$ direction can be represented as

$$\underline{\overline{E}}(z) = \underline{\overline{E}}_0 e^{-jkz} = \overline{E}_0 e^{j\phi_0 - jkz}$$

where the frequency dependence $e^{j\omega t}$ is separated from the spatial portion of the wave. Thus,

$$\overline{E}(\overline{r}, t) = \text{Re} \left\{ \underline{\overline{E}}_0 e^{-jkz} e^{j\omega t} \right\} = \overline{E}_0 \cos(\omega t - kz + \phi_0)$$

Because Maxwell's equations are linear, a solution that satisfies all of the boundary conditions for a given problem may be constructed by *superposing* a possibly infinite number of time-harmonic solutions of differing frequencies. Therefore, understanding the behavior of a single-frequency wave has greater generality than might at first be expected.

In both notations, the *polarization* of a uniform plane wave at frequency ω can be described by a single complex number $\underline{A} = A\, e^{j\phi}$. If the electric field phasor is $\overline{E} = \hat{x} + A\, e^{j\phi}\hat{y}$, then circular polarization is described by $A = 1, \phi = \pm\pi/2$ and linear polarization by $\phi = 0, \pi$. These are special cases of the more general elliptical polarization. Waves of arbitrary spectral content (i.e., containing more than one frequency) can be described as linear superpositions of the waves with differing values of ω.

A very useful theorem derived from Maxwell's equations relates dissipation and changes in energy storage in a volume to the total flux of electromagnetic energy across a surface bounding that volume. This flux is the *Poynting vector* $\overline{S} = \overline{E} \times \overline{H}$, or $\underline{S} = \underline{E} \times \underline{H}^*$ in complex notation. The *complex Poynting vector* is not the phasor part of the real Poynting vector; instead, the time-averaged real Poynting vector is one-half the real part of the complex Poynting vector. The *Poynting theorem* is essentially a statement that energy is conserved and that the Poynting vector represents power (W/m^2).

Another theorem derived from Maxwell's equations is the *energy theorem*, which relates the frequency derivatives of field quantities to the total energy stored in these fields. This theorem will prove especially useful for understanding how resonators couple to the external world in Section 8.8.

The generality of wave phenomena is illustrated by considering waves on strings, in gases, and in lines of automobile traffic, all of which obeyed *wave equations* for which the second derivatives of variables with respect to time and space were equal to within a constant factor. These waves propagate at different characteristic velocities and most of the properties of electromagnetic waves treated in the following chapters are also observed in nonelectromagnetic waves. In Chapter 10, we study the special case of acoustic wave propagation in gases, displaying the many analogies between acoustics and electromagnetics.

1.11 PROBLEMS

1.2.1 What is the force density (N/m) on a wire carrying 1 A at 45° from a magnetic field of 1 tesla (equal to 1 weber/m^2 = 10,000 gauss)?

1.2.2 If $\overline{J} = \hat{x}\,(1 + 2x)$ A/m^2, what is $\partial\rho/\partial t$?

1.2.3 A test charge of 1 C moving at 1 m/s experiences the following forces: (i) $\overline{f} = 3\hat{x} - \hat{y} + 2\hat{z}$ when $\overline{v} = \hat{x}$, (ii) $\overline{f} = 2\hat{x} - 2\hat{y} - \hat{z}$ when $\overline{v} = \hat{y}$, and (iii) $\overline{f} = 2\hat{x} + \hat{z}$ when $\overline{v} = \hat{z}$. What are the electric and magnetic fields $\overline{E}$ and $\overline{B}$ causing these forces?

1.3.1 The moon is about 400,000 km from Earth. How long would it take a radio wave to travel that distance? Assume that the space between the Earth and moon is essentially vacuum.

1.3.2 The known spectrum of electromagnetic waves covers a wide range of frequencies. Radio waves, television signals, radar beams, visible light, X-rays, and gamma rays are all examples of electromagnetic waves.

(a) Give in meters the wavelength corresponding to the following frequencies: (i) 60 Hz, (ii) AM radio (535–1605 kHz), (iii) FM radio (88–106 MHz), (iv) C-band (4–8 GHz, used in satellite communication), (v) visible light ($7.5 \times 10^{14} - 1.75 \times 10^{15}$ Hz), and (vi) x-rays ($\simeq 10^{18}$ Hz).

(b) Give in hertz the frequencies corresponding to the following wavelengths: (i) 1 km, (ii) 1 m, (iii) 1 mm, (iv) 1 μm, and (v) 1 Angstrom ($= 10^{-10}$ m).

1.3.3 An electromagnetic wave traveling in the positive $\hat{z}$ direction passes an observer at $z = 0$. He measures an electric field $\overline{E} = \hat{x} E_0 t / T$ for $0 < t < T$ and zero for all other times. Plot the magnitude of the field as function of z at two later times: (i) $2T$ and (ii) $3.5T$. Label axes clearly.

1.3.4 Consider the following complex field vectors in free space: $\overline{E}_1 = \hat{y} E_0 e^{-3jkz}$, $\overline{E}_2 = \hat{x} E_0 e^{jkz}$, $\overline{E}_3 = (\hat{x} + 2j\hat{z}) E_0 e^{jky}$.

(a) Which of these satisfy the wave equation $(\nabla^2 + \omega^2 \mu_0 \epsilon_0)\overline{E} = 0$? If they do, what is the relationship between k and ω?

(b) Which are electromagnetic waves? For those that are not, which of Maxwell's equations do they violate?

1.3.5 Consider an electromagnetic wave in free space having $\overline{E} = \hat{x} e^{j3\pi y}$.

(a) Compute $\nabla \times \overline{E}$ and $\nabla \cdot \overline{E}$ in Cartesian coordinates and show that they are consistent with Maxwell's equations.

(b) Find the corresponding expression for $\overline{H}$.

(c) What are the wavelength λ and the direction of propagation for this wave?

(d) Sketch the values of $\overline{E}(y, t = 0)$ and $\overline{H}(y, t = 0)$ as a function of position along the $\hat{y}$-axis.

1.3.6 A uniform plane wave in vacuum has a complex electric field given by $\overline{E} = \hat{x} e^{j2\pi y}$

(a) Compute $\nabla \times \overline{E}$ and $\nabla \cdot \overline{E}$ in Cartesian coordinates and use the results together with Gauss's law to show that $\overline{E}$ is consistent with Maxwell's equations.

(b) Find the corresponding $\overline{H}(x, y, z)$; show it too is consistent with Gauss's law.

(c) What is the wavelength and direction of propagation for this wave?

(d) For $t = 0$, sketch $\overline{E}$ and $\overline{H}$ along the $\hat{y}$-axis.

(e) How many volts per meter can be measured when this wave has a magnetic field comparable to that of the Earth, say 0.5 gauss (1 tesla = 10,000 gauss)?

1.4.1 (a) For $\overline{A} = \hat{x} + 3\hat{y} - \hat{z}$, and $\overline{B} = 2\hat{x} - \hat{y} + \hat{z}$, find: (i) $\overline{A} + \overline{B}$, (ii) $\overline{A} \cdot \overline{B}$, (iii) $\overline{A} \times \overline{B}$, (iv) the angle between $\overline{A}$ and $\overline{B}$, and (v) the vector perpendicular to both $\overline{A}$ and $\overline{B}$ and with the magnitude of $\overline{A}$.

(b) For $\overline{E}(x, t) = \hat{y} \cos(\omega t - kx) + \hat{z} \sin(\omega t - kx)$, find: (i) $\nabla \cdot \overline{E}$, (ii) $\nabla \times \overline{E}$, (iii) the magnetic flux density $\overline{B}$, and (iv) $\underline{E}(x)$, if $\overline{E}(x, t) = \text{Re}\{\underline{E}(x) e^{j\omega t}\}$.

1.4.2 Let $a = 1 - j$, $b = 1 + 3j$.
   (a) Compute $a + b$, $a\,b$ and a/b in real/imaginary form.
   (b) Evaluate $a\,b$ and a/b in magnitude and angle form.
   (c) Find the square roots of $2j$ and $3 + 4j$.
   (d) For $x + jy = e^{j7\pi/3}$, find x and y.

1.4.3 Define $v(t) = \text{Re}\{V\,e^{j\omega t}\}$.
   (a) Find $v(t)$, if possible, for (i) $V = j$, (ii) $V = e^{j\pi/4}$, (iii) $V = (1 + j)\,e^{-j\pi/2}$, and
       (iv) $V = 2\hat{x} + j\hat{y}$.
   (b) Find V, if possible, for (i) $v(t) = \cos(\omega t + \phi)$, (ii) $v(t) = \sin \omega t + \sqrt{2}\cos \omega t$,
       (iii) $v(t) = 2 \sin \omega t \cos \omega t$, and (iv) $v(t) = \hat{x}\,E_0 \cos(kz - \omega t)$.
   (c) Find the time average of $v(t)$ and $v^2(t)$ for $V = 1 - j$.

1.4.4 Let $a = 2 + j$ and $b = 1 - 3j$.
   (a) Compute $a + b$, $a - b$, $a\,b$, and a/b; give the answers in real/imaginary form.
   (b) Evaluate $a\,b$ and a/b; give the answers in magnitude/angle form.
   (c) Find two values of x that satisfy $x^2 = 3 + 4j$.
   (d) Represent $v(t)$ as a phasor $V\,e^{j\phi}$ for the case (i) $v(t) = \cos(\omega t - \pi/8)$, $v(t) = 4 \sin \omega t - 3 \cos \omega t$.
   (e) Find $w(t)$ if $W\,e^{j\phi} = 3 - 4j$.
   (f) Find $a(t)$, $b(t)$, and the time-average of $a(t)\,b(t)$.

1.4.5 (a) Let $\overline{A} = 2\hat{x} - \hat{y} + \hat{z}$ and $\overline{B} = \hat{x} + 2\hat{y} + \hat{z}$. Find (i) $\overline{A} + \overline{B}$, (ii) $\overline{A} \cdot \overline{B}$, (iii) $\overline{A} \times \overline{B}$,
       (iv) the magnitude of $\overline{A}$, and (v) the angle between $\overline{A}$ and $\overline{B}$.
   (b) Let $\overline{A} = \hat{x} + (1 - j)\hat{y} + j\hat{z}$. What is $\overline{a}(t)$ corresponding to this phasor?
   (c) If $\overline{b}(t) = 2\hat{x} \cos(\omega t - \pi/4) + \hat{y} \sin(\omega t + \pi/2)$, what is the corresponding phasor
       $\overline{B}$?
   (d) Sketch the locus of points traced by the tip of the vector $\overline{d}(t)$ if $\overline{D} = j\hat{x} + 2\hat{y}$.

1.5.1 For each of the uniform plane waves in vacuum characterized below, give the direction of propagation, the wavelength in meters, the frequency (Hz) and sketch the polarization ellipse: (i) $\overline{E} = \hat{x}\,3\,e^{jy}$, (ii) $\overline{E} = (j\hat{x} - \hat{y})\,e^{-j2\pi z}$, and (iii) $\overline{H} = (\hat{y} + \hat{z})\,e^{j2\pi x}$.

1.5.2 Find the polarization (linear, circular, or elliptical, and left-handed or right-handed) of the following fields: (i) $\overline{E} = (j\hat{y} + \hat{z})\,e^{-jkx}$, (ii) $\overline{E} = [\hat{x}(2 + j) + \hat{z}(3j + 1)]\,e^{jky}$, and (iii) $\overline{H} = (\hat{x} - j\hat{y})\,e^{jkz}$.

1.5.3 A uniform infinite medium with $\epsilon = 2\epsilon_0$, $\mu = 2\mu_0$ is propagating a 1-cm wavelength uniform plane wave in the $+\hat{z}$ direction; $\overline{E} \cdot \hat{x}$ is maximum at $x = y = z = t = 0$. The wave conveys 1 watt per square meter.
   (a) If this wave is $\hat{x}$-polarized, then what are (i) $\overline{E}(z, t)$? (ii) the complex electric field $\overline{E}(z)$? (iii) the complex magnetic field $\overline{H}(z)$?
   (b) If this wave is right-hand circularly polarized, then what is $\overline{E}(z)$?

1.5.4 Electromagnetic waves satisfy all of Maxwell's equations. Consider the following complex electric field vectors in free space: (i) $\overline{E}_1 = \hat{x}\,E_0\,e^{jkx}$, (ii) $\overline{E}_2 = \hat{z}\,E_0\,e^{-jkx}$,

(iii) $\overline{E}_3 = E_0(\hat{x} + \hat{z}) e^{-jk(x-z)/\sqrt{2}}$, (iv) $\overline{E}_4 = E_0(\hat{x} e^{-jkz} + \hat{y} j e^{jkz})$, and (v) $\overline{E}_5 = E_0(\hat{x} + j\hat{y}) e^{jkz}$.

(a) Which of these electric fields can satisfy the wave equation $(\nabla^2 + \omega^2 \mu_0 \epsilon_0)\overline{E} = 0$? If so, what is the relationship between ω and k?

(b) Which of these fields represent electromagnetic waves? For those that do not, specify which of Maxwell's equations are violated. For those that do: (i) find the corresponding complex magnetic field $\overline{H}$, (ii) determine the time-averaged Poynting vector, (iii) find the direction of propagation, and (iv) give the polarization and sketch the polarization ellipse.

1.5.5 Any monochromatic wave $\overline{E}$ can be represented as the linear combination of two opposite circularly polarized waves, such as $\overline{E}_r$ and $\overline{E}_\ell$ (the subscript r is for right circular, the subscript ℓ for left circular):

$$\overline{E}_r = (\hat{x} + j\hat{y}) e^{-jkz}$$
$$\overline{E}_\ell = (\hat{x} - j\hat{y}) e^{-jkz}$$

If $\overline{E} = a\overline{E}_r + b\overline{E}_\ell = 3j\hat{x} - 2\hat{y}$ at $z = 0$, then what are the complex constants a and b?

1.5.6 Wave motion can be viewed either by taking a series of still pictures at fixed times or by observing the time variation at a fixed point in space. We have discussed the definition of polarization from this second point of view; now let us look at polarization from the first point of view. Consider the wave solution for the electric field vector $\overline{E} = E_0(\hat{x} + j\hat{y}) e^{-jkz}$.

(a) Find the real space-time vector $\overline{E}(\bar{r}, t)$, sketch the time variation of the tip of $\overline{E}(\bar{r}, t)$ at $z = 0$, and show that it is left-hand circularly polarized.

(b) Sketch the electric field vectors at several points along the $\hat{z}$-axis at $\omega t = 0$ and $\omega t = \pi/2$.

(c) Find the corresponding space-time vector $\overline{H}(\bar{r}, t)$ and sketch the magnetic field vectors at $\omega t = 0$ and $\omega t = \pi/2$, as in (b). Note that the helices in (b) and (c) are right-handed. Thus, the left-hand circularly polarized wave is an advancing right-handed helix, which is why some texts define this wave as right-handed.

1.6.1 An AM radio receiver (1 MHz) can easily detect a wave with a peak electric field amplitude of 10 mV/m. (i) What is the power density (W/m^2) associated with such a wave? (ii) What is the peak magnitude of the associated magnetic field (A/m)?

1.6.2 A plane wave propagating in free space can be characterized by its complex electric field $\overline{E}$. If $\overline{E} = \hat{x} e^{-\hat{y} - j2\hat{z}}$,

(a) Is this a uniform plane wave?

(b) What is its frequency (Hz)?

(c) Evaluate the time-averaged power.

1.6.3 An electric field $\overline{E}(\bar{r}, t)$ is right-hand circularly polarized and propagating in vacuum in the $+\hat{y}$ direction. It has a wavelength λ and peak amplitude and direction specified by $\hat{x} E_0$ at $y = 0$, $t = 0$.

(a) Give the complex, time-harmonic field vector $\overline{E}(\bar{r})$ of this wave.

(b) What are the propagation constant k and frequency $f = \omega/2\pi$ in terms of the given parameters?

(c) Determine the corresponding $\overline{H}(\bar{r})$.

(d) Determine the time-averaged Poynting vector.

1.6.4 (a) Consider two plane waves with angular frequency ω propagating in the $+\hat{z}$ direction. One is $\hat{x}$-polarized with amplitude $\underline{E}_x$, and the other is $\hat{y}$-polarized with amplitude $\underline{E}_y$. Find the complex Poynting vector $\overline{\underline{S}}(z)$ for each wave and show whether the sum of the real parts of $\overline{\underline{S}}(z)$ for the two waves is equal to the real part of $\overline{\underline{S}}(z)$ for the combined wave $\underline{E}_x\hat{x} + \underline{E}_y\hat{y}$. Do the same for the imaginary parts.

(b) Repeat (a) for the case where the two waves are: $\underline{E}_1 = E_{x1}\,e^{-jkz}$ and $\underline{E}_2 = E_{x1}\,e^{-jkz-j\phi}$. Discuss briefly the difference in your answers between parts (a) and (b).

1.6.5 A uniform $\hat{x}$-polarized plane wave is propagating in the $-\hat{z}$ direction in free space at 1 GHz. The magnetic field is $\overline{H}(z,t) = \hat{y}\sin(\omega t + kz)$ (A/m).

(a) Sketch and dimension the instantaneous magnetic and electric energy densities $W_m(z)$ and $W_e(z)$ (J/m^3) for $t = 0$.

(b) Sketch and dimension $\overline{\underline{S}}(z, t = 0) \cdot \hat{z}$ (W/m^2).

(c) If instead $\overline{H}(z,t) = \hat{y}\,\sin(\omega t - kz) + 3\hat{y}\,\sin(\omega t + kz)$, sketch and dimension the real and imaginary parts of $\overline{\underline{S}}(z) \cdot \hat{z}$.

(d) For the waves of part (c), plot and dimension $W_e(z = 0, t)$ and $W_m(z = 0, t)$.

(e) For the fields of part (c), sketch and dimension $\overline{\underline{S}}(z, t) \cdot \hat{z}$ for $z = 0$ and also for $z = \lambda/4$. Can you see a relationship between $\overline{\underline{S}}(z) \cdot \hat{z}$ [part (c)] and $\overline{\underline{S}}(z, t) \cdot \hat{z}$ [part (e)]?

1.6.6 A certain plane wave in glass ($\epsilon = 2\epsilon_0$) is characterized by $\underline{E} = (\sqrt{2}\hat{x} + \hat{y} - \hat{z})\,e^{-j2\pi \cdot 10^6\,(y+z)}$ (V/m) where y and z are given in meters.

(a) In what direction is this wave propagating?

(b) What is its frequency (Hz)?

(c) What polarization does it have?

(d) Evaluate the complex Poynting vector $\overline{\underline{S}}$ at the origin.

(e) Repeat (a), (b), (c), and (d) for the case where $\underline{E} = \hat{x}\,e^{-\pi \cdot 10^6\,(z+2jy)}$.

1.6.7 A uniform $\hat{x}$-polarized plane wave is propagating in vacuum in the $+\hat{y}$ direction with a wavelength of 10 μm and a flux density of 10^{11} W/m^2 (e.g., a 10-W laser focused on a 100 μm^2 spot). The electric and magnetic fields are proportional to $\sin(\omega t - ky)$.

(a) What are $\overline{E}(\bar{r}, t)$ and $\overline{H}(\bar{r}, t)$ for this wave? Evaluate all constants numerically.

(b) Two such equal laser beams propagating in opposite directions are made to interfere so that $\overline{E}$ is proportional to $\sin(\omega t - ky) + \sin(\omega t + ky)$. For the resulting fields, evaluate and sketch quantitatively the electric and magnetic energy densities (J/m^3) $W_e(y)$ and $W_m(y)$ for $t = 0$ and $W_e(t)$ and $W_m(t)$ for $y = 0$.

(c) For the waves of parts (a) and (b), sketch and dimension $\overline{\underline{S}}(y = 0, t) \cdot \hat{y}$ (W/m^2) and $\overline{\underline{S}}(y, t = 0) \cdot \hat{y}$.

(d) For the waves of parts (a) and (b), evaluate the complex Poynting vector $\overline{\underline{S}}$.

(e) Sketch $\overline{E}(y, t = 0)$ and $\overline{H}(y, t = 0)$ along the $\hat{y}$-axis.

1.6.8* The Federal Communications Commission (FCC) of the United States requires a minimum of 25-mV/m field intensity for AM stations covering the commercial area of a city.

(a) What is the power density (W/m^2) associated with this minimum field?

(b) What is the intensity (A/m) of the minimum magnetic field $\overline{H}$?

(c) A small 100-turn coil with a 10-cm diameter is oriented so as to maximize its open circuit voltage V when excited by this 1-MHz 25-mV/m wave. What is V? Since the coil is small compared to the wavelength, quasistatics applies. Furthermore, since the loop is also open circuited, no significant currents flow in it and Faraday's law can be used in a simple way.

1.6.9 Two equal-strength 1-GHz $\hat{x}$-polarized waves propagate in vacuum along the $\hat{z}$-axis in opposite directions, producing a standing wave. Compute $\overline{S}(z)$ for the case where $\overline{S}(z = 0) = 0$ and $\overline{S}_{max} = 1$ W/m^2.

1.6.10 Complex power delivered to a circuit element is $\frac{1}{2}\underline{V}\,\underline{I}^*$ where $\underline{V}$ and $\underline{I}$ are complex voltage and current, respectively, across the element.

(a) A current source $I_0 \cos \omega t$ drives an ideal capacitor C. What is the complex power delivered to the capacitor? What is the real power delivered (converted to heat, for example)?

(b) The same current source drives an inductor L in series with the capacitor C. What is the complex power delivered to this circuit as a function of frequency?

(c) At what frequency ω_0 is the complex power zero?

(d) Briefly interpret your answers to (b) and (c) physically.

1.6.11 For the free-space wave $\overline{H} = \hat{x}\,\overline{H}_0\,e^{3y+j5z}$, what is the frequency (Hz)? In what directions are Re$\{\overline{S}\}$ and Im$\{\overline{S}\}$ pointing? Explain.

1.8.1* We have determined that uniqueness holds in free space if $\overline{E}$ and $\overline{H}$ are specified at an initial time and their tangential components are given along the boundaries surrounding a region of interest. Supposed the medium is not free space, but is instead a conductive dielectric (discussed further in Section 3.4), with

$$\overline{D} = \epsilon\overline{E}$$
$$\overline{B} = \mu\overline{H}$$
$$\overline{J} = \sigma\overline{E}$$

The material constants ϵ, μ, and σ are all positive. Show that Maxwell's equations give rise to a unique field solution for this medium.

1.9.1 A column of soldiers marching at 1 m/s can support density waves if each soldier acts to equalize the spacings between him and his neighbors immediately in front of and behind him. If this equalization is perfect once per step (once per second) and the average separation between soldiers is 2 meters, what are the velocities of these forward and backward density waves?

1.9.2 Acoustic disturbances in a gas are characterized by the sound pressure p (N/m^2) and the particle velocity $\overline{u}$ m/s, which refers to the time derivative of the particle displacement $\bar{\xi}$. These quantities are governed by the following equations:

$$-\nabla p = \rho_0 \frac{\partial \bar{u}}{\partial t}$$

$$\frac{\partial \bar{p}}{\partial t} = -\gamma P_0 \nabla \cdot \bar{u}$$

where ρ_0 is the average gas density (kg/m^3), P_0 is the average ambient pressure (N/m^2), and γ is the ratio of specific heat at constant pressure to that at constant volume.

(a) Show that the acoustic wave equation is given by

$$\nabla^2 p - \frac{\rho_0}{\gamma P_0} \frac{\partial^2 p}{\partial t^2} = 0$$

(b) What is the speed of wave propagation c_s for this wave equation?

(c) Assuming time-harmonic dependence, find the solutions for the pressure $p(x, t)$, the particle velocity $\bar{u}(x, t)$ and the particle displacement $\bar{\xi}(x, t)$ for a wave traveling in the positive $\bar{x}$ direction. Let the sound pressure amplitude be p_1 and the angular frequency be ω.

(d) The human ear canal can detect a 1-kHz signal with a peak pressure of 10^{-4} N/m^2. Find the corresponding maximum particle displacement in units of Angstroms. (Note: $\rho_0 c_s = 420$ kg/(m$^2 \cdot$s)). Also, one Angstrom is 10^{-10} meters, roughly the radius of an atom.)

2

RADIATION BY CURRENTS AND CHARGES IN FREE SPACE

2.1 STATIC SOLUTIONS TO MAXWELL'S EQUATIONS

In this chapter, we consider how electromagnetic disturbances are created and how they propagate in free space. In Chapter 1, we saw that Maxwell's equations predicted the existence of electromagnetic waves, even in vacuum where no currents or charges were present (though sources **outside** the region of interest were of course necessary to generate these waves). We shall now consider the case where the sources ρ and $\overline{J}$ are not zero, resulting in an *inhomogeneous wave equation*. Knowledge of the exact current and charge distributions on an object (antenna) located in free space is then sufficient to predict the field distributions at every point in space.

Before developing the inhomogeneous wave equation, we first discuss the simpler *static field* solutions to Maxwell's equations. Static fields do not change in time and therefore cannot produce waves. We shall see that the static field solutions to a particular problem are quite similar to the *dynamic field* solutions to the same problem with a time-dependent source, but the dynamic solutions must take into account the finite propagation velocity of electromagnetic waves. Because information does not propagate instantaneously from one point in space to another, this retardation effect must be included in the description of radiation. Section 2.2 describes the dynamics of radiating fields, but for now we consider only the simpler static fields.

If we let $\partial/\partial t \rightarrow 0$ in the time-varying Maxwell's equations (1.2.1)–(1.2.4), or let $\omega \rightarrow 0$ in the time-harmonic Maxwell's equations (1.4.4)–(1.4.7), we see that the four basic equations of electromagnetism immediately decouple into two pairs.

Faraday's law and Gauss's law in free space become

$$\nabla \times \overline{E} = 0 \tag{2.1.1}$$

$$\nabla \cdot \overline{E} = \rho/\epsilon_0 \tag{2.1.2}$$

and this pair of *electrostatic* equations, along with appropriate boundary conditions, uniquely determines the electric field. The charge density ρ is assumed to be specified.

Likewise, Ampere's law and Gauss's magnetic law in free space are

$$\nabla \times \overline{H} = \overline{J} \tag{2.1.3}$$

$$\nabla \cdot \mu_0 \overline{H} = 0 \tag{2.1.4}$$

where again this pair of *magnetostatic* equations with boundary conditions specifies $\overline{H}$ completely if $\overline{J}$ is known. Because $\overline{E}$ and $\overline{H}$ are decoupled, it is impossible to have wave propagation, since it is the interaction between the time derivatives of $\overline{E}$ and the space derivatives of $\overline{H}$ and vice versa which leads to electromagnetic radiation. We may solve (2.1.1) and (2.1.2) for the electric field by noticing that the vector identity $\nabla \times (\nabla\Phi) = 0$ holds for any scalar field Φ. Since $\nabla \times \overline{E} = 0$, we can write

$$\overline{E} = -\nabla\Phi \tag{2.1.5}$$

where Φ is called the *scalar electric potential*, and the negative sign is chosen so that electric field lines point from regions of high potential to regions of low potential (i.e., in the direction that a positive charge would move if it were placed in the field). The electric potential is a particularly useful quantity because it is a **scalar** which contains complete information about the three components of the **vector** electric field. Substitution of (2.1.5) into (2.1.2) yields the *scalar Poisson equation* in vacuum:

$$\nabla^2\Phi = -\rho/\epsilon_0 \tag{2.1.6}$$

It is Poisson's equation, an inhomogeneous second-order partial differential equation, which we should like to solve, and since it is a linear equation, we shall use the method of *superposition*.

Consider, first, the simplest charge distribution, namely a point charge q located at the origin. There are two ways to find the scalar electric potential; the first makes use of Gauss's law directly. The differential form of Gauss's law (2.1.2) may be converted to an integral representation by using *Gauss's divergence theorem* (1.6.7) discussed in Chapter 1:

$$\int_V (\nabla \cdot \overline{G}) \, dv = \oint_A \overline{G} \cdot \hat{n} \, da \tag{2.1.7}$$

The quantities V, A, $\hat{n}$, da, and dv have been defined in Section 1.6.

Taking the volume integral of Gauss's law (2.1.2) and applying the divergence theorem (2.1.7) yields

$$\oint_A \overline{E} \cdot \hat{n} \, da = \int_V \frac{\rho}{\epsilon_0} \, dv \qquad (2.1.8)$$

But $\int_V \rho \, dv$ is simply q, the total amount of charge enclosed by the volume V. If there is only a single point charge in space, then by symmetry the electric field must be radially directed away from the charge. Therefore, we choose the volume of integration to be a sphere of radius r centered at the origin. The electric field thus is normal to the surface of the sphere enclosing q, so $\overline{E}$ is parallel to $\hat{n}$. Hence, we just multiply E_r, constant at a given radius, by the surface area of a sphere, giving

$$\oint_A \overline{E} \cdot \hat{n} \, da = 4\pi r^2 E_r \qquad (2.1.9)$$

Combining (2.1.8) and (2.1.9) gives the static electric field solution for a point charge:

$$\overline{E}(r) = \hat{r} \frac{q}{4\pi \epsilon_0 r^2} \qquad (2.1.10)$$

Because $\overline{E}$ is radially directed and dependent only on r, we expect Φ to be a function of r alone (no angular dependence). Therefore, the equation $\overline{E} = -\nabla \Phi$ reduces to

$$E_r = -\frac{\partial \Phi}{\partial r} = \frac{q}{4\pi \epsilon_0 r^2} \qquad (2.1.11)$$

Integration over r yields

$$\Phi(r) = \frac{q}{4\pi \epsilon_0 r} + \Phi_0 \qquad (2.1.12)$$

where Φ_0 is an integration constant. We usually set Φ_0 to zero so that the potential will be zero infinitely far from the point charge.

The second method for finding $\Phi(r)$ from a point charge distribution is more mathematical, as it solves the Poisson equation (2.1.6) directly by writing down the Laplacian in spherical coordinates. But first we need a mathematical way to express the fact that we have a charge singularity at the origin. We write the charge distribution $\rho(\overline{r})$ as

$$\rho(\overline{r}) = q \delta^3(\overline{r}) \qquad (2.1.13)$$

where $\delta^3(\overline{r})$ is the three-dimensional Dirac delta function. This generalized function has the following properties:

(1) $\delta^3(\overline{r}) = 0$ for $r \neq 0$,

(2) $\delta^3(\overline{r}) \to \infty$ for $r = 0$,

(3) $\int_{-\infty}^{+\infty} \delta^3(\overline{r}) \, dv = 1$.

Property (3) implies that the delta function has dimensions of m^{-3}.

We can now rewrite (2.1.6) using the spherical form of the Laplacian operator and the charge distribution given by (2.1.13). Since ρ is spherically symmetric, the angular derivatives of Φ in the Laplacian are zero, and we find that only the radial derivative contributes:

$$\frac{1}{r}\frac{d^2}{dr^2}(r\Phi) = -\frac{q}{\epsilon_0}\delta^3(\overline{r}) \tag{2.1.14}$$

For $r \neq 0$, the delta function is zero by property (1):

$$\frac{d^2}{dr^2}(r\Phi) = 0; \quad r \neq 0 \tag{2.1.15}$$

By inspection, (2.1.15) has the solution $\Phi = K/r + \Phi_0$, where again Φ_0 may be set to zero so that $\Phi(r \rightarrow \infty) = 0$. The constant K is determined by integrating both sides of (2.1.14) over a sphere of radius $r > 0$

$$\int_0^r dr \int_0^\pi rd\theta \int_0^{2\pi} r\sin\theta d\phi \frac{1}{r}\frac{d^2}{dr^2}(r\Phi) =$$
$$-\int_0^r dr \int_0^\pi rd\theta \int_0^{2\pi} r\sin\theta d\phi \frac{q}{\epsilon_0}\delta^3(\overline{r}) = -\frac{q}{\epsilon_0} \tag{2.1.16}$$

where the last equality follows from properties (1) and (3) of the delta function. Performing the θ and ϕ integrations on the left side of (2.1.16) is trivial since the integrand is independent of both coordinates. Thus, (2.1.16) becomes

$$\int_0^r \frac{1}{r}\frac{d^2}{dr^2}(r\Phi) \, 4\pi r^2 dr = -\frac{q}{\epsilon_0} \tag{2.1.17}$$

Integrating the left side of (2.1.17) by parts and dividing by 4π gives

$$r\frac{d}{dr}(r\Phi)\Big|_{-\infty}^{+\infty} - \int_0^r \frac{d}{dr}(r\Phi) \, dr = -\frac{q}{4\pi\epsilon_0} \tag{2.1.18}$$

where the surface term [the first term on the left side of (2.1.18)] is zero because $d(r\Phi)/dr = d(K)/dr = 0$. The second term is an integral of a total derivative, so

$$r\Phi(r) = \frac{q}{4\pi\epsilon_0}$$

We conclude, then, that for a point charge located at the origin, the scalar potential is

$$\Phi(\overline{r}) = \frac{q}{4\pi\epsilon_0 r} \tag{2.1.19}$$

where $r = |\overline{r}|$ is the distance from the charge (at the origin) to an observer at $\overline{r}$. Both approaches therefore give the same result for the potential of a point charge located at the origin.

If the point charge is not located at the origin, but is instead found at $\bar{r}'$, then (2.1.19) becomes

$$\Phi(\bar{r}) = \frac{q}{4\pi\epsilon_0 |\bar{r} - \bar{r}'|}$$

where $|\bar{r} - \bar{r}'|$ measures the distance between the observer at $\bar{r}$ and the point charge at $\bar{r}'$. And if, instead of a point charge, we have an incremental volume $\delta v'$ filled with charge $\delta q' = \rho(\bar{r}')\delta v'$, then this incremental amount of charge will give rise to the incremental potential

$$\delta\Phi(\bar{r}) = \frac{\rho(\bar{r}')\delta v'}{4\pi\epsilon_0 |\bar{r} - \bar{r}'|} \qquad (2.1.20)$$

We now integrate the potential created by each infinitesimal amount of charge (2.1.20) over each of the infinitesimal volume elements $\delta v'$ where $\rho(\bar{r}') \neq 0$ to find the **total** electric potential:

$$\Phi(\bar{r}) = \int_{v'} \frac{\rho(\bar{r}')\,dv'}{4\pi\epsilon_0 |\bar{r} - \bar{r}'|} \qquad (2.1.21)$$

This integral is designated as the *scalar Poisson integral* or the *superposition integral*. The latter name comes from the fact that since Maxwell's equations are linear, we can simply sum the potentials contributed by each incremental amount of charge acting alone to get the total potential. Notice that all of the $\bar{r}'$ dependence disappears after integration over dv' in (2.1.21), so that Φ is only a function of the observer coordinate $\bar{r}$. (In performing this integral, $\bar{r}$ is kept constant.) The scalar potential may be computed by using the superposition integral if ρ is known, and then $\bar{E}$ can be determined by taking the negative gradient of Φ. Figure 2.1 is a pictorial representation of the superposition integral.

The static magnetic field equations (2.1.3) and (2.1.4) give rise to a similar *vector Poisson equation*. Because $\nabla \cdot B = 0$ and the identity $\nabla \cdot (\nabla \times A) = 0$ is obeyed by **any** vector $\bar{A}$, we can express $\bar{B} = \mu_0 \bar{H}$ as the curl of an unknown vector $\bar{A}$

$$\bar{B} = \mu_0\bar{H} = \nabla \times \bar{A} \qquad (2.1.22)$$

where $\bar{A}$ is called the *vector potential*. As usual, it is necessary ultimately to check that (2.1.22) does not conflict with any of the other Maxwell's equations. Substitution of (2.1.22) into (2.1.3) gives

$$\nabla \times (\nabla \times \bar{A}) = \mu_0\bar{J} \qquad (2.1.23)$$

which may be simplified by the vector identity (1.3.6)

$$\nabla \times (\nabla \times \bar{A}) = \nabla(\nabla \cdot \bar{A}) - \nabla^2\bar{A} \qquad (2.1.24)$$

Because only the curl of $\bar{A}$ is specified, we are free to choose its divergence to be anything we wish; this choice is known as setting the *gauge*. To see that (2.1.22)

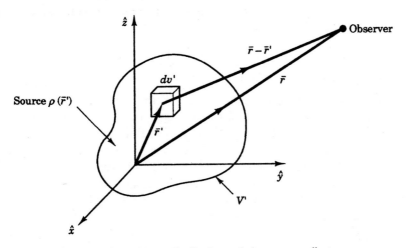

Figure 2.1   Charge distribution and observer coordinates.

is satisfied even if the divergence of $\overline{A}$ is arbitrary, we let $\overline{A}' = \overline{A} + \nabla \psi$ where ψ is any scalar. Taking the curl of both sides gives $\nabla \times \overline{A}' = \nabla \times \overline{A}$ since the curl of the gradient of any scalar is zero. Thus $\overline{A}$ is not unique. For static fields, we simplify (2.1.24) by letting

$$\nabla \cdot \overline{A} = 0 \tag{2.1.25}$$

which is called the *Coulomb gauge*. Combining (2.1.23), (2.1.24) and (2.1.25) gives the *vector Poisson equation* for vacuum

$$\nabla^2 \overline{A} = -\mu_0 \overline{J} \tag{2.1.26}$$

which may also be written as the three scalar equations

$$\nabla^2 A_i = -\mu_0 J_i$$

where $i = x, y, z$.[1] Because we can transform the previously derived scalar Poisson equation (2.1.6) into each of the Cartesian components of the vector Poisson equation (2.1.26) by making the substitutions $\Phi \to A_i$, $\rho \to J_i$, and $\epsilon_0 \to 1/\mu_0$, where i can be x, y, or z, we can immediately find the vector solution to (2.1.26) by similarly transforming the superposition integral (2.1.21), yielding a superposition integral for each Cartesian component of the static vector potential:

$$A_i(\overline{r}) = \int_{V'} \frac{\mu_0 J_i(\overline{r}') \, dv'}{4\pi |\overline{r} - \overline{r}'|}$$

---

1. Note that this separation of a vector differential equation into three scalar equations is straightforward **only** in Cartesian coordinates; for example, $\nabla^2 A_\theta \neq -\mu_0 J_\theta$ in spherical coordinates.

Again, the subscript i stands for a Cartesian component of the field, and we may make the notation more compact using vector notation:

$$\overline{A}(\overline{r}) = \int_{V'} \frac{\mu_0 \overline{J}(\overline{r'})\, dv'}{4\pi |\overline{r} - \overline{r'}|} \tag{2.1.27}$$

This means that if we know the current distribution in a problem, we can calculate $\overline{A}$ directly and then take its curl to get $\mu_0 \overline{H}$.[2] Therefore, in theory we can find the electric or magnetic fields in a statics problem if we know the charge or current distribution simply by performing an integration. Although this integral may be difficult to evaluate, a more serious problem also exists, as the fields themselves may alter the source distributions. Fortunately, for most of the simple problems solved in this text, the charge and/or current distributions are known exactly or may be approximated quite accurately.

2.2 RADIATION BY DYNAMIC CURRENTS AND CHARGES

If we now consider the fully dynamic form of Maxwell's equations in free space, we see that we can no longer let $\overline{E}$ be the gradient of a scalar potential because $\nabla \times \overline{E} = -\partial \overline{B}/\partial t \neq 0$. But since $\nabla \cdot \mu_0 \overline{H} = 0$, we may again express $\mu_0 \overline{H}$ as the curl of a vector potential; (2.1.22) is still a valid equation.

Substitution of (2.1.22) into Faraday's law (1.2.1) gives

$$\nabla \times \left(\overline{E} + \frac{\partial \overline{A}}{\partial t}\right) = 0 \tag{2.2.1}$$

which means that $\overline{E} + \partial \overline{A}/\partial t$ (instead of $\overline{E}$ alone) may be expressed as the gradient of a scalar

$$\overline{E} + \frac{\partial \overline{A}}{\partial t} = -\nabla \Phi \tag{2.2.2}$$

since $\nabla \times \nabla \Phi = 0$ for any scalar Φ. This equation has the correct static limit: as $\partial/\partial t \to 0$, (2.2.2) reduces to $\overline{E} = -\nabla \Phi$. Ampere's law (1.2.2) can now be rewritten with substitutions for $\overline{H}$ and $\overline{E}$ using (2.1.22) and (2.2.2):

$$\nabla \times (\nabla \times \overline{A}) = \mu_0 \overline{J} - \mu_0 \epsilon_0 \frac{\partial^2 \overline{A}}{\partial t^2} - \mu_0 \epsilon_0 \nabla \left(\frac{\partial \Phi}{\partial t}\right) \tag{2.2.3}$$

---

2. Using this procedure, we arrive at the useful *Biot-Savart law*, which relates the magnetic field $\overline{H}$ directly to the current $\overline{J}$ without the vector potential as an intermediary:

$$\overline{H}(\overline{r}) = \frac{\nabla \times \overline{A}(\overline{r})}{\mu_0} = \int_{V'} \frac{\overline{J}(\overline{r'}) \times (\overline{r} - \overline{r'})}{4\pi |\overline{r} - \overline{r'}|^2}\, dv'$$

After use of the identity (2.1.24) and rearrangement of the terms in (2.2.3), we find:

$$\nabla^2 \overline{A} - \mu_0\epsilon_0 \frac{\partial^2 \overline{A}}{\partial t^2} = -\mu_0 \overline{J} + \nabla \left(\nabla \cdot \overline{A} + \mu_0\epsilon_0 \frac{\partial \Phi}{\partial t} \right) \qquad (2.2.4)$$

Since we are free to choose the gauge (divergence of $\overline{A}$), we let

$$\nabla \cdot \overline{A} = -\mu_0\epsilon_0 \frac{\partial \Phi}{\partial t} \qquad (2.2.5)$$

which is called the *Lorentz gauge*. Combining (2.2.4) and (2.2.5) now gives the *inhomogeneous vector Helmholtz equation*:

$$\nabla^2 \overline{A} - \mu_0\epsilon_0 \frac{\partial^2 \overline{A}}{\partial t^2} = -\mu_0 \overline{J} \qquad (2.2.6)$$

If $\overline{J} = 0$, we have a wave equation (the homogeneous Helmholtz equation discussed in Section 1.3) which is consistent with the fact that both $\overline{E}$ and $\overline{H}$ obey wave equations in source-free vacuum. If $\overline{E}$ and $\overline{H}$ obey wave equations, then $\overline{A}$ and Φ will also.

In a similar fashion, we may combine Gauss's law (2.1.2) with (2.2.2) to yield

$$\nabla^2 \Phi + \frac{\partial}{\partial t}(\nabla \cdot \overline{A}) = -\frac{\rho}{\epsilon_0} \qquad (2.2.7)$$

Again using the Lorentz gauge (2.2.5) to eliminate $\nabla \cdot \overline{A}$ from (2.2.7), we obtain the *inhomogeneous scalar Helmholtz equation* for the scalar potential Φ:

$$\nabla^2 \Phi - \mu_0\epsilon_0 \frac{\partial^2 \Phi}{\partial t^2} = -\rho/\epsilon_0 \qquad (2.2.8)$$

If ρ and $\overline{J}$ are time-harmonic sources, then the scalar and vector Helmholtz equations can also be expressed as

$$\nabla^2 \underline{\overline{A}} + \omega^2 \mu_0\epsilon_0 \underline{\overline{A}} = -\mu_0 \underline{\overline{J}} \qquad (2.2.9)$$

$$\nabla^2 \underline{\Phi} + \omega^2 \mu_0\epsilon_0 \underline{\Phi} = -\underline{\rho}/\epsilon_0 \qquad (2.2.10)$$

where the phasor nature of $\underline{\overline{A}}$, $\underline{\Phi}$, $\underline{\overline{J}}$, and $\underline{\rho}$ is made explicit. Notice that both inhomogeneous Helmholtz equations (2.2.6) and (2.2.8) or (2.2.9)–(2.2.10) reduce to the static Poisson equations (2.1.26) and (2.1.6) respectively in the limit $\partial/\partial t \to 0$ ($\omega \to 0$). The dynamic solution to (2.2.9) or (2.2.10) is thus simply the static solution modified by a factor which accounts for the finite propagation time of the electromagnetic waves through space; this time delay or retardation is $|\overline{r} - \overline{r}'|/c$.

We would now like to find solutions to the inhomogeneous Helmholtz equations analogous to the superposition integrals (2.1.21) and (2.1.27). For simplicity, we shall consider time-harmonic fields since they form the basis for more arbitrary solutions. We first demonstrate that if the static point charge q at the origin starts to oscillate with time dependence $q(t) = q \cos \omega t$, then the static potential $\Phi_s = q/4\pi\epsilon_0 r$ can be replaced with the time-harmonic potential phasor $\underline{\Phi} = \Phi_s e^{-jkr}$. Using the chain rule and the spherical Laplacian, we find

$$\nabla^2 \underline{\Phi} = \left(\nabla^2 \Phi_s - 2jk \left(\frac{\partial \Phi_s}{\partial r} + \frac{\Phi_s}{r} \right) - k^2 \Phi_s \right) e^{-jkr}$$

The second term in this expression vanishes since Φ_s is inversely proportional to r, and we conclude that

$$\nabla^2 \underline{\Phi} = \left(\nabla^2 \Phi_s - k^2 \Phi_s \right) e^{-jkr} \tag{2.2.11}$$

If we now substitute (2.2.11) into the inhomogeneous time-harmonic wave equation (2.2.10), we find that

$$\left(\nabla^2 \Phi_s - k^2 \Phi_s \right) e^{-jkr} + \omega^2 \mu_0 \epsilon_0 \, \Phi_s \, e^{-jkr} = -\frac{q}{\epsilon_0} \delta^3(\overline{r})$$

which reduces to the simpler equation

$$\left(\nabla^2 \Phi_s \right) e^{-jkr} = -\frac{q}{\epsilon_0} \delta^3(\overline{r}) = -\frac{q}{\epsilon_0} e^{-jkr} \delta^3(\overline{r}) \tag{2.2.12}$$

if the dispersion relation in free space $k^2 = \omega^2 \mu_0 \epsilon_0$ holds. The second equality in (2.2.12) follows because e^{-jkr} is unity at $\overline{r} = 0$, the only value for which the delta function is nonzero. Therefore, (2.2.12) simply reduces to the static scalar Poisson equation for a point charge and we have shown that the potential phasor

$$\underline{\Phi} = \frac{q}{4\pi\epsilon_0 r} e^{-jkr} \tag{2.2.13}$$

is a valid solution to (2.2.10) in free space when the source ρ is just an oscillating point charge. Superposition of the incremental charge elements $\delta \underline{q}' = \rho(\overline{r}')\delta v'$ over the volume V', which encloses any arbitrary charge distribution, gives a *dynamic superposition integral* similar to the static version (2.1.21):

$$\underline{\Phi}(\overline{r}) = \int_{V'} \frac{\rho(\overline{r}')}{4\pi\epsilon_0 |\overline{r} - \overline{r}'|} e^{-jk|\overline{r} - \overline{r}'|} dv' \tag{2.2.14}$$

Therefore, (2.2.14) is a general solution to the inhomogeneous time-harmonic scalar Helmholtz equation (2.2.10). The inhomogeneous vector Helmholtz equation may be written down by inspection from (2.2.14) by noting that (2.2.9) is identical to (2.2.10) when the substitutions $\rho \rightarrow J_i$, $\Phi \rightarrow A_i$, and $\epsilon_0 \rightarrow 1/\mu_0$ ($i = x, y, z$) are made. Recombining the three scalar equations into the more compact vector notation gives the dynamic vector Poisson integral [3]

$$\underline{\overline{A}}(\overline{r}) = \int_{V'} \frac{\mu_0 \underline{\overline{J}}(\overline{r}')}{4\pi |\overline{r} - \overline{r}'|} e^{-jk|\overline{r} - \overline{r}'|} dv' \tag{2.2.15}$$

Notice that in the static limit ($\omega = k = 0$), the dynamic superposition integrals (2.2.14) and (2.2.15) reduce to the static Poisson integrals (2.1.21) and (2.1.27).

---

3. Although we have chosen to derive time-harmonic superposition integrals, it should be realized that time-dependent forms of these integrals also exist. We omit their lengthy derivations, but include

Once $\Phi(\bar{r})$ and $\overline{A}(\bar{r})$ are found from the superposition integrals (2.2.14) and (2.2.15), the electric and magnetic fields may be calculated by applying (2.1.22) and (2.2.2).

2.3 RADIATION FROM A HERTZIAN DIPOLE

The simplest radiating source is called a *Hertzian dipole* and is illustrated in Figure 2.2(a). It consists of two reservoirs separated by a distance d which contain equal and opposite amounts of charge; the charge oscillates back and forth between the reservoirs at frequency ω. The charge reservoirs are located on the $\hat{z}$-axis at $z = \pm d/2$, constituting an *electric dipole* with *dipole moment*   .

$$p = \hat{z}\, qd \tag{2.3.1}$$

In the limit that $d \to 0$ and $q \to \infty$ but the qd product remains finite, we have a Hertzian dipole.

Conservation of charge dictates that a current I must flow between the two charge reservoirs as the charge oscillates, where $I = \partial q/\partial t$ or $I = j\omega q$ in time-harmonic form. (We shall find the electric and magnetic fields for the Hertzian dipole at a single frequency ω, where we understand that for an arbitrary frequency spectrum we can superpose single-frequency solutions.) Therefore, the current density $\overline{J}$ is

$$\overline{J} = \hat{z}\, Id\, \delta^3(\bar{r}) \tag{2.3.2}$$

where we check the dimensions to make sure that (2.3.2) makes sense. Because the delta function has the dimensions of m^{-3}, $Id\,\delta^3(\bar{r})$ has dimensions of A/m^2 which is also the proper dimension for $\overline{J}$. We find (2.3.2) by considering the dipole to have finite (though small) width w, length l, and depth d as shown in Figure 2.2(b). Because the current flows in the $\hat{z}$ direction, $\overline{J} = \hat{z}\, I/lw = \hat{z}\, Id/lwd$. But lwd is the volume of the dipole, and as $l, w, d \to 0$, $1/lwd \to \delta^3(\bar{r})$, giving the result (2.3.2).

If we substitute the current distribution for a Hertzian dipole (2.3.2) into the

---

the time-dependent superposition integrals for $\Phi(\bar{r}, t)$ and $\overline{A}(\bar{r}, t)$ for completeness:

$$\Phi(\bar{r}, t) = \int_{v'} \frac{\rho(\bar{r}', t - |\bar{r} - \bar{r}'|/c)}{4\pi\epsilon_0|\bar{r} - \bar{r}'|}\, dv'$$

$$\overline{A}(\bar{r}, t) = \int_{v'} \frac{\mu_0 \overline{J}(\bar{r}', t - |\bar{r} - \bar{r}'|/c)}{4\pi|\bar{r} - \bar{r}'|}\, dv'$$

If the charge density oscillates sinusoidally as $\rho(\bar{r}', t) = \text{Re}\{\rho(\bar{r}')\, e^{j\omega t}\}$, then the retarded charge density $\rho(\bar{r}', t - |\bar{r} - \bar{r}'|/c) = \rho(\bar{r}')\, e^{j\omega t - jk|\bar{r} - \bar{r}'|}$ is observed in the integrand of (2.2.14) (with the $e^{j\omega t}$ factor implicit) since $k = \omega/c$. Therefore, the time-dependent equations above yield the correct time-harmonic equations when a sinusoidally varying source is chosen.

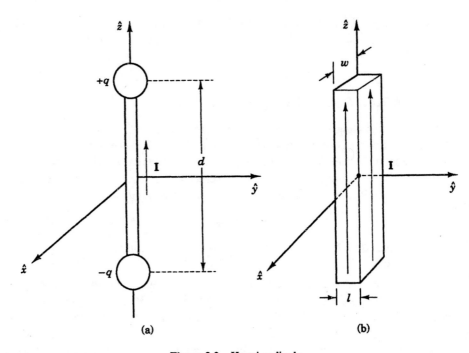

Figure 2.2   Hertzian dipole.

expression for the vector potential (2.2.15), we find that the vector potential for the dipole is

$$\overline{A}(\overline{r}) = \hat{z}\, \frac{\mu_0 I d}{4\pi r}\, e^{-jkr} \tag{2.3.3}$$

because all of the current is concentrated at the origin.[4] Since $\overline{A}$ varies with r, it is natural to use spherical coordinates r, θ, and ϕ as defined in Figure 2.3. The $\hat{z}$-unit vector may be written as

$$\hat{z} = \hat{r}\cos\theta - \hat{\theta}\sin\theta \tag{2.3.4}$$

which may also be seen from Figure 2.3, and therefore

$$\overline{A}(r, \theta) = (\hat{r}\cos\theta - \hat{\theta}\sin\theta)\, \frac{\mu_0 I d}{4\pi r}\, e^{-jkr} \tag{2.3.5}$$

---

4. Mathematically, the integration

$$\int_{-\infty}^{+\infty} f(\overline{r} - \overline{r}')\delta^3(\overline{r}')\, dv' = f(\overline{r})$$

is carried out to obtain (2.3.3).

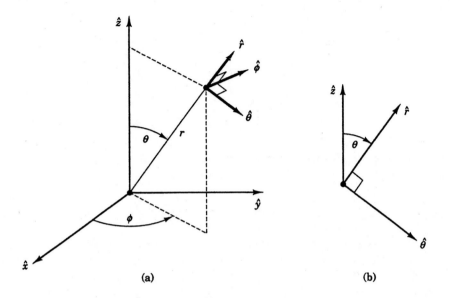

Figure 2.3 Spherical coordinates.

We can find $\overline{H}$ directly by taking the curl of $\overline{A}$ in spherical coordinates

$$\overline{H} = \frac{1}{\mu_0} \nabla \times \overline{A} = \frac{1}{\mu_0 r^2 \sin\theta} \begin{vmatrix} \hat{r} & r\hat{\theta} & r\sin\theta\hat{\phi} \\ \partial/\partial r & \partial/\partial\theta & \partial/\partial\phi \\ A_r & rA_\theta & r\sin\theta A_\phi \end{vmatrix} \qquad (2.3.6)$$

where $\partial/\partial\phi$ and A_ϕ are zero for the Hertzian dipole.

$$\overline{H} = \hat{\phi}\frac{1}{\mu_0 r}\left[\frac{\partial}{\partial r}(rA_\theta) - \frac{\partial}{\partial\theta}(A_r)\right]$$

$$= \hat{\phi}\frac{jkId}{4\pi r}e^{-jkr}\left[1 + \frac{1}{jkr}\right]\sin\theta \qquad (2.3.7)$$

We can find $\overline{E}$ from $\overline{H}$ using Ampere's law (1.2.2) where $\overline{J} = 0$ for $r \neq 0$:

$$\overline{E} = \frac{1}{j\omega\epsilon_0}\nabla \times \overline{H}$$

$$= \sqrt{\frac{\mu_0}{\epsilon_0}}\frac{jkId}{4\pi r}e^{-jkr}\left\{\hat{r}\left[\frac{1}{jkr} + \left(\frac{1}{jkr}\right)^2\right]2\cos\theta \qquad (2.3.8)\right.$$

$$\left. + \hat{\theta}\left[1 + \frac{1}{jkr} + \left(\frac{1}{jkr}\right)^2\right]\sin\theta\right\}$$

We can also find the complex Poynting vector for the Hertzian dipole from these electric and magnetic fields:

$$\overline{S} = \overline{E} \times \overline{H}^*$$

$$= \eta_0 \left| \frac{kId}{4\pi r} \right|^2 \left\{ \hat{r} \left[1 + \left(\frac{1}{jkr} \right)^3 \right] \sin^2 \theta \right.$$

$$\left. - \hat{\theta} \left[\left(\frac{1}{jkr} \right) - \left(\frac{1}{jkr} \right)^3 \right] \sin 2\theta \right\} \tag{2.3.9}$$

From the complex Poynting vector, we may find the time-averaged power in the usual way:

$$\langle \overline{S} \rangle = \frac{1}{2} \operatorname{Re} \{ \overline{S} \} = \hat{r} \frac{\eta_0}{2} \left| \frac{kId}{4\pi r} \right|^2 \sin^2 \theta \tag{2.3.10}$$

In the Hertzian dipole limit ($d \to 0$), these fields are valid for all r. For a short dipole where we can approximate the current as constant over the dipole length, (2.3.7)–(2.3.9) are valid for $r \gg d$. The electric field $\overline{E}(r, \theta, t)$ is displayed in Figure 2.4 at four instants of time. Were we to watch the field patterns evolve in time, it would appear that new dipole patterns were constantly being formed at $r = 0$ and that each pattern would begin to propagate outward once it was about a wavelength from the origin, until far from the source the fields would resemble plane waves. It is evident that for $r \ll \lambda/2\pi$ ($kr \ll 1$), the field resembles the static field of an electric dipole, and that for $r \gg \lambda/2\pi$ it resembles a uniform plane wave with gradually diminishing amplitude. In the region where $r \approx \lambda/2\pi$, the field lines appear to detach from the dipole field structure and begin propagating. We examine the exact electric and magnetic field solutions (2.3.8) and (2.3.7) in the limit where $kr \ll 1$ (the *nearfield zone*) and also where $kr \gg 1$ (the *farfield zone*).

First, we consider the nearfield case where $kr \ll 1$. Using the dispersion relation $k = \omega\sqrt{\mu_0\epsilon_0}$ and (2.3.1), we write

$$Id = j\omega q d = j\omega p = jkp/\sqrt{\mu_0\epsilon_0} \tag{2.3.11}$$

Substituting (2.3.11) into the expressions for $\overline{E}$ (2.3.8) and $\overline{H}$ (2.3.7) and keeping only the leading terms (highest powers of $1/kr$) gives the following expressions for the fields:

$$\overline{E} = \frac{p}{4\pi\epsilon_0 r^3} (\hat{r}2\cos\theta + \hat{\theta}\sin\theta) \quad (kr \ll 1) \tag{2.3.12}$$

$$\overline{H} = \hat{\phi} \frac{j\omega p}{4\pi r^2} \sin\theta \quad (kr \ll 1) \tag{2.3.13}$$

(Note that $e^{-jkr} \to 1$ in the nearfield zone.) Equation (2.3.12) is the nearfield expression for the electric field, and (2.3.13) is the induction field. We note that $\overline{H}$,

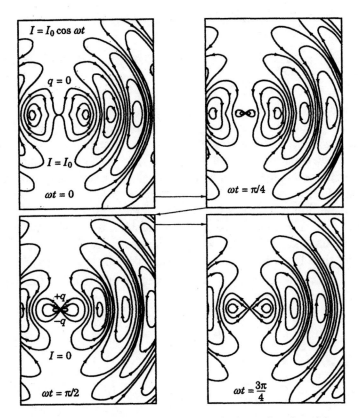

Figure 2.4   Electric fields for a Hertzian dipole as a function of time.

which is proportional to $1/r^2$, is much smaller than $\overline{E}$ (proportional to $1/r^3$) when $r \ll \lambda/2\pi$, so that the near field is dominated by electric field (capacitive) effects. The terms that dominate the complex Poynting vector (2.3.9) in the near field are proportional to $1/r^5$ and may be seen to be purely imaginary. This means that although electric energy is stored in the near field, there is no net power flow out of the nearfield region, consistent with the fact that there is no wave propagation near the dipole. All of these observations are completely consistent with the quasistatic solutions to the electric dipole, and we make the additional observation that in the completely static limit ($\omega = k = 0$), the magnetic field vanishes identically, and the dipole is completely capacitive.

By contrast, in the farfield zone ($kr \gg 1$), we neglect terms of order $1/(kr)^2$ (or smaller) in (2.3.7) and (2.3.8), resulting in electric and magnetic fields which both have $1/r$ dependence:

$$\overline{E} = \hat{\theta}\, \frac{j\eta_0 k I d}{4\pi r}\, e^{-jkr} \sin\theta \quad (kr \gg 1) \qquad (2.3.14)$$

$$\overline{H} = \hat{\phi}\, \frac{jkId}{4\pi r}\, e^{-jkr} \sin\theta \quad (kr \gg 1) \tag{2.3.15}$$

We note first that since the static electric field decays as $1/r^3$ away from the origin and the far fields decay as $1/r$, the near field rapidly becomes unimportant in comparison with the far field for large r. The electric and magnetic far fields have identical spatial dependencies, are orthogonal, and have an amplitude ratio $E_\theta / H_\phi = \eta_0$, the impedance of free space. These were just the properties of uniform plane waves discussed in Section 1.3, with the exception that the plane waves in Chapter 1 had constant amplitudes in space. For the farfield solutions (2.3.14) and (2.3.15), the amplitude gradually decreases as $1/r$ away from the origin, but if we are observing the wave at very large distances ($r \gg \lambda/2\pi$), we see what locally resembles a uniform plane wave propagating in the $+\hat{r}$ direction. (If we are far enough away, the $1/r$ factor doesn't change significantly as the wave propagates outward. We also look at the plane wave segment over a sufficiently small range of θ so that the $\sin\theta$ factor is relatively constant, especially near $\theta = \pi/2$.)

We can calculate the time-averaged power (W/m^2) found in this wave from the complex Poynting vector:

$$\overline{S} = \overline{E} \times \overline{H}^* = \hat{r}\, \eta_0 \left| \frac{kId}{4\pi r} \right|^2 \sin^2\theta$$

It follows that

$$\langle \overline{S} \rangle = \frac{1}{2} \operatorname{Re}\{\overline{S}\} = \hat{r}\, \frac{|\overline{E}|^2}{2\eta_0} = \hat{r}\, \frac{\eta_0}{2} \left| \frac{kId}{4\pi r} \right|^2 \sin^2\theta \tag{2.3.16}$$

which is **identical** with the exact expression (2.3.10) calculated from the complete Hertzian dipole fields. Thus (2.3.16) is valid everywhere in space, which is not surprising since power must be conserved. Integration of (2.3.16) over a spherical shell of radius r must be a constant independent of r, since this integration yields the total power emitted by the dipole. The $\sin^2\theta$ dependence of the Poynting vector means that most of the power is radiated to the sides of the dipole; there is no radiation along the dipole axis. Figure 2.5 illustrates the $\overline{E}$, $\overline{H}$, and $\overline{S}$ vectors for the far field of a Hertzian dipole, as well as the electric field and Poynting vector magnitudes as functions of θ.

The total power P radiated by the dipole may be calculated by integrating $\langle \overline{S} \rangle$, the average power radiated per unit area, over the surface of a sphere enclosing the dipole, as described above:

$$
\begin{aligned}
P = \oint_A \langle \overline{S} \rangle \cdot \hat{n}\, da &= \int_0^\pi r\, d\theta \int_0^{2\pi} r \sin\theta\, d\phi\, \langle S_r \rangle \\
&= \pi \eta_0 \left| \frac{kId}{4\pi} \right|^2 \int_0^\pi d\theta \sin^3\theta = \frac{\eta_0}{12\pi} |kId|^2
\end{aligned}
\tag{2.3.17}
$$

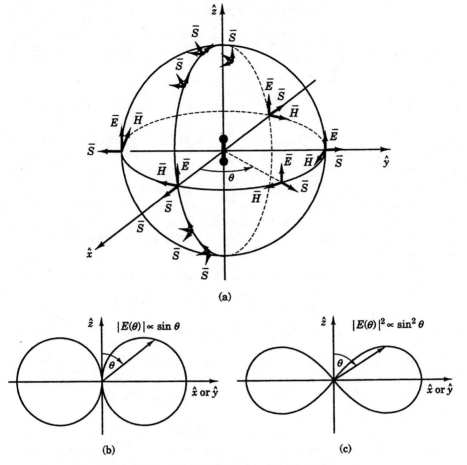

Figure 2.5 Radiation from a Hertzian dipole.

The total power radiated thus increases as the square of the frequency and as the square of the dipole moment.

Sometimes it is useful to think of the dipole element as part of an equivalent circuit; this concept is pursued in much greater depth in Chapter 9. We define the *radiation resistance* of any radiating element as the total power radiated by that element divided by one half the magnitude of the current squared; for the Hertzian dipole, we find

$$R_{rad} = \frac{P}{\frac{1}{2}|I|^2} = \frac{\eta_0}{6\pi}(kd)^2 \cong 20(kd)^2 \qquad (2.3.18)$$

if we approximate $\eta_0 = 377 \ \Omega = 120\pi \ \Omega$. Physically, this resistor R_{rad} would dissipate the same total power P that the dipole with current I radiates.

We also define the *gain* $G(\theta, \phi)$ of a radiating element as the ratio of the power density (W/m^2) radiated in the direction (θ, ϕ) to the power density radiated by an isotropic source emitting uniform power flux $\langle \overline{S}_{\text{isotropic}} \rangle = P/4\pi r^2$ in all directions. Clearly, if the radiating element is isotropic, the gain is 1 for all angles. For the Hertzian dipole,

$$G(\theta, \phi) = \frac{\langle S_r(\theta, \phi, r) \rangle}{P/4\pi r^2} = \frac{3}{2} \sin^2 \theta \qquad (2.3.19)$$

as can be seen by combining (2.3.16) and (2.3.17). The *radiation pattern* for a source is determined by the gain and is a plot of $G(\theta, \phi)$ over all angles but normalized so its maximum value is unity. Since $G(\theta, \phi)$ is independent of ϕ for the Hertzian dipole, the radiation pattern is symmetric around the $\hat{z}$-axis. The dipole radiates most strongly at $\theta = \pi/2$ (in the x–y plane), and has a null at $\theta = 0, \pi$ (along the $\hat{z}$-axis). Thus, the radiation pattern looks like a doughnut with a vanishingly small hole, oriented along the $\hat{z}$-axis. Figure 2.5 plots the gain for a Hertzian dipole.

If we integrate the gain over the solid angle $d\Omega = \sin \theta d\theta d\phi$, we should **always** find that

$$\int_0^{2\pi} d\phi \int_0^{\pi} \sin \theta d\theta G(\theta, \phi) = 4\pi \qquad (2.3.20)$$

no matter what type of radiating element we choose, since the integral of $\langle S_r \rangle$ over the surface of any sphere is equal to the total power P of the source. Equation (2.3.20) is readily derived from (2.3.17) and the definition of gain (2.3.19); it may be easily verified for the Hertzian dipole.

We also define G_0 to be the maximum gain of the antenna. For the Hertzian dipole, $G_0 = 1.5$ whereas for an isotropic radiator, $G_0 = 1$. Thus the dipole is 50 percent more directive than an isotropic antenna (which isn't directive at all). In Chapter 9 we shall discuss how to make antennas that are much more directive than the Hertzian dipole—useful if one wants to maximize the power transmitted by the antenna to a given region of space.

Example 2.3.1

What is the relation between the farfield $\overline{E}(r, \theta)$ radiated by a Hertzian dipole and the vector potential $\overline{A}(r, \theta)$?

Solution: The farfield $\overline{E}_{\text{ff}}(r, \theta)$ produced by a $\hat{z}$-oriented dipole is given by (2.3.14)

$$\overline{E}_{\text{ff}} = \hat{\theta} \sqrt{\frac{\mu_0}{\epsilon_0}} \frac{jkId}{4\pi r} e^{-jkr} \sin \theta$$

and the corresponding vector potential $\overline{A}$ (2.3.3) is

$$\overline{A}(r) = \hat{z} \frac{\mu_0 Id}{4\pi r} e^{-jkr}$$

Therefore

$$\overline{E}_{\text{ff}}(r, \theta) = \hat{\theta}\,\frac{j\eta_0 k}{\mu_0}(-\hat{\theta} \cdot \overline{A}(r))$$

$$= -j\omega A_\theta \hat{\theta}$$

which is easy to remember.

Example 2.3.2

A short dipole 0.5 m long carries $I = 2$ A at 1 MHz. What is the total power radiated into free space, and what is the maximum electric field $\overline{E}_{\text{max}}$ at 100 km?

Solution: The total power radiated by a Hertzian dipole is given by (2.3.17)

$$P = \frac{\eta_0}{12\pi}\,|kId|^2 = \frac{377}{12\pi}\left|\frac{2\pi}{\lambda} \cdot 2 \cdot 0.5\right|^2 = \frac{377\pi}{3 \times 300^2} \cong 4.4 \text{ mW}$$

assuming that the 2 A current is constant over the full length of the dipole. Here $\lambda = c/f$ is 300 m, or $600d$, and therefore the Hertzian dipole approximation ($d \ll \lambda$) is quite accurate. The case where d is not infinitesimal compared to λ is discussed in Chapter 9. Since 100 km is large compared to λ, we may use the farfield approximation (2.3.14):

$$\overline{E} = \hat{\theta}\,\frac{j\eta_0 kId}{4\pi r}\,e^{-jkr}\sin\theta$$

$$|\overline{E}_{\text{max}}| = 377\,\frac{2\pi \cdot 2 \cdot 0.5}{300 \cdot 4\pi \cdot 10^5}$$

$$\cong 6.3 \times 10^{-6} \text{ (V/m)}$$

This electric field is marginally detectable. Because it is proportional to Id/λ, the field would therefore increase with frequency.

Example 2.3.3

An electron moving at a nonrelativistic velocity $\overline{v}$ ($v \ll c$) perpendicular to a uniform magnetic field $\overline{H} = \hat{z}\,H_0$ experiences a force $\overline{f} = -e\,\overline{v} \times \mu_0\overline{H}$ newtons where $-e$ is the electron charge. If the electron moves in a circular orbit in the x–y plane, find the *electron cyclotron frequency* ω_c radians/s, which is the angular frequency of this orbit. How much power does the electron radiate?

Solution: Newton's law states that

$$\overline{f} = m\overline{a} = q\overline{v} \times \mu_0\overline{H}$$

or

$$m\frac{d^2\overline{r}}{dt^2} = -e\,\frac{d\overline{r}}{dt} \times \hat{z}\,\mu_0 H_0 \tag{1}$$

where the electron velocity $\overline{v}$ is $d\overline{r}/dt$ and acceleration $\overline{a}$ is $d^2\overline{r}/dt^2$. If the electron orbit is circular, $|\overline{r}|$ is constant so the time derivative of $\overline{r} = \hat{r}r$ only operates on the unit vector $\hat{r}$. For circular, motion,

$$\frac{d\hat{r}}{dt} = \hat{\phi}\,\frac{d\phi}{dt}$$

$$\frac{d\hat{\phi}}{dt} = -\hat{r}\,\frac{d\phi}{dt} \tag{2}$$

where the particle's position in the x–y plane is given by the angle ϕ (measured from the $\hat{x}$-axis). We substitute (2) into (1), yielding

$$\hat{\phi} m \frac{d^2\phi}{dt^2} - \hat{r} m \left(\frac{d\phi}{dt}\right)^2 = -\hat{r} e\mu_0 H_0 \frac{d\phi}{dt} \tag{3}$$

The $\hat{\phi}$ component of (3) states that $d^2\phi/dt^2 = 0$ so the angular velocity $d\phi/dt$ is just a constant defined to be the electron cyclotron frequency ω_c. The radial component of (3) then yields

$$\omega_c = \frac{e\mu_0 H_0}{m}$$

To find the power of the radiating electron, we recognize that an electron orbiting in a circle is a current loop with effective electric dipole moment $\int_{V'} \overline{J} dv' = qv = Id$ in each of two orthogonal directions that together define the plane of the loop. If we replace v by $\omega_c r$, substitute $Id = -e\omega_c r$ into (2.3.17), and double the radiated power because there are effectively two orthogonal dipoles radiating,[5] we find

$$P = \frac{\eta_0 k^2 e^2 \omega_c^2 r^2}{6\pi} = \frac{2}{3} \left(\frac{e^2}{4\pi\epsilon_0}\right) \frac{a^2}{c^3} \tag{4}$$

which is the Larmor formula for the power radiated by a nonrelativistic orbiting electron, and $a = -\omega_c^2 r$ is the acceleration of the electron. (The latter formula is commonly seen in physics texts with $e^2/4\pi\epsilon_0 \to e^2$ in cgs units). Note that $P \propto \omega_c^4 \propto H_0^4$, so electrons orbiting in strong magnetic fields rapidly radiate away their orbital energy.

2.4 RADIATION FROM ARRAYS OF HERTZIAN DIPOLES

In the previous section, we determined the farfield radiation pattern for a single $\hat{z}$-directed Hertzian dipole located at the origin to be

$$\overline{E}_{\text{ff}} = \hat{\theta} \frac{j\eta_0 k I d}{4\pi r} e^{-jkr} \sin\theta \tag{2.4.1}$$

where the subscript ff reminds us that (2.4.1) is valid only for $kr \gg 1$. Because Maxwell's equations are linear, we may superpose the electric fields radiated by individual dipoles in an array to find the total farfield radiation pattern. If a sufficient number of dipole elements, appropriately spaced and phased, are utilized in constructing the array, an arbitrary radiation pattern can be created that may be tailored to a specific engineering application. In Chapter 9 we shall discuss techniques used to determine the radiation pattern from an array of N identical radiating elements, but in this chapter we discuss radiation from just two Hertzian dipoles. Even with two dipoles, a wide variety of radiation patterns may be created, and two such elements are sufficient to understand the basic issues of phasor addition and interference effects between individual fields.

---

5. The total radiated powers of the two equal magnitude, perpendicular, 90°-out-of-phase dipoles can be summed only because they are orthogonal in space and time. Even in this case, superposition does not apply in specific directions or at specific times, but only on average.

For simplicity, we let both dipoles be $\hat{z}$-directed and consider radiation only in the x–y plane ($\theta = \pi/2$ in spherical coordinates). Because we are only interested in the radiation pattern in the farfield zone, we rewrite (2.4.1) for $\theta = \pi/2$:

$$\overline{E}_{\text{ff}} = -\hat{z}\, \frac{j\eta_0 k I d}{4\pi r}\, e^{-jkr} \qquad (\theta = \pi/2) \tag{2.4.2}$$

which is not a function of angle. A single Hertzian dipole radiates isotropically in any plane perpendicular to the dipole axis. If we now consider two $\hat{z}$-directed Hertzian dipoles located at $x = \pm D/2$ on the $\hat{x}$-axis having dipole moments $I_1 d_1$ and $I_2 d_2$, respectively, we may simply add the two $\hat{z}$ components of the electric field for each dipole to get

$$\overline{E}_{\text{ff}} = -\hat{z}\, \frac{j\eta_0 k}{4\pi}\left(I_1 d_1 \frac{e^{-jkr_1}}{r_1} + I_2 d_2 \frac{e^{-jkr_2}}{r_2} \right) \tag{2.4.3}$$

We note that I_1 and I_2 may be arbitrary complex numbers so that the dipoles do not necessarily have to oscillate in phase. The distance between the ith dipole and the observer is r_i ($i = 1, 2$), and the geometry of this two-dipole array is shown in Figure 2.6. For the farfield analysis to be valid, the distance between each dipole and the observer at $\overline{r}$ must be large compared to a wavelength ($kr \gg 1$). But note that no such constraints are placed on kD, as the kD product is not necessarily small (and is in fact on the order of unity if the dipoles are separated by roughly a wavelength).

We now make the *Fraunhofer farfield approximation* which says that when r is large, the denominators r_1 and r_2 in (2.4.3) may both be approximated by r. Far away from the origin, the function $1/r$ does not change appreciably as the wave propagates outward. On the other hand, the phase terms e^{-jkr_1} and e^{-jkr_2} may **not** be approximated by e^{-jkr}. Even small differences in path length contribute to a substantial change in the relative phases of the two individual electric fields in (2.4.3) because it is only the modulo 2π part of the phase that is important. Put another way, the phase is very sensitive to small changes in path length, leading to interference effects in the radiation pattern if kD is not small (and there is no guarantee that it is). If the observer is very far from the origin ($r \gg D$), then the vectors r_1 and r_2 will be almost parallel. In this *paraxial limit* , we may approximate

$$\begin{aligned} r_1 &\simeq r + \frac{D}{2}\cos\phi \\[4pt] r_2 &\simeq r - \frac{D}{2}\cos\phi \end{aligned} \tag{2.4.4}$$

where the difference in path length from each dipole to the observer ($D\cos\phi$) is found by inspecting Figure 2.6.

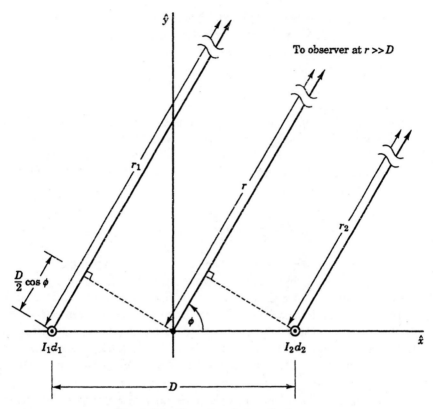

Figure 2.6　Geometry of a two-dipole array.

The far electric field for the two dipoles is thus approximated in the Fraunhofer limit by

$$\overline{E}_{ff} = -\hat{z}\,\frac{j\eta_0 k I_1 d_1}{4\pi r}\,e^{-jk(r+\frac{D}{2}\cos\phi)}\left(1 + \frac{I_2 d_2}{I_1 d_1}\,e^{jkD\cos\phi}\right) \qquad (2.4.5)$$

Because we would like to plot the radiation pattern, we are concerned here with finding the gain of the two dipoles. The time-averaged Poynting vector may be written in terms of the electric field alone by invoking (1.6.25). Using the formula for gain (2.3.19) for two Hertzian dipoles in free space, we see that in the x–y plane,

$$G(\phi) = K\left|1 + \frac{I_2 d_2}{I_1 d_1}\,e^{jkD\cos\phi}\right|^2 \qquad (2.4.6)$$

Because the overall magnitude of the radiation pattern is unimportant, we do not worry about the actual value of the proportionality constant K, which includes factors common to both dipole elements. Only the interference term is necessary to determine the radiation pattern. The constant K could be calculated by integrating the gain over all values of the solid angle $d\Omega = \sin\theta d\theta d\phi$; the result of this integration is always the constant 4π. Expanding (2.4.6) yields

$$G(\phi) = K(1 + A\,e^{j\alpha}\,e^{jkD\cos\phi})(1 + A\,e^{-j\alpha}\,e^{-jkD\cos\phi})$$

or

$$G(\phi) = K\left[1 + A^2 + 2A\cos(kD\cos\phi + \alpha)\right] \tag{2.4.7}$$

where $I_2 d_2 / I_1 d_1 \equiv A\,e^{j\alpha}$ and A and α are real numbers. The farfield radiation pattern is proportional to the gain $G(\theta, \phi)$ for any array, since the radiation pattern is just the gain $G(\theta, \phi)$ normalized to unity. From (2.4.7), we find

$$p(\phi) = \frac{1 + A^2 + 2A\cos(kD\cos\phi + \alpha)}{(1 + A)^2} \tag{2.4.8}$$

A few simple examples will help illustrate how a radiation pattern may be calculated, both analytically, by using (2.4.8), and physically.

Case 1: $A = 1$, $\alpha = 0$, $D = \lambda/2$

The dipoles are both the same strength and radiating in phase. Substituting the above parameters into (2.4.8) gives the radiation pattern

$$p(\phi) = \frac{1}{2} + \frac{1}{2}\cos(\pi\cos\phi) = \cos^2\left(\frac{\pi}{2}\cos\phi\right)$$

and $P(\phi)$ is plotted as a function of ϕ in Figure 2.7(a). Without actually computing $P(\phi)$, we can still describe the farfield radiation pattern based on physical arguments. If the two dipoles are located at $x = \pm\lambda/4$ and are radiating in phase, then we expect that the gain will have a maximum on the $\hat{y}$-axis ($\phi = \pm\pi/2$), since any point on the $\hat{y}$-axis is equidistant from the two dipoles. Therefore, the electric field radiation from both dipoles is said to interfere constructively at $\phi = \pm\pi/2$. On the other hand, the $\hat{x}$-axis is the location of a null in the radiation pattern because any point at $\phi = 0$ or $\phi = \pi$ is one-half wavelength closer to one dipole than to the other. This means that the two dipoles interfere destructively on the $\hat{x}$-axis, minimizing the farfield radiation pattern, and since the dipoles have the same magnitude, complete cancellation occurs. For any intermediate angle, $G(\phi)$ falls somewhere between the null at $\phi = 0$, π and the maximum at $\phi = \pm\pi/2$, and thus our physical analysis is consistent with the more formally computed gain plotted in Figure 2.7(a).

Case 2: $A = 1$, $\alpha = \pi$, $D = 3\lambda/2$

Again, the dipoles are the same strength, but this time they are radiating 180° out of phase. The radiation pattern is computed from (2.4.8) to be

$$p(\phi) = \frac{1}{2} + \frac{1}{2}\cos(3\pi\cos\phi + \pi) = \cos^2\left(\frac{3\pi}{2}\cos\phi + \frac{\pi}{2}\right)$$

and this radiation pattern is plotted in Figure 2.7(b). Again, we can locate the

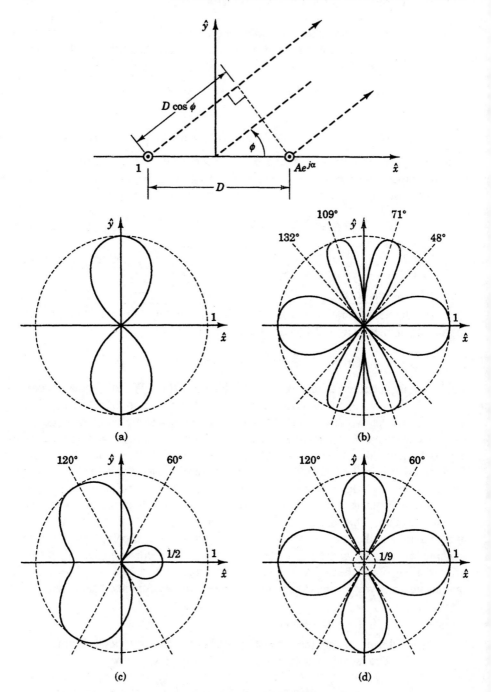

Figure 2.7   Two-dipole array radiation patterns.

maxima and minima of this pattern by physical argument. If the dipoles are the same strength and $180°$ out of phase, then there will be a null along the $\hat{y}$-axis where the path lengths are equal. In Figure 2.6, we see that nulls are present in the pattern when the excess path length $D \cos\phi$ is an integer multiple of wavelengths long, so that the out-of-phase dipoles interfere completely destructively:

$$\cos\phi_{null} = \frac{n\lambda}{3\lambda/2} = \frac{2n}{3} = 0, \pm\frac{2}{3}$$

Notice that $n = 0, \pm 1$ are the only integers possible because $|\cos\phi_{null}| \le 1$. Therefore, $\phi_{null} = \pm 48°, \pm 90°$, and $\pm 132°$. For completely constructive interference, corresponding to maxima in the radiation pattern, the path length difference must be a half-integer number of wavelengths

$$\cos\phi_{max} = \frac{(n+\frac{1}{2})\lambda}{3\lambda/2} = \frac{2n}{3} + \frac{1}{3} = \pm\frac{1}{3}, \pm 1$$

so $\phi_{max} = 0, \pm 71°, \pm 109°$, and $180°$. This is consistent with the gain pattern plotted in Figure 2.7(b).

Case 3: $A = 1$, $\alpha = \pi/2$, $D = \lambda/2$

The radiation pattern (2.4.8) for this case is given by

$$p(\phi) = \frac{1}{2} + \frac{1}{2}\cos\left(\pi\cos\phi + \frac{\pi}{2}\right) = \cos^2\left(\frac{\pi}{2}\cos\phi + \frac{\pi}{4}\right)$$

and is plotted in Figure 2.7(c). Because the dipoles are $90°$ out of phase, with the dipole on the $+\hat{x}$-axis leading the dipole on the $-\hat{x}$-axis, constructive interference is possible only when the excess path length is $n\lambda - \lambda/4$, so

$$\cos\phi_{max} = \frac{n\lambda - \lambda/4}{\lambda/2} = 2n - \frac{1}{2} = -\frac{1}{2}$$

and $\phi_{max} = \pm 120°$. Complete destructive interference occurs when the excess path length is $n\lambda + \lambda/4$

$$\cos\phi_{null} = \frac{n\lambda + \lambda/4}{\lambda/2} = 2n + \frac{1}{2} = \frac{1}{2}$$

so $\phi_{null} = \pm 60°$. Along the $\hat{x}$-axis, the magnitude of the gain pattern is half of the maximum gain, most easily seen by setting $\phi = 0$ or π in the expression for $P(\phi)$, and thus we expect a radiation pattern like that seen in Figure 2.7(c).

Case 4: $A = 2$, $\alpha = 0$, $D = \lambda$

This time, the dipoles are of unequal magnitude, but we can immediately determine the maximum and minimum intensities in the farfield radiation pattern. If the dipole interfere constructively, $P(\phi) = |1 + 2|^2/9 = 1$, but if they interfere destructively, then $P(\phi) = |1 - 2|^2/9 = 1/9$. Therefore, there are no nulls in the pattern, and we suspect that pattern nulls exist only when the dipoles have equal magnitudes. In general, (2.4.8) has a maximum when $\cos(kD\cos\phi + \alpha) = 1$, in which case $P(\phi) = 1$, and a minimum when $\cos(kD \cos \phi + \alpha) = -1$, forcing $P(\phi)$ to be $(1 - A)^2/(1 + A)^2$. From this last expression, we see that $P(\phi)$ is indeed zero only when $A = 1$. The radiation pattern for this configuration of dipoles is

$$p(\phi) = \frac{5}{9} + \frac{4}{9} \cos(2\pi \cos \phi)$$

Geometrically, the minima and maxima are still found by considering the difference in path lengths of the two dipoles. For dipoles in phase, the excess path length should be an integer multiple of wavelengths for constructive interference, so

$$\cos \phi_{max} = \frac{n\lambda}{\lambda} = n = 0, \pm 1$$

or $\phi_{max} = 0, \pm 90°$, and $180°$. The pattern minima are located at

$$\cos \phi_{min} = \frac{(n + \frac{1}{2})\lambda}{\lambda} = n + \frac{1}{2} = \pm\frac{1}{2}$$

and therefore $\phi_{min} = \pm 60°$ and $\pm 120°$. Thus, the physical argument agrees exactly with the pattern derived from the gain equation (2.4.8).

These four simple examples demonstrate how phasors may be added to produce interference patterns. We shall return to the subject of dipole arrays in Chapter 9, but already we can see that even two dipoles may produce a variety of radiation patterns.

Example 2.4.1

Two orthogonal Hertzian dipoles located at $x = y = z = 0$ are excited $\pi/2$ out of phase with dipole moments $I_x d$ and $I_y d$, respectively, where $I_y = j I_x$. Describe the polarization of the resulting radiation and calculate the power radiated along the $+\hat{z}$-axis.

Solution: The total far field is the linear vector sum of the farfield radiation from each dipole acting alone. Equation (2.3.14) indicates that the electric field from a single dipole is always $\hat{\theta}$-directed, where θ is referenced to the dipole axis. Therefore, using (2.3.14)

$$\overline{E}_{ff} = \hat{\theta}_1 \frac{j\eta_0 k I_x d}{4\pi r} e^{-jkr} \sin\theta_1 + \hat{\theta}_2 \frac{j\eta_0 k I_y d}{4\pi r} e^{-jkr} \sin\theta_2$$

$$= \frac{j\eta_0 k I_x d}{4\pi r} e^{-jkr} \left[\hat{\theta}_1 \sin\theta_1 + j\hat{\theta}_2 \sin\theta_2 \right] \tag{1}$$

where $\hat{\theta}_1$ and $\hat{\theta}_2$ are shown in Figure 2.8.

In the $x-y$ plane, $\hat{\theta}_2 = -\hat{\theta}_1$ and $\sin\theta_2 = \cos\theta_1$; both these relations may be found by examining Figure 2.8. Thus, in the $x-y$ plane, (1) simplifies to

$$\overline{E}_{ff} = \frac{j\eta_0 k I_x d}{4\pi r} e^{-jkr} \left[\hat{\theta}_1 \sin\theta_1 - j\hat{\theta}_1 \cos\theta_1 \right]$$

$$= \hat{\theta}_1 \frac{\eta_0 k I_x d}{4\pi r} e^{-jkr} e^{j\theta_1} \tag{2}$$

where $I_y = j I_x$ has been used. The polarization of (2) is therefore linear and $\hat{\theta}_1$-directed.

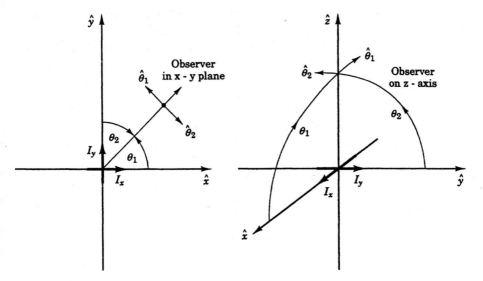

Figure 2.8   Turnstile antenna.

The total radiated power in the x–y plane is given by $\langle \overline{S} \rangle = |\overline{E} \cdot \overline{E}^*|/2\eta_0$, or

$$\langle \overline{S}(z = 0) \rangle = \hat{r} \, \frac{\eta_0}{2} \left| \frac{kI_x d}{4\pi r} \right|^2 \quad (\text{W/m}^2)$$

Along the $\hat{z}$-axis, (1) reduces to

$$\overline{E}_{\text{ff}} = -\frac{j\eta_0 k I_x d}{4\pi r} \, e^{-jkr} \left(\hat{x} + j\hat{y} \right)$$

because $\hat{\theta}_1 = -\hat{x}$ and $\hat{\theta}_2 = -\hat{y}$; $\theta_1 = \theta_2 = 90°$. Therefore, the polarization is left circular for $+\hat{z}$-directed propagation and right circular for $-\hat{z}$-directed propagation. The total radiated power is

$$\langle \overline{S}(x = 0, y = 0) \rangle = \frac{\eta_0}{2} \left| \frac{kI_x d}{4\pi r} \right|^2 (\hat{x} + j\hat{y}) \cdot (\hat{x} - j\hat{y})$$

$$= \eta_0 \left| \frac{kI_x d}{4\pi r} \right|^2 \quad (\text{W/m}^2)$$

The power radiated along the $\hat{z}$-axis is thus twice the power radiated in the x–y plane. This result is not surprising because on the $\hat{x}$- or $\hat{y}$-axes, one of the two dipoles has a null in its radiation pattern (and the other is maximally radiating), but on the $\hat{z}$-axis, **both** dipoles are maximally radiating. Such an orthogonal pair of dipoles is sometimes called a *turnstile antenna*.

2.5 SUMMARY

The primary source of all electromagnetic fields is electric charge. Maxwell's equations for a stationary charge density ρ (C/m^3) and charge in steady motion (current density $\overline{J}$ (A/m^2)) quickly yield simple static expressions for $\overline{E}$ and $\overline{H}$ everywhere in terms of the *scalar electric potential* Φ (V) and the *vector magnetic potential* $\overline{A}$ (webers/m = T·m). We define these potentials below:

$$\overline{E}(\overline{r}) = -\nabla\Phi$$

$$\overline{B}(\overline{r}) = \nabla \times \overline{A}$$

where

$$\Phi(\overline{r}) = \int_{v'} \frac{\rho(\overline{r}')}{4\pi\epsilon_0 |\overline{r} - \overline{r}'|} \, dv'$$

$$\overline{A}(\overline{r}) = \int_{v'} \frac{\mu_0 \overline{J}(\overline{r}')}{4\pi |\overline{r} - \overline{r}'|} \, dv'$$

Here we assume there are no media or boundaries present, only charge density $\rho(\overline{r}')$ and current density $\overline{J}(\overline{r}')$ in free space within some source region V'.

In general $\rho(\overline{r}')$ and $\overline{J}(\overline{r}')$ may vary with time, and therefore these equations must be modified. In particular, allowance must be made for the finite propagation velocity of electromagnetic signals in vacuum. In the sinusoidal steady state, the scalar and vector potentials become

$$\Phi(\overline{r}) = \int_{v'} \frac{\rho(\overline{r}')}{4\pi\epsilon_0 |\overline{r} - \overline{r}'|} e^{-jk|\overline{r} - \overline{r}'|} \, dv'$$

$$\overline{A}(\overline{r}) = \int_{v'} \frac{\mu_0 \overline{J}(\overline{r}')}{4\pi |\overline{r} - \overline{r}'|} e^{-jk|\overline{r} - \overline{r}'|} \, dv'$$

where

$$\overline{E} = -j\omega\overline{A} - \nabla\Phi$$

$$\overline{B} = \nabla \times \overline{A}$$

The significance of these equations was first seen in terms of a *Hertzian dipole* consisting of a current I flowing over a straight path of length d in the limit where $d \to 0$. The current density of a Hertzian dipole oriented along the $\hat{z}$-axis is therefore

$$\overline{J}(\overline{r}') = \hat{z} \, I d \, \delta^3(\overline{r}')$$

For this special source, the expressions for $\overline{E}$ and $\overline{H}$ have particularly simple forms in the limiting cases where $r \ll \lambda/2\pi$ (the *nearfield zone*) and $r \gg \lambda/2\pi$ (the *farfield zone*). In the near field, relevant distances are sufficiently small compared to a wavelength that signal propagation times are negligible compared to one cycle time $(1/f)$, and the field solutions approximate the static solutions. In the far field, the static solutions ($\propto 1/r^3$) are negligible compared to the radiated fields that decay only as $1/r$. These radiated fields propagate outward from a dipole with spherical wave fronts that, on a local microscopic scale, appear to be uniform plane waves; $\overline{S} = \overline{E} \times \overline{H}$ points radially outward, and $\overline{H} = \eta_0 \hat{r} \times \overline{E}$. The *impedance of free space* η_0 is approximately 377 Ω. For a $\hat{z}$-directed dipole with *dipole moment* Id,

$$\overline{E} = \hat{\theta} \frac{j\eta_0 k Id}{4\pi r} e^{-jkr} \sin\theta$$

where $k = \omega\sqrt{\mu_0\epsilon_0}$.

The *gain* of an antenna is found to be proportional to the square of the electric far field; for the Hertzian dipole,

$$G(\theta, \phi) = \frac{3}{2}\sin^2\theta$$

The total radiated power is found by integrating $\langle S_r \rangle$ over the surface of a sphere; for a Hertzian dipole,

$$P = \frac{\eta_0}{12\pi}|kId|^2$$

and its *radiation resistance* is therefore

$$R_{\text{rad}} = 2P/|I|^2 \cong 20(kd)^2$$

The farfield radiation pattern (proportional to gain) was plotted for a single Hertzian dipole and also for a two-dipole array. Even with only two dipoles, a wide variety of radiation patterns are observed, dependent on the relative magnitudes and phases of the two dipoles as well as the spacing between them. An analytic expression for the gain was derived, but more intuitive ways of understanding phasor addition also exist.

2.6 PROBLEMS

2.1.1 A straight wire 1 m long connects two charge reservoirs that pass one ampere DC through the wire. (i) What is the vector potential $\overline{A}(r)$ in a plane perpendicular to the wire and bisecting it, where r is the radial distance? (ii) What is $\overline{H}(r)$ in the same plane?

2.1.2 (a) What charge Q centered inside a sphere 1 cm in diameter would produce an electric field strength at radius 0.5 cm that equals a nominal air breakdown voltage of 3×10^7 V/m?

(b) What is the electric potential Φ (volts) at this radius?

2.2.1 The charge Q of Problem 2.1.2 is centered at the origin and varies as $Q = \cos 2 \times 10^{11} t$.

    (a) What is $\overline{E}(t)$ at r = 1 cm?

    (b) What is the potential $\Phi(t)$?

2.2.2* The charge q on an electron is -1.602×10^{-19} coulombs.

    (a) What is $\overline{E}(r)$ at a distance of 1 Angstrom (10^{-10} m), approximately the radius of an atom?

    (b) Electrons orbit nuclei at $r \approx 1$ Angstrom at a frequency f corresponding to a wavelength of 10^{-7} m, so approximately qf C/s flows at any point on the orbit. If we model an orbiting electron as a dipole element with current magnitude qf, what is the magnetic field at $r = 1$ Angstrom from this current?

2.3.1 Assume a short-dipole element is driven at its center with current $I_0 \sin \omega t$; the current is uniform over the 1-m long element.

    (a) What is $\overline{A}(r, \theta, t)$ in the far field, where θ is the angle between the dipole axis and the dipole-observer line?

    (b) What is $\overline{E}(r, \theta, t)$ (V/m)?

    (c) If 1 kW of power is radiated by this dipole and a given radio receiver can detect field strengths of 10^{-4} V/m, at what range r_{max} from the transmitter will this receiver work?

    (d) At what frequency f (Hz) would this 1-m dipole radiate 1 W if the current I were 1 A?

2.3.2 Sun navigation was first observed in 1911 when it was found that some species of ants, horseshoe crabs, and honeybees are sensitive to polarized light. As long as a small patch of blue sky exists, the animals can navigate well. The sky polarization depends on the angle ϕ between the sun's rays to a particular point in the sky and an observer's line of sight to the same point. This is because sunlight is scattered by air molecules that, when irradiated, behave like small dipole antennas. Consider an induced dipole oscillating at the same frequency as the incident sunlight. The scattered electric field $\overline{E}_s$ of the dipole is polarized parallel to planes containing the induced dipole: the magnitude of the scattered field is maximum in the plane perpendicular to the dipole axis and zero along the dipole axis. Draw a picture to show that when looking directly at the sun, the light is unpolarized (randomly polarized) and becomes more linearly polarized when looking at the sky farther away from the sun. Show that as ϕ approaches $90°$, the polarization will become linear, and specify the direction of polarization. (See *Scientific American*, July 1955 for further details.)

2.3.3* A uniform plane wave at frequency ω and intensity S (W/m^2) will cause a free electron to move with velocity v and radiate power P_{rad} as a Hertzian dipole. The velocity v can be found from Newton's law $\overline{f} = m\overline{a} = q\overline{E}$.

    (a) What are v and P_{rad}?

    (b) What is the effective scattering cross-section σ_T (the *Thompson scattering cross-section* of the electron, defined so that $\sigma_T \overline{S} = P_{rad}$)?

2.3.4* What is the approximate radiation force $\overline{f}$ of sunlight (having intensity ~ 1.5 kW/m^2 near Earth) upon a free electron in space? Approximate the average wavelength of sunlight as 500 nm and assume monochromaticity. Assess this force by finding the length of time before an electron at rest accelerates to 10 m/s. Hint: We may either compute $\bar{v} \times \bar{B}$ or use the Thompson scattering cross-section plus photon momentum $p = hf/c$ (see Example 4.2.4 and Problem 2.3.3). This force drives the solar wind and is much more intense near the sun.

2.4.1 Two $\hat{z}$-directed equal magnitude Hertzian dipoles are excited in phase and are spaced $D = 2\lambda$ apart along the $\hat{z}$-axis. Sketch the antenna power gain $G(\theta)$ in the y–z plane, and find the angle θ of each null.

2.4.2 A turnstile antenna consists of two Hertzian dipoles oscillating at an angular frequency ω and positioned at right angles to each other. The dipoles may be represented by the current distributions $\bar{J}_1 = \hat{z} I_0 \ell \delta(\bar{r}')$ and $\bar{J}_2 = \hat{y} j I_0 \ell \delta(\bar{r}')$.

 (a) Using spherical coordinates, find the total electric and magnetic field in the far field ($kr \gg 1$) in the x–y plane, i.e., at $\theta = \pi/2$.

 (b) Find and sketch the radiation power pattern in the x–y plane. Compare the power radiated in the $+\hat{x}$ direction to that in the $+\hat{y}$ direction.

 (c) What is the polarization of the radiated wave along the $+\hat{x}$-axis in the far field?

2.4.3 For two $\hat{z}$-directed Hertzian dipoles spaced a distance D apart along the $\hat{y}$-axis with the right dipole driven at $A e^{j\alpha}$ with respect to the left, sketch the radiation patterns in the x–y and y–z planes for the cases: (i) $D = \lambda$, $A e^{j\alpha} = -1$, (ii) $D = \lambda/2$, $A e^{j\alpha} = 1 + j\sqrt{3}$, and (iii) $D = 2\lambda$, $A e^{j\alpha} = j$. In which direction(s) is the gain maximum for each case?

2.4.4* A 1-MHz AM radio station wants to broadcast only to an area west of its two transmitting towers, since ocean lies to the east. Assume each tower is 3 m tall (where reflections from the ground will cause it to appear 6 m electrically) and that current is uniform along each tower length.

 (a) Show that the two towers should be separated by 75 m and be driven 90° out of phase to produce a pattern (called a cardioid) with a single null to the east. Should the eastern dipole lead or lag?

 (b) If the radio station is granted 50 kW power by the FCC, how far away can this radio station be heard? (For this calculation, assume that all the power is radiated isotropically into the above-ground, western half plane. Note that actual radio stations use half-wave dipoles, not short dipoles, since they have better impedance properties. See Section 9.7 for more details.)

3

WAVES IN MEDIA

3.1 TYPES OF MEDIA

Maxwell's equations (1.2.1)–(1.2.4) are generally expressed in terms of the four field quantities $\overline{E}$, $\overline{H}$, $\overline{D}$, and $\overline{B}$, where in vacuum

$$\overline{D} = \epsilon_0 \overline{E} \tag{3.1.1}$$

$$\overline{B} = \mu_0 \overline{H} \tag{3.1.2}$$

with $\epsilon_0 = 8.8542 \times 10^{-12}$ (F/m) and $\mu_0 = 4\pi \times 10^{-7}$ (H/m).

Equations (3.1.1) and (3.1.2) are the *constitutive relations* for free space, which must be modified to account for the presence of matter, consisting of molecules, atoms, ions, or even just electrons occupying the vacuum. All material properties enter Maxwell's equations through the constitutive relations, as Maxwell's equations themselves are media independent. This chapter will try to explain both the physics underlying each of the several types of media discussed and also how solutions to Maxwell's equations are modified by the more complicated inter-relationships between the fields.

There are many subtleties inherent in moving from a microscopic description of matter at the level of atoms or molecules to the macroscopic vantage point necessary to establish the constitutive relations. Once the connections between $\overline{E}$, $\overline{H}$, $\overline{D}$, and $\overline{B}$ are understood, however, then the actual medium may be characterized solely by its constitutive relations. In general, the relations between the fields depend upon the detailed physical structure of the medium, and this structure may

TABLE 3.1   Representative Types of Media

Designation	Dependency	Examples
polarizable		water, plastic
anisotropic	direction	polarizers, mica
inhomogeneous	position	atmosphere
nonstationary	time	acousto-optical modulators
dispersive	frequency	ionized plasma, human tissue
nonlinear	$\overline{E}$ or $\overline{H}$	iron
temperature dependent	temperature	iron
compressive	pressure	plastic
stationary	time-independent	stable dielectric
amnesic	history-independent	vacuum

further depend on direction, temperature, frequency, or other parameters. We review the nomenclature and give examples of several of the more common types of media in Table 3.1. In the remainder of this chapter, we discuss the character of wave propagation within anisotropic, lossy, and dispersive media.

3.2 POLARIZABLE MEDIA

One simple medium consists of atoms or molecules whose internal charge distributions are rearranged by the application of an electric field. Media in which charge separation may be induced are called *polarizable*; examples include water, oil, and plastic. Figure 3.1(a) shows how the application of an electric field causes the electron cloud to be displaced from the nucleus of an atom, and Figure 3.1(b) suggests that the permanent dipole moments of water molecules align with an applied field. The alignment is statistical in nature because the water molecules are in constant thermal motion; an applied electric field simply makes certain molecular orientations more probable in direct proportion to the field strength. In both cases, the charge distribution of a single atom or molecule may be modeled by a positive charge q displaced a directed distance $\overline{x}$ from its negative counterpart $-q$. A schematic view of this microscopic polarization is shown in Figure 3.1(c). The dipole moment for this idealized bit of polarized matter is a vector given by

$$\overline{p} = q\overline{x} \qquad (3.2.1)$$

We construct the macroscopic *polarization* vector $\overline{P}$ by multiplying $\overline{p}$ by the density of atoms in the material N

$$\overline{P} = Nq\overline{x} \qquad (3.2.2)$$

where N is expressed in m^{-3}, q is the charge of an electron (-1.602×10^{-19} C),

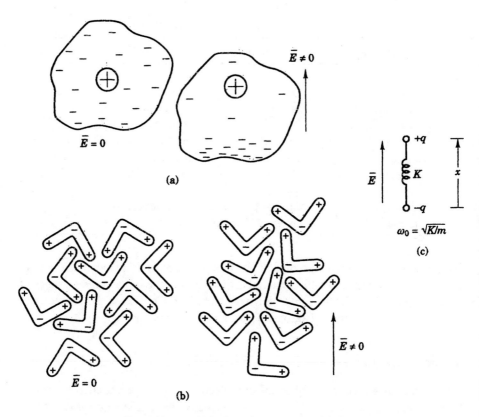

Figure 3.1 Polarization of atoms and molecules.

and $\overline{x}$ has units of meters. Therefore, both $\overline{D}$ and $\overline{P}$ have units of C/m^2, and it is natural to add $\overline{P}$ to $\epsilon_0\overline{E}$ to get the total displacement field:

$$\overline{D} = \epsilon_0\overline{E} + \overline{P} \qquad (3.2.3)$$

Equation (3.2.3) is actually a rather general empirical equation, where $\overline{P}$ can be determined in various types of media. For simple polarizable media, we relate $\overline{P}$ to the electric field by exploring the analogy between the restoring force (Hooke's law) that tends to bring the positive and negative charge centers together, and the restoring force in a simple mechanical spring. Inertial effects can be ignored if $\omega \ll \omega_0$, where ω_0 is the resonant frequency of the spring. Therefore, we can equate the restoring force on the spring, $K\overline{x}$, to the force produced by the electric field $q\overline{E}$ to find

$$\overline{x} = \frac{q}{K}\overline{E} = \frac{q}{m\omega_0^2}\overline{E} \qquad (3.2.4)$$

where K is related to the natural frequency of the spring ω_0 by $\omega_0 = \sqrt{K/m}$.

In this model, ω_0 corresponds to the resonant frequency of a polarizable atom or molecule. Combining (3.2.2) and (3.2.4) yields a polarization proportional to the electric field

$$\overline{P} = \frac{Nq^2}{m\omega_0^2}\overline{E} = \epsilon_0 \chi_e \overline{E} \qquad (3.2.5)$$

where $\chi_e = Nq^2/m\epsilon_0\omega_0^2$ is called the *electric susceptibility*. This expression suggests that media for which the charges are tightly bound (such that ω_0 and K are large) will have small susceptibilities because it is difficult to polarize such media. Equations (3.2.3) and (3.2.5) may be combined to yield

$$\overline{D} = \epsilon_0(1 + \chi_e)\overline{E} = \epsilon\overline{E} \qquad (3.2.6)$$

where $\epsilon = \epsilon_0(1 + \chi_e)$ is the *dielectric constant* of the medium. Such media are isotropic, homogeneous, and linear.

For *magnetizable media*, the general constitutive relation

$$\overline{B} = \mu_0(\overline{H} + \overline{M}) \qquad (3.2.7)$$

holds, where $\overline{M}$ is called the *magnetization*. The magnetization may be found by considering the physics of materials in which the moving charges in an atom or molecule can be acted upon by a magnetic field, causing their orbits or spins to shift. For simple media, $\overline{M}$ is linearly proportional to $\overline{H}$ and therefore

$$\overline{B} = \mu\overline{H} \qquad (3.2.8)$$

The more complicated gyromagnetic behavior of ferrites is considered in Section 3.7.

The permittivities ϵ and permeabilities μ of some common materials are given in Tables 3.2 and 3.3. Maxwell's equations for these simple media may be solved simply by replacing ϵ_0 by ϵ and μ_0 by μ in all of the derivations that have been already presented in Chapters 1 and 2. Of course, these derivations only apply

TABLE 3.2 Permittivities of some common materials [Von Hippel, 1954]

Material	ϵ/ϵ_0 at 1 MHz
vacuum	1
water	78.2
ice	4.15
sandy soil	2.59
teflon	2.1
ethyl alcohol	24.5
vaseline	2.16

TABLE 3.3   Permeabilities of some
common materials [Chegwedden, 1948]

Material	μ/μ_0
vacuum	1
biological tissue	1
cold steel	2,000
iron (99.91%)	5,000
purified iron (99.95%)	180,000
mu metal (FeNiCrCu)	100,000
supermalloy (FeNiMoMn)	800,000

to media that are unbounded and homogeneous, because we are not yet ready to explore the consequences of boundaries between media with differing properties. The subject of piecewise-uniform media will be taken up in the next chapter.

3.3 ANISOTROPIC MEDIA

For *anisotropic* media, the permittivity and/or permeability is not a single number, but depends instead on the orientation of the medium with respect to the direction of propagation of electromagnetic waves. The constitutive relations for anisotropic media may no longer be represented by a simple constant of proportionality between $\overline{D}$ and $\overline{E}$; instead, a tensor (matrix) equation is necessary:

$$\overline{D} = \overline{\overline{\epsilon}} \cdot \overline{E} \tag{3.3.1}$$

$$\overline{B} = \overline{\overline{\mu}} \cdot \overline{H} \tag{3.3.2}$$

The permittivity and permeability tensors $\overline{\overline{\epsilon}}$ and $\overline{\overline{\mu}}$ (a double overbar signifies a tensor) may be purely real or may be complex. The significance of complex constitutive relations is postponed until Section 3.4. In the discussion that follows, we consider an anisotropic dielectric medium where $\overline{B} = \mu_0 \overline{H}$ and $\overline{D} = \overline{\overline{\epsilon}} \cdot \overline{E}$. For clarity, we write out the components of (3.3.1):

$$\begin{aligned}
D_x &= \epsilon_{xx} E_x + \epsilon_{xy} E_y + \epsilon_{xz} E_z \\
D_y &= \epsilon_{yx} E_x + \epsilon_{yy} E_y + \epsilon_{yz} E_z \\
D_z &= \epsilon_{zx} E_x + \epsilon_{zy} E_y + \epsilon_{zz} E_z
\end{aligned} \tag{3.3.3}$$

We see immediately that $\overline{D}$ is no longer always parallel to $\overline{E}$, as shown in Figure 3.2, and that the permittivity tensor is a function of nine numbers. (Since we are still considering homogeneous media, each ϵ_{ij} is a constant.) But on physical grounds, we might expect that the permittivity tensor will be symmetric (that is, $\overline{\overline{\epsilon}} = \overline{\overline{\epsilon}}^T$ or $\epsilon_{ij} = \epsilon_{ji}$), which means that at most only six different constants are

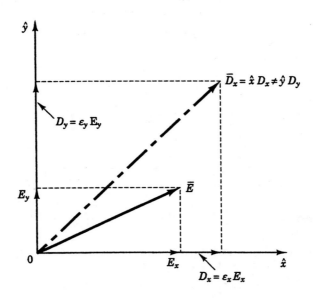

Figure 3.2 $\overline{D}$ and $\overline{E}$ in an anisotropic medium.

necessary to specify $\overline{\overline{\epsilon}}$. (Interesting exceptions in which $\overline{\overline{\epsilon}} \neq \overline{\overline{\epsilon}}^T$ will be discussed in Sections 3.7 and 3.8.) Moreover, an important theorem of linear algebra states that any symmetric matrix may be diagonalized through a proper choice of basis vectors. This means that if we choose our coordinate system intelligently, $\overline{\overline{\epsilon}}$ may be expressed as a diagonal tensor with only three independent constants

$$\overline{\overline{\epsilon}} = \begin{bmatrix} \epsilon_x & 0 & 0 \\ 0 & \epsilon_y & 0 \\ 0 & 0 & \epsilon_z \end{bmatrix} \tag{3.3.4}$$

where in general $\epsilon_x \neq \epsilon_y \neq \epsilon_z$ and the medium is *biaxial*. The coordinate system in which the permittivity tensor is diagonal is called the *principal system* and the coordinate axes are the *principal axes*. If the electric field lies along one of the principal axes, $\overline{D}$ is parallel to $\overline{E}$ in that special case. We therefore see that diagonalizing $\overline{\overline{\epsilon}}$ by choosing the principal coordinate system will greatly simplify the mathematical analysis that follows. It is evident from referring to Figure 3.2 that the principal axes of that medium ($\hat{x}$ and $\hat{y}$) are the only directions in which $\overline{D}$ and $\overline{E}$ are parallel; i.e., $D_x = \epsilon_x E_x$ and $D_y = \epsilon_y E_y$. No other direction in the x–y plane has this property.

Before we undertake the analysis of an anisotropic medium, we can partially explain anisotropy with a simple model. We consider a slab of dielectric ϵ with dimensions d, $d/2$, and W. In the first case, we orient a pair of parallel conducting plates as in Figure 3.3(a) and measure the capacitance of the dielectric/vacuum medium. Because the dielectric slab is connected in parallel with the vacuum, the total capacitance is

$$C_x = \frac{\epsilon(d/2)W}{d} + \frac{\epsilon_0(d/2)W}{d} = (\epsilon + \epsilon_0)\frac{W}{2}$$

where fringing effects have been neglected. For the configuration in Figure 3.3(b), the capacitors add in series, and

$$C_y = \left[\left(\frac{\epsilon dW}{(d/2)}\right)^{-1} + \left(\frac{\epsilon_0 dW}{(d/2)}\right)^{-1}\right]^{-1} = \left(\frac{\epsilon\epsilon_0}{\epsilon + \epsilon_0}\right)2W$$

Clearly C_x and C_y are not equal, and if $\epsilon \gg \epsilon_0$, then $C_x \simeq \epsilon W/2$ is much larger than $C_y \simeq 2\epsilon_0 W$.

For the crystalline structure in Figure 3.3(c), which is oriented so that its atoms are aligned along the $\hat{x}$-axis, we therefore expect that $\epsilon_x > \epsilon_y$, ϵ_z by similar reasoning. If the crystal arrangement is such that only the $\hat{x}$ direction is singled out, and therefore the $\hat{y}$- and $\hat{z}$-axes are indistinguishable, then the permittivity tensor may be written

$$\overline{\overline{\epsilon}} = \begin{bmatrix} \epsilon^e & 0 & 0 \\ 0 & \epsilon^o & 0 \\ 0 & 0 & \epsilon^o \end{bmatrix} \qquad (3.3.5)$$

where $\hat{y}$ and $\hat{z}$ are the *ordinary axes* and $\hat{x}$ is the *extraordinary* or *optic axis*. This type of medium is given the name *uniaxial*. Many crystals including mica, calcite, and quartz have such a structure. We can also fabricate a uniaxial medium using Figure 3.3(c) as a guide, provided that the wavelength of the electromagnetic radiation in this uniaxial medium is much larger than structural inhomogeneities of this pseudo-crystal.

We may derive the wave behavior of anisotropic media in a fashion similar to that employed for wave propagation in vacuum. Consider a source-free medium characterized by a permittivity tensor $\overline{\overline{\epsilon}}$ and uniform permeability μ. For simplicity, we choose the anisotropic medium to be uniaxial and the coordinate system to be oriented along the principal axes of the medium, so that $\overline{\overline{\epsilon}}$ is given by (3.3.5). We first derive a new wave equation from the source-free Maxwell's equations:

$$\nabla \times \overline{E} = -j\omega\mu\overline{H} \qquad (3.3.6)$$

$$\nabla \times \overline{H} = j\omega\overline{D} \qquad (3.3.7)$$

$$\nabla \cdot \overline{D} = 0 \qquad (3.3.8)$$

$$\nabla \cdot \overline{B} = 0 \qquad (3.3.9)$$

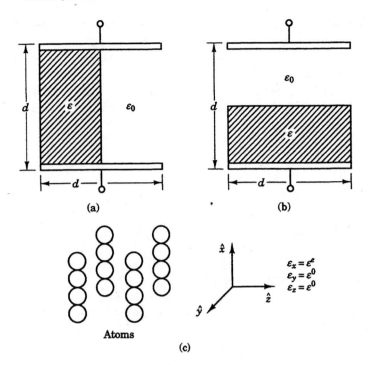

Figure 3.3   Anisotropically filled capacitors and a uniaxial crystal.

The curl of Faraday's law (3.3.6) combined with (3.3.7) yields

$$\nabla \times (\nabla \times \overline{E}) = \nabla(\nabla \cdot \overline{E}) - \nabla^2 \overline{E} = \omega^2 \mu \overline{D} \qquad (3.3.10)$$

But it is not immediately clear from (3.3.8) whether $\nabla \cdot \overline{E} = 0$. If we assume that we may choose a uniform plane wave propagating in the $+\hat{z}$ direction as one possible solution to Maxwell's equations, then $\partial/\partial x = \partial/\partial y = 0$ because there can be no variation in the plane perpendicular to the direction of propagation for a uniform plane wave. Therefore,

$$\nabla \cdot \overline{D} = \frac{\partial}{\partial x} D_x + \frac{\partial}{\partial y} D_y + \frac{\partial}{\partial z} D_z = \frac{\partial}{\partial z} D_z = 0$$

But by making use of (3.3.5),

$$\frac{\partial}{\partial z} D_z = \epsilon^o \frac{\partial}{\partial z} E_z = 0 \qquad (3.3.11)$$

where the mathematics is made much more tractable since the $\hat{x}$-, $\hat{y}$-, and $\hat{z}$-axes are coincident with the principal axes. Then, since $\partial E_z / \partial z = 0$ (3.3.11), and $\partial/\partial x = \partial/\partial y = 0$, it follows that

$$\frac{\partial}{\partial x} E_x + \frac{\partial}{\partial y} E_y + \frac{\partial}{\partial z} E_z = \nabla \cdot \overline{E} = 0$$

That $\nabla \cdot \overline{E} = 0$ is indeed zero can ultimately be tested by examining the final solution. Since we suppose $\nabla \cdot \overline{E}$ is zero, (3.3.10) becomes

$$\nabla^2 \overline{E} + \omega^2 \mu \overline{D} = 0$$

which can be reduced to three independent equations describing the wave behavior along each axis. Only the $\hat{x}$ and $\hat{y}$ components of the electric field are nonzero here for a $\hat{z}$-directed wave:

$$\left[\frac{\partial^2}{\partial z^2} + \omega^2 \mu \epsilon^e\right] E_x = 0 \tag{3.3.12}$$

$$\left[\frac{\partial^2}{\partial z^2} + \omega^2 \mu \epsilon^o\right] E_y = 0 \tag{3.3.13}$$

The wave equation (3.3.12) describes $\hat{x}$-polarized waves propagating in the $\hat{z}$ direction with propagation constant $k^e = \omega\sqrt{\mu\epsilon^e}$ and phase velocity $1/\sqrt{\mu\epsilon^e}$. Likewise, (3.3.13) describes $\hat{y}$-polarized waves also propagating in the $\hat{z}$ direction with $k^o = \omega\sqrt{\mu\epsilon^o}$ and phase velocity $1/\sqrt{\mu\epsilon^o}$. The polarization axis ($\hat{x}$ or $\hat{y}$ here) for which the wave propagates more slowly is called the "slow" axis, and the other is the "fast" axis. The solutions to (3.3.12) and (3.3.13) are written in phasor form as

$$E_x = E_{x0}\, e^{-jk^e z}$$
$$E_y = E_{y0}\, e^{-jk^o z}$$

and because E_x and E_y are not functions of x or y, we verify that $\nabla \cdot \overline{E} = 0$, which was assumed in the derivation of (3.3.12) and (3.3.13).

The phenomenon wherein $\hat{x}$- and $\hat{y}$-polarized waves propagate at different velocities in an anisotropic medium is called *birefringence*. Many natural crystals are birefringent, as are most transparent materials that are compressed along any one axis perpendicular to the direction of propagation. For example, because the extent of birefringence can be observed optically, pressure or stress patterns in plastic or glass are routinely measured using such techniques.

One of the major uses of birefringent media is the lossless conversion of one wave polarization to another. Consider a $\hat{z}$-propagating wave that is linearly polarized at 45° to the $\hat{x}$-axis at $z = 0$, with an electric field that may be represented by

$$\overline{E}(z = 0) = E_0 (\hat{x} + \hat{y}) \tag{3.3.14}$$

As the wave propagates in the uniaxial medium, the $\hat{x}$ and $\hat{y}$ components travel at different velocities, and the electric field within the medium is thus

$$\overline{E}(z) = E_0 \left(\hat{x}\, e^{-jk^e z} + \hat{y}\, e^{-jk^o z}\right) \tag{3.3.15}$$

which reduces to (3.3.14) at $z = 0$ as expected. If the crystal has thickness d in the $\hat{z}$ direction, then the wave leaves the uniaxial medium with the polarization it had at $z = d$:

$$\overline{E}(z = d) = E_0 \left(\hat{x}\, e^{-jk^e d} + \hat{y}\, e^{-jk^o d} \right) = E_0\, e^{-jk^e d} \left(\hat{x} + \hat{y}\, e^{j(k^e - k^o)d} \right)$$

From our discussion of polarization in Section 1.5, we note that if

$$\Delta\phi = (k^o - k^e)d = (n + \frac{1}{2})\pi, \qquad n = 0, \pm 1, \pm 2, \ldots$$

then the emerging wave is circularly polarized, and in fact is right circularly polarized if n is an even integer, and left circularly polarized if n is odd. We can also make elliptical polarization for intermediate values of $\Delta\phi$, and we may even use the birefringent medium to create the orthogonal linear polarization at $-45°$ to the $\hat{x}$-axis if $\Delta\phi = (2n + 1)\pi$. Figure 3.4 shows the types of polarization that may be established by varying the thickness of the crystal when the incident wave is linearly polarized at $45°$ to the extraordinary axis. This figure is the same as Figure 1.5 for $A = 1$ and arbitrary angle ϕ, where A and ϕ are defined in Section 1.5. If d is chosen such that the differential phase $\Delta\phi$ is $(n + \frac{1}{2})\pi$, then the $\hat{x}$ and $\hat{y}$ components of the electric field are out of phase by a quarter of a wavelength when the field emerges from the birefringent material. Such a slab is called a *quarter-wave plate* and is routinely used in optical systems to convert linear to circular polarization and vice versa. If $\Delta\phi = (2n + 1)\pi$, then the slab is called a *half-wave plate*, and when properly oriented it will convert a given polarization

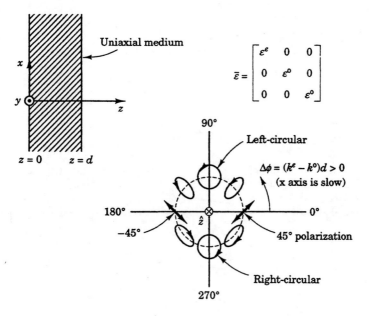

Figure 3.4   Birefringent medium for polarization conversion.

to its orthogonal partner (i.e., horizontal polarization will be converted to vertical polarization, or left circular to right circular.)

Our presentation of anisotropic media in this section has focused on the anisotropic permittivity tensor $\overline{\overline{\epsilon}}$, but a similar analysis could be made for a medium in which $\overline{D} = \epsilon_0 \overline{E}$ and $\overline{B} = \overline{\overline{\mu}} \cdot \overline{H}$. Because of the symmetry of Maxwell's equations, an exactly analogous derivation of birefringence could be carried out. Examples of magnetically birefringent media include ionized plasmas in static magnetic fields and many magnetic materials such as ferrites.

Example 3.3.1

A certain dielectric has

$$\overline{\overline{\epsilon}} = \epsilon_0 \begin{bmatrix} 9 & 0 & 0 \\ 0 & 4 & 0 \\ 0 & 0 & 4 \end{bmatrix}$$

What should its dimensions and orientation be in order to convert an $\hat{x}$-polarized $\hat{z}$-directed uniform infrared plane wave into a $\hat{y}$-polarized wave? The frequency f is 10^{14} Hz.

Solution: Since $\epsilon^e = 9\epsilon_0$ and $\epsilon^o = 4\epsilon_0$, the extraordinary phase velocity is $c/3$ and the ordinary phase velocity is $c/2$. Thus, the extraordinary axis of the dielectric is the slow axis, and the ordinary axis is the fast axis. To convert to the orthogonal linear polarization, it is necessary to project the incident electric field at $45°$ to the fast and slow axes, so that the components of the incident field along the fast and slow axes will have the same magnitude. The slow axis wave is then delayed by $180°$ with respect to the fast axis, as illustrated in Figure 3.5. A relative phase shift of $180°$ between the two components of $\overline{E}$ requires a half-wave plate with

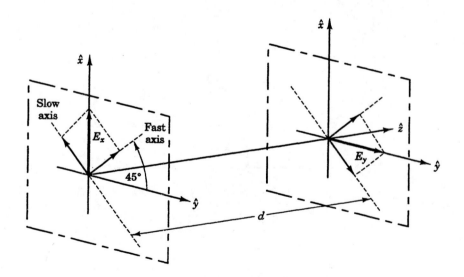

Figure 3.5 Half-wave plate.

$$(k^e - k^o)d = (2n + 1)\pi$$

$$(\omega\sqrt{\mu_0 9\epsilon_0} - \omega\sqrt{\mu_0 4\epsilon_0})d = (2n + 1)\pi$$

$$(\frac{3\omega d}{c} - \frac{2\omega d}{c}) = \frac{\omega d}{c} = (2n + 1)\pi$$

$$d = \frac{c(2n + 1)\pi}{\omega} = \frac{3 \times 10^8 (2n + 1)\pi}{2\pi \cdot 10^{14}} = 1.5, \ 4.5, \ 7.5, \ \dots \ \mu m$$

We generally choose the smallest value of d that works because of inevitable losses in the crystal.

3.4 WAVES IN LOSSY MEDIA

Before we go on to discuss more sophisticated versions of the constitutive relations (3.2.3) and (3.2.7), we pause to consider a heretofore unmentioned relationship between the current density $\overline{J}$ and the electric field $\overline{E}$. For a wide variety of materials, *Ohm's law* has been experimentally verified

$$\overline{J} = \sigma\overline{E} \tag{3.4.1}$$

where σ is the *conductivity* of the media, measured in siemens/m; a siemen (S) is equivalent to an inverse ohm. (Inverse ohms were formerly known as mhos, or "ohm" spelled backward.) Table 3.4 lists the conductivities of a variety of materials, and we note that almost no other material property exhibits as large a range of values.

Ohm's law (3.4.1), Gauss's law (1.2.3), and the charge continuity equation (1.2.5) may all be combined to yield a first-order differential equation for the free charge ρ:

TABLE 3.4   Conductivities of some common materials [Englund, 1944]

Material	σ (S/m)
vacuum	0
ocean water	3-5
physiological saline (0.15 $\underline{M}$)	4
distilled water	2×10^{-4}
silver	6.14×10^7
copper	5.80×10^7
iron	1×10^7
dry earth	$10^{-4} - 10^{-5}$
glass	10^{-12}
paraffin	$10^{-14} - 10^{-16}$

$$-\frac{\partial \rho}{\partial t} = \nabla \cdot \overline{J} = \sigma \nabla \cdot \overline{E} = \frac{\sigma}{\epsilon} \nabla \cdot \epsilon \overline{E} = \frac{\sigma}{\epsilon} \rho$$

The final differential equation for ρ is thus

$$\frac{\partial \rho}{\partial t} + \frac{\sigma}{\epsilon} \rho = 0 \tag{3.4.2}$$

which has the solution

$$\rho(\overline{r}, t) = \rho(\overline{r}, t = 0) \, e^{-t/\tau_c} \tag{3.4.3}$$

where $\tau_c = \epsilon/\sigma$ is called the *charge relaxation time*. We note that for our analysis to hold exactly, ϵ and σ must be constants in an unbounded, homogeneous medium. The physical significance of (3.4.3) is that a charge distribution that is imposed on a conductor at $t = 0$ will decay away essentially to zero after a few charge relaxation times. This charge will eventually end up on the boundaries of the conducting medium as *surface charge*, which could be shown analytically by considering a bounded medium, and the surface charge will arrange itself in such a way as to eliminate the electric fields inside the conductor.

That $\overline{E} = 0$ inside a good conductor is easily seen. For a "perfect" conductor, $\sigma \to \infty$. If we wish to ensure that the current density $\overline{J} = \sigma \overline{E}$ is finite everywhere, then the electric field must vanish inside the conductor by Ohm's law. Even if the conductivity is not actually infinite, we can consider the conductor to be essentially perfect if we are willing to allow the conduction of free charge out of the bulk of the medium to occur; in other words, if we are willing to wait several charge relaxation times. The charge relaxation time for physiological saline[1] is about 0.1 nanosecond (10^{-10} s), and for a good conductor like copper, it is less than a femtosecond (10^{-15} s), so we don't have to wait very long! For insulating material like paraffin, charge relaxation can take hours.[2]

---

1. Physiological saline is about 0.15 $\underline{M}$ (moles/liter) salt solution, which is the approximate salinity of blood.

2. The magnetic analog to charge relaxation is *magnetic diffusion*, in which nonuniform magnetic fields in a conductor relax toward uniformity with a *magnetic diffusion time* τ_m, provided that $\overline{J} \gg \partial \overline{D}/\partial t$. In this limit, Ampere's law becomes

$$\nabla \times \overline{H} = \overline{J} = \sigma \overline{E} \tag{1}$$

If we now take the curl of (1), using the vector identity (1.3.6) and Gauss's law (1.2.4) to simplify the left side and Faraday's law to simplify the right side, we are left with the magnetic diffusion equation

$$\nabla^2 \overline{H} = \sigma \mu \frac{\partial \overline{H}}{\partial t} \tag{2}$$

A simple solution to (2) is $\overline{H}(z, t) = \hat{y} H_0 e^{-t/\tau_m} \cos kz$ where the magnetic diffusion time τ_m is seen to equal $\sigma \mu / k^2$ after $\overline{H}$ is substituted into (2). Therefore, the magnetic field diffuses away slowly in a good conductor or high permeability material, or when the wavelength $\lambda = 2\pi/k$ is large.

We wish now to consider the effects of a conducting medium on the time-harmonic form of Maxwell's equations. We have already shown that in the steady state $\rho = 0$, since charge relaxation is a purely transient response. Ampere's law may then be written

$$\nabla \times \overline{H} = \overline{J} + j\omega\epsilon\overline{E} = \sigma\overline{E} + j\omega\epsilon\overline{E}$$

by applying Ohm's law (3.4.1), or

$$\nabla \times \overline{H} = j\omega\epsilon_{\text{eff}}\overline{E} \tag{3.4.4}$$

where the effective permittivity ϵ_{eff} is now a complex number

$$\epsilon_{\text{eff}} = \epsilon \left(1 - \frac{j\sigma}{\omega\epsilon}\right) \tag{3.4.5}$$

This mathematical construction of a complex dielectric constant allows us to substitute ϵ_{eff} for ϵ in all previous derivations to include the possibility of loss. For example, a uniform wave propagating in a lossy medium can be understood by reconsidering the free space dispersion relation (1.3.10) with ϵ_{eff} in place of ϵ_0:

$$k^2 = \omega^2\mu\epsilon \left(1 - \frac{j\sigma}{\omega\epsilon}\right) \tag{3.4.6}$$

Clearly, the wavenumber k is now a complex number and we express k as $k' - jk''$ where k' and k'' are both real numbers. Substituting $k = k' - jk''$ into (3.4.6) leads to a complex expression with real and imaginary parts given by

$$(k')^2 - (k'')^2 = \omega^2\mu\epsilon \tag{3.4.7}$$

and

$$2k'k'' = \omega\mu\sigma \tag{3.4.8}$$

respectively.

The quantity $\sigma/\omega\epsilon$ is called the *loss tangent* of the medium and is a measure of how successfully the medium conducts. Mathematically, the conductor is considered good if

$$\sigma \gg \omega\epsilon$$

and poor (insulating) if the reverse is true. Thus we can see that a material that might be an excellent conductor at low frequencies will become increasingly insulating as the frequency is raised.

For a wave propagating in the $+\hat{z}$ direction we have z dependence given by

$$e^{-jkz} = e^{-jk'z - k''z} \tag{3.4.9}$$

Such waves attenuate exponentially in the **same** direction as they propagate, and a sketch of a wave propagating in a lossy medium is seen in Figure 3.6. The distance over which the wave amplitude decays by $1/e$ is called the *penetration depth* Δ, equal to $1/k''$ m.

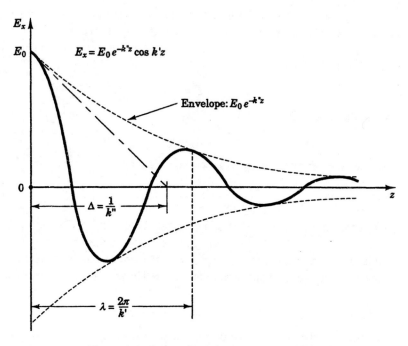

Figure 3.6 Wave propagation in a lossy medium.

The real and imaginary parts of the wavenumber, k' and k'', may be computed exactly from (3.4.7) and (3.4.8), but it is more useful to consider how k behaves in the limits of either very high or very low conductivity. For media with low conductivity, the loss tangent $\sigma/\omega\epsilon$ is much smaller than 1, and we can write

$$k = \omega\sqrt{\mu\epsilon}\sqrt{1 - \frac{j\sigma}{\omega\epsilon}}$$

$$\simeq \omega\sqrt{\mu\epsilon}\,(1 - \frac{j\sigma}{2\omega\epsilon}) \qquad (\sigma/\omega\epsilon \ll 1)$$

(3.4.10)

by using the Taylor expansion $\sqrt{1+x} \cong 1 + x/2$ for small x. Thus, $k' \simeq \omega\sqrt{\mu\epsilon}$ is basically unaffected by the slight conductivity of the medium; the wavelength $\lambda = 2\pi/k'$ is approximately the same as it would be if σ were zero. The imaginary part of k is given by

$$k'' = \frac{\sigma}{2}\sqrt{\frac{\mu}{\epsilon}} \qquad (\sigma/\omega\epsilon \ll 1)$$

(3.4.11)

so the penetration depth is $2(\sqrt{\epsilon/\mu})/\sigma$, which may be quite large, and

$$\eta = \sqrt{\frac{\mu}{\epsilon(1 - j\sigma/\omega\epsilon)}} \simeq \sqrt{\frac{\mu}{\epsilon}}\left(1 + \frac{j\sigma}{2\omega\epsilon}\right) \qquad (\sigma/\omega\epsilon \ll 1)$$

(3.4.12)

if the Taylor expansion for small loss tangent is again used.

For media that are highly conducting, $\sigma/\omega\epsilon \gg 1$, and k may also be simplified in this limit:

$$k = \omega\sqrt{\mu\epsilon}\sqrt{\left(1 - \frac{j\sigma}{\omega\epsilon}\right)}$$

$$\simeq \omega\sqrt{\mu\epsilon}\sqrt{\frac{-j\sigma}{\omega\epsilon}} = \sqrt{-j\omega\mu\sigma} \qquad (\sigma/\omega\epsilon \gg 1)$$

The square root of $-j$ may easily be computed by writing $-j$ as a phasor: $-j = e^{-j\pi/2}$, so $\sqrt{-j} = \pm e^{-j\pi/4} = \pm(1-j)/\sqrt{2}$. We choose the positive root of $\sqrt{-j}$ so that the wave will propagate in the $+\hat{z}$ direction; the negative root of $\sqrt{-j}$ propagates and attenuates in the $-\hat{z}$ direction. Note that if the conductivity were negative, $k \propto \sqrt{+j} = \pm(1+j)/\sqrt{2}$ and the wave would propagate in the same direction as it exponentially increased in magnitude. Some media (e.g., bulk lasers and masers), do amplify a propagating signal over finite distances. Such active media have negative conductivity and the energy for amplification must be supplied externally. However, we shall only be concerned with passive media in this text. Thus

$$k \simeq \sqrt{\frac{\omega\mu\sigma}{2}}\,(1 - j) \qquad (\sigma/\omega\epsilon \gg 1) \qquad (3.4.13)$$

for a highly conducting medium and both the real and imaginary parts of the wave vector k are the same: $k' = k'' = \sqrt{\omega\mu\sigma/2}$.

The penetration depth, also called the *skin depth* in a good conductor because it is very short, is given by

$$\delta = \frac{1}{k''} = \sqrt{\frac{2}{\omega\mu\sigma}} \quad \text{(m)} \qquad (3.4.14)$$

The wavelength in the highly conducting medium is $\lambda = 2\pi/k' = 2\pi\delta$, which is much shorter than the wavelength in the same medium with $\sigma = 0$.[3] For example, copper has a conductivity $\sigma \simeq 5.8 \times 10^7$ S/m at $f = 1$ GHz and $\mu = \mu_0$. The skin depth δ of copper is then approximately 2 μm and the wavelength is about 13 μm at 1 GHz, extremely short compared to the 1 GHz free-space wavelength of 30 cm. We also note that the characteristic wave impedance η is complex for a lossy medium:

---

3. We also recognize that

$$\omega^{-1} = \frac{\delta^2\sigma\mu}{2} = \frac{\sigma\mu}{2(k')^2}$$

is closely related to the quasistatic magnetic diffusion time τ_m: $\omega^{-1} = \tau_m/2$.

$$\eta = \sqrt{\frac{\mu}{\epsilon_{\text{eff}}}} = \sqrt{\frac{\mu}{\epsilon\,(1 - j\sigma/\omega\epsilon)}}$$

$$\cong \sqrt{\frac{j\omega\mu}{\sigma}} = \sqrt{\frac{\omega\mu}{2\sigma}}\,(1 + j) \qquad (\sigma/\omega\epsilon \gg 1)$$

(3.4.15)

and therefore the Poynting vector $\overline{S} = \overline{E} \times \overline{H}^* = |\overline{E}|^2/\eta^*$ is complex too. A more general treatment of waves in lossy media appears in Chapter 4.

Example 3.4.1

A certain uniaxial medium has permittivity

$$\overline{\overline{\epsilon}} = \epsilon_0 \begin{bmatrix} 9 & 0 & 0 \\ 0 & 4(1 - j10^{-3}) & 0 \\ 0 & 0 & 4(1 - j10^{-3}) \end{bmatrix}$$

Explain how this can be used to make Polaroid sunglasses that absorb one linear polarization but transmit most of the other. Find the appropriate thickness, assuming the light frequency is 6.3×10^{14} Hz.

Solution: The $\hat{y}$- and $\hat{z}$-axes are lossy, while the $\hat{x}$-axis is not. If we choose to propagate the light in the $\hat{x}$ direction, then we will have attenuation in both the $\hat{y}$ and $\hat{z}$ directions and the glass will not preferentially filter one of the two linear polarizations. Since Polaroid sunglasses work by filtering horizontally polarized light (most reflected light is horizontally polarized, as we shall see in Chapter 4), we should choose either $\hat{y}$ or $\hat{z}$ to be the direction of propagation. If the glasses are fabricated such that light propagates in the $\hat{z}$ direction, then $E_z = 0$, and the $\hat{z}$-axis conductivity is irrelevant. Assume the thickness d is chosen so that the $\hat{y}$-polarized power propagating in the $\hat{z}$ direction, $S_y(z)$, is attenuated 20 dB or more. That is,

$$\frac{|S_y(z = d)|}{|S_y(z = 0)|} = \frac{\dfrac{1}{2\eta}\,|\overline{E}_y(z = d)|^2}{\dfrac{1}{2\eta}\,|\overline{E}_y(z = 0)|^2} = \left|e^{-jk^o d}\right|^2 \le 10^{-2} \tag{1}$$

The ordinary wavenumber is

$$k^o = \omega\sqrt{\mu_0 4\epsilon_0(1 - j10^{-3})}$$

$$\cong \frac{2\omega}{c}\left(1 - \frac{j}{2}\cdot 10^{-3}\right)$$

But

$$\left|e^{-jk^o d}\right|^2 = e^{-2(k^o)'' d} = \exp\left(-\frac{2\omega}{c}10^{-3}d\right) \lesssim 10^{-2}$$

and therefore $d \gtrsim 0.18$ mm. The sunglasses are oriented so that the $\hat{y}$-axis is horizontal and the $\hat{x}$-axis is vertical.

Example 3.4.2

A uniform plane wave is propagating in the $+\hat{z}$ direction in a good conductor having conductivity σ S/m. The permittivity and permeability in the conductor are the same as in free space and the electric field is $\hat{x} E_0$ at $z = 0$. What power (W/m^2) is dissipated in this medium for $z > 0$? Assume $\sigma \gg \omega\epsilon$.

Solution: The time-averaged power is

$$\langle \overline{S} \rangle = \frac{1}{2} \operatorname{Re} \left\{ \overline{E} \times \overline{H}^* \right\}$$

where we use the fields

$$\overline{E} = \hat{x} \, E_0 \, e^{-jkz}$$

$$\overline{H} = -\frac{\nabla \times \overline{E}}{j\omega\mu_0} = \hat{y} \, \frac{k}{\omega\mu_0} E_0 e^{-jkz} = \hat{y} \, \frac{E_0}{\eta} e^{-jkz}$$

The high-conductivity-limit complex wave admittance $1/\eta$ is

$$\frac{1}{\eta} = \sqrt{\frac{\epsilon}{\mu_0}} = \sqrt{\frac{\epsilon_0}{\mu_0} \left(1 - j \frac{\sigma}{\omega\epsilon_0} \right)}$$

$$\cong \sqrt{\frac{\sigma}{2\omega\mu_0}} (1 - j) \qquad (\sigma \gg \omega\epsilon_0)$$

Thus,

$$\langle \overline{S}(z = 0) \rangle = \frac{1}{2} \operatorname{Re} \left\{ \hat{x} \, E_0 \times \hat{y} \left(\sqrt{\frac{\sigma}{2\omega\mu_0}} (1 - j) E_0 \right)^* \right\}$$

$$= \hat{z} \sqrt{\frac{\sigma}{8\omega\mu_0}} |E_0|^2 \quad \text{(W/m}^2\text{)}$$

Example 3.4.3
An $\hat{x}$-polarized beam of light passes through an ideal absorbing polarizer with its transmission axis rotated by an angle ϕ from the $\hat{x}$-axis in the x–y plane. We neglect any reflections at the surface of the polarizer and assume that no light is transmitted in the direction orthogonal to the transmission axis. What value of ϕ maximizes the fraction of power converted to $\hat{y}$-polarization, and what is this fraction? If N such ideal polarizers are stacked sequentially such that the transmission axis of the $k + 1^{st}$ polarizer is rotated by an angle ϕ with respect to the transmission axis of the kth polarizer, and the Nth polarizer has a transmission axis at $\phi = \pi/2$, then find the amount of power that passes through the series of polarizers. (A stack of $N = 4$ such polarizers is shown in Figure 3.7.) Show that as $N \to \infty$, 100 percent of the power that is originally $\hat{x}$-polarized is transmitted with $\hat{y}$-polarization.

Solution: If the original light beam is $\hat{x}$-polarized, then

$$\overline{E}_i = \hat{x} \, E_0$$

If the transmitting axis of the polarizer is in the $\hat{x}'$ direction, then the incident wave can be resolved along the $\hat{x}'$-$\hat{y}'$ axes as

$$\overline{E}_i = [(\hat{x} \cdot \hat{x}')\hat{x}' + (\hat{x} \cdot \hat{y}')\hat{y}']E_0 = \hat{x}' \, E_0 \cos\phi + \hat{y}' \, E_0 \sin\phi$$

so the only light that is transmitted is

$$\overline{E}_t = \hat{x}' \, E_0 \cos\phi$$

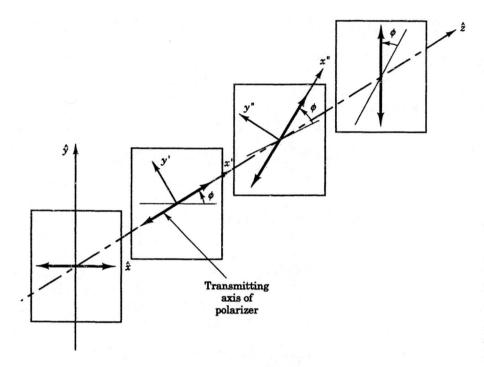

Figure 3.7   Transmission through a stack of $N = 4$ polarizers.

As we wish to find out how much of this transmitted wave is parallel to the $\hat{y}$-axis, we now re-express $\overline{E}_t$ in terms of components along the $\hat{x}$- and $\hat{y}$-axes:

$$\overline{E}_t = [(\hat{x}' \cdot \hat{x})\hat{x} + (\hat{x}' \cdot \hat{y})\hat{y}]E_0 \cos\phi = \hat{x}\, E_0 \cos^2\phi + \hat{y}\, E_0 \sin\phi \cos\phi$$

Therefore, the amount of light transmitted along the $\hat{y}$-axis is $E_{ty} = E_0 \sin\phi \cos\phi$, which is maximized when $\phi = \pi/4$. The transmitted field strength is thus one-half the incident field strength when $\phi = \pi/4$ and so the transmitted power is only one-fourth of the incident power.

If we have light that is polarized in the $\hat{x}'$ direction and it passes through a polarizer that has a transmitting axis in the $\hat{x}''$ direction, then the light that gets through is $\cos\phi$ times the original field and points in the $\hat{x}''$ direction, where ϕ is the angle between $\hat{x}'$ and $\hat{x}''$. Thus, if the incident wave is $\hat{x}$-polarized, we see that the electric field just after the first polarizer is

$$\overline{E}_1 = \hat{x}_1(\hat{x}_1 \cdot \hat{x})E_i = \hat{x}_1\, E_i \cos\phi$$

and after the second is

$$\overline{E}_2 = \hat{x}_2(\hat{x}_2 \cdot \hat{x}_1)E_1 = \hat{x}_2\, E_i \cos^2\phi$$

In general,

$$\overline{E}_n = \hat{x}_n(\hat{x}_n \cdot \hat{x}_{n-1})E_{n-1} = \hat{x}_n\, E_i \cos^n\phi$$

where $\overline{E}_n$ is the light passing through the nth polarizer. But if there are N polarizers with transmitting axes evenly spaced between $\phi = 0$ and $\phi = \pi/2$, then the field passing through the Nth polarizer is

$$\overline{E}_N = \hat{y}\, E_i \cos^N(\pi/2N)$$

and therefore the fraction of transmitted power is $[\cos(\pi/2N)]^{2N}$. As $N \to \infty$, this fraction may be expanded in a Taylor series for the small argument $\pi/2N$

$$\text{fraction} \simeq \left(1 - \frac{\pi^2}{8N^2}\right)^{2N} \simeq 1 - \frac{\pi^2}{4N}$$

which approaches unity when N is sufficiently large. Thus we can losslessly convert $\hat{x}$- to $\hat{y}$-polarization using lossy polarizers that transmit only along one axis if we keep the angle between adjacent polarizers sufficiently small!

Example 3.4.4

A slab of thickness L in the $\hat{x}$-direction and of infinite extent in the $\hat{y}$ and $\hat{z}$ directions has conductivity $\sigma(x) = \sigma_0/(1 + x/L)$ and dielectric constant ϵ_0. What is the electric field $\overline{E}$, displacement field $\overline{D}$, free charge density ρ_f, and polarization (bound) charge density ρ_p if current density $\overline{J} = \hat{x}\, J_0$ A/m² flows in the slab? If we replace the slab with a nonconducting dielectric having the same shape where $\epsilon(x) = \epsilon_0(1 + x/L)$, what are $\overline{E}$, $\overline{D}$, ρ_f and ρ_p if the potential difference across the slab is V_0 volts? Finally, if this slab has both the conductivity and dielectric function $\sigma(x)$ and $\epsilon(x)$ specified above, what are $\overline{E}$, $\overline{D}$, ρ_f, and ρ_p?

Solution: Case 1: $\sigma = \sigma_0/(1 + x/L)$, $\epsilon = \epsilon_0$
The electric field is just given by $\overline{J}/\sigma$ so

$$\overline{E}(x) = \hat{x}\, \frac{J_0}{\sigma_0}\left(1 + \frac{x}{L}\right) \quad \text{(V/m)}$$

and $\overline{D} = \epsilon \overline{E}$:

$$\overline{D}(x) = \hat{x}\, \frac{\epsilon_0 J_0}{\sigma_0}\left(1 + \frac{x}{L}\right) \quad \text{(C/m²)}$$

Because the material is not a dielectric, the polarization vector $\overline{P}$ defined in (3.2.2) is zero so the bound charge density is also zero

$$\rho_p \equiv -\nabla \cdot \overline{P} = 0$$

and the free charge density is

$$\rho_f \equiv \nabla \cdot \overline{D} = \frac{\epsilon_0 J_0}{\sigma_0 L} \quad \text{(C/m³)}$$

Therefore, nonuniform conductors typically possess free-charge distributions.

Case 2: $\sigma = 0$, $\epsilon = \epsilon_0(1 + x/L)$
The insulating slab has no free charge so $\rho_f = \nabla \cdot \overline{D} = 0$. Therefore $D_x(x) = \epsilon(x)E_x(x)$ cannot be a function of x, so $\overline{E} = \hat{x}\, E_0/(1 + x/L)$ where E_0 is an unknown constant. If we now integrate E_x across the slab, we find the slab voltage

$$V_0 = -\int_0^L dx\, E_x = -\int_0^L dx\, \frac{E_0}{1 + x/L} = -LE_0 \ln 2 \quad \text{(V)}$$

so $E_0 = -V_0/(L \ln 2)$. The electric field for the insulating slab is therefore

$$\overline{E} = -\hat{x}\, \frac{V_0}{\ln 2(x+L)} \quad \text{(V/m)}$$

and the displacement field is

$$\overline{D} = \epsilon(x)\overline{E} = -\hat{x}\, \frac{\epsilon_0 V_0}{L \ln 2} \quad \text{(C/m}^2)$$

The polarization charge density is found from

$$\rho_p = -\nabla \cdot \overline{P} = \nabla \cdot \epsilon_0 \overline{E} - \nabla \cdot \overline{D} = \frac{\epsilon_0 V_0}{\ln 2\,(x+L)^2} \quad \text{(C/m}^3)$$

where $\overline{D}$, $\overline{E}$, and $\overline{P}$ are related by (3.2.3). The dielectric slab therefore has polarization charge but no free charge.

    Case 3: $\sigma = \sigma_0/(1 + x/L)$, $\epsilon = \epsilon_0(1 + x/L)$
Once again, $\overline{E} = \overline{J}/\sigma$ so

$$\overline{E}(x) = \hat{x}\, \frac{J_0}{\sigma_0} \left(1 + \frac{x}{L}\right) \quad \text{(V/m)}$$

and

$$\overline{D}(x) = \epsilon(x)\overline{E}(x) = \hat{x}\, \frac{\epsilon_0 J_0}{\sigma_0} \left(1 + \frac{x}{L}\right)^2 \quad \text{(C/m}^2)$$

The free charge is found from

$$\rho_f = \nabla \cdot \overline{D} = \frac{2\epsilon_0 J_0}{\sigma_0 L} \left(1 + \frac{x}{L}\right) \quad \text{(C/m}^3)$$

and the polarization charge from

$$\rho_p = -\nabla \cdot \overline{P} = \nabla \cdot \epsilon_0 \overline{E} - \nabla \cdot \overline{D} = -\frac{\epsilon_0 J_0}{\sigma_0 L} \left(1 + \frac{2x}{L}\right) \quad \text{(C/m}^3)$$

Note that ρ_f and ρ_p are now each influenced by **both** the nonuniform conductivity and dielectric function, so the effects of each are no longer separable.

3.5 WAVES IN DISPERSIVE MEDIA: PHASE AND GROUP VELOCITIES

Dispersive media have values of μ, ϵ, or σ that depend on frequency. As a result, the wave velocity is generally frequency-dependent. To understand what frequency-dependent propagation means physically, we will use an approach that has been helpful previously. In Section 3.3 we saw how a wave at one point in a uniaxial medium could be divided into two constituent parts, each of which propagated with a different velocity. Then, after allowing each wave to propagate a finite distance, the two parts could be added back together, typically with altered polarization. This decomposition of waves into components is useful any time the propagation characteristics for the components are different. Decomposition and reconstitution work because Maxwell's equations are linear for linear media, where linear media have values of ϵ, μ, and σ which are independent of the field strengths. We may use this approach to understand dispersive media.

Consider a simple wave composed of only two sinusoids at very slightly differing frequencies, $\omega \pm \delta\omega$, where $\delta\omega \ll \omega$. The wave numbers at these two frequencies are $k \pm \delta k$ where $\delta k \ll k$, and so the total wave is the sum of the two independent solutions to the wave equation (1.3.7)

$$\cos\left[(\omega + \delta\omega)t - (k + \delta k)z\right] + \cos\left[(\omega - \delta\omega)t - (k - \delta k)z\right]$$

$$\equiv 2\cos(\omega t - kz)\cos(\delta\omega \cdot t - \delta k \cdot z)$$

where we used the identity

$$\cos\alpha + \cos\beta = 2\cos\left(\frac{\alpha + \beta}{2}\right)\cos\left(\frac{\alpha - \beta}{2}\right)$$

These two waves beat together, as suggested in Figure 3.8. The first factor, $\cos(\omega t - kz)$, is a sinusoid at the average frequency ω of the wave; it propagates at velocity $v_p = \omega/k$, which is designated the *phase velocity*. The modulation envelope of the wave, $\cos(\delta\omega \cdot t - \delta k \cdot z)$, has a period of $2\pi/\delta\omega$ seconds and

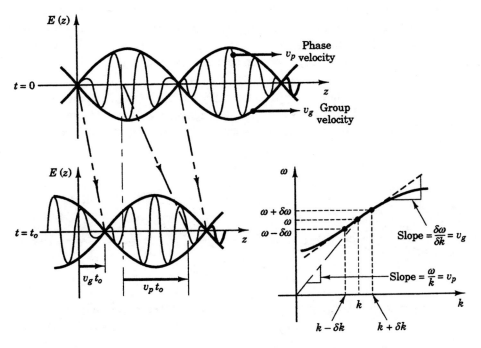

Figure 3.8   Group and phase velocities for two interfering sinusoids.

propagates at a velocity v_g of $\delta\omega/\delta k$, designated the *group velocity*. Thus

$$v_p = \frac{\omega}{k} \tag{3.5.1}$$

and

$$v_g = \left(\frac{\partial k}{\partial \omega}\right)^{-1} \tag{3.5.2}$$

In free space, $v_p = v_g = c$ because $\omega = ck$.

In general, the velocity of the envelope of any modulated sinusoid is the group velocity v_g derived here. "Packets" of energy or information therefore travel at v_g, the velocity of the modulation envelope, and this group velocity can never exceed the velocity of light c. The phase velocity v_p is not the velocity at which either energy or information propagates, and therefore it may exceed c, as it does in a plasma for a certain frequency range.

The phase velocity resembles the apparent velocity of a breaking wave at a beach, because the intersection point between the ocean wave and the shore can move at very high velocities as the wave fronts become increasingly parallel to the shore line. Figure 3.9 depicts this phenomenon. While the actual wave front has moved a distance $\overline{BC} = d$ in time Δt, the intersection point of the wave with the shore has moved a distance $\overline{AC} = d/\sin\theta$ in the same time, where θ is the angle between the wave fronts and the shore. So we see an apparent "intersection" velocity $v/\sin\theta$ moving along the shore, which becomes infinitely large as $\theta \to 0$. Notice that no information of energy in the wave can propagate with the intersection velocity, however, so it does not matter that it may eventually exceed the speed of light!

Any finite-duration wave packet actually has all frequencies present, as can be seen from the Fourier transform of such a signal. Therefore, the group velocity v_g generally differs for each frequency comprising the wave packet, and the wave

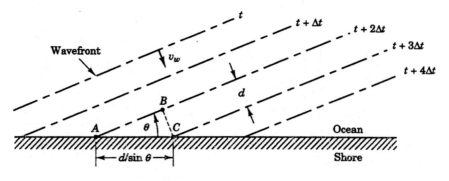

Figure 3.9 Phase velocity at a beach.

envelope will become progressively more distorted as the wave travels. For this reason, dispersive media are avoided in communications systems, or are accommodated by reducing the signal bandwidth to the point where the variation of v_g over the band is acceptably small for the given propagation path length. Since dispersive delay is a linear phenomenon, inverse dispersion filters can be employed in either the receiver or transmitter to cancel these distortions if necessary.

3.6 WAVES IN PLASMAS

A *plasma* is an assembly of positive and negative charged particles in which the time-averaged charge density is zero. The ionosphere, the surface of the sun, hydrogen in a fusion reactor, and even electrons in a highly conducting metal are all examples of plasmas. To find the effective dielectric constant ϵ of a plasma, we first find the polarization vector $\overline{P}$, where

$$\overline{D} = \epsilon\overline{E} = \epsilon_0\overline{E} + \overline{P} \tag{3.6.1}$$

$$\overline{P} \cong Nq\overline{x} \tag{3.6.2}$$

The charge on each electron is q ($q = -1.602 \times 10^{-19}$ C), and we assume N electrons per m^3 are each displaced a directed distance $\overline{x}$ from their origin by the electric field as previously described in Section 3.2 for the model of isotropic polarization. If the mass of the positive ions is not infinite, then they too respond to the field $\overline{E}$ and will increase the polarization. Here we make the reasonable assumption that the ion mass is so much greater than that of the electrons that their contribution to $\overline{P}$ can be neglected.

To find $\overline{P}$, we solve for the electron motion $\overline{x}$ by assuming there are no collisions. We recall that Newton's law may be written for a single electron as

$$\overline{f} = m\overline{a}$$

where the acceleration $\overline{a}$ is given by $\overline{a} = d^2\overline{x}/dt^2 = (j\omega)^2\overline{x}$ in time-harmonic form. Because the Lorentz force on the electron in the absence of magnetic fields is $\overline{f} = q\overline{E}$ (1.2.6), we find that

$$(j\omega)^2 m\overline{x} = q\overline{E} \tag{3.6.3}$$

where m is the electron mass.[4] Only the inertial term has been kept because the electrons are not strongly bound to the ions. Therefore, the restoring term $K\overline{x}$ discussed in Section 3.2 for bound charges is neglected here.

---

4. More exactly, m is the reduced mass of the electron-ion system

$$m = \frac{m_e M}{m_e + M}$$

where m_e and M are the masses of the electron and ion, respectively. For $M \gg m_e$, $m \simeq m_e$.

Solving (3.6.3) yields

$$\overline{x} = -\frac{q}{m\omega^2}\overline{E}$$

$$\overline{P} = -\frac{Nq^2}{m\omega^2}\overline{E}$$

$$\overline{D} = \epsilon_0\overline{E} + \overline{P} = \epsilon_0\left(1 - \frac{Nq^2}{\epsilon_0 m\omega^2}\right)\overline{E}$$

or

$$\overline{D} = \epsilon_0\left(1 - \frac{\omega_p^2}{\omega^2}\right)\overline{E} = \epsilon\overline{E} \tag{3.6.4}$$

where

$$\epsilon = \epsilon_0\left(1 - \frac{\omega_p^2}{\omega^2}\right) \tag{3.6.5}$$

and we have defined the *plasma frequency* ω_p to be

$$\omega_p = \sqrt{\frac{Nq^2}{m\epsilon_0}} \tag{3.6.6}$$

The plasma frequency is the natural frequency or resonance of a displaced electron (or cluster of electrons) in a uniform plasma oscillating about equilibrium. As we might expect, interesting things occur when we drive the plasma at its resonant frequency ω_p, as will be discussed shortly. It is clear from (3.6.5) that since the function ϵ depends on frequency, a plasma is a dispersive medium.

If we allow uniform plane waves to propagate in a simple plasma, we will find two basic types of behavior, depending on whether the frequency ω at which we propagate is above or below the plasma frequency. For $\omega > \omega_p$, the permittivity (3.6.5) is a positive number and normal wave propagation results. The dispersion relation for a plasma may be written

$$k^2 = \omega^2\mu_0\epsilon_0\left(1 - \frac{\omega_p^2}{\omega^2}\right) \tag{3.6.7}$$

The phase velocity of the plasma is then

$$v_p = \frac{\omega}{k} = \frac{c}{\sqrt{1 - (\omega_p/\omega)^2}} \tag{3.6.8}$$

which is greater than c for $\omega > \omega_p$. It is perfectly reasonable for the phase velocity to exceed the speed of light, since no information or energy is transported at the phase velocity. The group velocity, on the other hand, must normally be less than

c. Noting from (3.6.7) that $\omega^2 = c^2 k^2 + \omega_p^2$, we find that

$$v_g = \frac{\partial \omega}{\partial k} = c^2 \frac{k}{\omega} = c\sqrt{1 - (\omega_p/\omega)^2} \qquad (3.6.9)$$

which is always less than c.[5] We note that since a plasma has a quadratic dispersion relation, $v_p v_g = c^2$. The dispersion relation for the electron plasma is shown in Figure 3.10.

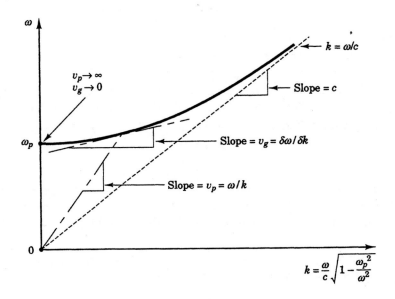

Figure 3.10   Dispersion relation for a simple plasma.

For $\omega < \omega_p$, the permittivity is a negative number; consequently, both the wave number k and the wave impedance η are purely imaginary in this frequency range. For a $\hat{z}$-directed uniform plane wave with $\omega < \omega_p$, we can write

$$k = \pm \frac{j}{c}\sqrt{\omega_p^2 - \omega^2} = \pm j\alpha \qquad (3.6.10)$$

where α is a real number, and the negative sign of α is chosen so that the waves will decay into the plasma. For this underdriven plasma,

---

5. Physically, we see that the plasma reduces the wave group velocity below c when $\omega > \omega_p$ and also reduces the wave energy from its free space value of $\epsilon_0|\overline{E}|^2/2$ to $\epsilon_0(1 - \omega_p^2/\omega^2)|\overline{E}|^2/2$. The "lost" wave energy resides partly in the kinetic energy of the plasma particles.

$$\overline{E}(z) = \hat{x}\, E_0\, e^{-jkz} = \hat{x}\, E_0\, e^{-\alpha z}$$

$$\overline{H}(z) = -\frac{\nabla \times \overline{E}}{j\omega\mu_0} = \hat{y}\,\frac{\alpha}{j\omega\mu_0} E_0\, e^{-\alpha z} \tag{3.6.11}$$

$$\overline{S} = \overline{E} \times \overline{H}^* = \hat{z}\,\frac{j\alpha}{\omega\mu_0}|E_0|^2\, e^{-2\alpha z}$$

Since $\overline{k}$ is purely imaginary, the wave oscillates in time and decays exponentially in space without any sinusoidal spatial component. The electric and magnetic fields are 90° out of phase, so the Poynting vector is purely imaginary, corresponding only to stored magnetic energy with no time-averaged power flow.[6] This type of wave, attenuated without power being dissipated, is called an *evanescent wave*, and the frequency at which propagating waves start to become evanescent is called the *cut-off frequency*. For a simple plasma, the cut-off frequency is the same as the plasma frequency. That the stored energy in the evanescent wave is predominantly magnetic can be seen by noting that the complex power into an inductor L is

$$P = \frac{j\omega L}{2}|I|^2$$

which is proportional to $+j$, just as is $\overline{S}$ in (3.6.11). We always associate $+j$ reactive power with an excess of average magnetic stored energy. A more complicated plasma which includes the effects of applying a DC magnetic field is considered below.

The permittivity of a plasma containing a DC magnetic field $\overline{B}_0$ must in general be expressed by a tensor $\overline{\overline{\epsilon}}$. We assume the electron density n, current $\overline{J}$, and $\overline{E}$ and $\overline{H}$ may be expressed as

$$n(t) = N + \mathrm{Re}\left\{n\, e^{j\omega t}\right\}$$

$$\overline{J}(t) \cong Nq\overline{v}(t)$$

$$\overline{E}(t) = \mathrm{Re}\left\{\overline{E}\, e^{j\omega t}\right\} \tag{3.6.12}$$

$$\overline{B}(t) = \overline{B}_0 + \mathrm{Re}\left\{\overline{B}\, e^{j\omega t}\right\}$$

where $\overline{v}$ is the electron velocity, $|n| \ll N$, and $|\overline{B}| \ll |\overline{B}_0|$. The time-harmonic equation of motion for an electron in such a plasma is approximately

$$j\omega m\overline{v} + mv\overline{v} = q(\overline{E} + \overline{v} \times \overline{B}_0) \tag{3.6.13}$$

---

6. In the underdriven plasma ($\omega < \omega_p$), the plasma kinetic energy density exceeds the level that a propagating electromagnetic wave can supply, leading to attenuation and nonpropagation of the wave itself. (Recall that $W_k \simeq \frac{1}{2}Nm|dx/dt|^2 = Nq^2|\overline{E}|^2/(2m\omega^2) = (\omega_p^2/\omega^2)\frac{1}{2}\epsilon_0|\overline{E}|^2$, so the kinetic energy of the plasma is equal to the total energy of the wave at cut-off, when $\omega = \omega_p$.)

where the Lorentz force on the right side of (3.6.13) has been equated to the inertial term $m \partial \overline{v} / \partial t \rightarrow j \omega m \overline{v}$ plus a dissipative term ($m \nu \overline{v}$), which accounts for collisions within the plasma. The plasma collision frequency is ν and $m \overline{v}$ is the momentum lost per collision, so $m \nu \overline{v}$ is the rate of momentum lost per electron due to collisions and has units of newtons. Rearranging (3.6.13) gives

$$q \overline{E} = j \omega m \left(1 - \frac{j \nu}{\omega} \right) \overline{v} + q \overline{B}_0 \times \overline{v} \tag{3.6.14}$$

and we may substitute (3.6.12) into (3.6.14) to yield

$$N q^2 \overline{E} = j \omega m \left(1 - \frac{j \nu}{\omega} \right) \overline{J} + q \overline{B}_0 \times \overline{J} \tag{3.6.15}$$

In order to solve for $\overline{J}$, it is useful to convert (3.6.15) into a matrix equation. We may write

$$\overline{B}_0 \times \overline{J} = \begin{bmatrix} 0 & -B_{0z} & B_{0y} \\ B_{0z} & 0 & -B_{0x} \\ -B_{0y} & B_{0x} & 0 \end{bmatrix} \cdot \begin{bmatrix} J_x \\ J_y \\ J_z \end{bmatrix} = \overline{\overline{B}}_0 \cdot \overline{J}$$

Using this notation, (3.6.15) becomes

$$\overline{E} = \left[\frac{j \omega m}{N q^2} \left(1 - \frac{j \nu}{\omega} \right) \overline{\overline{I}} + \frac{1}{N q} \overline{\overline{B}}_0 \right] \cdot \overline{J} \equiv \overline{\overline{M}} \cdot \overline{J} \tag{3.6.16}$$

where $\overline{\overline{I}}$ is the identity matrix

$$\overline{\overline{I}} = \begin{bmatrix} 1 & 0 & 0 \\ 0 & 1 & 0 \\ 0 & 0 & 1 \end{bmatrix}$$

Inverting (3.6.16) gives us

$$\overline{J} = \overline{\overline{M}}^{-1} \cdot \overline{E} \equiv \overline{\overline{\sigma}} \cdot \overline{E} \tag{3.6.17}$$

where $\overline{\overline{\sigma}}$ is the complex *plasma conductivity tensor*.

Now we may combine $\overline{\overline{\sigma}}$ with ϵ_0 to yield a *plasma permittivity tensor* $\overline{\overline{\epsilon}}$ for magnetized plasma, in the same fashion that we found an equivalent complex permittivity ϵ for a lossy medium in (3.4.5). That is, we use Ampere's law

$$\nabla \times \overline{H} = \overline{J} + j \omega \epsilon_0 \overline{E} = \left[\overline{\overline{\sigma}} + j \omega \epsilon_0 \overline{\overline{I}} \right] \cdot \overline{E} = j \omega \overline{\overline{\epsilon}} \cdot \overline{E} \tag{3.6.18}$$

where $j \omega \overline{\overline{\epsilon}} = \overline{\overline{\sigma}} + j \omega \epsilon_0 \overline{\overline{I}}$. If $\overline{B}_0 = B_0 \hat{z}$, this complex permittivity becomes

$$\overline{\overline{\epsilon}} = \epsilon_0 \begin{bmatrix} K' & j K'' & 0 \\ -j K'' & K' & 0 \\ 0 & 0 & K_0 \end{bmatrix} \tag{3.6.19}$$

where

$$K' = 1 - \frac{UX}{U^2 - Y^2}$$

$$K'' = -\frac{XY}{U^2 - Y^2}$$

$$K_0 = 1 - \frac{X}{U}$$

$$U = 1 - \frac{j\nu}{\omega} \tag{3.6.20}$$

$$X = \frac{\omega_p^2}{\omega^2}$$

$$Y = \frac{qB_0}{m\omega} = \frac{\omega_c}{\omega}$$

A tensor of the form (3.6.19) is given the name *gyrotropic* and has the interesting property that $\bar{\bar{\epsilon}} = (\bar{\bar{\epsilon}}^T)^*$. Matrices whose transposes are the same as their complex conjugates are called *Hermitian tensors*.

The natural frequencies here are the plasma frequency

$$\omega_p = \sqrt{\frac{Nq^2}{m\epsilon_0}}$$

and the *electron cyclotron frequency* ω_c

$$\omega_c = \frac{qB_0}{m} \tag{3.6.21}$$

The cyclotron frequency is the frequency at which a free electron circulates in the magnetic field B_0. In the limit where the electron collision frequency ν is zero, we find

$$K' = 1 - \frac{\omega_p^2}{\omega^2 - \omega_c^2}$$

$$K'' = -\frac{\omega_p^2 \omega_c}{\omega(\omega^2 - \omega_c^2)} \tag{3.6.22}$$

$$K_0 = 1 - \frac{\omega_p^2}{\omega^2}$$

Note that K' and K'' both become infinite at the cyclotron frequency when $\nu = 0$. Also, if B_0 is zero (and $\nu = 0$), then $K' = K_0 = 1 - \omega_p^2/\omega^2$, and K'' is zero. Therefore, $\bar{\bar{\epsilon}}$ is an isotropic tensor equivalent to the expression for ϵ derived in (3.6.5) for a field-free ($\bar{B}_0 \equiv 0$) collisionless plasma. As $\bar{B}_0 \to \infty$, the cyclotron frequency goes to infinity as well, and therefore $K' \to 1$, $K'' \to 0$, and

$K_0 = 1 - \omega_p^2/\omega^2$. The permittivity tensor therefore becomes

$$\bar{\bar{\epsilon}} = \epsilon_0 \begin{bmatrix} 1 & 0 & 0 \\ 0 & 1 & 0 \\ 0 & 0 & K_0 \end{bmatrix} \tag{3.6.23}$$

so we have a uniaxial plasma at sufficiently high magnetic field strengths.

A medium characterized by a Hermitian permittivity tensor ($\bar{\bar{\epsilon}}$ is equal to its complex conjugate transposed) is called an *electrically gyrotropic* medium. We defer the question of how waves propagate in such a medium to Section 3.8, after we derive the gyrotropic permeability tensor for a ferrite.

Example 3.6.1

A radio wave from outer space passes through the terrestrial ionosphere on its way to Earth. If the electron density is $N = 10^{12}$ m^{-3}, below what frequency f (Hz) can we no longer receive these signals? If the ionosphere were 100 km thick and $f = 10$ MHz, what would be the group delay T (seconds) of signals passing through it relative to the delay T_0 over the same path if N were zero?

Solution: The plasma frequency (Hz) is

$$f_p = \frac{\omega_p}{2\pi} = \frac{1}{2\pi} \sqrt{\frac{Nq^2}{\epsilon_0 m}} = 8.97 \text{ MHz}$$

where $m = 9.11 \times 10^{-31}$ kg is the mass of an electron.

The group delay depends on the group velocity v_g

$$v_g = \left(\frac{\partial k}{\partial \omega} \right)^{-1}$$

where

$$\frac{\partial k}{\partial \omega} = \frac{\partial}{\partial \omega} \left[\frac{\omega}{c} \sqrt{1 - \frac{\omega_p^2}{\omega^2}} \right] = \frac{1}{c \sqrt{1 - \frac{\omega_p^2}{\omega^2}}}$$

The group delay T for 100 km would be

$$T = \frac{10^5}{v_g} = \frac{10^5}{c \sqrt{1 - \frac{\omega_p^2}{\omega^2}}}$$

whereas the delay for 100 km of free space would be

$$T_0 = \frac{10^5}{c}$$

The differential delay is then

$$T - T_0 = \frac{10^5}{c} \left[\frac{1}{\sqrt{1 - \frac{\omega_p^2}{\omega^2}}} - 1 \right]$$

$$= 0.42 \text{ ms}$$

3.7* GYROMAGNETIC MEDIA

In a ferrite, the magnetization $\overline{M}$ roughly obeys the relation

$$\frac{\partial \overline{M}}{\partial t} = g\mu_0 \overline{M} \times \overline{H} \tag{3.7.1}$$

where g is called the *gyromagnetic ratio*.[7] We assume that a large $\hat{z}$-directed DC magnetic field is present in the ferrite, with a small-signal AC component added to it, so that

$$\overline{H}(t) = \hat{z} H_0 + \text{Re}\{\overline{H}_1 e^{j\omega t}\} \tag{3.7.2}$$

$$\overline{M}(t) = \hat{z} M_0 + \text{Re}\{\overline{M}_1 e^{j\omega t}\} \tag{3.7.3}$$

where $|\overline{H}_1| \ll |H_0|$ and $|\overline{M}_1| \ll |M_0|$. As usual, the constitutive relation for $\overline{B}$ and $\overline{H}$ is written

$$\overline{B} = \mu_0(\overline{H} + \overline{M}) = \overline{\overline{\mu}} \cdot \overline{H} \tag{3.7.4}$$

We should like to find the complex permeability tensor $\overline{\overline{\mu}}$ for the ferrite. Substituting (3.7.2) and (3.7.3) into (3.7.1), we find

$$\frac{\partial}{\partial t}(M_0\hat{z} + \text{Re}\{\overline{M}_1 e^{j\omega t}\}) = g\mu_0(M_0\hat{z} + \text{Re}\{\overline{M}_1 e^{j\omega t}\}) \times$$

$$(H_0\hat{z} + \text{Re}\{\overline{H}_1 e^{j\omega t}\})$$

or

$$j\omega \overline{M}_1 \simeq g\mu_0 \left[M_0(\hat{z} \times \overline{H}_1) - H_0(\hat{z} \times \overline{M}_1) \right] \tag{3.7.5}$$

where $\overline{H}_1$ and $\overline{M}_1$ are time harmonic fields. We have neglected the term $g\mu_0(\overline{M}_1 \times \overline{H}_1)$ because it is a product of two small quantities and is thus second-order compared to the two first-order terms on the right side of (3.7.5). The time derivative of

---

7. The gyromagnetic ratio has the units of charge/mass and is defined as the ratio of the magnetic moment to the angular momentum of an orbiting particle. For an electron, classical electromagnetic theory predicts $g = e/2m_e$, which is almost exactly half of the correct answer $g = -1.759 \times 10^{11}$ C/kg $\simeq e/m_e$ predicted by quantum mechanics.

$\hat{z}\, M_0$ is zero because M_0 is a DC field, and the $e^{j\omega t}$ dependence has been eliminated from (3.7.5) because it is common to all the terms in the equation.

If we explicitly express (3.7.5) in Cartesian coordinates, we find that

$$j\omega M_{1x} = -g\mu_0 M_0 H_{1y} + g\mu_0 H_0 M_{1y}$$

and

$$j\omega M_{1y} = g\mu_0 M_0 H_{1x} - g\mu_0 H_0 M_{1x}$$

We can rewrite these two equations in matrix form as:

$$\begin{bmatrix} M_{1x} \\ M_{1y} \end{bmatrix} = \frac{g\mu_0 M_0}{j\omega} \begin{bmatrix} 0 & -1 \\ 1 & 0 \end{bmatrix} \begin{bmatrix} H_{1x} \\ H_{1y} \end{bmatrix} + \frac{jg\mu_0 H_0}{\omega} \begin{bmatrix} 0 & -1 \\ 1 & 0 \end{bmatrix} \begin{bmatrix} M_{1x} \\ M_{1y} \end{bmatrix} \quad (3.7.6)$$

We know that M_{1z} is a second-order quantity because it arises from the cross-product of $\overline{M}_1$ and $\overline{H}_1$, which we have neglected. The dominant contribution to the $\hat{z}$ component of $\overline{M}$ is instead $\hat{z}\, M_0$, given in (3.7.3). The matrix equation (3.7.6) may be rewritten more compactly by combining terms

$$\begin{bmatrix} 1 & j\alpha \\ -j\alpha & 1 \end{bmatrix} \begin{bmatrix} M_{1x} \\ M_{1y} \end{bmatrix} = \frac{M_0}{H_0} \begin{bmatrix} 0 & j\alpha \\ -j\alpha & 0 \end{bmatrix} \begin{bmatrix} H_{1x} \\ H_{1y} \end{bmatrix} \quad (3.7.7)$$

where $\alpha \equiv g\mu_0 H_0/\omega$. We can multiply the matrix which premultiplies the vector $\overline{M}$ by its inverse on both sides of (3.7.7) to isolate $\overline{M}$, where the inverse matrix is

$$\begin{bmatrix} 1 & j\alpha \\ -j\alpha & 1 \end{bmatrix}^{-1} = \frac{1}{1-\alpha^2} \begin{bmatrix} 1 & -j\alpha \\ j\alpha & 1 \end{bmatrix}$$

Therefore, we find that

$$\begin{bmatrix} M_{1x} \\ M_{1y} \end{bmatrix} = \frac{M_0/H_0}{1-\alpha^2} \begin{bmatrix} 1 & -j\alpha \\ j\alpha & 1 \end{bmatrix} \begin{bmatrix} 0 & j\alpha \\ -j\alpha & 0 \end{bmatrix} \begin{bmatrix} H_{1x} \\ H_{1y} \end{bmatrix}$$

$$= \frac{\alpha M_0/H_0}{1-\alpha^2} \begin{bmatrix} -\alpha & j \\ -j & -\alpha \end{bmatrix} \begin{bmatrix} H_{1x} \\ H_{1y} \end{bmatrix} \quad (3.7.8)$$

We now substitute (3.7.8) into the constitutive relation (3.7.4):

$$\begin{bmatrix} B_x \\ B_y \end{bmatrix} = \mu_0 \begin{bmatrix} H_{1x} \\ H_{1y} \end{bmatrix} + \mu_0 \frac{\alpha M_0/H_0}{1-\alpha^2} \begin{bmatrix} -\alpha & j \\ -j & -\alpha \end{bmatrix} \begin{bmatrix} H_{1x} \\ H_{1y} \end{bmatrix} \quad (3.7.9)$$

$$B_z = \mu_0(H_0 + M_0)$$

This set of equations (3.7.9) may finally be combined to form a complex permeability tensor $\overline{\overline{\mu}}$ which relates $\overline{B}$ to $\overline{H}$: $\overline{B} = \overline{\overline{\mu}} \cdot \overline{H}$, where

$$\overline{\overline{\mu}} = \begin{bmatrix} \mu & j\mu_g & 0 \\ -j\mu_g & \mu & 0 \\ 0 & 0 & \mu_z \end{bmatrix} \quad (3.7.10)$$

and

$$\mu = \mu_0 \left[1 - \frac{(g\mu_0/\omega)^2 H_0 M_0}{1 - (g\mu_0 H_0/\omega)^2} \right]$$

$$\mu_g = \mu_0 \left[\frac{(g\mu_0 M_0/\omega)}{1 - (g\mu_0 H_0/\omega)^2} \right] \qquad (3.7.11)$$

$$\mu_z = \mu_0 \left[1 + \frac{M_0}{H_0} \right]$$

We see that the permeability matrix (3.7.10) is a gyrotropic permeability tensor, just as the permittivity matrix (3.6.19) for a plasma in a DC magnetic field also had the form of a gyrotropic tensor. We have thus demonstrated that ferrites exhibit gyromagnetism and also shown that both permeabilities and permittivities may be gyrotropic. The next section discusses the consequences of wave propagation in gyrotropic media.

3.8* FARADAY ROTATION

We next consider how waves propagate in a gyrotropic medium, and we choose to illustrate this by the example of gyromagnetic ferrites, with $\overline{\overline{\mu}}$ given by (3.7.10). We could equally well study gyrotropic media with the example of a plasma in a DC magnetic field, which yields a gyrotropic permittivity tensor $\overline{\overline{\epsilon}}$ (3.6.19). We choose the $\hat{z}$-axis to be the direction of wave propagation, so Faraday's law may then be written

$$k\hat{z} \times \overline{E} = \omega\overline{B}$$

where $k = |\overline{k}|$. Written in component form, this equation states that $E_y = -vB_x$ and $E_x = vB_y$; $v \equiv \omega/k$. Ampere's law for a $\hat{z}$-propagating wave is

$$k\hat{z} \times \overline{H} = -\omega\epsilon_0\overline{E}$$

or $H_y = v\epsilon_0 E_x$ and $H_x = -v\epsilon_0 E_y$. Combining the component forms of Ampere's and Faraday's laws for a gyrotropic magnetic medium (3.7.10) yields

$$H_x = v^2\epsilon_0 B_x = v^2\epsilon_0(\mu H_x + j\mu_g H_y) \qquad (3.8.1)$$

$$H_y = v^2\epsilon_0 B_y = v^2\epsilon_0(-j\mu_g H_x + \mu H_y) \qquad (3.8.2)$$

We can rearrange (3.8.1) and (3.8.2) to give

$$H_x(1 - v^2\epsilon_0\mu) = jv^2\epsilon_0\mu_g H_y \qquad (3.8.3)$$

$$H_y(1 - v^2\epsilon_0\mu) = -jv^2\epsilon_0\mu_g H_x \qquad (3.8.4)$$

Equation (3.8.3) can be substituted into (3.8.4) to find that

$$(1 - v^2\epsilon_0\mu)^2 H_x = (-jv^2\epsilon_0\mu_g)(jv^2\epsilon_0\mu_g) H_x \qquad (3.8.5)$$

But (3.8.5) has only a nontrivial solution when

$$1 - v^2 \mu \epsilon_0 = \pm v^2 \mu_g \epsilon_0 \qquad (3.8.6)$$

which is the dispersion relation for a gyromagnetic medium. This means that

$$v = \frac{\omega}{k} = \sqrt{\frac{1}{\epsilon_0(\mu \pm \mu_g)}} \qquad (3.8.7)$$

gives the velocities of the two waves that can propagate in this medium. We wish to know the polarizations of the waves that are able to propagate, and they are found by calculating H_y/H_x from (3.8.4):

$$\frac{H_y}{H_x} = \frac{-jv^2 \mu_g \epsilon_0}{1 - v^2 \mu \epsilon} = \pm j \qquad (3.8.8)$$

where the last equality holds by virtue of the dispersion relation (3.8.6) for a gyrotropic medium. Thus, the wave traveling in the $+\hat{z}$ direction with velocity $v = [\epsilon_0(\mu + \mu_g)]^{-1/2}$ has magnetic field components $H_y = jH_x$ and is therefore left circularly polarized, while the wave traveling in the $+\hat{z}$ direction with velocity $v = [\epsilon_0(\mu - \mu_g)]^{-1/2}$ is right circularly polarized, since $H_y = -jH_x$ (H_y lags H_x by $\pi/2$).

In order to understand what is happening to an arbitrary wave propagating in the $+\hat{z}$ direction, it must first be decomposed into two orthogonal circular polarizations, because we can only assign a meaningful propagation velocity to the two circular polarizations described above. We have just shown that each of the circular polarizations travels at a different velocity. We now investigate what happens to a $+\hat{z}$-propagating, linearly polarized wave as it passes through a gyromagnetic medium. We suppose that initially

$$\overline{H} = \hat{x} H_0 = (\hat{x} + j\hat{y})\frac{H_0}{2} + (\hat{x} - j\hat{y})\frac{H_0}{2} \qquad (3.8.9)$$

where we have decomposed the linearly polarized wave into two orthogonal circular polarizations. (We showed in Example 1.5.1 that it is possible to decompose **any** polarization into two orthogonal polarizations.) The wave on the left is left circularly polarized, and propagates with a wave vector k_L

$$k_L = \omega\sqrt{\epsilon_0(\mu + \mu_g)}$$

while the wave on the right is right circularly polarized and has a wave vector k_R

$$k_R = \omega\sqrt{\epsilon_0(\mu - \mu_g)}$$

The values k_L and k_R result from the dispersion relation (3.8.7). Therefore, after the wave has propagated a distance d into the ferrite, the magnetic field can be written

$$\overline{H} = (\hat{x} + j\hat{y})\frac{H_0}{2} e^{-jk_L d} + (\hat{x} - j\hat{y})\frac{H_0}{2} e^{-jk_R d}$$

and we regroup the $\hat{x}$ and $\hat{y}$ components of the $\overline{H}$ field to find (after some algebra), that

$$\overline{H} = \frac{H_0}{2}(e^{-jk_Ld} + e^{-jk_Rd})\left\{\hat{x} + \hat{y}\tan\left(\frac{(k_L - k_R)d}{2}\right)\right\} \qquad (3.8.10)$$

From our discussion of polarization in Section 1.5, we see that the outgoing wave (3.8.10) is **still** linearly polarized, but the polarization axis has been rotated counterclockwise by an angle $(k_L - k_R)d/2$. The rotation of a linearly polarized field to another linear polarization of differing orientation by a gyrotropic medium is called *Faraday rotation*.

3.9 SUMMARY

Electromagnetic waves interact with *media* by influencing the motions of the electric charges that compose the matter. Electric fields typically polarize molecules slightly such that each molecule produces a small dipole-like *polarization* field, which in turn influences the behavior of neighboring molecules. Similarly, magnetic fields can alter the orbits or spins of electrons and protons so as to induce small magnetic dipole fields that perturb the incident field. The cumulative effect of all these mutual interactions can usually be macroscopically characterized by the simple *constitutive relations*

$$\overline{D} = \epsilon\overline{E} = \epsilon_0\overline{E} + \overline{P}$$
$$\overline{B} = \mu\overline{H} = \mu_0(\overline{H} + \overline{M})$$

where $\epsilon = \epsilon_0$ and $\mu = \mu_0$ in vacuum.

Several properties of media are of particular interest. If the permittivity and/or permeability of a medium is dependent on the direction of the electric or magnetic fields, then the medium is said to be *anisotropic*. Waves travel with differing velocities along the different *principal axes* of an anisotropic medium, and thus anisotropic crystals may be used to losslessly convert one polarization to another; this phenomenon is known as *birefringence*.

If the permittivity possesses an imaginary part, then the medium exhibits loss, and waves attenuate as they propagate through this *lossy medium*. Loss is incorporated into the permittivity via the *conductivity* σ which relates $\overline{J}$ to $\overline{E}$ by *Ohm's law*: $\overline{J} = \sigma\overline{E}$, and $\epsilon_{\text{eff}} = \epsilon - j\sigma/\omega$.

Another important property of some media, including *plasmas*, is *dispersion*. A dispersive medium has frequency-dependent constitutive laws, and therefore waves of different frequencies propagate with different velocities. This leads to waveform distortion as a wave-packet propagates in such a medium. In a dispersive medium, the *group velocity* $v_g = (\partial k/\partial\omega)^{-1}$ is not equal to the *phase velocity* $v_p = \omega/k$. The group velocity, which is the speed at which energy and information travels, is only equal to the phase velocity, the speed at which the phase fronts travel, when $\omega/k = \text{constant}$.

The medium may also be *gyrotropic*, in which case the two orthogonal circular polarizations propagate at different velocities. A wave of any other polarization must be decomposed into right and left circular polarizations before it can be meaningfully analyzed. A linearly polarized wave propagating through a gyrotropic medium is rotated as it travels, all the while maintaining linear polarization; this phenomenon is known as *Faraday rotation*.

3.10 PROBLEMS   ·

3.2.1 Energy storage densities are limited by breakdown voltages for electric energy and mechanical collapse for magnetic energy. If $\overline{E}$ and $\overline{H}$ cannot exceed 5×10^6 V/m and 10 T, respectively, what are the maximum energy storage densities (J/m^3) for vacuum, alcohol, iron, and supermalloy (see Tables 3.2 and 3.3)? What are the electric susceptibilities χ_e and $\overline{M}/\overline{H}$ for each of these cases?

3.2.2 Find the approximate equivalent resonant frequency of an idealized water molecule [see Equation (3.2.4)] if its mass is equal to 18 amu (1 atomic mass unit = 1.67×10^{-27} kg). Distilled water (55 $\underline{M}$) has a dielectric constant of about $80\epsilon_0$. Estimate the electric charge magnitude $|q|$ to be $2|e|$, where e is the charge of an electron.

3.3.1 A wave has a magnetic field given by

$$\overline{H} = (\hat{y} - 2j\hat{z}) e^{2jx}$$

where x is measured in meters.
   (a) What is the polarization of this wave?
   (b) If this wave is passed through a quarter-wave plate with $\epsilon_y > \epsilon_z$, what is the resulting polarization of the emerging wave?
   (c) If $\epsilon_y = 4\epsilon_0$ and $\epsilon_z = \epsilon_0$, what is the minimum thickness of the quarter-wave plate?

3.3.2 A uniaxial dielectric has $\mu = \mu_0$ and

$$\overline{\overline{\epsilon}}/\epsilon_0 = \begin{bmatrix} 2 & 0 & 0 \\ 0 & 2 & 0 \\ 0 & 0 & 2.001 \end{bmatrix}$$

   (a) Design a quarter-wave plate for use at $\lambda = 0.5$ μm (green light).
   (b) A certain optical disk memory can be rewritten by changing the surface magnetic state. The state is read by reflecting a 0.5 μm laser from the surface. State 0 yields a right-hand polarization ellipse with an axial ratio of 2:1 oriented in the $\hat{y}$ direction in the y–z plane. Show how the quarter-wave plate of part (a) can be combined with a polaroid so as to absorb all the laser light for state 0, but pass some light for any other polarization. Find the necessary angles of the fast and slow axes and the angle of the polaroid axis, with respect to $\hat{y}$ and $\hat{z}$.

3.4.1* Consider a slab of material of thickness $w = d$ described by the following permittivity tensor:

$$\overline{\overline{\epsilon}} = \begin{bmatrix} 12\epsilon_0 & 0 & 0 \\ 0 & \epsilon_0 & 0 \\ 0 & 0 & 4(1 - j0.05)\epsilon_0 \end{bmatrix} \qquad \begin{array}{l} \epsilon_x = 12\epsilon_0 \\ \epsilon_y = \epsilon_0 \\ \epsilon_z = 4(1 - j0.05)\epsilon_0 \end{array}$$

This material can be used to make a polarizer, a quarter-wave plate, and a half-wave plate. Let the incident electric field propagate in the $\hat{w}$ direction.

Polarizer:

(a) Assign the dielectric axes $\hat{x}$, $\hat{y}$, $\hat{z}$ to the dimensions of the slab $\hat{u}$, $\hat{v}$, $\hat{w}$ (not necessarily in that order) so that for any arbitrarily polarized incident electric field and sufficiently thick slab, the transmitted field is linearly polarized.

(b) Determine the minimum thickness $w = d$ in free space wavelengths such that the undesirable component of the incident field is attenuated by $1/e$.

Quarter-wave plate (QWP):

(c) Assign $\hat{x}$, $\hat{y}$, $\hat{z}$ to the slab dimensions $\hat{u}$, $\hat{v}$, $\hat{w}$ (not necessarily in that order) so that for a given linearly polarized incident electric field, the transmitted field may be circularly polarized by appropriately positioning the QWP. Specify the axes so that there is no power absorption.

(d) Give an expression for the electric field incident on the QWP of part (c) so that the transmitted electric field is circularly polarized.

(e) For the QWP of part (c) and the incident field of part (d), determine the smallest thickness d in free space wavelengths such that the transmitted electric field is left circularly polarized.

Half-wave plate (HWP):

(f) Assign $\hat{x}$, $\hat{y}$, $\hat{z}$ to $\hat{u}$, $\hat{v}$, $\hat{w}$ (not necessarily in that order) so that for a given linearly polarized incident electric field, the transmitted field may be polarized in a direction orthogonal to the incident wave by appropriately positioning the HWP.

(g) For the HWP of part (f), determine the minimum thickness d in free space wavelengths. Specify the axes so that there is no power absorption. **Note:** Throughout the problem, neglect reflections at the slab interfaces. There may be more than one correct answer for each of the parts of the problem. Give one answer for each part; however, the answers must be consistent.

3.4.2 (a) The complex permittivity for bottom round steak is about $\epsilon = 40(1 - j0.3)\epsilon_0$ at the operating frequency (2.5 GHz) of a microwave oven. What is the wave penetration depth (skin depth) δ? Compare this penetration depth to that of polystyrene foam, which has complex permittivity $\epsilon = 1.03(1 - j0.3 \times 10^{-4})\epsilon_0$.

(b) Earth is considered to be a good conductor when $\sigma/\omega\epsilon \gg 1$. Determine the highest frequency for which Earth can be considered a good conductor if $\sigma/\omega\epsilon = 10$. Assume $\sigma = 5 \times 10^{-3}$ S/m and $\epsilon = 10\epsilon_0$.

(c) Aluminum has $\epsilon = \epsilon_0$, $\mu = \mu_0$ and $\sigma = 3.54 \times 10^7$ S/m. If an antenna for VHF (100 MHz) reception is made of wood coated with a layer of aluminum and if its thickness should be five times greater than the skin depth of the aluminum at that frequency, determine the thickness of the aluminum layer. Is ordinary aluminum foil thick enough for that purpose? (Ordinary aluminum foil is approximately 1/1000 inch thick.)

(d) Calculate loss tangents and skin depths for sea water at 100 Hz and 5 MHz. Sea water can be characterized by conductivity $\sigma = 4$ S/m, permittivity $\epsilon = 80\epsilon_0$, and permeability $\mu = \mu_0$ at those frequencies.

(e) A ship at the ocean surface wishes to communicate electromagnetically with a deeply submerged vehicle 100 meters below the surface. Consider a ULF signal at 1 kHz propagating down into the sea water. What fraction of the incident power density (just below the sea surface) reaches the submerged vehicle?

3.4.3 A uniform 1-GHz $\hat{x}$-polarized plane wave is propagating in copper ($\sigma = 5 \times 10^7$ S/m). The field $\overline{E}(z, t)$ is maximum at $t = 0$, $z = 0$.

(a) Sketch $\overline{E}(z, t = 0)$ and $\overline{H}(z, t = 0)$. Dimension the figure carefully, indicating quantitatively the positions of any field nulls and the approximate relative magnitudes of peaks in field strength.

(b) Calculate the imaginary part of $\overline{S}$ for this wave as a function of z; assume $\overline{E}$ at $z = 0$ is $\hat{x} E_0$. Is the reactive power inductive or capacitive?

(c) Express the total power dissipated (W/m^2) in the copper for $z > 0$ in terms of E_0, σ, μ, ϵ, and ω.

3.4.4 The electric field in a certain isotropic nonmagnetic material ($\mu = \mu_0$) is

$$\overline{E} = \hat{y}\, e^{-j(z-x)-z)/100}$$

where x and z have units of meters. The frequency is $\omega = \mu_0^{-1/2}$ rad/s.

(a) Find the magnetic field $\overline{H}$.

(b) What is the complex dielectric constant ϵ_{eff} for this medium?

3.4.5 Water vapor in the atmosphere has a molecular absorption band near 22 GHz which causes the power P in a $\hat{z}$-propagating wave to be attenuated as $P = P_0\, e^{-z/1000}$ where z is measured in meters.

(a) If we characterize the permittivity of the atmosphere as $\epsilon_{\text{eff}} = \epsilon_0 - j\epsilon''$, what is ϵ''?

(b) If the permittivity is characterized by ϵ_0 and the medium is (incorrectly) assumed to have an equivalent bulk conductivity σ (S/m), what is σ?

3.5.1 What are the group and phase velocities v_g and v_p as a function of f (Hz) for a material with $\epsilon = \epsilon_0$, $\mu = 4\mu_0(1 - (f_0/f)^3)$? Does v_p ever exceed c?

3.5.2 Figure 3.9 (in Section 3.5) illustrates the phase velocity of a wave along the shore (i.e., greater than v_w, the wave velocity in the ocean). For wave angle θ, what are v_p and v_g, the projected phase and group velocities along the shore line? What is the product $v_p v_g$?

3.6.1* Pulsars are rapidly rotating neutron stars composed primarily of neutrons at nuclear densities. When observed on Earth, pulsars emit periodic RF pulses each lasting approximately 10 ms. As these sharp pulses propagate through the interstellar medium, dispersion introduced by the interstellar plasma slows the pulse envelope more at lower RF frequencies that at higher frequencies. The first pulsar to be seen visually is located in the Crab Nebula, and was discovered in 1968 using the data shown in Figure 3.11(a).[8] The sloping lines formed by dense dots represent the envelopes of received pulse signals. Optical astronomers estimate that the Crab Nebula is approximately 6×10^{19} m ($\sim$ 6350 light years) from the Earth. In the following analysis, use the permittivity for plasma $\epsilon = \epsilon_0(1 - \omega_p^2/\omega^2)$.

(a) For $\omega \gg \omega_p$, find expressions for the phase and group velocities.

(b) In Figure 3.11(a), note the 1.5 s dispersion between frequencies 110 MHz and 115 MHz; i.e., the relative time delay between the 110-MHz and 115-MHz frequency components arriving at Earth is 1.5 s. Use this fact plus the known distance and the expression for v_g to calculate the interstellar electron density (m^{-3}).

(c) Aliens in the Crab Nebula wish to send messages to Earth using on-off-keyed binary signals (see Figure 3.11(b)). These signals occupy a bandwidth W (Hz) near 110 MHz, where $W \cong 1/T$. They want the group (envelope) delays at the high and low ends of the band W to differ less than $T/10$ s after their $\sim$ 6350-year journey from the Crab Nebula to Earth. If they differ much more, the bits are smeared and hard to detect. (However, a linear inverse filter at the transmitter or receiver can compensate for this.) What is the approximate maximum useable data rate R (bits/second) using this modulation scheme in the absence of compensation? Use appropriate approximations to simplify the algebra.

3.6.2 (a) Derive its group velocity v_g of a dispersive ionized plasma if $\mu = \mu_0$ and $\epsilon = \epsilon_0(1 - \omega_p^2/\omega^2)$, where $\omega_p = \sqrt{Ne^2/m\epsilon_0}$, N is the number of free electrons per cubic meter, e is the charge of an electron (coulombs), and m is the mass of an electron (kg).

(b) What is the difference in arrival times between a flash of light ($\lambda = 0.5$ μm) and a simultaneous radio pulse ($f = 10$ MHz) seen through an idealized homogeneous ionosphere where $\omega_p = 2\pi \times 8$ MHz along a path of 100 km?

3.6.3 X-ray lasing results from the electron plasma generated by a high-intensity 1.06 μm neodynium-glass pump laser. What is the *critical density* of this plasma, where most of the pump energy is absorbed? (The critical density is that density for which the pump laser frequency is equal to the plasma frequency.)

3.6.4 Superconductivity was first observed by Kamerlingh Onnes in 1911. In 1933, Meissner and Ochsenfeld discovered that superconducting metals cannot be penetrated by magnetic fields. Magnetic fields are expelled from a normal metal when it is cooled to the superconducting state. A macroscopic theory of superconductivity was developed by London and London in 1935, followed by the microscopic theory of Bardeen, Cooper and Schrieffer in 1957. A simple model of superconductivity calls for a collisionless electron plasma with a very high electron density N. The electrons avoid collisions if they are paired.

(a) Show that the wave penetration depth of a plasma with very large N is

$$d_p = \sqrt{\frac{m}{Ne^2\mu_0}}$$

---

8. Reprinted from *Science*, D. H. Staelin and E. C. Reifenstein, "Pulsating Radio Sources Near the Crab Nebula," Vol. 162, p. 1481, Dec. 27, 1968.

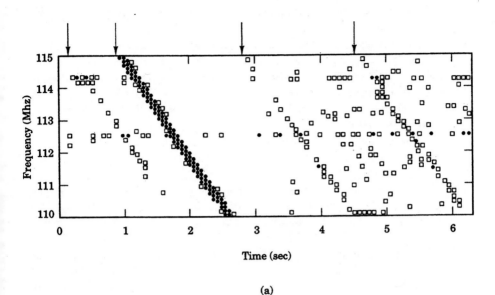

(a)

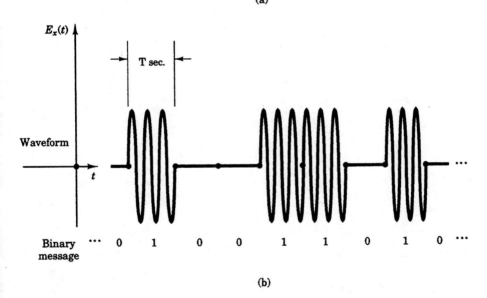

(b)

Figure 3.11   Crab Nebula data and alien binary signals.

(b) If $N = 7 \times 10^{28}$ m^{-3}, calculate d_p.

(c) Compare the above result with the skin depth of a good conductor. Explain why a very slowly varying magnetic field can penetrate a good conductor but not a superconductor. To the extent fields do penetrate superconductors microscopically, they interact with the conducting, unpaired, collision-prone electrons, introducing small losses at high frequencies.

3.7.1* (a) For very high frequencies, show that gyromagnetic media become uniaxial, just as plasmas become uniaxial at sufficiently high magnetic fields.

   (b) Find the dispersion relation (ω as a function of k) for a gyromagnetic medium.

3.8.1* A certain Faraday rotation device consists of a ferrite slab magnetized along the direction of propagation with $\overline{B} = 1$ T. The gyromagnetic ratio g for this material is -10^{10} C/kg and $\mu = 10^5 \mu_0$.

   (a) How thick must this slab be to yield 90° Faraday rotation at 10 GHz?

   (b) Over what frequency band (approximately) does the rotation angle lie between 89° and 91°?

3.8.2* If the ionized plasma of interstellar space has $N = 0.03$ electrons per cm^3 and $\overline{B} \approx 10^{-6}$ T in a particular region of interest, find the distance 1-GHz light must travel to rotate 360°. Express your answer in light years, where one light year is 9.46×10^{15} m.

3.8.3* The magnetic field of the Earth (approximately 0.6 gauss) is sufficient to cause electromagnetic fields to rotate when passing through the ionospheric plasma. If a linearly polarized, 2-GHz wave propagates 100 km through the ionosphere ($N = 10^{12}$ m^{-3}), what is the maximum rotation possible? Assume that the magnetic field and electron density are uniform (which isn't a very good approximation), and ignore electron-ion collisions in the plasma.

4

WAVES AT PLANAR BOUNDARIES

4.1 BOUNDARY CONDITIONS

In this chapter, we extend our interest to uniform plane waves propagating in regions of space containing two media separated by an infinite planar boundary. We find solutions to Maxwell's equations both for plane wave-generating sources infinitely far from the boundary and also for radiating current sources located on the boundary itself. In general, we solve Maxwell's equations in each of the two media separately and then use boundary conditions derived from Maxwell's equations to match these solutions at the planar interface. Once the boundary conditions are satisfied in a well-posed problem, we can be sure that the solutions in both media are unique. Therefore, we must first derive the boundary conditions at a surface (not necessarily planar) between two different media. To do this, we need to express Maxwell's equations in integral form.

Gauss's divergence theorem relates the volume integral of the divergence of an arbitrary vector $\overline{G}$ to the integral of that vector over the surface enclosing the volume:

$$\int_V (\nabla \cdot \overline{G})\, dv = \oint_S \overline{G} \cdot \hat{n}\, da \tag{4.1.1}$$

The outwardly pointing unit vector normal to the closed surface S spanning the volume V is denoted by $\hat{n}$, and da and dv are differential area and volume elements, respectively. These mathematical quantities are illustrated in Figure 4.1(a).

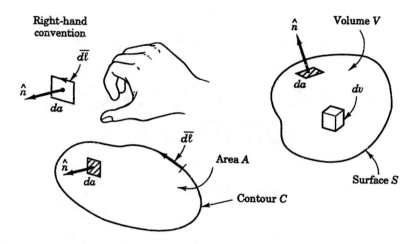

Figure 4.1   Gauss's and Stokes's theorems: Contours, surfaces, and volumes.

If we perform an integration of Gauss's laws (1.2.3) and (1.2.4) over an arbitrary volume before using (4.1.1), we find that

$$\oint_S \overline{D} \cdot \hat{n}\, da = \int_V \rho\, dv \quad (Gauss's\ law) \tag{4.1.2}$$

and

$$\oint_S \overline{B} \cdot \hat{n}\, da = 0 \qquad (Gauss's\ law) \tag{4.1.3}$$

Similarly, we can use *Stokes's theorem* to relate the area integral of the curl of an arbitrary vector $\overline{G}$ to the integral of that vector along the closed contour C encircling the area A:

$$\int_A (\nabla \times \overline{G}) \cdot \hat{n}\, da = \oint_C \overline{G} \cdot d\overline{\ell} \tag{4.1.4}$$

Again, $\hat{n}$ is the unit vector normal to the differential area da, and $d\overline{\ell}$ is a differential length tangent to the contour C. The direction of $d\overline{\ell}$ and the sign of $\hat{n}$ are related by the right-hand rule. With fingers of the right hand pointing in the direction $d\overline{\ell}$ of the contour, the thumb points in the direction $\hat{n}$. This convention is illustrated in Figure 4.1(b), along with the symbols used in Gauss's and Stokes's theorems. Faraday's law (1.2.1) and Ampere's law (1.2.2) may be written in integral form with the help of Stokes's theorem (4.1.4), if both sides of the differential laws are first integrated over an arbitrary surface:

$$\oint_C \overline{E} \cdot d\overline{\ell} = -\frac{d}{dt} \int_A \overline{B} \cdot \hat{n}\, da \qquad (Faraday's\ law) \tag{4.1.5}$$

and

$$\oint_C \overline{H} \cdot d\overline{\ell} = \frac{d}{dt} \int_A \overline{D} \cdot \hat{n}\, da + \int_A \overline{J} \cdot \hat{n}\, da \quad (\textit{Ampere's law}) \quad (4.1.6)$$

The total time derivative may be taken outside the integrals in (4.1.5) and (4.1.6) only if the area is not moving or deforming with time. We will ignore the complications that arise when the boundary conditions are nonstationary.

We may derive boundary conditions from Gauss's laws (4.1.2) and (4.1.3) by choosing a differential volume V shaped like a pillbox as illustrated in Figure 4.2(a). The top and bottom faces have area a and are oriented such that the top face is in one medium and the bottom face is in the other medium. Both faces are parallel to the boundary surface, and the thickness δ of the box is allowed to approach zero faster than the area does. Therefore, the contributions to the left side of (4.1.2) come only from the top and bottom surfaces of the pillbox

$$\oint_S \overline{D} \cdot \hat{n}\, da = (D_{1n} - D_{2n})a \quad (4.1.7)$$

where D_{1n} is the component of $\overline{D}$ in medium 1, which is normal to the boundary surface, and D_{2n} is the normal component of $\overline{D}$ in medium 2. The surface normal is arbitrarily defined to point **from** medium 2 **to** medium 1 as shown in Figure 4.2(a), which accounts for the negative sign in (4.1.7). The surface normal in medium 2 is in the opposite direction from the outward-pointing normal to the box surface a.

The right side of (4.1.2) can also be simplified, as $\int_V \rho\, dv$ is just the total charge inside the pillbox. As δ approaches zero, no volume charge ρ remains inside the pillbox. The only charge that survives the limit-taking process is a possible *surface charge* σ_s, which has dimensions of C/m^2. (The subscript s is added to distinguish the symbol for surface charge from that of conductivity.) Surface charge is actually a mathematical idealization in which the charge density at the surface goes to infinity at the same time the thickness of the charge layer goes to zero:

$$\sigma_s = q/a = \lim_{\substack{\rho \to \infty \\ \delta \to 0}} \rho\delta$$

But for surfaces on which the charge layer is very thin (exemplified by a good conductor that may have a skin depth of only a few micrometers), the charge distribution may be accurately approximated as a surface charge singularity.

The total charge in the pillbox is thus $\sigma_s a$, so (4.1.2) and (4.1.7) are combined to give

$$D_{1n} - D_{2n} = \sigma_s$$

which is also written

$$\hat{n} \cdot (\overline{D}_1 - \overline{D}_2) = \sigma_s \quad (4.1.8)$$

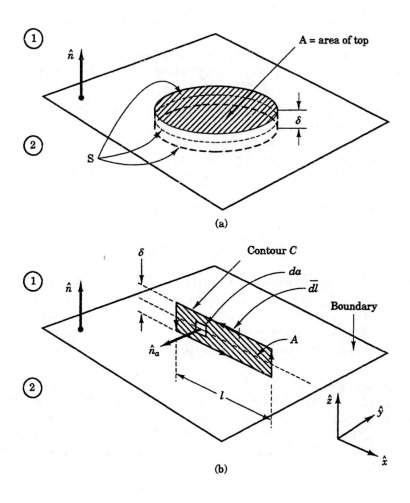

Figure 4.2 Elemental surfaces and volumes for deriving boundary conditions.

Exactly analogous reasoning for Gauss's magnetic law yields

$$\hat{n} \cdot (\overline{B}_1 - \overline{B}_2) = 0 \tag{4.1.9}$$

This means that the normal component of the displacement field $\overline{D}$ is continuous across a boundary between media, unless surface charge exists on that boundary, and the normal component of the magnetic flux $\overline{B}$ is always continuous. The charge continuity equation (1.2.5) also has a corresponding boundary condition that is included here for completeness. A similar infinitesimal pillbox volume used in conjunction with Gauss's theorem (4.1.1) can be used to show that

$$\hat{n} \cdot (\overline{J}_1 - \overline{J}_2) = \frac{\partial \sigma_s}{\partial t} \tag{4.1.10}$$

Faraday's law (4.1.5) and Ampere's law (4.1.6) yield two more boundary conditions if we choose a differential area a with contour C encircling the boundary between media as shown in 4.1.2(b). Two sides of the contour, each of length ℓ, run parallel to the boundary and the other two sides of length δ pierce it. Again, δ goes to zero faster than ℓ does. The left side of (4.1.5) may be written

$$\oint_C \overline{E} \cdot d\overline{\ell} = (E_{1t} - E_{2t})\ell \tag{4.1.11}$$

because we can ignore the contribution from the electric field along the two sides of length δ in the limit that $\delta/\ell \to 0$. The subscript t stands for tangential; E_{1t} is the component of the electric field in medium 1 tangential to the boundary surface, and a similar notation is used for medium 2. The negative sign in (4.1.11) comes from the fact that the direction of integration along the contour is opposite for the two segments of length ℓ.

The right side of Faraday's law (4.1.5) is written

$$-\frac{d}{dt} \int_A \overline{B} \cdot \hat{n}_a \, da = -\frac{d}{dt}(B_n \ell \delta) \tag{4.1.12}$$

where $\hat{n}_a$ is the unit vector perpendicular to the loop area. Equation (4.1.12) vanishes as δ goes to zero because the magnetic flux density $\overline{B}$ must be finite. If $\overline{B}$ were infinite in any finite region of space, then the magnetic energy density would also be infinite there, which is an impossibility. Therefore, we combine (4.1.11) and (4.1.12) with (4.1.5) to yield the boundary condition

$$E_{1t} - E_{2t} = 0$$

or

$$\hat{n} \times (\overline{E}_1 - \overline{E}_2) = 0 \tag{4.1.13}$$

where $\hat{n}$ again points from region 2 to region 1 (and is not to be confused with $\hat{n}_a$). Equation (4.1.13) states that the tangential components of the electric field are always continuous across a boundary.

Similar reasoning from Ampere's law (4.1.6) yields

$$(H_{1t} - H_{2t})\ell = \int_A \overline{J} \cdot \hat{n}_a \, da \tag{4.1.14}$$

where again the displacement current term $\dfrac{d}{dt} \displaystyle\int_A \overline{D} \cdot \hat{n}_a \, da$ goes to zero because $\overline{D}$ is finite and $\delta \to 0$. But the $\int_A \overline{J} \cdot \hat{n}_a \, da$ term does **not** vanish, because $\overline{J}$ will be infinite at the boundary if a surface current is flowing there. Like the surface charge singularity, the surface current $\overline{J}_s$ is also a mathematical convenience, where the magnitude of $\overline{J}_s$ is given by

$$|\overline{J}_s| = I/\ell = \lim_{\substack{J \to \infty \\ \delta \to 0}} |\overline{J}|\delta$$

The surface current has the dimensions of A/m and is the amount of current flowing along the boundary surface per unit length. Because $I = \int_A \overline{J} \cdot \hat{n}_a \, da$ has the units of amperes, we can write

$$\left| \int_A \overline{J} \cdot \hat{n}_a \, da \right| = |\overline{J}_s|\ell \tag{4.1.15}$$

and thus combine (4.1.14) and (4.1.15) to give

$$|H_{1t} - H_{2t}| = |\overline{J}_s|$$

or

$$\hat{n} \times (\overline{H}_1 - \overline{H}_2) = \overline{J}_s \tag{4.1.16}$$

where $\hat{n}$ again points from medium 2 to medium 1. We recall that magnetic fields circulate around the direction of current flow in a right-hand sense, and therefore the directions of the vectors in (4.1.16) may be easily confirmed.

The four boundary conditions (4.1.8), (4.1.9), (4.1.13) and (4.1.16) may easily be remembered from the differential form of Maxwell's equations. Each ∇ is converted to a $\hat{n}$, and each field $\overline{E}, \overline{H}, \overline{D},$ or $\overline{B}$ is replaced by the difference of the fields at the boundary: $\overline{E}_1 - \overline{E}_2, \overline{H}_1 - \overline{H}_2, \overline{D}_1 - \overline{D}_2,$ or $\overline{B}_1 - \overline{B}_2$. The volume charge density ρ is replaced by the surface charge σ_s, the current density $\overline{J}$ by the surface current $\overline{J}_s$, and the $\partial \overline{D}/\partial t$ and $-\partial \overline{B}/\partial t$ terms are ignored.

In the special case of a perfectly conducting medium, it has already been shown that $\overline{E} = 0$ (Section 3.4). If $\overline{J}$ is to be finite in the perfect conductor ($\sigma \to \infty$), then by Ohm's law (3.4.1), $\overline{E}$ must be zero inside the conductor. The fact that $\overline{J}$ approaches infinity on a perfectly conducting surface isn't a contradiction because the thickness of the current-containing surface δ approaches zero such that the product $|\overline{J}|\delta = |\overline{J}_s|$ is finite. If $\overline{E} = 0$, then $\omega\overline{B}$ is also zero by Faraday's law (1.2.1), so the magnetic field inside the conductor vanishes for time-harmonic fields.[1] Therefore, inside a perfect conductor,

$$\overline{E} = 0 \qquad (\sigma \to \infty) \tag{4.1.17}$$

and

$$\overline{B} = 0 \qquad (\sigma \to \infty) \tag{4.1.18}$$

This means that electric field lines in a medium adjacent to a perfect conductor

---

1. If $\omega = 0$, then $\overline{B}$ is not forced to vanish, which implies that a constant DC magnetic field can exist inside a perfect conductor.

must be perpendicular to the conducting surface, and the magnetic field lines must be purely tangential to the conducting surface. The normal displacement field $\overline{D}$ and tangential magnetic field $\overline{H}$ are both discontinuous at the conductor surface, giving rise to surface charge σ_s and surface current $\overline{J}_s$, respectively. We can therefore summarize the boundary conditions for fields inside a linear medium (ϵ, μ) bounded by a perfect conductor:

$$\hat{n} \times \overline{E} = 0 \tag{4.1.19}$$

$$\hat{n} \cdot \overline{E} = \sigma_s/\epsilon \tag{4.1.20}$$

$$\hat{n} \times \overline{H} = \overline{J}_s \tag{4.1.21}$$

$$\hat{n} \cdot \overline{H} = 0 \tag{4.1.22}$$

Example 4.1.1

Two thin, parallel metal plates of area A are separated by a small distance ℓ. If the plates are oppositely charged to $\pm Q$ coulombs, what are the static fields in the gap between them? What is the force on the plates? If both plates are charged to $+Q$ coulombs, how does the force change?

Solution: Because no charges are moving, $\overline{H} = \overline{B} = 0$, and the electric field in the charge-free region between the plates satisfies the electrostatic equations $\nabla \cdot \overline{E} = \nabla \times \overline{E} = 0$. Therefore, a uniform electric field perpendicular to the plates exists in the gap, consistent with the boundary condition (4.1.19), which says that only a normal electric field component is present at the surface of a perfect conductor. We neglect fringing fields if ℓ is small compared to the size of the plates. From (4.1.20), we find the surface charge $\sigma_s = Q/A$ is equal to $\hat{n} \cdot \epsilon_0 \overline{E}$. Since $\overline{E}$ is parallel to $\hat{n}$, $\overline{E} = \hat{n} Q/\epsilon_0 A$ where $\hat{n}$ is normal to the plates and directed from the positive to the negative plate.

If we let $\hat{n} = \hat{x}$ and assume that the plates are not quite perfectly conducting, a charge density $\rho(x)$ (C/m^3) exists on the positive plate, which is only appreciable over a small thickness δ (i.e., the charge is almost a surface charge). Therefore, $\sigma_s = Q/A = \int_{-\delta}^0 \rho(x)dx$. The Lorentz force law (1.2.6) thus becomes

$$\overline{f} = q\overline{E} = \hat{x} A \int_{-\delta}^0 \rho(x)\overline{E}(x) \, dx$$

where the electric field drops from $\hat{x} Q/\epsilon_0 A$ at the surface of the plate ($x = 0$) to $\overline{E} \approx 0$ at $x = -\delta$. Because the electric field is only a function of x, Gauss's law (1.2.3) becomes $\rho(x) = \epsilon_0 \partial E_x/\partial x$ and we find the attractive force

$$\overline{f} = \hat{x} \epsilon_0 A \int_{E_x(x=-\delta)=0}^{E_x(x=0)=Q/\epsilon_0 A} E_x(x) \, dE_x$$

$$= \hat{x} \frac{1}{2}\epsilon_0 A \left(\frac{Q}{\epsilon_0 A} \right)^2$$

even as $\delta \to 0$. An equal and opposite force is exerted on the negative plate.

We can also compute the Lorentz force by finding the total electric energy stored between the plates: $w = \frac{1}{2}\epsilon_0 E^2 A\ell$ (J), which may also be expressed as $w = \frac{1}{2}CV^2 = \frac{1}{2}Q^2/C$

where $C = \epsilon_0 A/\ell$ is the capacitance between the plates. If the charge is held constant on the plates, then

$$f_x = \left.\frac{\partial w}{\partial \ell}\right|_Q = -\frac{Q^2}{2C^2}\frac{\partial C}{\partial \ell} = \frac{1}{2}\epsilon_0 A \left(\frac{Q}{\epsilon_0 A}\right)^2 \tag{1}$$

as before. If the voltage between the plates were held constant instead, then the force calculated would be $f_x = (\partial w/\partial \ell)|_V = \frac{1}{2}V^2(\partial C/\partial \ell) = -\frac{1}{2}\epsilon_0 A(Q/\epsilon_0 A)^2$, which is oppositely directed and incorrect! In this calculation, we have ignored the energy of the power supply needed to maintain a constant voltage drop between the plates (which changes as a function of plate spacing). If the charge on the plates is held constant as originally specified, the power supply energy does not change with ℓ, so (1) is the correct procedure for computing the force between the plates.

If both plates are charged to $+Q$ coulombs, mutual repulsion forces the charges to the outsides of the two plates. As before, no magnetic field exists, and the electric field can be found by superposition. Were only one positively charged plate in the problem, it would produce the uniform field $\overline{E} = \hat{n}\, Q/2\epsilon_0 A$ pointing away from each side of the plate. With two such plates, the fields cancel in the gap between the plates and add outside of them, so $\overline{E} = \pm\hat{n}\, Q/\epsilon_0 A$ pointing away from the capacitor on both sides. The magnitude of the Lorentz force is the same as in the previous calculation but the force is repulsive (so that the plates tend to fly apart if free to move). The force computation is more subtle in this case because energy is stored throughout all space.

4.2 REFLECTION AND TRANSMISSION OF NORMALLY INCIDENT PLANE WAVES

Consider a semi-infinite, nonconducting medium characterized by ϵ_t and μ_t for $z \geq 0$; a different medium with ϵ_i and μ_i exists for $z < 0$. A uniform plane wave is propagating in the $+\hat{z}$ direction within the incident medium, having been generated by sources infinitely far from the boundary at $z = 0$. This plane wave is incident upon the medium from the left, as shown in Figure 4.3. If the electric and magnetic fields immediately to the left of the boundary at $z = 0^-$ are nonzero (which is generally the case), then the fields to the right ($z = 0^+$) are also nonzero because the tangential components of $\overline{E}$ and $\overline{H}$ must be continuous across a nonconducting boundary. Thus, a transmitted wave generally exists for $z > 0$. Usually the incident and transmitted waves alone cannot satisfy all of the boundary conditions; consequently a reflected wave is generated that travels back toward the incident sources.

The superposition of an incident $\hat{x}$-polarized wave and a reflected wave in the incident medium can be represented as

$$\overline{E}(z < 0) = \hat{x}\, E_i\, e^{-jk_iz} + \hat{x}\, E_r\, e^{+jk_iz} \tag{4.2.1}$$

$$\overline{H}(z < 0) = -\frac{\nabla \times \overline{E}}{j\omega\mu_i} = \hat{y}\, \frac{1}{\eta_i}\left(E_i\, e^{-jk_iz} - E_r\, e^{+jk_iz}\right) \tag{4.2.2}$$

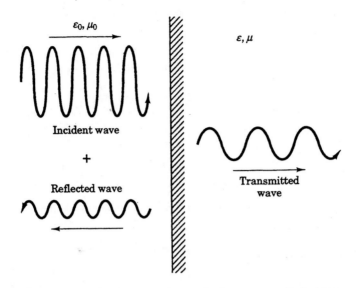

Figure 4.3   Incident, reflected, and transmitted waves normally incident at a planar boundary.

where E_i is the complex amplitude of the incident wave, E_r is the complex amplitude of the reflected wave, $k_i = \omega\sqrt{\mu_i \epsilon_i}$ is the wave number, and $\eta_i = \sqrt{\mu_i/\epsilon_i}$ is the impedance of the incident medium. We expect that the reflected wave will have the same $\hat{x}$-polarization as the incident wave in order to satisfy the boundary conditions; there is no source that can generate any other polarization.[2] We note also that the reflected wave propagates in the $-\hat{z}$ direction.

For $z > 0$, only a transmitted wave exists

$$\overline{E}(z > 0) = \hat{x}\, E_t\, e^{-jk_t z} \tag{4.2.3}$$

$$\overline{H}(z > 0) = \hat{y}\, \frac{1}{\eta_t} E_t\, e^{-jk_t z} \tag{4.2.4}$$

where E_t is the complex amplitude of the transmitted wave, $k_t = \omega\sqrt{\mu_t \epsilon_t}$ is the wave number, and $\eta_t = \sqrt{\mu_t/\epsilon_t}$ is the impedance of the transmitting medium. We do not expect to find a wave traveling in the $-\hat{z}$ direction for $z > 0$, as this would necessitate the existence of sources somewhere in the right half space. The only source in this system is the one that generates the original incident wave.

Because the tangential components of the electric field must be continuous at

---

2. If we attempt to add a $\hat{y}$-polarized reflected electric field to (4.2.1), we would also have to add a $\hat{y}$-polarized transmitted electric field in order to maintain continuity of the tangential electric field. The resulting $\hat{x}$-polarized magnetic field would be discontinuous at $z = 0$, giving rise to a nonzero surface current. But as the boundary between two nonconducting dielectric media does not support a surface current, we are forced to conclude that $E_y = 0$ everywhere in space.

the boundary between the incident and transmitting media, (4.2.1) and (4.2.3) may be equated at $z = 0$ to yield

$$\hat{x}\,(E_i + E_r) = \hat{x}\,E_t \qquad (4.2.5)$$

Similarly, the tangential magnetic field is continuous at $z = 0$, since neither dielectric media is conducting, no surface currents can exist. The tangential magnetic fields (4.2.2) and (4.2.4) are set equal at $z = 0$:

$$\hat{y}\,\frac{1}{\eta_i}(E_i - E_r) = \hat{y}\,\frac{1}{\eta_t}E_t \qquad (4.2.6)$$

Dividing (4.2.5) and (4.2.6) by E_i and E_i/η_i, respectively, yields two equations in the two unknowns E_r/E_i and E_t/E_i:

$$1 + \frac{E_r}{E_i} = \frac{E_t}{E_i} \qquad (4.2.7)$$

$$1 - \frac{E_r}{E_i} = \frac{\eta_i}{\eta_t}\frac{E_t}{E_i} \qquad (4.2.8)$$

We define E_r/E_i to be Γ, the complex *reflection coefficient*, E_t/E_i to be T, the complex *transmission coefficient*, and η_t/η_i to be η_n, the *normalized wave impedance*. Solving (4.2.7) and (4.2.8) for T and Γ yields

$$\Gamma = \frac{E_r}{E_i} = \frac{\eta_n - 1}{\eta_n + 1} \qquad (4.2.9)$$

$$T = \frac{E_t}{E_i} = 1 + \Gamma = \frac{2\eta_n}{\eta_n + 1} \qquad (4.2.10)$$

The time-averaged Poynting vector may be calculated for this $\hat{x}$-polarized wave in each of the two regions. For $z < 0$,

$$\langle \overline{S}(z < 0) \rangle = \frac{1}{2}\,\text{Re}\left\{\overline{E} \times \overline{H}^*\right\}$$

$$= \frac{1}{2}\,\text{Re}\left\{ \hat{x}\,E_i\,\left(e^{-jk_iz} + \Gamma\,e^{+jk_iz}\right) \right. \qquad (4.2.11)$$

$$\left. \times \hat{y}\,\left(\frac{E_i}{\eta_i}\,\left(e^{-jk_iz} - \Gamma\,e^{+jk_iz}\right)\right)^* \right\}$$

$$= \hat{z}\,\frac{|E_i|^2}{2\eta_i}\,\text{Re}\left\{\left(1 - |\Gamma|^2 + \Gamma\,e^{2jk_iz} - \Gamma^*\,e^{-2jk_iz}\right)\right\}$$

Because the difference of a number and its complex conjugate is purely imaginary, the third and fourth terms of (4.2.11) do not survive taking the real part of the complex Poynting vector, so

$$\langle \overline{S}(z < 0) \rangle = \hat{z} \frac{|E_i|^2}{2\eta_i}(1 - |\Gamma|^2) \tag{4.2.12}$$

where $\langle \overline{S} \rangle$ is the difference between the incident and reflected power. For a passive (nonamplifying) medium, $|\Gamma|^2 \leq 1$ because the amount of reflected power $|\Gamma|^2|E_i|^2/2\eta_i$ must not exceed the amount of incident power $|E_i|^2/2\eta_i$.

For the transmitting medium,

$$\langle \overline{S}(z > 0) \rangle = \frac{1}{2} \text{Re} \left\{ \hat{x} \, T E_i \, e^{-jk_t z} \times \hat{y} \left(\frac{T E_i}{\eta_t} e^{-jk_t z} \right)^* \right\}$$

$$= \hat{z} \frac{|E_i|^2 |T|^2}{2} \text{Re} \left\{ \frac{1}{\eta_t^*} \right\} \tag{4.2.13}$$

Notice that the time-averaged power flowing on either side of the planar interface is independent of coordinate z. Equations (4.2.12) and (4.2.13) are equal because power must be conserved across the $z = 0$ interface; hence

$$1 - |\Gamma|^2 = |T|^2 \text{Re} \left\{ \frac{\eta_i}{\eta_t^*} \right\}$$

If the magnitude of the reflection coefficient is unity ($|\Gamma| = 1$), then from (4.2.12) we see that no power flows across the planar interface.

From (4.2.9) and (4.2.10), we see that for real values of the normalized wave impedance η_n, the complex reflection and transmission coefficients are purely real. Therefore, there can be no phase shift upon the reflection and transmission of a normally incident plane wave at such a boundary, except for a possible sign change in Γ. If $\mu_t = \mu_i$ and $\epsilon_t = \epsilon_i$, then the normalized wave impedance has the value $\eta_n = 1$ and the reflection coefficient Γ is zero. Thus, there is no reflected wave and all of the incident power is transmitted across the boundary at $z = 0$, consistent with the fact that no difference exists between the two media in this case.

If the normalized wave impedance approaches $+\infty$, which can happen for a highly magnetizable material ($\mu_t \gg \mu_i$) or for very low values of ϵ_t (as in a collisionless plasma, for ω only slightly greater than the plasma frequency), then Γ approaches $+1$. In this infinite-impedance or open-circuit situation, all of the incident power is reflected, consistent with the fact that no power flows across the $z = 0$ boundary when $|\Gamma| = 1$. In this case, the reflected electric field is exactly in phase with the incident electric field and so the total $\overline{E}$ field doubles at the boundary while the total $\overline{H}$ field goes to zero there. If η_n is close to zero, which could mean that the transmitting medium is highly conducting ($\epsilon_t \gg \epsilon_i$, $\mu_t \simeq \mu_i$), then Γ is approximately -1. Again, all of the incident power is reflected at this zero-impedance or short-circuit medium, and the reflected $\overline{E}$ field is 180° out of phase with the incident $\overline{E}$ field. In this case, the total electric field is zero at the boundary, and the total magnetic field is twice the incident magnetic field.

Figure 4.4 gives plots of the electric and magnetic fields at several instants in time for the case $\Gamma = +1$. We note that the resulting wave pattern in the vacuum does not propagate; the pattern of nulls in the wave envelope is completely stationary when $|\Gamma| = 1$. A wave that does not propagate at all is called a pure *standing wave* and is created by superposing forward-traveling and backward-traveling plane waves of equal magnitude. The electric field $\overline{E}_t$ in the transmitting medium propagates to the right, although since $|\overline{H}_t| = |\overline{E}_t/\eta_t| = 0$, no power propagates for $z > 0$.

In general, the incident and reflected waves superpose to form a wave pattern

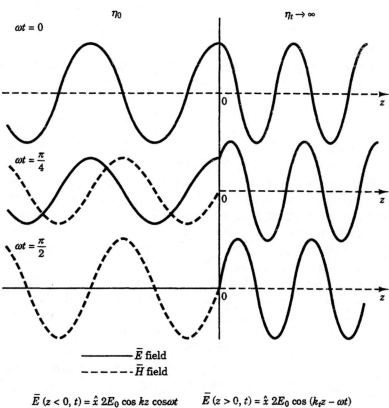

$\overline{E}\,(z < 0, t) = \hat{x}\, 2E_0 \cos kz \cos \omega t$ $\overline{E}\,(z > 0, t) = \hat{x}\, 2E_0 \cos (k_t z - \omega t)$

$\overline{H}\,(z < 0, t) = \hat{y}\, \dfrac{2E_0}{\eta_0} \sin kz \sin \omega t$ $\overline{H}\,(z > 0, t) = 0$

Figure 4.4   Electric and magnetic fields for $\Gamma = +1$.

of the type illustrated in Figure 4.5 at various instants of time for the particular case $\eta_n = \frac{1}{2}$ $\left(\Gamma = -\frac{1}{3}\right)$. This general wave pattern can be created by superposing a pure standing wave on a traveling wave. The maximum value of $E_x(t)$ at any position z defines the *standing wave envelope*, which is generally **not** a sinusoid; this envelope is shown in Figure 4.5 for $\Gamma = -\frac{1}{3}$.

For a boundary at which $|\Gamma| = +1$, we have already seen that the incident and reflected waves superpose to form a pure standing wave, and the resulting standing wave envelope can be seen to possess nulls. It is useful to define the *standing wave ratio* or SWR for the incident plus reflected waves on one side of a boundary. The standing wave ratio is the ratio of the maximum to minimum values of the stationary envelope for the electric or magnetic fields:

$$
\begin{aligned}
\mathrm{SWR} &= \frac{|E(z)|_{\max}}{|E(z)|_{\min}} = \frac{|H(z)|_{\max}}{|H(z)|_{\min}} \\
&= \frac{|\overline{E}_i| + |\overline{E}_r|}{|\overline{E}_i| - |\overline{E}_r|} = \frac{1 + |\Gamma|}{1 - |\Gamma|}
\end{aligned}
\tag{4.2.14}
$$

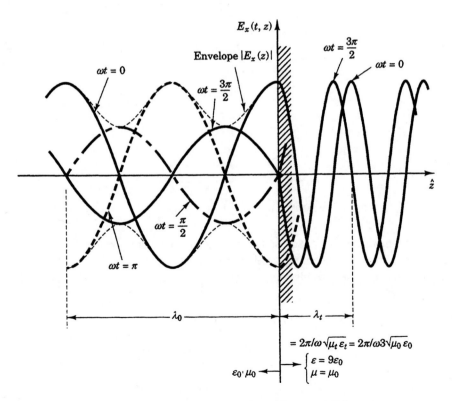

Figure 4.5   Standing-wave pattern for $\eta_n = 1/2$.

We see that for a perfectly reflecting boundary for which a pure standing wave results, the SWR is infinite, whereas for a perfectly transmitting boundary ($\Gamma = 0$), the SWR is unity. An intermediate type of wave, which is neither purely standing nor purely propagating, results when $1 < \text{SWR} < \infty$. We can invert (4.2.14) to find $|\Gamma|$ in terms of the SWR

$$|\Gamma| = \frac{\text{SWR} - 1}{\text{SWR} + 1} \tag{4.2.15}$$

which is a useful relation because of the simplicity with which the SWR can often be measured. The quantity $|\Gamma|$ is of interest because the ratio of reflected to incident power is $|\Gamma|^2$.

If the medium in the right half space is a perfect conductor ($\sigma_t \to \infty$), it can neither dissipate nor transmit power, so 100 percent of the incident power must be reflected. This is easily shown by calculating the wave impedance of the conductor

$$\eta_t = \sqrt{\mu_t / \left(\epsilon_t - \frac{j\sigma_t}{\omega} \right)} = 0 \qquad (\sigma_t \to \infty)$$

where (3.4.5) has been used. Evaluating equations (4.2.9) and (4.2.10) then implies that $\Gamma = -1$ and $T = 0$, or equally, $E_r = -E_i$ and $E_t = 0$. The total magnetic field at $z = 0$ is thus

$$\overline{H} = \hat{y} \frac{1}{\eta_i}(E_i - E_r) = \hat{y}\, 2E_i/\eta_i = \hat{y}\, 2H_i \tag{4.2.16}$$

where $E_i = \eta_i H_i$. Using the boundary condition derived from Ampere's law (4.1.21) at the surface of a perfect conductor,

$$\overline{J}_s = \hat{n} \times \overline{H} = -\hat{z} \times \hat{y}\, 2H_i = \hat{x}\, 2H_i \tag{4.2.17}$$

where $-\hat{z}$ is the surface normal pointing from the perfect conductor to the incident medium.

The fields, energies, and standing wave envelope are given in Figure 4.6 for the case of waves incident on a perfectly conducting medium. We note that the electric and magnetic fields are 90° out of phase in both space and time when $\Gamma = -1$, with

$$\overline{E}(z, t) = \hat{x}\, 2E_i \sin kz \sin \omega t$$
$$\overline{H}(z, t) = \hat{y}\, \frac{2E_i}{\eta_i} \cos kz \cos \omega t \tag{4.2.18}$$

Therefore, the time-averaged Poynting vector is zero because $\langle \sin \omega t\ \cos \omega t \rangle = 0$, and the total energy in the field simply oscillates between electric and mag-

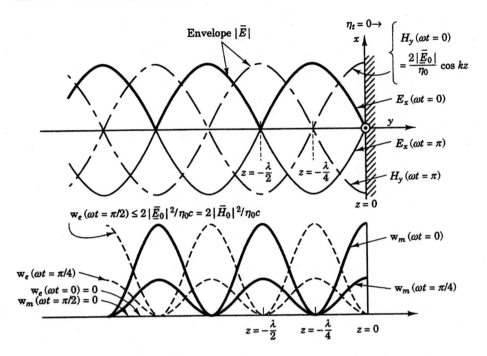

Figure 4.6 Fields, energies, standing wave envelope for a perfectly conducting medium.

netic energy. The time-averaged electric and magnetic energy densities are

$$\langle W_e \rangle = \frac{1}{4}\epsilon_i(2E_i)^2 \sin^2 kz$$

$$\langle W_m \rangle = \frac{1}{4}\mu_i\left(\frac{2E_i}{\eta_i}\right)^2 \cos^2 kz = \frac{1}{4}\epsilon_i(2E_i)^2 \cos^2 kz \tag{4.2.19}$$

so we see that $\langle W_e \rangle + \langle W_m \rangle = \frac{1}{4}\epsilon_i(2E_i)^2$ is constant in space for $z < 0$.

A more interesting case occurs when the loss tangent is much greater than one but still finite $(1 \ll \sigma_t/\omega\epsilon_t < \infty)$. Plane wave propagation in this case was examined in Section 3.4, and it was shown that the complex wave number k_t in a highly conductive medium is given by (3.4.13):

$$k_t \cong \sqrt{\frac{\omega\mu_t\sigma_t}{2}}(1 - j) \qquad (\sigma_t \gg \omega\epsilon_t) \tag{4.2.20}$$

Thus the wave decays sharply away from the $z = 0$ boundary with the exponential dependence $e^{-k''z} = e^{-z/\delta}$, where δ is the skin depth given by $\delta = \sqrt{2/\omega\mu_t\sigma}$ (3.4.14). The impedance of a highly conducting medium was derived in (3.4.15):

$$\eta_t = \sqrt{\frac{\omega\mu_t}{2\sigma_t}}(1 + j) \qquad (\sigma_t \gg \omega\epsilon_t) \tag{4.2.21}$$

Therefore,

$$\eta_n = \sqrt{\frac{\eta_t}{\eta_i}} \cong \sqrt{\frac{\omega \mu_t \epsilon_i}{2\sigma_t \mu_i}}(1 + j) \quad (\sigma_t \gg \omega \epsilon_i) \qquad (4.2.22)$$

Using (4.2.9) and noting that $|\eta_n| \ll 1$, we find

$$\Gamma = \frac{\eta_n - 1}{\eta_n + 1} \cong (\eta_n - 1)(1 - \eta_n)$$

$$\cong -1 + 2\eta_n = \sqrt{\frac{2\omega \mu_t \epsilon_i}{\mu_i \sigma_t}}(1 + j) - 1 \qquad (4.2.23)$$

$$T = \Gamma + 1 \cong \sqrt{\frac{2\omega \mu_t \epsilon_i}{\mu_i \sigma_t}}(1 + j)$$

where the Taylor expansion $(1 + \eta_n)^{-1} \cong 1 - \eta_n$ for small η_n has been used and terms of order η_n^2 dropped.

We recognize that $\frac{1}{2} \operatorname{Re}\{\overline{S}\}|_{z=0}$ is the average power flow into the transmitting medium at $z = 0$. Because the conductor is lossy, the electric and magnetic fields are attenuated essentially to zero intensity by the time the fields have penetrated several skin depths into the medium. Thus, nearly all of the power that flows into the conducting surface at $z = 0$ is dissipated within a distance δ. We can thus express P_d, the power dissipated per unit area (W/m^2) in the conductor, as

$$P_d = \frac{1}{2} \operatorname{Re}\{\overline{E} \times \overline{H}^*\} \cdot \hat{z}\Big|_{z=0} \qquad (4.2.24)$$

where $\overline{E}$ and $\overline{H}$ are evaluated at the $z = 0$ interface. We now seek an expression for the dissipated power P_d in terms of $\overline{H}(z = 0)$, the total magnetic field at $z = 0$. The electric and magnetic fields at $z = 0$ may be found from (4.2.3) and (4.2.4), where the fields in the transmitting medium are used for simplicity. [Equations (4.2.1) and (4.2.2) could just as easily be used instead because $\overline{E}$ and $\overline{H}$ are continuous at $z = 0$.] The power dissipated is

$$P_d = \frac{1}{2} \operatorname{Re}\left\{ \hat{x} E_t \times \hat{y} \left(\frac{E_t}{\eta_t} \right)^* \right\} \cdot \hat{z} = \operatorname{Re}\left\{ \frac{1}{2\eta_t^*}|E_t|^2 \right\}$$

But since $E_t = T E_i$ from (4.2.10), we find

$$P_d = \frac{\eta_i^2 |H_i|^2 |T|^2}{2} \operatorname{Re}\left\{ \frac{1}{\eta_t^*} \right\} \qquad (4.2.25)$$

where $E_i = \eta_i H_i$ has been used.

Since $\sigma_t \gg \omega\epsilon_t$ and we have already found in (4.2.16) that the total magnetic field to the left of the interface $\overline{H}(z=0)$ is approximately $2\overline{H}_i$ for a good conductor, (4.2.21) and (4.2.23) may then be substituted into (4.2.25) to yield

$$P_d = \frac{1}{2}\eta_i^2|H_i|^2\frac{4\omega\mu_t\epsilon_i}{\mu_i\sigma_t}\sqrt{\frac{\sigma_t}{2\omega\mu_t}} \cong \frac{|H(z=0)|^2}{4}\sqrt{\frac{2\omega\mu_t}{\sigma_t}} \tag{4.2.26}$$

A simple way to remember the result (4.2.26) is to suppose that the power dissipated in the conductor is approximately the same as if the surface current $\overline{J}_s$ flowed uniformly through a slab of conductivity σ_t and thickness equal to the skin depth δ (3.4.14). In this case, $|\overline{J}_s| = |\overline{H}(z=0)|$ by virtue of (4.1.21), and

$$P_d = \frac{1}{2}\,\mathrm{Re}\{\overline{E}\cdot\overline{J}^*\}\delta = \frac{1}{2\sigma_t}|\overline{J}|^2\delta = \frac{|\overline{J}_s|^2}{2\sigma_t\delta} = \frac{|\overline{H}(z=0)|^2}{4}\sqrt{\frac{2\omega\mu_t}{\sigma_t}} \tag{4.2.27}$$

which is identical to (4.2.26). Notice that the power dissipated is proportional to $\sqrt{\omega/\sigma_t}$ if $\sigma_t \gg \omega\epsilon_t$, so that in the limit that $\sigma_t \to \infty$, no power at all is dissipated in the perfect conductor!

Here, the surface current is distributed over a thickness on the order of the skin depth δ. The skin depth is identically zero only for perfect conductors where $\sigma_t \to \infty$. Since this conduction current flows through a nonzero skin depth for finite σ_t, there is no actual surface current. Therefore, by (4.1.16), the tangential magnetic field is continuous across the boundary $z=0$ for a real, imperfect conductor. However, $\overline{H}_t$ is still equal to $\hat{n} \times \overline{J}_s$, where $\overline{J}_s$ is now given by

$$\overline{J}_s = \int_0^\infty \overline{J}(z)\,dz$$

Example 4.2.1

What fraction F of the power in a 100-GHz uniform plane wave traveling in vacuum is lost within a polished metal surface after the wave is reflected at normal incidence? Here, $\sigma = 3 \times 10^7$ (S/m), $\mu_t = \mu_0$, and $\epsilon_t = \epsilon_0$? What is the standing wave ratio (SWR)?

Solution: Since $|\Gamma|^2$ is the ratio of the amount of power reflected by the surface to the amount of incident power, the fraction of power lost in the polished metal is $F = 1 - |\Gamma|^2$. Because $\sigma/\omega\epsilon_t \gg 1$, we can use (4.2.23) to approximate Γ by

$$\Gamma \simeq \sqrt{\frac{2\omega\epsilon_0}{\sigma}}(1+j) - 1$$

Therefore,

$$|\Gamma|^2 = \Gamma\cdot\Gamma^* \simeq 1 - 2\sqrt{\frac{2\omega\epsilon_0}{\sigma}}$$

where terms of order $\omega\epsilon_0/\sigma$ are neglected, and

$$F \simeq 2\sqrt{\frac{4\pi f\epsilon_0}{\sigma}} = 1.2 \times 10^{-3}$$

So, using (4.2.14), we find that

$$\text{SWR} = (1 + |\Gamma|)/(1 - |\Gamma|)$$
$$= (1 + \sqrt{1 - F})/(1 - \sqrt{1 - F}) \simeq 4/F$$
$$\simeq 3282$$

Example 4.2.2

A uniform plane wave $\overline{E}(z) = \hat{x}\, E_i\, e^{-jkz}$ traveling in vacuum is normally incident upon a good conductor with $\sigma \gg \omega\epsilon$. This plane wave induces conduction currents $\overline{J}_c$ in the conductor that correspond to an equivalent surface conduction current $\overline{J}_{sc} = \int_0^\infty \overline{J}_c(z)\,dz$. What is $\overline{J}_{sc}$ and how does it relate to the surface current $\overline{J}_s$ found in Ampere's boundary condition (4.1.16)?

Solution: We start by considering the integral form of Ampere's law for the contour C shown in Figure 4.7(a), which is at least a few skin depths wide and lies in the y–z plane so as to encircle all of the $\hat{x}$-directed current $\overline{J}_c$. (We can decide what direction the current flows from the direction $\hat{n} \times \overline{H}$). The magnitude of $|\overline{J}_c|$ is shown in Figure 4.7(b) and is seen to be primarily concentrated within the skin depth. We recall that Ampere's law is written

$$\oint_C \overline{H} \cdot d\overline{\ell} = j\omega \int_A \overline{D} \cdot d\overline{a} + \int_A \overline{J}_c \cdot d\overline{a}$$

and the left side is easily seen to be

$$\oint_C \overline{H} \cdot d\overline{\ell} \simeq H_y \Big|_{z=0} \ell = \frac{2E_i}{\eta_0} \ell$$

because only the segment of the contour outside the medium contributes to the integral, and the total magnetic field at a good conducting surface is approximately twice the incident $\overline{H}_i$. (The opposite segment is chosen far enough into the conductor that the magnetic field is zero there, and the $\hat{z}$-directed contour lengths are perpendicular to the $\hat{y}$-directed magnetic field so that $\overline{H} \cdot d\overline{\ell} = 0$ along those paths.) The displacement field $\overline{D}$ can be expressed in terms of $\overline{E}$ by using the constitutive law for a linear medium $\overline{D} = \epsilon \overline{E}$, where the dielectric constant does not include conductivity because the conduction current term is treated separately.

The first term on the right side of Ampere's law is written

$$j\omega \int_A \overline{D} \cdot d\overline{a} = j\omega\ell \int_0^\infty \epsilon T E_i\, e^{-jkz} dz = \frac{\omega \epsilon \ell T E_i}{k}$$

This last equality results because the wave number k has a negative imaginary part in lossy media, which forces the integral to converge as $z \to \infty$. Using the approximate expressions for T (4.2.23) and k (4.2.20) in a good conductor yields

$$j\omega \int_A \overline{D} \cdot d\overline{a} = \omega \epsilon \ell \left(\sqrt{\frac{2\omega\mu\epsilon_0}{\sigma\mu_0}}(1 + j) \right) E_i \left(\sqrt{\frac{\omega\mu\sigma}{2}}(1 - j) \right)^{-1}$$

$$= j \frac{2\omega\epsilon E_i \ell}{\sigma\eta_0}$$

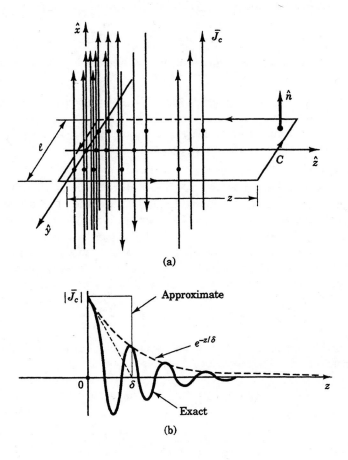

Figure 4.7 Conduction currents at the surface of a good conductor.

The second term on the right side of Ampere's law is the area integral of the conduction current, given by

$$\int_A \overline{J}_c \cdot d\overline{a} = \ell \int_0^\infty J_c(z)\,dz = \ell J_{sc}$$

so we may combine terms to simplify our original equation:

$$\frac{2E_i}{\eta_0}\ell = j\frac{2\omega\epsilon\ell}{\eta_0\sigma}E_i + \ell J_{sc}$$

or

$$\overline{J}_{sc} = \hat{x}\left(\frac{2E_i}{\eta_0} - j\frac{2\omega\epsilon}{\eta_0\sigma}E_i\right)$$

For $\sigma \gg \omega\epsilon$, the second term becomes negligible compared to the first term, and

$$\overline{J}_{sc} \simeq \hat{x}\,\frac{2E_i}{\eta_0}$$

which is exactly what we would find if the current were confined to the surface of the conductor

$$\overline{J}_s = \hat{n} \times \overline{H}\Big|_{z=0} = -\hat{z} \times \hat{y}\,\frac{2E_i}{\eta_0} = \hat{x}\,\frac{2E_i}{\eta_0} \cong \overline{J}_{sc}$$

even though $\overline{J}_{sc}$ is distributed over a thin layer on the order of the skin depth δ, has a phase dependence on depth, and is not a surface impulse.

Example 4.2.3

The reflectivity $|\Gamma|^2$ of a metal was calculated earlier in terms of σ. As the frequency increases, what additional physical mechanisms must we consider in estimating $|\Gamma|^2$ and the wave penetration depth Δ? Estimate Δ for silver at $\lambda = 2$ mm and $\lambda = 2$ μm.

Solution: Free electrons in a metal behave like a plasma that has a very high electron collision frequency v, where the time-harmonic equation of motion for such a plasma is given by (3.6.13). In the absence of a strong magnetic field, the complex permittivity matrix (3.6.19) reduces to that of an isotropic medium with an effective dielectric constant given by

$$\epsilon_{\text{eff}} = \epsilon_0 K' = \epsilon_0 K_0 = \epsilon_0 \left(1 - \frac{\omega_p^2/\omega^2}{1 - jv/\omega}\right) \tag{1}$$

where K' and K_0 are defined in (3.6.20) and the cyclotron frequency $\omega_c = qB_0/m$ is zero for zero magnetic field. The expression (1) reduces to

$$\epsilon_{\text{eff}} \simeq \epsilon_0 \left(1 - \frac{j\omega_p^2}{v\omega}\right) \qquad (\omega \ll 2\pi v)$$

$$\epsilon_{\text{eff}} \simeq \epsilon_0 \left(1 - \frac{\omega_p^2}{\omega^2}\right) \qquad (\omega \gg 2\pi v)$$

Therefore, for frequencies $\omega \ll 2\pi v$, the conductivity of a metal is real;

$$\sigma = j\omega(\epsilon_{\text{eff}} - \epsilon_0) = \epsilon_0 \omega_p^2/v$$

where (3.4.5) has been used. The conductivity is dominated by the conversion of electron energy to heat via collisions with the ions. However, when $\omega \gg 2\pi v$, the metal behaves more like a collisionless plasma with a different $|\Gamma|^2$ and Δ. Above the plasma frequency, photons can in principle penetrate metals deeply, limited by residual collisions and by various quantum effects; x-rays are quite penetrating.

To estimate ω_p and v, we assume one free electron per atom in silver (specific gravity 10.5 g/cm^3, molecular weight 107.88 g/mole, conductivity 5×10^7 S/m). Other useful constants are $q = -1.602 \times 10^{-19}$ C, $m = 9.11 \times 10^{-31}$ kg, and Avogadro's number (6.02×10^{23} electrons/mole). The number density of free electrons is therefore

$$N = 6.02 \times 10^{23} \text{ electrons/mole} \cdot \frac{1}{107.88} \text{ moles/g} \cdot 10.5 \text{ g/cm}^3 \cdot 10^6 \text{ cm}^3/\text{m}^3$$

$$= 5.86 \times 10^{23} \text{ m}^{-3}$$

The plasma frequency is then found from (3.6.6):

$$\omega_p = \sqrt{Nq^2/m\epsilon_0} = 1.366 \times 10^{16} \text{ s}^{-1}$$

When collisions dominate, (3.6.13) becomes $m\nu\overline{\upsilon} \cong q\overline{E}$, and since $\overline{J} = \sigma\overline{E} = Nq\overline{\upsilon}$, we find

$$\nu = Nq^2/m\sigma = 3.30 \times 10^{13} \text{ Hz}$$

Thus for wavelengths longer than $\lambda_c \cong c/\nu \cong 9 \ \mu\text{m}$, the conductivity is real and dominated by collisions. The penetration depth in this case is the traditional skin depth δ

$$\delta \cong \sqrt{2/\omega\mu\sigma} \qquad (\omega \ll 2\pi\nu)$$

For wavelengths between $\sim 9 \ \mu\text{m}$ and $\lambda_p = 2\pi c/\omega_p \cong 0.14 \ \mu\text{m}$, collisionless plasma effects dominate and ϵ_{eff} is negative. The penetration depth Δ is then:

$$\Delta = 1/k'' = \frac{1}{\omega\sqrt{\mu_i\epsilon_i\left((\omega_p/\omega)^2 - 1\right)}}$$

$$= \frac{c}{\sqrt{\omega_p^2 - \omega^2}} \qquad (2\pi\nu \ll \omega < \omega_p)$$

At $\lambda = 2\text{mm}$, $\lambda > \lambda_c$ and $\delta \cong 0.18 \ \mu\text{m}$. At $\lambda = 2 \ \mu\text{m}$, plasma effects dominate and $\Delta \cong 0.022 \ \mu\text{m}$. For frequencies greater than ω_p, Thompson scattering and various quantum effects dominate and this simple model is no longer valid.

Example 4.2.4

What pressure p (N/m^2) does a uniform plane wave exert on a planar mirror when all of the wave is reflected at normal incidence? If only 80 percent of the incident wave is reflected, how does the pressure change?

Solution: The pressure on the mirror may be found by integrating the force density F_z from the surface of the mirror at $z = 0$ to $z = \infty$:

$$p = \int_0^\infty F_z \, dz$$

The instantaneous $\hat{z}$-directed force is $\rho E_z + (\overline{J} \times \mu_0\overline{H})_z$, but since $\overline{E}$ is transverse to the direction of propagation of the wave, $E_z = 0$. The current $\overline{J}$ arises predominantly from conduction electrons if the mirror is a good conductor, so $\overline{J} \simeq \nabla \times \overline{H}$. We therefore find the time-averaged pressure

$$p = \frac{1}{2}\text{Re}\left\{\mu_0\int_{z=0}^\infty \left[(\nabla \times \overline{H}) \times \overline{H}\right]_z dz\right\} = -\frac{1}{2}\mu_0\text{Re}\left\{\int_{z=0}^\infty \frac{\partial H_y}{\partial z} H_y \, dz\right\}$$

$$= \frac{1}{2}\mu_0\text{Re}\left\{\int_{H_y(z=0)=H_0}^{H_y(z=\infty)=0} H_y \, dH_y\right\} = \frac{1}{4}\mu_0|\overline{H}_0|^2$$

where H_0 is the magnetic field at the mirror surface and is twice the magnetic field magnitude $|\overline{H}_i|$ of the incident wave. The factor of $1/2$ comes from taking the time-average of the sinusoidally varying fields. This magnetic pressure $p = \frac{1}{4}\mu_0|\overline{H}_0|^2 = \mu|\overline{H}_i|^2$ is the pressure that the incident wave exerts on the mirror. If only 80 percent of the wave is reflected, then $p_T = 0.8\mu_0|\overline{H}_i|^2$.

We can also find the total pressure by considering the uniform plane wave to be composed of photons, each having energy hf (J), where h is Planck's constant and f is the photon frequency. The momentum of an individual photon is therefore hf/c (J·s/m) and the total pressure is just $p_T = 2n(hf/c)$ (N/m^2), where n is number of photons per square meter per second transported by the incident beam. The factor of two results because the photon momentum is not absorbed by the mirror but is instead reversed in direction. The time-averaged power density in the forward-propagating wave is $\frac{1}{2}\eta_0|\overline{H}_i|^2$ (W/m^2), which is also equal to nhf, so we finally can compute the pressure resulting from 100 percent reflection: $p_T = 2(\frac{1}{2}\eta_0|\overline{H}_i|^2/hf)(hf/c) = \mu_0|\overline{H}_i|^2$. If only 80 percent of the wave is reflected, then the pressure is reduced by 20 percent as before.

Example 4.2.5

What is the force per unit length $\overline{f}'$ (N/m) acting on a long stationary metal bar (μ_0, ϵ_0) having cross-section 1 cm × 10 cm in the $\hat{x}$ and $\hat{y}$ directions, respectively, when it is subject to a static magnetic field $\overline{H} = \hat{x} H_0$ while a static current density $\overline{J} = \hat{y} J_0$ flows? Assume that $\mu_0\overline{H}_0 = 1$ tesla and $J_0 = 10^7$ A/m. What is $\overline{f}'$ if the bar has permeability μ rather than μ_0? If the bar is open-circuited and moves at velocity $\overline{v} = \hat{z} 100$ m/s, what happens?

Solution: The Lorentz force density (1.2.7) is $\overline{F} = \rho\overline{E} + \overline{J} \times \mu_0\overline{H}$ (N/m^2) where $\rho = \overline{E} = 0$. Therefore, the force per unit length $\overline{f}'$ is

$$\overline{f}' = (\overline{J} \times \mu_0\overline{H})A = -\hat{z} 10,000 \text{ (N/m)} \tag{1}$$

which is roughly equivalent to 2,250 pounds force per meter. Because it is difficult to built an efficient power source yielding high currents at low voltages (source resistances are too high), such a bar would normally be composed of parallel, insulated wires formed in a coil with the return wires outside the region of uniform magnetic field to avoid force cancellation.

If the bar were made of a highly permeable medium, the force per unit length would remain as given in (1). From the Lorentz force density given in (1.2.7), we might expect to replace $\overline{J} \times \overline{B}$ with $\overline{J} \times \mu\overline{H}$, but the expression in (1.2.7) is actually correct only for media with permeability $\mu = \mu_0$. If the medium is magnetizable, then we modify the force density by adding the *Kelvin magnetization force density* $\mu_0\overline{M} \cdot \nabla\overline{H}$ to (1.2.7), where $\mu_0\overline{M} = (\mu - \mu_0)\overline{H}$; in this case,

$$\overline{F} = \overline{J} \times \mu_0\overline{H} + \mu_0\overline{M} \cdot \nabla\overline{H}$$

Because $\overline{H}$ is uniform, $\nabla\overline{H}$ is equal to zero and the Kelvin force density does not contribute to the total force.

The conduction electrons in an open-circuited, moving bar experience no net force (or they would leave the bar), so $\overline{f} = q(\overline{E} + \overline{v} \times \mu_0\overline{H}) = 0$, and the electric field $\overline{E}$ is therefore equal to $-\overline{v} \times \mu_0\overline{H} = -\hat{y} 100$ V/m. The open-circuit voltage across the 10-cm width bar is thus 10 V; the bar acts as an electrical generator. If current is drawn from the bar, then the force $\overline{f}$ required to keep the bar moving increases and mechanical power $\overline{f} \cdot \overline{v}$ is converted to electrical power $V I$. The field $\overline{E} = \hat{y} E_0$ inside the bar terminates on surface charge σ_s on the sides of the bar, and these moving charges create a small additional magnetic field if the bar is observed from a stationary viewpoint.

Example 4.2.6

The motor illustrated in Figure 4.8 is made of an iron toroid (stator) wrapped by an N-turn coil and interrupted by a movable iron rotor free to turn. The permeability of the iron is $\mu = 10,000\mu_0$. The thin gaps between rotor and stator are d_1 and d_2 meters and have areas A_1 and A_2, which vary as the rotor spins. If the gap d_2 were zero, what would be the approximate static fields $\overline{H}$ and $\overline{B}$ in the iron and in the gap d_1? What is the inductance L seen at the motor terminals? What is L if d_2 is nonzero? What is the total energy w_T in this motor? If $d_2 = 0$, what is the force acting close to the gap? What is the torque on the rotor?

Solution: We first note that the normal component of $\overline{B}$ and the tangential components

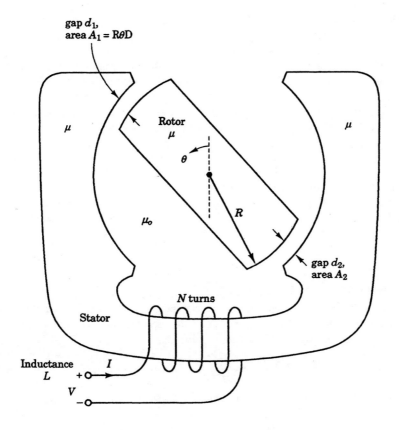

gap d_1,
area $A_1 = R\theta D$

Rotor
μ

θ

μ

μ

μ_0

R

gap d_2,
area A_2

N turns

Stator

Inductance
L I

V

Figure 4.8   Stator and rotor in an iron motor.

of $\overline{H}$ must be continuous across the gap, as no surface current flows on the stator or rotor. If we consider just the gap d_1 ($d_2 = 0$), then the gap fields are related to the magnetic fields in the stator by

$$H_{1t} = H_{\text{stat},t}$$

$$H_{1n} = \frac{\mu}{\mu_0} H_{\text{stat},n}$$

The normal $\overline{H}$ field is thus approximately $\mu/\mu_0 = 10{,}000$ times the tangential field in the gap, and we shall approximate the magnetic field in the gap as essentially perpendicular to the stator and rotor surfaces with magnitude $(\mu/\mu_0)H_{\text{stat},n}$. The $\overline{B}$ field in the gap is therefore approximately $\mu H_{\text{stat},n} = B_{\text{stat},n}$. The inductance of the motor is the amount of flux linked by the current flowing in the windings. The flux Λ (T$\cdot$ m^2) is given by

$$\Lambda = \int_A \overline{B} \cdot \hat{n}\, da$$

which is constant across any cross-section of the motor since Gauss's law $\oint_A \overline{B} \cdot \hat{n}\, da = 0$ is always valid. For the case where $d_2 = 0$, it is easiest to evaluate the flux in the air gap d_1: $\Lambda \simeq \mu_0 H_{1n} A_1$. We now evaluate Ampere's law $\oint_C \overline{H} \cdot d\overline{\ell} = \int_A \overline{J} \cdot \hat{n}\, da$ by choosing a closed contour C of length G meters perpendicular to the air gaps and passing through the body of the motor. Because the normal component of the magnetic field $\overline{H}$ is 10,000 times greater in the air gap than in the stator or rotor, we approximate the left side of Ampere's law as $H_{1n}d_1$ if $\mu d_1 \gg \mu_0 G$. This quantity must be equal to the current passing through the N-turn coil, so $H_{1n}d_1 \simeq NI$. The inductance can now be found

$$L = \frac{N\Lambda}{I} = \frac{\mu_0 N^2 A_1}{d_1} \quad \text{(henries)} \tag{2}$$

where the extra factor of N arises because the current I links N times the flux Λ. If d_2 were not zero, then Ampere's law would yield $H_{1n}d_1 + H_{2n}d_2 \simeq NI$ and the flux would be found as before to be $\Lambda = \mu_0 H_{1n} A_1 = \mu_0 H_{2n} A_2$. With these equations, the inductance would be

$$L = \frac{N\Lambda}{I} = \frac{\mu_0 N^2 A_1 A_2}{A_2 d_1 + A_1 d_2}$$

which reduces to (2) when $d_2 = 0$.

To calculate the energy in the inductor, we can integrate the power flowing into it over time, since power is just the derivative of energy with respect to time. The voltage measured across the inductor terminals can be found from Faraday's law

$$V = -\int \overline{E} \cdot d\overline{\ell} = \frac{d}{dt} \int_A \overline{B} \cdot \hat{n}\, da = \frac{d\Lambda}{dt} = L \frac{dI}{dt}$$

for linear media. The total energy in the inductor is then

$$w_T = \int_{-\infty}^t V(t)I(t)\, dt = \int_{-\infty}^t L \frac{dI}{dt} I\, dt = \frac{1}{2} L I^2$$

which is analogous to the energy density $W_m = \frac{1}{2}\mu_0 |\overline{H}|^2$ of a magnetic field. If $d_2 = 0$ and the current I is flowing losslessly around the loop formed by a short-circuited source, then the total energy in the motor is just that in the inductor, and the force acting to close the gap d_1 is

$$f = \frac{dw}{dd_1}\bigg|_\Lambda = \frac{d}{dd_1}\left(\frac{\Lambda^2}{2L}\right) = -\frac{\Lambda^2}{2L^2}\frac{dL}{dd_1} = \frac{\mu_0 N^2 A_1 I^2}{2d_1^2}$$

where $\Lambda = LI$. To find the torque, we again differentiate the stored energy, but instead with respect to θ, the rotational angle of the rotor. The areas A_1 and A_2 are both equal to $R\theta D$, where R is the radius of the rotor and D is its width. The torque τ is thus

$$\tau = \frac{dw}{d\theta}\bigg|_\Lambda = -\frac{\Lambda^2}{2L^2}\frac{dL}{d\theta} = -\frac{\mu_0 N^2 R D I^2}{2(d_1 + d_2)}$$

where the torque tries to align the rotor with the stator, maximizing A. Note that the torque is inversely proportional to the size of the gaps between the stator and rotor, so for high torque, the gap sizes should be made as small as possible.

4.3 NON-NORMAL INCIDENCE OF PLANE WAVES IN LOSSLESS MEDIA

We first consider the representation of plane waves propagating at arbitrary angles in free space. A $\hat{y}$-polarized wave propagating in the $+z$ direction can be characterized by the phasor

$$\overline{E} = \hat{y}\, E_i\, e^{-jkz}$$

where the wave number is $k = 2\pi/\lambda = \omega/c$, and k is the rate of phase change along the $\hat{z}$-axis. The same wave propagating at an angle θ in the x–z plane would exhibit phase variations with respect to both the $\hat{x}$ and $\hat{z}$ directions and would thus be expressed by

$$\overline{E} = \hat{y}\, E_i\, e^{-jk_x x - jk_z z} \qquad (4.3.1)$$

as illustrated in Figure 4.9

We may write (4.3.1) more compactly by letting

$$\overline{k} = \hat{x}\, k_x + \hat{z}\, k_z$$
$$\overline{r} = \hat{x}\, x + \hat{z}\, z \qquad (4.3.2)$$

so that (4.3.1) becomes

$$\overline{E} = \hat{y}\, E_i\, e^{-j\overline{k}\cdot\overline{r}}$$

where $\overline{k}$ and $\overline{r}$ may in general have components in the $\hat{x}$, $\hat{y}$, and $\hat{z}$ directions. For the discussion that follows, it is sufficient to consider $\overline{k}$ and $\overline{r}$ vectors that lie only in the x–z plane.

Positions in space having the same value of $\overline{k} \cdot \overline{r}$ have the same phase and so define a *phase front*. Constant phase fronts are perpendicular to $\overline{k}$, as illustrated in Figure 4.9. A wave having planar phase fronts is called a *plane wave*, and if the

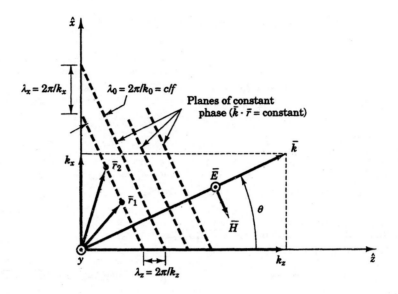

Figure 4.9   Uniform plane wave propagating in the x–z plane.

amplitude of the wave is constant for any particular phase front, then the wave is also a *uniform plane wave*.

Substituting (4.3.1) into the free space, homogeneous wave equation (1.4.9), which we recall is $(\nabla^2 + \omega^2 \mu_0 \epsilon_0)\overline{E} = 0$, yields the dispersion relation

$$k_x^2 + k_z^2 = \omega^2 \mu_0 \epsilon_0 \equiv k_0^2 \tag{4.3.3}$$

Equation (4.3.2) then yields

$$\overline{k} \cdot \overline{k} = \left|\overline{k}\right|^2 = k_x^2 + k_z^2 = k_0^2$$

From (4.3.3) and Figure 4.9 we also see that

$$k_x = \overline{k} \cdot \hat{x} = k_0 \sin\theta$$
$$k_z = \overline{k} \cdot \hat{z} = k_0 \cos\theta$$

We may interpret the *propagation vector* or *wave number* $\overline{k}$ for a uniform plane wave in free space as a vector that points in the direction of wave propagation and has the magnitude $\left|\overline{k}\right| = \omega\sqrt{\mu_0 \epsilon_0}$.

Note that we may define the *projected wavelengths* λ_x and λ_z along the $\hat{x}$- and $\hat{z}$-axes, respectively, to be $\lambda_x = 2\pi/k_x$ and $\lambda_z = 2\pi/k_z$. These projected wavelengths λ_x and λ_z are generally longer than $\lambda_0 = 2\pi/k_0$, and in view of

(4.3.3) must satisfy

$$(1/\lambda_x)^2 + (1/\lambda_z)^2 = (1/\lambda_0)^2$$

The magnetic field $\overline{H}$ for this uniform plane wave is readily calculated from Faraday's law to be

$$\overline{H} = -\frac{\nabla \times \overline{E}}{j\omega\mu_0} = \frac{\overline{k} \times \overline{E}}{\omega\mu_0} = (\hat{x}\, k_x + \hat{z}\, k_z) \times \hat{y}\, \frac{E_i}{\omega\mu_0}\, e^{-j\overline{k}\cdot\overline{r}}$$

$$= \left(\hat{z}\, \frac{k_x}{k_0} - \hat{x}\, \frac{k_z}{k_0}\right) \frac{E_i}{\eta_i}\, e^{-j\overline{k}\cdot\overline{r}}$$

where $k_x/k_0 = \sin\theta$, and $k_z/k_0 = \cos\theta$. Since $\overline{k} \times \overline{E}$ is proportional to $\overline{H}$ for a uniform plane wave, we know immediately that $\overline{H} \cdot \overline{E} = \overline{H} \cdot \overline{k} = 0$, so $\overline{H}$ is perpendicular to both $\overline{E}$ and $\overline{k}$. But Gauss's law for a uniform plane wave in free space forces $\overline{k} \cdot \overline{E} = 0$. Therefore, it is apparent that $\overline{E}$, $\overline{H}$, and $\overline{k}$ are **mutually orthogonal** for a uniform plane wave, as illustrated in Figure 4.9. The vectors $\overline{E}$, $\overline{H}$, and $\overline{k}$ are related by a right-hand coordinate system, which means that $\overline{E} \times \overline{H}$ points in the direction of $\overline{k}$.

Suppose the uniform plane wave (4.3.1) is in a medium characterized by ϵ_i, μ_i, and is incident at angle θ upon the planar boundary of a medium having permittivity ϵ_t and permeability μ_t for $z > 0$. We define the *plane of incidence* to be that plane containing the vector $\overline{k}$ and the surface normal $\hat{n}$, which in this case is the x–z plane. Since the electric field given by (4.3.1) is in the $\hat{y}$ direction and hence is everywhere perpendicular to the plane of incidence, we say that $\overline{E}$ has *perpendicular* or *transverse polarization*. This plane wave is called a *TE (Transverse Electric) wave*. If $\overline{E}$ were everywhere parallel to the plane of incidence, it would have *parallel polarization*.   In this case, $\overline{H}$ would be transverse to the plane of incidence because $\overline{E}$ and $\overline{H}$ are orthogonal. Such a transverse magnetic field wave is called a *TM wave*. As an arbitrary wave is the superposition of *TE* and *TM* components, it is sufficient to understand the propagation of both types of waves to completely understand wave propagation at a planar boundary.

For the case of normal incidence, we saw that Maxwell's equations were satisfied by one reflected and one transmitted wave in addition to the incident wave. We assume the same to be true for obliquely incident waves, where the wave vectors $\overline{k}_r$ and $\overline{k}_t$ for the reflected and transmitted waves are initially unknown. In addition, the complex transmission and reflection coefficients T and Γ must be determined. If Maxwell's equations are all satisfied by the two hypothesized waves in addition to the incident wave, then we have found the correct solution. The directions of propagation of the incident, reflected, and transmitted waves are

$$\overline{k}_i = \hat{x}\, k_{ix} + \hat{z}\, k_{iz}$$

$$\overline{k}_r = \hat{x}\, k_{rx} - \hat{z}\, k_{rz}$$

$$\overline{k}_t = \hat{x}\, k_{tx} + \hat{z}\, k_{tz}$$

where the negative sign is explicitly included in the reflected wave vector so that the real part of each component of the three wave numbers is positive. The electric fields for the incident, reflected, and transmitted waves thus become

$$\overline{E}_i = \hat{y}\, E_i\, e^{-jk_{ix}x - jk_{iz}z}$$
$$\overline{E}_r = \hat{y}\, \Gamma E_i\, e^{-jk_{rx}x + jk_{rz}z} \qquad (4.3.4)$$
$$\overline{E}_t = \hat{y}\, T E_i\, e^{-jk_{tx}x - jk_{tz}z}$$

as illustrated in Figure 4.10 for real $\overline{k}_i$, $\overline{k}_r$, and $\overline{k}_t$. Therefore, the electric field for $z < 0$ may be written

$$\overline{E}(z < 0) = \hat{y}\, E_i\, (e^{-jk_{ix}x - jk_{iz}z} + \Gamma\, e^{-jk_{rx}x + jk_{rz}z}) \qquad (4.3.5)$$

and the electric field for $z > 0$ is

$$\overline{E}(z > 0) = \hat{y}\, E_i (T\, e^{-jk_{tx}x - jk_{tz}z}) \qquad (4.3.6)$$

The tangential component of $\overline{E}$ must be continuous across the boundary at $z = 0$, forcing (4.3.5) and (4.3.6) to be equal at $z = 0$:

$$E_i(e^{-jk_{ix}x} + \Gamma\, e^{-jk_{rx}x}) = E_i(T\, e^{-jk_{tx}x}) \qquad (4.3.7)$$

Since this equation must be satisfied for all values of x, it follows that

$$k_{ix} = k_{rx} = k_{tx} \equiv k_x \qquad (4.3.8)$$

so the tangential components of all three $\overline{k}$ vectors are equal along the boundary.

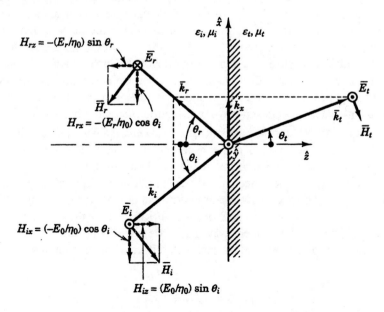

Figure 4.10 *TE wave incident upon a planar boundary.*

This is called the *phase-matching condition* and may also be written

$$\hat{n} \times (\overline{k}_i - \overline{k}_t) = 0$$

Therefore, we may express these k_x components in terms of the angles of incidence, reflection, and transmission to yield

$$k_x = k_i \sin \theta_i = k_r \sin \theta_r = k_t \sin \theta_t \qquad (4.3.9)$$

where $k_i = k_r = \omega\sqrt{\mu_i\epsilon_i}$ is the magnitude of the propagation vectors $\overline{k}_i$ and $\overline{k}_r$. Both the incident and reflected waves obey the same dispersion relation because they propagate in the same medium. The magnitude of the transmitted wave vector is $k_t = \omega\sqrt{\mu_t\epsilon_t}$, which is in general not equal to k_i. From (4.3.9) we find that

$$\theta_r = \theta_i \qquad (4.3.10)$$

and

$$\frac{\sin \theta_t}{\sin \theta_i} = \frac{k_i}{k_t} = \frac{\sqrt{\mu_i\epsilon_i}}{\sqrt{\mu_t\epsilon_t}} = \frac{n_i}{n_t} \qquad (4.3.11)$$

which are the *reflection law* and *Snell's law*, respectively. Equation (4.3.10) says that the *angle of reflection* θ_r equals the *angle of incidence* θ_i. Snell's law, which relates the angle of incidence to the angle of transmission θ_t, is often expressed as shown in (4.3.11) using n, the *index of refraction*, which is the ratio of the velocity of light c to the phase velocity within the medium of interest:

$$n = c\sqrt{\mu\epsilon} \qquad (4.3.12)$$

For all nondispersive media, $n \geq 1$, though for some dispersive media, such as a plasma for $\omega > \omega_p$, the index of refraction may be less than unity. In general, $\theta_i \neq \theta_t$.

Snell's law can be visualized graphically by using the phase matching condition (4.3.8), as shown in Figure 4.11. In general, a light ray traveling from a medium of higher index of refraction to a medium with a lower index of refraction is bent farther away from the normal to the boundary surface.

Figure 4.11(a) illustrates the case where the transmitting medium ϵ_t, μ_t has a higher refractive index n_t than the medium in which the incident wave travels n_i. For any angle of incidence θ_i, there exists some real value of θ_t for which k_x is identical for all three waves. This is also true for the opposite case ($n_t < n_i$) illustrated in Figure 4.11(b), **provided that** the angle of incidence θ_i is less than some *critical angle* θ_c. Figures 4.11(c) and (d) are pictorial representations of wave propagation at an interface, both for $|\overline{k}_t|$ greater than and less than $|\overline{k}_i|$. At the critical angle of incidence, the transmitted wave is maximally bent away from the surface normal and so propagates parallel to the boundary surface ($\theta_t = 90°$). Snell's law (4.3.11) then yields

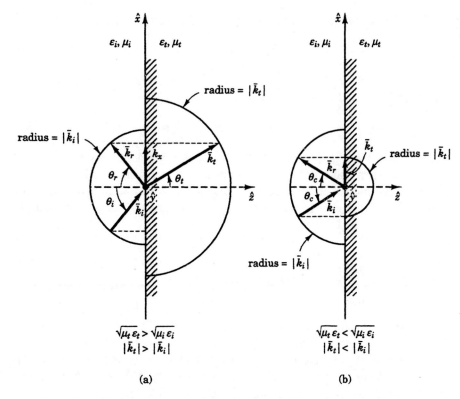

Figure 4.11(a,b)   Snell's law and the critical angle θ_c.

$$\frac{\sin \theta_t}{\sin \theta_c} = \frac{1}{\sin \theta_c} = \frac{n_i}{n_t}$$

and therefore

$$\theta_c = \sin^{-1}(n_t/n_i) \tag{4.3.13}$$

For angles of incidence θ_i greater than the critical angle θ_c, the $\hat{x}$ component of $\bar{k}_t$ is required to be larger than the magnitude of $\bar{k}_t$, which is not possible for real values of $\bar{k}_t$. The dispersion relation (4.3.3) for the transmitting medium is

$$k_t^2 = \omega^2 \mu_t \epsilon_t = k_{tx}^2 + k_{tz}^2 \tag{4.3.14}$$

At the critical angle θ_c, we observe that $\theta_t = 90°$, $k_{tz} = 0$, and $k_x = k_{tx} = \omega\sqrt{\mu_t \epsilon_t}$. For $\theta_i > \theta_c$, the dispersion relation yields

$$k_{tz}^2 = k_t^2 - k_x^2 < 0$$

$$k_{tz} = \pm j\sqrt{k_x^2 - k_t^2} = \pm j\alpha_z$$

where α_z is a real, positive quantity.

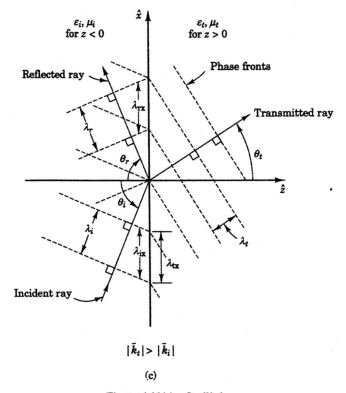

Figure 4.11(c)   Snell's law

We choose that sign in front of α_z that corresponds to a wave exponentially decaying away from the boundary. For a passive medium, any exponentially growing wave would have to be generated somewhere on the right side of the boundary ($z > 0$) and would thus not satisfy our assumption of a source only on the left of the boundary. (For an active medium, which amplifies waves propagating through it, exponential growth would be permissible.) Since we want $\overline{E}_z \propto e^{-\alpha_z z}$, it follows that

$$k_{tz} = -j\alpha_z$$

$$\overline{E}_t = \hat{y}\, T\, E_i\, e^{-jk_x x - \alpha_z z} \tag{4.3.15}$$

$$\alpha_z = \sqrt{k_x^2 - k_t^2} > 0$$

This is a new type of wave because $\overline{E}_t$, given by (4.3.15), propagates in the $+\hat{x}$ direction by virtue of the $e^{-jk_x x}$ factor, but exponentially decays in the $+\hat{z}$ direction as $e^{-\alpha_z z}$. Thus the amplitude of the wave is not uniform along the constant-phase surfaces (perpendicular to the $\hat{x}$-axis) where $e^{-jk_x x}$ is constant. Because the

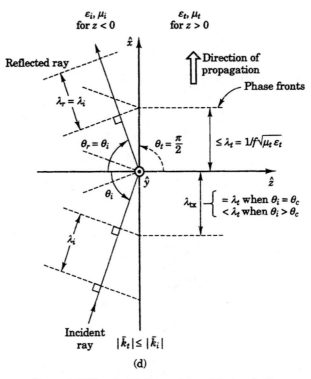

Figure 4.11(d) Snell's law and the critical angle θ_c.

phase fronts are planar but the amplitude is nonuniform, this is called a *nonuniform plane wave* or *inhomogeneous plane wave*. Since the amplitude is significant only near the surface of the medium, it is also called a *surface wave*, and is illustrated in Figure 4.12. For angles of incidence greater than the critical angle, the medium is said to exhibit *total internal reflection*, as no average power is transmitted into the second medium.

Note that the vector wave number $\bar{k}$ for a nonuniform plane wave is complex; here

$$\bar{k}_t = \hat{x}\, k_{tx} + \hat{z}\, k_{tz}$$
$$= \hat{x}\, k_x - j\hat{z}\, \alpha_z \tag{4.3.16}$$

In general, an arbitrary wave vector $\bar{k}$ may be written

$$\bar{k} = \bar{k}' - j\bar{k}'' \tag{4.3.17}$$

where the real part $\bar{k}'$ is perpendicular to the wave phase fronts and points in the direction of wave propagation, and the imaginary part $\bar{k}''$ is perpendicular to the planes of constant wave amplitude and points in the direction of wave attenuation. Here $\bar{k}'$ and $\bar{k}''$ are orthogonal because the medium ϵ_t, μ_t is lossless. This issue is

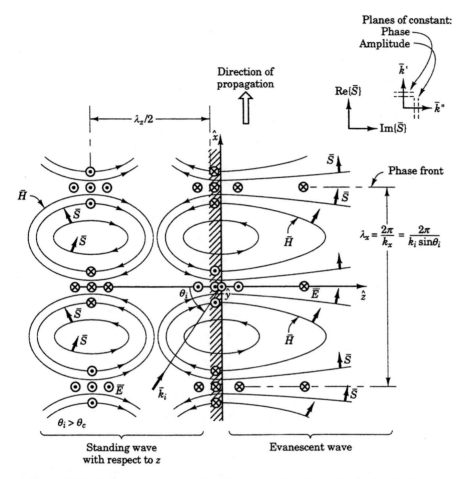

Figure 4.12 Evanescent wave at a planar boundary at $t = 0$ traveling in the $+\hat{x}$ direction.

discussed further in Section 4.4, and may be contrasted with waves propagating in lossy media where $\overline{k}'$ and $\overline{k}''$ point in the same direction, discussed in Section 3.4.

The Poynting vector for $z > 0$ is

$$\overline{S}_t(z > 0) = \overline{E}_t \times \overline{H}_t^*$$

where

$$\overline{E}_t = \hat{y}\, T E_i\, e^{-jk_x x - \alpha_z z}$$

and

$$\overline{H}_t = -\frac{\nabla \times \overline{E}_t}{j\omega\mu_t} = \left(\hat{z}\,\frac{k_x}{\omega\mu_t} + \hat{x}\,\frac{j\alpha_z}{\omega\mu_t}\right) T E_i\, e^{-jk_x x - \alpha_z z}$$

Thus,

$$\overline{S}_t(z > 0) = \left(\hat{x}\, \frac{k_x}{\omega\mu_t} + \hat{z}\, \frac{j\alpha_z}{\omega\mu_t} \right) |T|^2 E_i^2\, e^{-2\alpha_z z}$$

The real part of $\overline{S}_t$ points in the $+\hat{x}$ direction, which is the direction of propagation for this surface wave, and the imaginary part is positive (inductive) and points in the $+\hat{z}$ direction, corresponding to storage of excess magnetic energy in the transmitted wave. Since no real power is transmitted in the direction in which the wave decays ($+\hat{z}$), the wave is said to be *evanescent*.

Once $\overline{k}_r$ and $\overline{k}_t$ are found by matching phases at the boundary, we may then solve for Γ and T, the complex reflection and transmission coefficients. These two unknowns may be found by supplementing the boundary condition for continuity of the electric field (4.3.7) with a similar equation for the tangential magnetic field H_x. Re-examining (4.3.7) gives

$$1 + \Gamma = T \tag{4.3.18}$$

because $k_{ix} = k_{rx} = k_{tx} \equiv k_x$ from (4.3.8). To obtain H_x, we use Faraday's law for plane waves: $\overline{H} = -\nabla \times \overline{E}/j\omega\mu$. Applying this equation to the electric fields at $z = 0$ given by (4.3.4) yields the following tangential components of the incident, reflected, and transmitted magnetic fields at $z = 0$:

$$H_{ix}(z = 0) = -\frac{k_{iz}}{\omega\mu_i} E_i\, e^{-jk_x x} \tag{4.3.19}$$

$$H_{rx}(z = 0) = \frac{k_{rz}}{\omega\mu_i} \Gamma E_i\, e^{-jk_x x} \tag{4.3.20}$$

$$H_{tx}(z = 0) = -\frac{k_{tz}}{\omega\mu_t} T E_i\, e^{-jk_x x} \tag{4.3.21}$$

We can equate the total tangential magnetic fields on both sides of the boundary because no surface currents exist on an interface between two nonconducting media, so

$$H_{ix}(z = 0) + H_{rx}(z = 0) = H_{tx}(z = 0)$$

or

$$\frac{k_{iz}}{\omega\mu_i}(1 - \Gamma) = \frac{k_{tz}}{\omega\mu_t} T \tag{4.3.22}$$

Combining (4.3.18) and (4.3.22) gives us expressions for the reflection and transmission coefficients for *TE* waves:

$$\Gamma = \frac{Z_n^{TE} - 1}{Z_n^{TE} + 1} \tag{4.3.23}$$

$$T = \frac{2Z_n^{TE}}{Z_n^{TE} + 1} \tag{4.3.24}$$

The normalized wave impedance Z_n^{TE} is defined here to be

$$Z_n^{TE} \equiv \frac{\mu_t/k_{tz}}{\mu_i/k_{iz}} \qquad (4.3.25)$$

and

$$k_{iz} = \omega\sqrt{\mu_i\epsilon_i}\cos\theta_i$$

$$k_{tz} = \sqrt{k_t^2 - k_x^2} = \sqrt{\omega^2\mu_t\epsilon_t - k_i^2\sin^2\theta_i}$$

We note that for the case of normal incidence $(\theta_i = 0)$, (4.3.25) reduces to $Z_n^{TE} = \sqrt{\mu_t\epsilon_i/\mu_i\epsilon_t} = \eta_t/\eta_i$, which is the normalized wave impedance η_n described in Section 4.2. If k_{tz} is real, which occurs if both media are lossless and the angle of incidence θ_i is less than the critical angle θ_c, then Γ is real. Beyond θ_c we found that $k_{tz} = -j\alpha_z = -j\sqrt{k_x^2 - k_t^2}$, and therefore Z_n^{TE} is imaginary and Γ is generally complex. The phase angle of Γ is called the *Goos-Hänschen phase shift*.

The only example considered so far has been that of an incident *TE* wave, for which $\overline{E}$ is transverse to the plane of incidence, but the same methods lead to comparable results for an incident *TM* wave. An alternative method for deriving the results for *TM* waves uses the principle of *duality*. Duality enables us to construct a new, physically different solution to Maxwell's equations from a known solution, using the symmetry properties of these equations. Suppose we recall Maxwell's equations for a charge-free medium

$$\nabla \times \overline{E} = -j\omega\mu\overline{H} \qquad (4.3.26)$$

$$\nabla \times \overline{H} = j\omega\epsilon\overline{E} \qquad (4.3.27)$$

$$\nabla \cdot \overline{E} = 0 \qquad (4.3.28)$$

$$\nabla \cdot \overline{H} = 0 \qquad (4.3.29)$$

and make the following replacements:

$$\overline{E} \rightarrow \overline{H}$$

$$\overline{H} \rightarrow -\overline{E}$$

$$\mu \rightarrow \epsilon$$

$$\epsilon \rightarrow \mu$$

We then note that Faraday's and Ampere's laws, (4.3.26) and (4.3.27), are interchanged, and Gauss's laws, (4.3.28) and (4.3.29), are also interchanged. The result is the exact same four Maxwell's equations! Thus, if we find solutions $\overline{E}$ and $\overline{H}$ that satisfy Maxwell's equations in some medium characterized by μ and ϵ, then the solution to the dual problem can be written down immediately by making the indicated replacements, **provided that** the boundary conditions also exhibit duality.

Consider the duality inherent in the *TE* and *TM* reflection problems, as illustrated in Figure 4.13. In both the *TE* and *TM* problems, the boundary conditions require that the tangential electric and magnetic fields be continuous at $z = 0$, and therefore these boundary conditions remain unchanged when $\overline{E}$ and $\overline{H}$ are replaced by their duals $\overline{H}$ and $-\overline{E}$, respectively. If, however, the transmitting medium were a perfect conductor, the tangential electric field would vanish at the boundary, but the tangential magnetic field would not, and thus the *TE* and *TM* reflection problems would not be dual to each other.

Since *TM* reflection at the boundary between two dielectrics is dual to *TE* reflection at the same boundary, we may immediately write the solutions to the *TM* problem by making the appropriate replacements of variables in the *TE* solutions (4.3.4)

$$\overline{H}_i = \hat{y}\, H_i\, e^{-jk_{i_x}x - jk_{i_z}z}$$

$$\overline{H}_r = \hat{y}\, \Gamma^{TM} H_i\, e^{-jk_{r_x}x + jk_{r_z}z}$$

$$\overline{H}_t = \hat{y}\, T^{TM} H_i\, e^{-jk_{t_x}x - jk_{t_z}z}$$

where the k vectors are the same as before. Equations (4.3.23)–(4.3.25) become

$$\Gamma^{TM} = \frac{Y_n^{TM} - 1}{Y_n^{TM} + 1}$$

$$T^{TM} = \frac{2Y_n^{TM}}{Y_n^{TM} + 1}$$

where the normalized impedance Z_n^{TE} is transformed into the normalized admittance $Y_n^{TM} = \epsilon_t k_{iz}/\epsilon_i k_{tz}$ because the roles of μ and ϵ are interchanged via duality. We have defined Γ^{TM} as the ratio of the reflected to incident magnetic fields to be the dual *TE* reflection coefficient, which is defined as the ratio of the reflected to incident electric fields. However, it is customary to define the reflection coefficient as the ratio of reflected to incident electric fields for **both** polarizations,

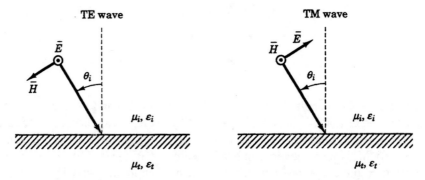

Figure 4.13 Duality of *TE* and *TM* reflections from a planar boundary.

so $\Gamma = E_r/E_i = -\Gamma^{TM}$. Likewise, the transmission coefficient T is always defined as the ratio of transmitted to incident electric fields (not magnetic fields), so $T = E_t/E_i = T^{TM}/Y^{TM}$. The TM reflection and transmission coefficients are therefore

$$\Gamma = -\frac{Y_n^{TM} - 1}{Y_n^{TM} + 1} \tag{4.3.30}$$

$$T = \frac{2}{Y_n^{TM} + 1} \tag{4.3.31}$$

$$Y_n^{TM} = \frac{\epsilon_t/k_{tz}}{\epsilon_i/k_{iz}} \tag{4.3.32}$$

For normally incident TM waves ($\theta_i = 0$), the wave impedance Z_n^{TM} reduces to $\sqrt{\mu_t \epsilon_i / \mu_i \epsilon_t} = \eta_t/\eta_i$, which is the exactly the same as the expression for Z_n^{TE} at normal incidence. The two expressions are identical because normal incidence does not differentiate between TE and TM waves; there is no uniquely defined plane of incidence for $\theta = 0$. We have seen that for normal incidence, there is no reflected wave if the characteristic impedances η_i and η_t are equal. It is interesting to explore the circumstances for which reflections are zero for non-normal incidence.

Zero reflection for TE or TM waves requires that the reflection coefficient Γ be zero. Therefore, we require that

$$Z_n^{TE} = 1 = \frac{\mu_t/k_{tz}}{\mu_i/k_{iz}} \quad \text{for } TE \text{ waves} \tag{4.3.33}$$

and

$$Z_n^{TM} = 1 = \frac{\epsilon_i/k_{iz}}{\epsilon_t/k_{tz}} \quad \text{for } TM \text{ waves} \tag{4.3.34}$$

If we assume $\mu_t = \mu_i = \mu$, then for TE waves, (4.3.33) implies that $k_{iz} = k_{tz}$ for zero reflection to occur, which may be written

$$k_i \cos \theta_i = k_t \cos \theta_t \tag{4.3.35}$$

Combining (4.3.35) with Snell's law $k_i \sin \theta_i = k_t \sin \theta_t$, we find that

$$\tan \theta_i = \tan \theta_t$$

which means that $\theta_t = \theta_i$ and $\epsilon_t = \epsilon_i$. Therefore, TE waves are fully transmitted at the boundary between two dielectric media ($\mu_i = \mu_t$) only when the permittivities of the media are the same; this is a physically uninteresting case.

For 100 percent transmission of TM waves to occur when $\mu_t = \mu_i$, (4.3.34) yields

$$k_i \cos \theta_t = k_t \cos \theta_i \tag{4.3.36}$$

A solution to (4.3.36) and Snell's law (4.3.11) exists when

$$\cos \theta_t = \sin \theta_i$$
$$\sin \theta_t = \cos \theta_i$$

or $\tan \theta_t = \cot \theta_i$, which is an equivalent way of stating

$$\theta_i + \theta_t = \pi/2 \qquad (4.3.37)$$

Since $\sin \theta_t = \cos \theta_i$, Snell's law can also be expressed as

$$\tan \theta_i = k_t/k_i = \sqrt{\epsilon_t/\epsilon_i}$$

We note that there always exists some angle $0 \le \theta_i \le \pi/2$ for which there are no *TM* reflections and 100 percent of the incident power is transmitted. This angle is given the name *Brewster's angle*, θ_B:

$$\theta_B = \tan^{-1} \sqrt{\epsilon_t/\epsilon_i} \qquad (4.3.38)$$

Brewster's angle exists when the incident and transmitting permeabilities are equal, and there is no more than one such angle for a given boundary. If the permeabilities are mismatched at the boundary ($\mu_i \neq \mu_t$) and the permittivities are equal, then a Brewster's angle $\theta_B = \tan^{-1} \sqrt{\mu_t/\mu_i}$ exists for *TE* waves only, easily observed by duality.

Figure 4.14 is a plot of $|\Gamma|^2$ for both *TE* and *TM* waves when $\mu_i = \mu_t$; $|\Gamma|^2$ is called the *surface reflectivity*. In this case, we see that *TE* waves are always more reflective than *TM* waves, even for angles other than θ_B.

The curious relation (4.3.37) for Brewster's angle, $\theta_B + \theta_t = \pi/2$, is physically significant. In the case of mismatched dielectrics, the reflected field arises from fields produced by polarized, dipole-like atoms in response to the incident field. Because the transmitted *TM* wave is propagating at an angle θ_t with its electric field parallel to the plane of incidence, any radiation emitted in the direction $\theta_r = \theta_B = \pi/2 - \theta_t$ would have to have been radiated along the polarization axis of the atoms. From Section 2.3, we recall that there is a null in the Hertzian dipole radiation pattern along the polarization axis ($\theta = 0$), and thus no reflected radiation is observed at this special angle. The geometry of Brewster's angle is suggested in Figure 4.15. For *TE* waves, the induced electric dipole moments produced by the incident electric field are perpendicular to the plane of incidence with no nulls in the dipole radiation pattern in the plane of incidence. Therefore, these dipoles can radiate at any angle consistent with $k_{rz} = k_{iz}$, or $\theta_r = \theta_i$, and there is no Brewster's angle in this case.

Example 4.3.1

A 20-MHz shortwave radio transmitter is broadcasting intercontinental programs by bouncing signals from the ionosphere. Assume the ionospheric plasma frequency f_p is 10 MHz and the ionosphere lies above an altitude of 100 km. What is the minimum distance from a ground-based transmitter to a ground-based receiver for which a single-bounce signal could be sent if it is 100 percent reflected at the ionosphere? What is the ionospheric

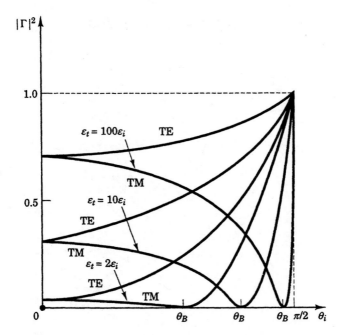

Figure 4.14   Reflectivity of a planar dielectric surface:
angular dependence of *TE* and *TM* waves.

reflection coefficient $|\Gamma|^2$ at zenith (normal incidence)? Assume that the curvature of the Earth is negligible.

Solution: Figure 4.16 illustrates the single-bounce signal path for the shortwave radio transmitter. For $\theta > \theta_c$ there is 100 percent reflection, where (4.3.13) gives the critical angle

$$\theta_c = \sin^{-1}(n_t/n_i)$$

From (4.3.12) and (3.6.5), the indices of refraction in the two media are

$$n_i = n_0 = 1$$

$$n_t = \sqrt{1 - \frac{\omega_p^2}{\omega^2}} = \sqrt{\frac{3}{4}}$$

and therefore

$$\theta_c = \sin^{-1}(n_t/n_i) = \sin^{-1}\sqrt{0.75} = 60°$$

The distance to a receiver for a wave bouncing at the critical angle $\theta_c = 60°$ is $2 \cdot 100$ km $\cdot$ tan $60° = 346$ km. If the ionospheric boundary were abrupt, then the normalized wave impedance of the *TE* and *TM* signals at normal incidence would just be the ratio

$$\eta_n = \eta_t/\eta_i = \sqrt{\epsilon_i/\epsilon_t} = n_i/n_t = \sqrt{4/3}$$

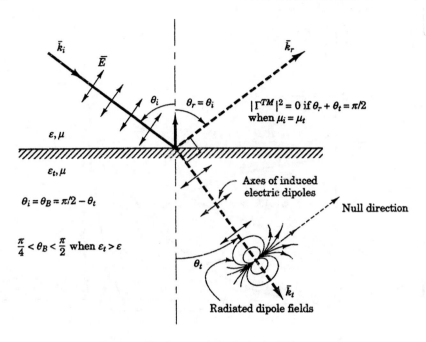

Figure 4.15   Brewster's angle θ_B for *TM* waves.

and the reflectivity would then be

$$|\Gamma|^2 = \left| \frac{\eta_n - 1}{\eta_n + 1} \right|^2 = 5.15 \times 10^{-3}$$

If the boundary were gradual with altitude because the electron density varied slowly compared to a wavelength, then the reflected power received for bounce distances less than 346 km would be negligible. In this case, the problem could be modeled as a series of smaller wave impedance transitions as discussed in Chapter 6. For angles greater than

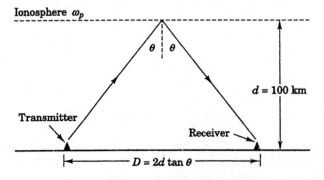

Figure 4.16   Reflection of radio waves from the ionosphere.

θ_c, reflection is always 100 percent, whether the boundary is sharp or not, and the wave penetrates the ionosphere roughly to the altitude where $n_t = n_i \sin \theta_i$.

Example 4.3.2

A beam of light is propagating inside glass with an index of refraction $n_i = 2$ and is incident upon a planar interface with vacuum ($n_0 = 1$) at an angle $\theta > \theta_c$. The resulting evanescent wave in vacuum decays as $e^{-z/\Delta}$, where the $\hat{z}$-axis is perpendicular to the boundary. What is the smallest possible value for the penetration depth Δ?

   Solution: The field strengths $\overline{E}$ and $\overline{H}$ in an evanescent wave decay as

$$e^{-jk_z z} = e^{-\alpha z} = e^{-z/\Delta}$$

where $\Delta = 1/\alpha$, and (4.3.15) implies that

$$1/\Delta = \sqrt{k_x^2 - k_t^2} > 0$$

To minimize Δ, we choose $k_x = k_i \sin \theta_i$ as large as possible, so $\theta_i = \pi/2$ and $k_x = k_i = 2\pi/\lambda_i$. Therefore,

$$\Delta \geq 1 \Big/ \sqrt{\left(\frac{2\pi}{\lambda_i}\right)^2 - \left(\frac{2\pi}{\lambda_0}\right)^2} = \frac{\lambda_i}{2\pi}\left[1 - \left(\frac{n_0}{n_i}\right)^2\right]^{-1/2} = \frac{\lambda_i}{\sqrt{3}\pi}$$

where $\lambda_i/\lambda_0 = k_0/k_i = n_0/n_i$. Thus, the penetration depth Δ of an evanescent wave cannot be less than $1/\sqrt{3}\pi$ times the wavelength λ_i inside glass with index of refraction $n_i = 2$.

4.4* NON-NORMAL INCIDENCE OF PLANE WAVES IN LOSSY MEDIA

A general plane wave in an isotropic, lossy medium may be represented by

$$\overline{E} = \overline{E}_0 e^{-j\overline{k}\cdot\overline{r}} \tag{4.4.1}$$

where $\overline{k}$ is complex, and $\overline{k} \cdot \overline{E} = 0$ because Gauss's law (1.3.3) in a source-free region must be satisfied. Substituting (4.4.1) into the homogeneous wave equation $(\nabla^2 + \omega^2 \mu \epsilon_{\text{eff}})\overline{E} = 0$ yields

$$\left[(-j\overline{k}) \cdot (-j\overline{k}) + \omega^2 \mu \epsilon_{\text{eff}}\right] \overline{E}_0 = 0$$

For nonzero $\overline{E}_0$,

$$\overline{k} \cdot \overline{k} = \omega^2 \mu \epsilon_{\text{eff}} \tag{4.4.2}$$

which is the general dispersion relation for a plane wave in an isotropic, lossy media. The permeability μ is usually a real, positive number and ϵ_{eff} is given by (3.4.5):

$$\epsilon_{\text{eff}} = \epsilon - \frac{j\sigma}{\omega} \tag{4.4.3}$$

If the conductivity of the medium is nonzero, then $\omega^2\mu\epsilon_{\text{eff}}$ is complex, so in general $\overline{k}$ is complex as well. If we represent

$$\overline{k} = \overline{k}' - j\overline{k}'' \tag{4.4.4}$$

where $\overline{k}'$ and $\overline{k}''$ are both real vectors that do not necessarily point in the same direction, then (4.4.2) can be rewritten as two real equations using (4.4.4):

$$\left|\overline{k}\right|'^2 - \left|\overline{k}\right|''^2 = \text{Re}\{\omega^2\mu\epsilon_{\text{eff}}\} = \omega^2\mu\epsilon \tag{4.4.5}$$

and

$$2\overline{k}' \cdot \overline{k}'' = \text{Im}\{\omega^2\mu\epsilon_{\text{eff}}\} = \mu\sigma\omega \tag{4.4.6}$$

For a lossless medium with $\sigma = 0$, it follows from (4.4.6) that $\overline{k}' \cdot \overline{k}'' = 0$. This means that the direction of propagation of the wave ($\overline{k}'$) is perpendicular to the direction in which the wave attenuates ($\overline{k}''$), which is the characteristic signature of evanescent waves.[3] In general, $\sigma \neq 0$ and therefore $\overline{k}' \cdot \overline{k}'' \neq 0$. Because lossy media absorb energy as waves propagate through them, there must be some attenuation in the direction of wave propagation $\overline{k}'$ and hence a component of $\overline{k}''$ that is parallel to $\overline{k}'$.

We can also compute the Poynting vector for the wave described by (4.4.1)

$$\overline{S} = \overline{E} \times \overline{H}^* = \overline{E} \times \left(\frac{\overline{k} \times \overline{E}}{\omega\mu}\right)^* \tag{4.4.7}$$

where Faraday's law for plane waves (1.6.19) has been used. The vector identity $\overline{A} \times (\overline{B} \times \overline{C}) = \overline{B}(\overline{A} \cdot \overline{C}) - \overline{C}(\overline{A} \cdot \overline{B})$ allows us to simplify (4.4.7), yielding

$$\overline{S} = \frac{1}{\omega\mu}\left(\overline{k}^* \left|\overline{E}\right|^2 - \overline{E}^*(\overline{k}^* \cdot \overline{E})\right) \tag{4.4.8}$$

Thus, $\overline{S}$ has both real and imaginary parts that generally point in different directions. For isotropic, charge-free, lossy media for which $\rho = -j\overline{k} \cdot \epsilon\overline{E} = 0$, the direction in which power flows (given by the real part of $\overline{S}$) is the same as the direction of wave propagation $\overline{k}'$. For more complicated media, power flow and propagation need not be parallel, as may be seen from (4.4.8).

If we wish to compute the reflection coefficient from a planar interface where the media on one or both sides of the boundary are lossy, we may utilize the same Maxwell boundary conditions as before. If the geometry of the surface is given in Figure 4.17, where the planar boundary between the two media (ϵ_i, μ_i, σ_i and ϵ_t, μ_t, σ_t) is placed at $z = 0$, we see that the incident, reflected, and

---

3. The lossless condition $\overline{k}' \cdot \overline{k}'' = 0$ is also satisfied when $\overline{k}'' = 0$, producing normal uniform plane wave propagation with no attenuation. Alternatively, $\overline{k}' = 0$, producing purely attenuated, nonpropagating evanescent waves as a result, such as those exhibited by a plasma excited below the plasma frequency.

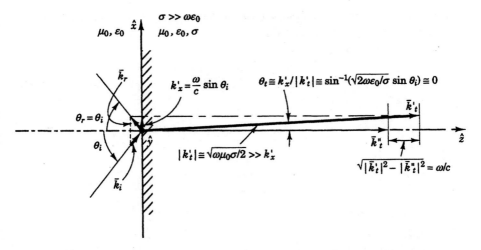

Figure 4.17 Reflection and transmission at a highly conducting boundary.

transmitted waves must all have the same tangential components of the wave vector $\bar{k}$. This phase matching condition, discussed in Section 4.3 for nonlossy media, implies that

$$k'_{ix} = k'_{rx} = k'_{tx} \tag{4.4.9}$$

and also that

$$k''_{ix} = k''_{rx} = k''_{tx} \tag{4.4.10}$$

The real and imaginary parts of the dispersion relation (4.4.2), given by (4.4.5) and (4.4.6), are

$$(k'_{rx})^2 + (k'_{rz})^2 - (k''_{rx})^2 - (k''_{rz})^2 = \omega^2 \mu_i \epsilon_i \tag{4.4.11}$$

$$k'_{rx}k''_{rx} + k'_{rz}k''_{rz} = \frac{1}{2}\omega\mu_i\sigma_i \tag{4.4.12}$$

for the reflected wave, and

$$(k'_{tx})^2 + (k'_{tz})^2 - (k''_{tx})^2 - (k''_{tz})^2 = \omega^2 \mu_t \epsilon_t \tag{4.4.13}$$

$$k'_{tx}k''_{tx} + k'_{tz}k''_{tz} = \frac{1}{2}\omega\mu_t\sigma_t \tag{4.4.14}$$

for the transmitted wave. So if the incident wave vector $\bar{k}_i = (k'_{ix} - jk''_{ix})\hat{x} + (k'_{iz} - jk''_{iz})\hat{z}$ is known, then the eight equations contained in (4.4.9)–(4.4.14) may be used to solve for the eight unknowns: k'_{rx}, k''_{rx}, k'_{rz}, k''_{rz}, k'_{tx}, k''_{tx}, k'_{tz} and k''_{tz}. With $\bar{k}_r$ and $\bar{k}_t$ determined, Γ and T are found from (4.3.18), (4.3.23)–(4.3.25) for TE waves and (4.3.30)–(4.3.32) for TM waves.

We now consider the special case of reflection from a highly conducting medium ($\sigma \gg \omega\epsilon$) as an example of how to find Γ and T for both TE and TM waves. There is vacuum to the left of the boundary $(z < 0)$ and a conducting medium characterized by μ_0, ϵ_0 and $\sigma \gg \omega\epsilon_0$ to the right $(z > 0)$, shown in Figure 4.17. In the vacuum, $k_i = \omega\sqrt{\mu_0\epsilon_0} = \omega/c$ so

$$k'_{ix} = \frac{\omega}{c}\sin\theta_i$$

and

$$k'_{iz} = \frac{\omega}{c}\cos\theta_i$$

where θ_i is the angle of incidence. Because vacuum is lossless, $k''_{ix} = k''_{iz} = 0$. In the lossy medium, we use (4.4.13) and (4.4.14) to find

$$(k'_{tx})^2 + (k'_{tz})^2 - (k''_{tx})^2 - (k''_{tz})^2 = \omega^2\mu_0\epsilon_0 = \left(\frac{\omega}{c}\right)^2 \tag{4.4.15}$$

and

$$k'_{tx}k''_{tx} + k'_{tz}k''_{tz} = \frac{1}{2}\omega\mu_0\sigma \tag{4.4.16}$$

But phase-matching at $z = 0$ allows us to set $k''_{tx} = 0$ by (4.4.10) and $k'_{tx} = k'_{ix} = (\omega/c)\sin\theta_i$ by (4.4.9). Thus (4.4.15) and (4.4.16) become

$$k'^2_{tz} - k''^2_{tz} = \left(\frac{\omega}{c}\right)^2\cos^2\theta_i \tag{4.4.17}$$

and

$$k'_{tz}k''_{tz} = \frac{1}{2}\omega\mu_0\sigma \tag{4.4.18}$$

We notice that $\frac{1}{2}\omega\mu_0\sigma \gg (\omega/c)^2\cos^2\theta_i$ because $\sigma \gg \omega\epsilon_0$, which means that the difference of the squares of k'_{tz} and k''_{tz} is a small number, while the product of k'_{tz} and k''_{tz} is a large number. To first order, this means that $k'_{tz} \simeq k''_{tz}$, if we neglect the right side of (4.4.17), and we substitute this into (4.4.18) to yield

$$k'_{tz} \simeq k''_{tz} \simeq \sqrt{\frac{\omega\mu_0\sigma}{2}} \tag{4.4.19}$$

We could also establish this result by substituting (4.4.18) into (4.4.17), solving the resulting quadratic equation, and keeping only the lowest order terms. The angle at which the transmitted wave propagates is given by

$$\tan\theta_t = \frac{k'_{tx}}{k'_{tz}} \simeq \frac{(\omega/c)\sin\theta_i}{\sqrt{\frac{1}{2}\omega\mu_0\sigma}} = \sqrt{\frac{2\omega\epsilon_0}{\sigma}}\sin\theta_i \tag{4.4.20}$$

which is very close to zero because the conductor is good. Therefore, no matter what the angle of incidence of the incoming wave, the transmitted wave propagates in a direction essentially normal to the conducting surface. (Though since the conductor is good, the transmitted wave amplitude decays essentially to zero within a skin depth.) Propagation in a highly conducting medium is illustrated in Figure 4.17.

We would next like to calculate the fraction of power transmitted across a highly conducting interface for a *TE* wave that is initially propagating in vacuum at an angle θ_i from the normal to the interface. This problem may be solved in at least two ways. The first method makes direct use of the expression (4.3.23) for Γ and then computes the transmitted fraction of power $1 - |\Gamma|^2$. For a *TE* wave, we recall from (4.3.25) that

$$Z_n^{TE} = \frac{\mu_t k_{iz}}{\mu_i k_{tz}} \tag{4.4.21}$$

We have already seen that $k_{iz} = k'_{iz} - jk''_{iz} = (\omega/c)\cos\theta_i$, and $k_{tz} = k'_{tz} - jk''_{tz} = \sqrt{\omega\mu_0\sigma/2}(1-j)$ from (4.4.19). Since $\mu_i = \mu_t = \mu_0$, (4.4.21) may be rewritten

$$Z_n^{TE} = \sqrt{\frac{\omega\epsilon_0}{2\sigma}}(1+j)\cos\theta_i \tag{4.4.22}$$

which is much smaller than unity if $\sigma \gg \omega\epsilon_0$. The *TE* reflection coefficient is given by (4.3.23) and may be approximated by a Taylor series because $Z_n^{TE} \ll 1$; thus

$$\Gamma = \frac{Z_n^{TE} - 1}{Z_n^{TE} + 1} \simeq -(1 - Z_n^{TE})(1 - Z_n^{TE})$$

$$\simeq -1 + 2Z_n^{TE}$$

where terms of order $(Z_n^{TE})^2$ have been neglected. Therefore,

$$\Gamma \simeq -1 + \sqrt{\frac{2\omega\epsilon_0}{\sigma}}(1+j)\cos\theta_i \tag{4.4.23}$$

We note that Γ is very close to -1, which is what we expect for reflection from a highly conducting surface: the reflected wave almost exactly cancels the incident wave at the boundary so that $\overline{E} \simeq 0$ inside the conductor. The fraction of power that is transmitted across the boundary for the *TE* wave is then

$$1 - |\Gamma|^2 \simeq \sqrt{\frac{8\omega\epsilon_0}{\sigma}}\cos\theta_i \tag{4.4.24}$$

where terms of order $\omega\epsilon_0/\sigma$ have been neglected.

We may also calculate the transmitted power by computing the total power dissipation in the conducting medium, as all of the power that passes into the conductor at $z = 0$ is eventually dissipated. If the conductor is sufficiently good, power is dissipated mostly within the first few skin depths. We have already shown in Section 4.2 that the power dissipated in an almost-perfect conductor is the same as the power that would be dissipated if a surface current $\overline{J}_s$ flowed uniformly through a slab of conductivity σ and thickness equal to its skin depth δ. This dissipated power P_d (W/m^2) is

$$P_d = \frac{1}{2} \text{Re}\{\overline{E} \cdot \overline{J}^*\}\delta = \frac{|\overline{J}_s|^2}{2\sigma\delta} = \frac{|\overline{H}(z=0)|^2}{4} \sqrt{\frac{2\omega\mu_0}{\sigma}} \qquad (4.4.25)$$

where $\overline{H}(z=0)$ is the total **tangential** magnetic field at the surface of the conductor.

We recall from (4.3.5) that the *TE* wave at $z = 0$ has an electric field given by

$$\overline{E}(z=0) = \hat{y}\, E_i(1+\Gamma)\, e^{-jk_x x} \qquad (4.4.26)$$

From (4.3.19) and (4.3.20), the tangential component of $\overline{H}$ is

$$\overline{H}_t(z=0) = -\hat{x}\, \frac{E_i}{\eta_0}(1-\Gamma)\cos\theta_i\, e^{-jk_x x} \qquad (4.4.27)$$

The amount of power transmitted into the conductor is then

$$\langle \overline{S}(z=0)\rangle = \frac{1}{2}\text{Re}\left\{\overline{E}\times\overline{H}^*\right\}\cdot\hat{z}\,\Big|_{z=0}$$

$$= \frac{|E_i|^2}{2\eta_0}(1-|\Gamma|^2)\cos\theta_i \qquad (4.4.28)$$

after (4.4.26) and (4.4.27) are used. Notice that we don't need to calculate the normal component of $\overline{H}$ (H_z) to find the time-averaged power flow across the conducting surface. Equation (4.4.28) may be compared with (4.2.12), the time-averaged Poynting vector at normal incidence. We see that except for the $\cos\theta_i$ factor, which accounts for oblique incidence, the two are the same.

The tangential magnetic field in (4.4.27) may be approximated by $\overline{H}_t(z=0) = -\hat{x}\,(2E_i/\eta_0)\,e^{-jk_x x}\cos\theta_i$ because we already know that Γ will be close to -1 for an almost-perfect conductor. This allows us to rewrite (4.4.25) as

$$P_d \simeq \frac{1}{4}\left|\frac{2E_i}{\eta_i}\cos\theta_i\right|^2 \sqrt{\frac{2\omega\mu_0}{\sigma}} \qquad (4.4.29)$$

We now equate (4.4.28) and (4.4.29), because power must be conserved at the interface, yielding

$$1 - |\Gamma|^2 \simeq \sqrt{\frac{8\omega\epsilon_0}{\sigma}}\cos\theta_i$$

which is exactly the same result as that given by (4.4.24).

For the *TM* case, the power transmitted into the conductor is still given by (4.4.28), because duality simply interchanges the roles of $\overline{E}$ and $\overline{H}$. The power dissipated in the conductor is also given by (4.4.25), but the tangential $\overline{H}$ field at the conducting surface is now

$$\overline{H}_t = \hat{y}\, \frac{E_i}{\eta_i}(1-\Gamma) \simeq \hat{y}\, \frac{2E_i}{\eta_i}$$

with no angular dependence since the magnetic field is wholly tangential. The reflection coefficient is still approximately -1. Therefore, equating (4.4.25) and (4.4.28) yields a transmitted power fraction equal to

$$1 - |\Gamma|^2 \simeq \sqrt{\frac{8\omega\epsilon_0}{\sigma}} \frac{1}{\cos\theta_i} \qquad (4.4.30)$$

which is **not** the same as the power transmitted (dissipated) for *TE* waves. Equation (4.4.30) may also be derived by computing the *TM* reflection coefficient directly from (4.3.30) and (4.3.32), where k_{iz} and k_{tz} are the same as for the *TE* case, and the permittivities are given by $\epsilon_i = \epsilon_0$ and $\epsilon_t = \epsilon_0 - j\sigma/\omega \simeq -j\sigma/\omega$.

The expression (4.4.30) for the fraction of power transmitted by a *TM* wave is only valid for incident angles not close to $\theta_i = \pi/2$ where (4.4.30) has a singularity. For normal incidence, where the distinction between *TE* and *TM* waves becomes arbitrary, both (4.4.24) and (4.4.30) reduce to the same expression for transmitted power, which is a check on the validity of the two formulas. A comparison of (4.4.24) and (4.4.30) indicates that the *TM* radiation is preferentially absorbed by a good conductor, and that unpolarized light reflected from a conducting surface is increasingly *TE*-polarized as the angle of incidence is increased.

4.5* CURRENT SHEETS AT PLANAR BOUNDARIES

The previous sections in this chapter have discussed reflection and transmission from a planar boundary with plane wave-generating sources located infinitely far from the interface. This section discusses radiation from a sheet of current located on the boundary itself. In principle, current sheets may be thought of as aggregates of Hertzian dipoles arrayed in a plane and excited with appropriate magnitudes and phases, but this approach is postponed until Chapter 9 when radiation from apertures is discussed. The boundary conditions derived from Maxwell's equations provide an alternative, elegant method for solving current sheet problems. We gain insight into these current sheet radiation fields as an almost trivial consequence of deriving the Maxwell boundary conditions.

We begin by considering a $\hat{y}$-directed traveling wave surface current $\overline{J}_s$ (A/m) flowing in the x–y plane at $z = 0$, illustrated in Figure 4.18. This surface current has the form

$$\overline{J}_s(x, y, t) = \hat{y} \, J_{sA} \cos(\omega t - k_A x)$$

expressed in phasor notation as

$$\overline{J}_s = \hat{y} \, J_{sA} e^{-jk_A x} \qquad (4.5.1)$$

where J_{sA} and k_A are specified constants. To the left (region 1, $z < 0$) and right (region 2, $z > 0$) of the current sheet, there is vacuum. Simple reasoning based on Maxwell's equations suggests the form of the electric field. Magnetic fields

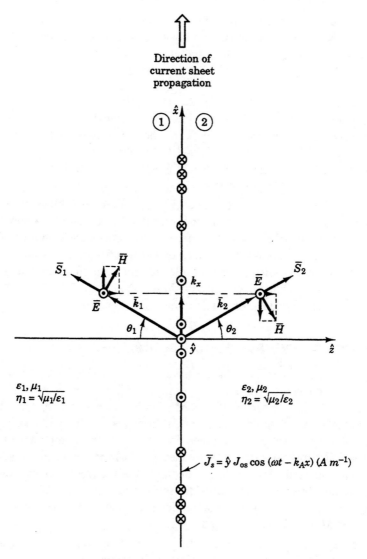

Figure 4.18 A radiating current sheet.

circulate around currents by Ampere's law, so we expect that $\overline{H}$ will lie in the x–z plane. Because the electric field circulates about this magnetic field by Faraday's law, the electric field must point in the $\hat{y}$ direction on both sides of the current sheet. But because $\overline{E} \times \overline{H}$ points in the direction of wave propagation, we also know that $\overline{k}$ must lie in the x–z plane. The expressions

$$\overline{E}_1(z < 0) = \hat{y}\, E_1\, e^{-jk_{1x}x - jk_{1z}z} \tag{4.5.2}$$

$$\overline{E}_2(z > 0) = \hat{y} \, E_2 \, e^{-jk_{2x}x - jk_{2z}z} \tag{4.5.3}$$

are thus the most general expressions for the electric field. Equations (4.5.2) and (4.5.3) are equal at $z = 0$ because the tangential components of the electric field are always continuous, and this leads to $E_1 = E_2 \equiv E_0$ and $k_{1x} = k_{2x} \equiv k_x$.

We can construct the magnetic field in each region using Faraday's law for plane waves (1.6.20):

$$\overline{H}_1 = \frac{\overline{k}_1 \times \overline{E}_1}{\omega \mu_0} = \frac{1}{\omega \mu_0}(k_x \hat{x} + k_{1z}\hat{z}) \times \hat{y} \, E_0 \, e^{-jk_x x - jk_{1z}z}$$

$$= \frac{E_0}{\omega \mu_0}(k_x \hat{z} - k_{1z}\hat{x}) \, e^{-jk_x x - jk_{1z}z} \tag{4.5.4}$$

$$\overline{H}_2 = \frac{\overline{k}_2 \times \overline{E}_2}{\omega \mu_0} = \frac{1}{\omega \mu_0}(k_x \hat{x} + k_{2z}\hat{z}) \times \hat{y} \, E_0 \, e^{-jk_x x - jk_{2z}z}$$

$$= \frac{E_0}{\omega \mu_0}(k_x \hat{z} - k_{2z}\hat{x}) \, e^{-jk_x x - jk_{2z}z} \tag{4.5.5}$$

The requirement that the normal component of $\overline{B} = \mu_0 \overline{H}$ be continuous at $z = 0$ is already satisfied by the $\hat{z}$ components of (4.5.4) and (4.5.5). The boundary condition derived from Ampere's law (4.1.16) relates the magnitude of the current to the tangential component of the magnetic field:

$$\overline{J}_s = \hat{y} \, J_{sA} \, e^{-jk_A x} = \hat{z} \times (\overline{H}_2 - \overline{H}_1)\Big|_{z=0} = \hat{y} \, \frac{(k_{1z} - k_{2z})}{\omega \mu_0} \, E_0 \, e^{-jk_x x} \tag{4.5.6}$$

We note that $k_{1z}^2 = k_{2z}^2$ because the dispersion relations for $z > 0$ and $z < 0$ are

$$k_x^2 + k_{1z}^2 = k_x^2 + k_{2z}^2 = \omega^2 \mu_0 \epsilon_0 \tag{4.5.7}$$

(Both waves travel in vacuum, and we have already shown $k_{1x} = k_{2x} \equiv k_x$.) Therefore, $k_{1z} = -k_{2z}$ in order that (4.5.6) be nonzero. Because (4.5.6) must be valid for all x, we note that $k_x = k_A$ and hence $J_{sA} = 2k_{1z}E_0/\omega\mu_0$. We can finally express all of the unknowns in (4.5.2)–(4.5.5) in terms of ω, μ_0, ϵ_0, J_{sA}, and k_A, resulting in

$$E_0 = \frac{\omega \mu_0 J_{sA}}{2k_{1z}}$$

$$k_{1x} = k_{2x} = k_A \tag{4.5.8}$$

$$k_{1z} = -k_{2z} = -\sqrt{\omega^2 \mu_0 \epsilon_0 - k_A^2}$$

where the sign of k_{1z} and k_{2z} is chosen to correspond to propagation **away** from the sheet. If $k_A^2 > \omega^2 \mu_0 \epsilon_0$, the waves generated are evanescent, and

$$k_{1z} = -k_{2z} = j\sqrt{k_A^2 - \omega^2 \mu_0 \epsilon_0} \tag{4.5.9}$$

where the sign of (4.5.9) is chosen so that the waves exponentially attenuate away from the boundary in both directions. For nonevanescent waves, the direction of wave propagation is given by the angle θ from the surface normal, where

$$\cos\theta = \frac{k_{2z}}{\sqrt{k_{2x}^2 + k_{2z}^2}} = \sqrt{1 - \left(\frac{ck_A}{\omega}\right)^2} \qquad (4.5.10)$$

The time-averaged power P radiated by the current sheet (W/m^2) can be evaluated by integrating the volume power density $-\frac{1}{2}\,\text{Re}\{\overline{E}\cdot\overline{J}^*\}$ over the thickness of the sheet. The negative sign in this expression is needed because $\frac{1}{2}\overline{E}\cdot\overline{J}^*$ corresponds to power expended by $\overline{E}$ on $\overline{J}$, but here we are computing the power expended by $\overline{J}$ on $\overline{E}$. Since $\overline{J} = \overline{J}_s\delta(z)$ is the mathematical expression for the relation between current density and surface current, we have

$$P = -\int_{-\infty}^{\infty}\frac{1}{2}\,\text{Re}\left\{\overline{E}\cdot\overline{J}^*\right\}dz = -\frac{1}{2}\overline{E}\cdot\overline{J}_s^*\bigg|_{z=0} \qquad (4.5.11)$$

Substituting (4.5.1) and (4.5.2) into (4.5.11), and using the values given in (4.5.8) and (4.5.10) to simplify the expression for radiated power yields

$$P = -\frac{1}{2}E_0 J_{sA}^* = -\frac{\omega\mu_0|J_{sA}|^2}{4k_{1z}} = \frac{1}{4}\eta_0\,|J_{sA}|^2\,\frac{1}{\cos\theta} \qquad (4.5.12)$$

For $k_A < \omega\sqrt{\mu_0\epsilon_0}$, the power radiated is real and positive, whereas for $k_A > \omega\sqrt{\mu_0\epsilon_0}$, the complex power is a $j\omega\mu_0|J_{sA}|^2/(4\sqrt{k_A^2 - \omega^2\mu_0\epsilon_0})$. No time-averaged power is radiated by the current sheet in this case; instead, there is an excess of stored magnetic energy (because the complex power is a positive imaginary quantity).

The time-averaged power radiated into the right half space ($z > 0$) is half that given by (4.5.12). It may also be found by calculating the real part of the $\hat{z}$ component of the Poynting vector:

$$\langle S_z\rangle = \frac{1}{2}\,\text{Re}\{\overline{E}\times\overline{H}^*\}\cdot\hat{z}$$

$$= \frac{1}{2}\,\text{Re}\{\hat{y}E_0 e^{-jk_Ax-jk_{2z}z}\}\times\left\{\frac{E_0^*}{\omega\mu_0}(k_0\hat{z} - k_{2z}^*\hat{x})\,e^{+jk_Ax+jk_{2z}^*z}\right\}\cdot\hat{z}$$

$$= \frac{1}{2}\,\text{Re}\left\{\frac{|E_0|^2}{\omega\mu_0}k_{2z}^*\,e^{-j(k_{2z}-k_{2z}^*)z}\right\} \qquad (4.5.13)$$

If k_{2z} is real ($\omega^2\mu_0\epsilon_0 > k_A^2$), then $k_{2z} = k_{2z}^*$ and the time-averaged, $\hat{z}$-directed Poynting vector is

$$\langle S_z\rangle = \frac{\omega\mu_0|J_{sA}|^2}{8\sqrt{\omega^2\mu_0\epsilon_0 - k_A^2}} = \frac{1}{8}\eta_0\,|J_{sA}|^2\,\frac{1}{\cos\theta} \qquad (4.5.14)$$

which confirms that half the radiated power given by (4.5.12) is indeed radiated into the right half space. The other half of the radiated power is carried by the wave propagating in the negative half space. For k_{2z} imaginary ($\omega^2 \mu_0 \epsilon_0 < k_A^2$), the $\hat{z}$ component of the complex Poynting vector is completely imaginary and therefore no average power is radiated into the vacuum, consistent with the fact that the waves are evanescent: $\langle S_z \rangle = 0$ in this case.

We conclude Section 4.5 by presenting a few more examples of Maxwell's boundary conditions involving current sheets.

Example 4.5.1

A planar current sheet at $z = 0$ has a surface current given by $\overline{J}_s(t) = \hat{y} \cos x \cos \omega t$ A/m. For $z < 0$ (medium 1), $\epsilon_1 = 9\epsilon_0$, and for $z > 0$ (medium 2), $\epsilon_2 = \epsilon_0$; $\mu_1 = \mu_2 = \mu_0$. The frequency $\omega/2\pi$ is 30 MHz. In what directions are waves radiated? If any of the waves are attenuated, give their penetration depths.

Solution: We start by writing $\overline{J}_s(x, t)$ as a phasor:

$$\overline{J}_s(x, t) = \text{Re}\left\{ \hat{y} \frac{1}{2} \left(e^{-jx} + e^{+jx} \right) e^{j\omega t} \right\}$$

so

$$\overline{J}_s(x) = \hat{y} \frac{1}{2} \left(e^{-jx} + e^{+jx} \right)$$

Because this problem is linear, we can consider the two oppositely directed current sheet waves separately. We initially choose to find the fields for $\overline{J}_{s1} = \hat{y} \frac{1}{2} e^{-jx}$. Because the tangential wave vectors must be continuous at the $z = 0$ boundary, we immediately apply the phase-matching condition to find

$$k_{1x} = k_{2x} = k_A$$

where $\left\{ {k_{1x} \atop k_{2x}} \right\}$ is the $\hat{x}$ component of $\overline{k}$ for $\left\{ {z<0 \atop z>0} \right\}$ and $k_A = 1$ m^{-1} is the wavenumber of the current sheet.

But we also know the dispersion relations in both regions 1 and 2 are

$$k_{1x}^2 + k_{1z}^2 = k_1^2$$
$$k_{2x}^2 + k_{2z}^2 = k_2^2$$

and since $k_i = \omega\sqrt{\mu_i \epsilon_i}$ where i may be 1 or 2, we find that

$$k_{1z}^2 = \left(\frac{3\omega}{c} \right)^2 - 1^2$$

$$k_{2z}^2 = \left(\frac{\omega}{c} \right)^2 - 1^2$$

But

$$\frac{\omega}{c} = \frac{2\pi \cdot 30 \times 10^6}{3 \times 10^8} = \frac{\pi}{5}$$

so we see that $k_{1z}^2 > 0$ but $k_{2z}^2 < 0$. Therefore, the waves in medium 1 (the dielectric) propagate with an angle θ above the $\hat{z}$-axis (where θ is exactly the same angle defined in Figure 4.18):

$$\sin \theta = \frac{k_{1x}}{k_1} = \frac{5}{3\pi}; \qquad \theta = 32°$$

The waves in the vacuum are evanescent, with penetration depth given by

$$\Delta = 1/\text{Im}\{k_{1z}\} = \left[1 - \left(\frac{\pi}{5}\right)^2\right]^{-1/2} = 1.29 \text{ m}$$

The part of the current sheet given by $\overline{J}_{s2} = \hat{y} \frac{1}{2} e^{+jx}$ gives exactly the same results discussed above for $\overline{J}_{s1}$ except that the current now propagates in the $-\hat{x}$ direction. This means that the angle of propagation of the waves radiated into medium 1 is now θ degrees **below** the $\hat{z}$-axis, as shown in Figure 4.19. The waves in medium 2 are still evanescent, with the same penetration depth, since only the magnitude of the current sheet wave vector is important in determining Δ. The sum of these two solutions for $\overline{J}_{s1}$ and $\overline{J}_{s2}$ has a standing wave structure in the $\hat{x}$ direction. If we wished to find all of the fields associated with this current sheet, we could find $\overline{E}_1$ and $\overline{H}_1$ from $\overline{J}_{s1}$ as already shown earlier in the section, and then find $\overline{E}_2$ and $\overline{H}_2$ from $\overline{J}_{s2}$ by letting $k_A \to -k_A$. The total solution is given by $\overline{E}_1 + \overline{E}_2$ and $\overline{H}_1 + \overline{H}_2$.

Example 4.5.2

Currents can be induced by plane waves incident upon conducting sheets that are thin compared to the skin depth $\delta = \sqrt{2/\omega\mu\sigma}$. A uniform plane wave in vacuum is normally incident from the left upon a membrane at $z = 0$ of conductance $\beta = \sigma d$ S, where d is the

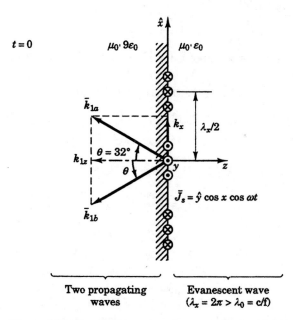

Figure 4.19 Propagating and evanescent waves generated by a current sheet.

thickness of the membrane and the surface current (A/m) is:

$$\overline{J}_s = \beta\overline{E}\big|_{z=0} \quad \text{(A/m)}$$

If $\overline{E}$ lies in the plane of the conductor (i.e., *TE* polarization), what conductivity β maximizes the power absorbed, and what waves result?

Solution: Physically, we expect that the incident wave $\overline{E}_i\,e^{-jkz}$ ($z < 0$) will induce current on the conducting membrane at $z = 0$ and that this current will in turn give rise to waves that propagate away from the sheet. We suppose that the incident $\overline{E}$ field is $\hat{y}$-directed and the incident and induced fields to the left of the boundary are

$$\overline{E} = \hat{y}\,E_i\,e^{-jkz} + \hat{y}\,E_r\,e^{+jkz} \qquad (z < 0)$$

while the induced field to the right of the boundary is

$$\overline{E} = \hat{y}\,E_t\,e^{-jkz} \qquad\qquad (z > 0)$$

The wave numbers are all identical because each wave is propagating at normal incidence in the same free space medium: $k = \omega/c$. Continuity of the tangential electric field at $z = 0$ means that $E_i + E_r = E_t$. The magnetic fields may be found by Ampere's law for plane waves (1.6.20)

$$\overline{H}(z < 0) = \frac{\overline{k}\times\overline{E}}{\omega\mu_0} = -\hat{x}\,\frac{E_i}{\eta_0}\,e^{-jkz} + \hat{x}\,\frac{E_r}{\eta_0}\,e^{+jkz}$$

$$\overline{H}(z > 0) = \frac{\overline{k}\times\overline{E}}{\omega\mu_0} = -\hat{x}\,\frac{E_t}{\eta_0}\,e^{-jkz}$$

where $\eta_0 = \omega\mu_0/k$. The discontinuity in the tangential magnetic field at $z = 0$ gives rise to a surface current $\overline{J}_s$ and we apply (4.1.16) to find

$$\overline{J}_s = \hat{z}\times[\overline{H}(z > 0) - \overline{H}(z < 0)]\big|_{z=0} = \hat{y}\,\frac{1}{\eta_0}(-E_t + E_i - E_r)$$

But since $\overline{J}_s = \beta\overline{E}\big|_{z=0}$, we see that

$$\frac{1}{\eta_0}(-E_t + E_i - E_r) = \beta E_t$$

Recalling that $E_i + E_r = E_t$, we now have two equations in the two unknowns E_t and E_r, which are solved to give

$$E_t = \frac{E_i}{1 + \beta\eta_0/2}$$

$$E_r = \frac{(-\beta\eta_0/2)E_i}{1 + \beta\eta_0/2}$$

The power per unit area dissipated in the conducting sheet is given by

$$P_d = \frac{1}{2}\,\text{Re}\left\{\overline{E}\cdot\overline{J}_s^*\right\}\big|_{z=0}$$

$$= \frac{1}{2}\beta\,|E|^2\big|_{z=0} = \frac{1}{2}\beta\,|E_t|^2$$

Therefore,

$$P_d = \frac{\frac{\beta}{2}|E_i|^2}{(1+\beta\eta_0/2)^2}$$

which we can maximize by setting $\partial P_d/\partial\beta = 0$; we find that $\beta = 2/\eta_0$. Plugging $\beta = 2/\eta_0$ back into the expression for dissipated power yields $P_d = |E_i|^2/4\eta_0$, which is half the incident power density.

4.6 SUMMARY

The integral form of Maxwell's equations is found by judicious use of *Gauss's* and *Stokes's theorems* on the differential form of Maxwell's equations. Boundary conditions at an interface follow by applying the integral form of Maxwell's equations to the boundary between two media. The Maxwell boundary conditions relate the fields on either side of a surface and are compactly expressed by

$$\hat{n} \times (\overline{E}_1 - \overline{E}_2) = 0$$
$$\hat{n} \times (\overline{H}_1 - \overline{H}_2) = \overline{J}_s$$
$$\hat{n} \cdot (\overline{D}_1 - \overline{D}_2) = \sigma_s$$
$$\hat{n} \cdot (\overline{B}_1 - \overline{B}_2) = 0$$

The unit normal $\hat{n}$ points from medium 2 to medium 1, and the subscripts on the fields refer to each of the two media. The fields are evaluated right at the boundary edge. The *surface current* is given the symbol $\overline{J}_s$ (A/m) and σ_s is the *surface charge* (C/m^2). These boundary conditions state that the tangential electric field and normal $\overline{B}$ field are always continuous, while discontinuities in normal $\overline{D}$ give rise to surface charge, and discontinuities in tangential $\overline{H}$ give rise to surface currents.

If the waves on either side of the boundary are planar and the boundary surface is planar, then the field expressions are quite simple. For a *plane wave* propagating at an arbitrary direction $\overline{k}$, we can write

$$\overline{E}(\overline{r}) = \overline{E}_i\, e^{-j\overline{k}\cdot\overline{r}} = \overline{E}_i\, e^{-jk_x x - jk_y y - jk_z z}$$

The wave number $\overline{k}$ may be a complex vector $\overline{k}' - j\overline{k}''$ where both $\overline{k}'$ and $\overline{k}''$ are real but not necessarily parallel. The real part of the wave vector $\overline{k}'$ points in the direction of wave propagation with $\lambda = 2\pi/|\overline{k}'|$, and the imaginary part $\overline{k}''$ points in the direction of maximum attenuation with wave *penetration depth* equal to $1/|\overline{k}''|$. If the medium is lossless, then $\overline{k}' \cdot \overline{k}'' = 0$. Substitution of $\overline{E}(\overline{r})$ into the wave equation yields

$$\overline{k} \cdot \overline{k} = \omega^2\mu\epsilon = k_x^2 + k_y^2 + k_z^2$$

where μ and ϵ may be complex. Such waves must possess the same phase variation along the boundary, so the tangential components of $\overline{k}$ must be continuous across the boundary:

$$\hat{n} \times (\overline{k}_1 - \overline{k}_2) = 0$$

This *phase-matching condition* in isotropic media requires that

$$\theta_i = \theta_r$$

and

$$\frac{\sin \theta_i}{\sin \theta_t} = \frac{n_t}{n_i} = \sqrt{\frac{\mu_t \epsilon_t}{\mu_i \epsilon_i}}$$

where θ_i, θ_r, and θ_t are the *angles of incidence, reflection*, and *transmission* at a boundary. This last equation is known as *Snell's law*. In a medium where $n_i > n_t$, it is possible that Snell's law cannot be satisfied for real values of θ_t. This happens when $\theta_i > \theta_c = \sin^{-1} \sqrt{n_t/n_i}$, where θ_c is called the *critical angle*. If the angle of incidence exceeds the critical angle, then *total internal reflection* results and no waves propagate into the second medium. Instead, waves propagate in the transmitting medium parallel to the boundary surface with an amplitude that decays exponentially away from the boundary. The rate of exponential attenuation α is

$$\alpha = \sqrt{k_{t_x}^2 - k_t^2} = \sqrt{k_i^2 \sin^2 \theta_i - k_t^2}$$

and the wave in the transmitting medium has the form $e^{-jk_{t_x}x - \alpha z}$, where $\hat{z}$ is the surface normal and the waves propagate in the $\hat{x}$ direction near the surface of the boundary. If the transmitting medium is lossless, these surface waves are said to be *evanescent*.

Once the phase-matching condition is met, the rest of Maxwell's boundary conditions may be applied to determine the *reflection coefficient* Γ, and the *transmission coefficient* T at a planar interface. These are generally different functions of θ_i for TE and TM waves, where for $\begin{Bmatrix} TE \\ TM \end{Bmatrix}$ waves the $\begin{Bmatrix} \text{electric} \\ \text{magnetic} \end{Bmatrix}$ field is perpendicular to the plane containing the direction of propagation $\overline{k}$ and the surface normal. We find that

$$\Gamma = \frac{Z_n^{TE} - 1}{Z_n^{TE} + 1}$$

$$T = \frac{2Z_n^{TE}}{Z_n^{TE} + 1}$$

$$Z_n^{TE} = \frac{\mu_t / k_{tz}}{\mu_i / k_{iz}}$$

and

$$\Gamma = -\frac{Y_n^{TM} - 1}{Y_n^{TM} + 1}$$

$$T = \frac{2}{Y_n^{TM} + 1}$$

$$Y_n^{TM} = \frac{\epsilon_t / k_{tz}}{\epsilon_i / k_{iz}}$$

where $Y_n^{TM} = 1/Z_n^{TM}$.

At *Brewster's angle*, the reflection coefficient is zero. For media in which $\epsilon_i \neq \epsilon_t$ but $\mu_i = \mu_t$, the Brewster's angle exists only for *TM* modes and is equal to

$$\theta_B = \tan^{-1} \sqrt{\epsilon_t / \epsilon_i}$$

For media in which $\mu_i \neq \mu_t$ but $\epsilon_i = \epsilon_t$, Brewster's angle exists only for *TE* modes.

Current sheets may also be located at boundaries where they may radiate, scatter, or absorb power. For surfaces along which the phase variation of the current is sinusoidal [i.e., $\overline{J} = \overline{J}_{sA} e^{-jk_A x} \delta(z)$], it is particularly simple to find the $\overline{E}$ and $\overline{H}$ fields on either side of the surface because they have the form of plane waves. Boundary conditions are then used to determine the magnitudes and directions of each of the fields.

4.7 PROBLEMS

4.1.1 An exotic material has the properties $\mu = \infty$, $\sigma = 0$, and $\epsilon = 0$. What boundary conditions does such a surface impose on $\overline{E}$, $\overline{H}$, $\overline{B}$, and $\overline{D}$ in an adjacent free space region at nonzero frequencies?

4.1.2 Repeat Problem 4.1.1 for $\epsilon = \infty$, $\mu = 0$, and $\sigma = 0$.

4.2.1 Ambient light in a room can produce annoying reflections from CRT displays. Prepare a simple plot showing the percent power reflected at normal incidence from a flat plate of permittivity ϵ for $\epsilon_0 < \epsilon < 10\epsilon_0$. Note that ϵ/ϵ_0 is approximately 1.0006, 2.5, 3.75, 6.00, and 9.5 for air, lucite, fused quartz, Pyrex 1710, and Corning 8879, respectively.

4.2.2 Narcissus looked at his reflection in a pool of water with $\mu = \mu_0$, $\epsilon = 1.8\epsilon_0$. What fraction of the light flux density ($\lambda = 0.5 \times 10^{-6}$ m) incident upon the water was reflected at normal incidence? What standing wave ratio would be measured at this boundary?

4.2.3 A uniform plane wave of intensity 1 W/m^2 in free space is normally incident on a planar perfect conductor located at $z = 0$. The total electric field is

$$\overline{E} = (\hat{x} + j\hat{y}) E_0 e^{-j\pi z} + \overline{E}_r$$

(a) What is the frequency f of this wave?
(b) Evaluate E_0 quantitatively.

(c) What is the reflected term $\overline{E}_r$ in the expression for the total electric field? Give your answer in the same format as that of the incident wave.

(d) What is the polarization of the reflected wave?

(e) At $z = -\lambda/2$, what is the time-dependent magnetic field $\overline{H}(x, y, z, t)$?

4.2.4 What is the highest frequency at which 99.9 percent of the energy of an incident plane wave would be reflected from a flat copper slab (thick compared to a skin depth) having conductivity $\sigma = 5 \times 10^7$ S/m? What is the frequency for 99 percent reflection?

4.2.5 Complex optical systems require low-loss mirrors because the signals may be reflected many times. Consider a polished silver mirror with $\sigma = 5 \times 10^7$ S/m.

(a) What is the skin depth δ in the silver for infrared light of 10 μm wavelength?

(b) What is the wavelength λ inside the silver?

(c) What fraction of the power in this infrared beam would be reflected at normal incidence?

4.3.1 A uniform laser beam in free space is normally incident on a glass surface that reflects nine percent of the power and transmits the rest.

(a) What is the permittivity ϵ of the glass?

(b) If the beam is incident at $60°$, what is the angle of transmission θ_t?

4.3.2 A 10-W $\hat{y}$-polarized uniform plane wave in free space is incident upon a planar interface at $\theta_i = 60°$; the plane of incidence is the x-z plane. The medium on the other side of the interface ($z > 0$) is characterized by $\epsilon = \epsilon_0/2$ and $\mu = \mu_0$.

(a) What is the direction of the transmitted electric field?

(b) Give the magnitude of the transmitted electric field and its phase relative to the incident field.

4.3.3 A plane wave is incident at $45°$ upon a perfectly conducting surface in the y-z plane. The plane wave consists of both TE and TM components as follows:

$$\overline{E}^{TE} = \hat{y} E_0 e^{jk(x-z)/\sqrt{2}}$$

$$\overline{E}^{TM} = (\hat{x} + \hat{z}) \frac{j E_0}{\sqrt{2}} e^{jk(x-z)/\sqrt{2}}$$

(a) Write the TE and TM components of the electric fields of the reflected wave in the same form used above in describing the incident wave, and show that the tangential electric fields satisfy the boundary conditions.

(b) What are the polarizations of the incident and the reflected waves?

4.3.4 A 1-MHz uniform plane wave is incident on a planar plasma boundary at $60°$ with $\mu = \mu_0$.

(a) If no wave is reflected, what is the dielectric constant of the transmitting medium?

(b) If no wave is transmitted, what is the minimum plasma frequency of the transmitting medium?

4.3.5 Light can be trapped inside optical fibers if it is reflected internally at angles θ greater than the critical angle θ_c. If the wavelength is small compared to the fiber radius, the reflection is approximately planar. The field strength in the evanescent region decays exponentially away from a planar boundary as $e^{-\alpha z}$, where $\hat{z}$ is perpendicular to the surface. What is α for

(a) $\theta = \theta_c$?

(b) $\theta = \theta_c + \frac{1}{2}(\frac{\pi}{2} - \theta_c)$

(c) the limit where $\theta = \pi/2$? The angle θ is measured in radians in parts (a)–(c).

4.3.6 Some gas lasers use "Brewster angle" quartz windows on the gas discharge tube in order to minimize reflection losses. They are simply thin pieces of glass set at angle θ with respect to the laser beam. Determine the angle θ if the index of refraction for quartz at the wavelength of interest is $n = 1.46$. Because of these windows, the laser output is almost completely linearly polarized. What is the direction of this polarization, Why?

4.3.7 A uniform plane wave in free space is incident at angle θ on a flat dielectric ϵ, where $\epsilon > \epsilon_0$. The planar boundary of the dielectric lies at $z = 0$ and $\hat{z}$ is normal to the surface, pointing into space. The electric field vector of the incident wave may be expressed as

$$\overline{E}_i = E_0\left(\hat{x}\cos\theta + \hat{y} + \hat{z}\sin\theta\right)e^{-jk_x x + jk_z z}$$

where E_0 is a real constant.

(a) Show that this incident electric field satisfies Gauss's Law.

(b) Decompose the incident electric field into its TE and TM components.

(c) What is the polarization of the incident wave?

(d) Determine the expression for the reflected electric field in terms of Γ^{TE} and Γ^{TM}.

(e) For incident angles less than the critical angle θ_c, what are the polarizations of the reflected and transmitted waves?

(f) For incident angles greater than the critical angle θ_c, show that the phase difference between the TE and TM components of the reflected electric field is given by

$$\Delta\phi = \phi^{TM} - \phi^{TE} = 2\tan^{-1}\left[\frac{\cos\theta\left[\sqrt{\sin^2\theta - n^{-2}}\right]}{\sin^2\theta}\right]$$

where n is the refractive index defined by $c\sqrt{\mu\epsilon}$. You may find the following identity useful:

$$\tan(A + B) = \frac{\tan A + \tan B}{1 - \tan A \tan B}$$

(g) If $\epsilon = 6\epsilon_0$, for what angle θ is the reflected wave circularly polarized?

4.3.8 Assume that planet Rigel-3 has a very thick ionosphere with an electron density N_e (electrons/m^3) $= 10^7 \cdot h$, where h is the altitude in meters above the planet surface.

(a) Is there an altitude h_0 above which a 9-MHz signal cannot propagate? If so, find h_0.

(b) A 9-MHz signal is transmitted upward from the planet's surface at 45°. It is then gradually refracted by the ionosphere and eventually returns to Rigel-3. What is the highest altitude h_1 which this signal can reach before returning?

4.3.9 The ionosphere extends from approximately 50 km above the Earth to altitudes many times higher where it merges with the magnetosphere. The maximum ionization density is at about 300 km. The ionization density profile shows ledges where it varies more slowly with altitude, labelled the C, D, E, F_1, and F_2 layers, as shown in Figure 4.20. These maxima arise because both the solar radiation and the composition of the atmosphere change with altitude. The heights and the intensities of ionization of these layers change with the hour of the day, the season of the year, the sunspot cycle, an so on. The electron density varies from approximately 10^7 m^{-3} to 10^{12} m^{-3} from the lowest to the highest layer. For simplicity, assume that the ionosphere consists of a 40-km thick E-layer with electron density $N = 10^{11}$ m^{-3} below a 200-km thick F-layer with $N = 6 \times 10^{11}$ m^{-3}.

(a) What are the plasma frequencies of the E and F layers?
(b) Consider a plane wave of 10 MHz incident at angle θ upon the ionosphere from below the E-layer. What are the transmitted angles of the ionospheric wave in the E and F layers?
(c) Let $\theta = 30°$. Below what frequency will the wave be totally reflected by the E-layer, and below what frequency will it be totally reflected by the F-layer?

4.4.1 A *TE* wave is incident at angle θ upon a thick flat copper sheet of $\sigma = 4 \times 10^7$ S/m.

(a) If $\theta = 45°$, approximately what is θ_t, the angle (from normal) of the transmitted wave? Assume ϵ_0, μ_0 everywhere, and $\omega = 10^7$ rad/s.

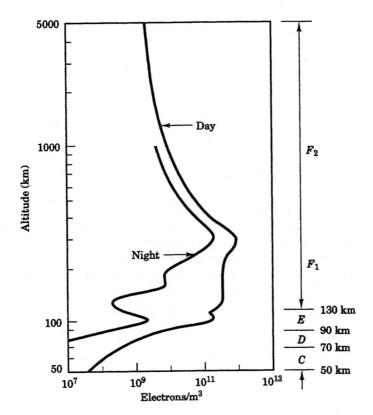

Figure 4.20   Ionization density profile.

(b) In this high-conductivity case, what is the approximate value of $|\overline{H}_z|$ just inside the slab if the original $\hat{y}$-polarized uniform wave in free space has power density 377 W/m^2? What is $\overline{J}_s$, the effective surface current density (A/m)?

(c) What is the approximate power (W/m^2) absorbed by the copper and what fraction of the incident power does this represent?

4.4.2 Sea water can be characterized at frequencies between 60 Hz and 10 MHz by a conductivity $\sigma = 4$ S/m, permittivity $\epsilon = 80\epsilon_0$, and permeability $\mu = \mu_0$.

(a) An incident, vertically polarized 100-Hz plane wave from a distant transmitter propagates parallel to the ocean surface. At what approximate angle does this wave propagate into the ocean?

(b) What is the penetration depth of this wave in meters?

(c) What is $|\overline{E}|$ (V/m) above the surface and also at a depth of 60 m, where an under-sea mining vehicle is trailing a long antenna? This receiver requires 10^{-5} V/m in order to work. Will the vehicle receive these 100-Hz messages if $|\overline{E}|$ above the surface is 1 V/m?

4.4.3 A flat steel radar reflector has conductivity $\sigma = 0.6 \times 10^5$ S/m. At $f = 100$ GHz, what fraction of the incident power is absorbed by the reflector (i) at normal incidence? (ii) for a *TE* wave at $\theta_i = 80°$?

4.4.4* A uniform plane wave in a magnetic medium (with $\mu = 6.25\mu_0$, $\epsilon = \epsilon_0$) is incident upon a planar interface at $z = 0$ at incidence angle θ_i. The magnetic field $\overline{H}$ for $z > 0$ in free space is

$$\overline{H} = \hat{y}\, e^{-6z - j10x} \qquad (z > 0)$$

(a) Is the incident wave *TE* or *TM*? Explain your reasoning.

(b) What is the frequency ω (rad/s)?

(c) What is the numerical value of θ_i?

(d) If the medium for $z > 0$ is slightly lossy (μ_0, ϵ_0; $\sigma \ll \omega\epsilon_0$), the unperturbed propagation vector $\overline{k}_0$ for $z > 0$ would become $\overline{k} = \overline{k}_0 + \overline{k}_p$, where $\overline{k}_p = \overline{k}' - j\overline{k}''$ is the perturbation wave vector and $|\overline{k}_p| \ll |\overline{k}_0|$. In what direction would the imaginary part of the perturbation vector $\overline{k}''$ point? Explain your reasoning (a simple argument suffices).

4.5.1 A certain phased-array antenna can be modeled as a current sheet $\overline{J}_s = \hat{x}\, e^{-jk_z z}$ (A/m) in the x–z plane where $k_z = 2k_0 = 8\pi$ where $k_0 = \omega/c$ is the free space wavenumber.

(a) In what direction(s) does the antenna radiate?

(b) What are $\overline{E}$ and $\overline{H}$ for the wave radiated on the $+\hat{y}$ side of the antenna?

4.5.2* A $\hat{z}$-directed current sheet is placed into a conducting dielectric ($\epsilon = 4\epsilon_0$, μ_0, $\sigma \gg \omega\epsilon$) at $x = 0$. The electric field for $x > 0$ is

$$\overline{E}(x > 0) = \hat{z}\, E_0\, e^{-jk_x x}$$

(a) Find the magnitude and phase of the current sheet surface current $\overline{J}_s$.

(b) How far does the electric field penetrate into the medium before decaying to $1/e$ of its amplitude at $x = 0$?

5

TRANSMISSION LINES

5.1 INTRODUCTION

Transmission lines are structures used to transmit energy or information from one place to another and therefore find many applications in fields ranging from power electronics to computers and communications. The electromagnetic waves supported by transmission lines generally propagate only in the $\pm \hat{z}$ direction, where z is the coordinate along the length of the structure, and thus have $e^{\mp jkz}$ dependence. Usually, the transmission line has a uniform cross-section (i.e., geometry independent of z), which makes the mathematical analysis of transmission line theory much more tractable. More than one transmission line may be interconnected and other types of circuit elements (e.g., capacitors, inductors, transistors) may be part of a transmission line network.

In the following two chapters, we restrict ourselves to solutions of Maxwell's equations in transmission lines for which $E_z = H_z = 0$. Both the electric and magnetic fields must then be transverse (perpendicular) to the direction of wave propagation $\hat{z}$; such waves are called *transverse electromagnetic* or *TEM* waves.

Chapter 5 describes the propagation of *TEM* waves on transmission line structures—the simplest example being a pair of parallel conducting strips—where we shall find that a voltage $V(z)$ and current $I(z)$ can be meaningfully defined. The differential equations that relate $V(z)$ and $I(z)$, known as the *transmission line equations*, are obeyed by waves on all other *TEM* structures. Wave propagation on arbitrary *TEM* lines of infinite extent are also discussed for both the time and frequency domains. Signals that vary sinusoidally in time are easiest to analyze, and these single-frequency waves form the backbone of *TEM* line theory. On the other hand, transient (nonsinusoidally time-dependent) behavior is important in many practical systems, including computers and power lines, and is often more straightforwardly analyzed in the time domain. Periodic structures incorporating passive lumped elements, such as capacitors, inductors, and resistors, can support *TEM* waves and complete the description of transmission lines in this chapter. Chapter 6 extends the boundaryless transmission line theory described in Chapter 5 through an exploration of the behavior of multiply connected transmission line systems, which may also incorporate passive elements.

5.2 PARALLEL-PLATE LINES

Perhaps the simplest *TEM* structure is the *parallel-plate transmission line* consisting of two long, perfectly conducting parallel plates, as illustrated in Figure 5.1. Any field solution that exists between the conductors must satisfy Maxwell's equations for a uniform medium characterized by ϵ and μ, as well as satisfying the boundary conditions imposed by the conducting plates, (4.1.19)–(4.1.22):

$$\hat{n} \times \overline{E} = 0 \tag{5.2.1}$$

$$\hat{n} \times \overline{H} = \overline{J}_s \tag{5.2.2}$$

$$\hat{n} \cdot \overline{D} = \sigma_s \tag{5.2.3}$$

$$\hat{n} \cdot \overline{B} = 0 \tag{5.2.4}$$

We recall that $\hat{n}$ is the unit vector normal to the conducting surface, where $\hat{n} = \hat{x}$ at $x = 0$ and $\hat{n} = -\hat{x}$ at $x = d$.

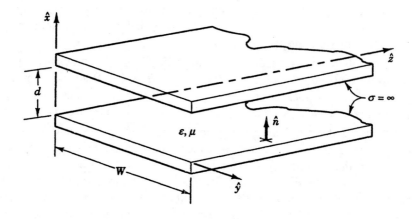

Figure 5.1 Parallel-plate transmission line.

We may neglect the fringing fields at the edges of the structure if the plate separation d is much smaller than the plate width W, which greatly simplifies the mathematical analysis. The directions in which $\overline{E}$ and $\overline{H}$ point are easily determined by reasoning on physical grounds. Because the electric field can have no tangential components at the plates, we guess that $\overline{E}$ is $\hat{x}$-directed. If the wave is to propagate in the $\hat{z}$ direction, then $\overline{H}$ must be $\hat{y}$-directed, which is also consistent with having a surface current on each of the plates in the $\hat{z}$ direction. Therefore, we can see that both Maxwell's equations and the boundary conditions (5.2.1)–(5.2.4) are satisfied by the fields

$$\overline{E} = \hat{x}\, E_0\, e^{-jkz} \tag{5.2.5}$$

and

$$\overline{H} = -\frac{\nabla \times \overline{E}}{j\omega\mu} = \frac{\overline{k} \times \overline{E}}{\omega\mu} = \hat{y}\,\frac{k}{\omega\mu}\, E_0\, e^{-jkz} \tag{5.2.6}$$

$$\overline{k} = \hat{z}\, k$$

This wave is simply an $\hat{x}$-polarized uniform plane wave bounded by the conducting plates, where

$$\frac{k}{\omega\mu} = \frac{\omega\sqrt{\mu\epsilon}}{\omega\mu} = \sqrt{\epsilon/\mu} = 1/\eta \tag{5.2.7}$$

Here, η is the characteristic wave impedance of the medium, which is approximately 377 Ω for free space.

The boundary conditions (5.2.2) and (5.2.3) yield the surface charge density σ_s (C/m^2) and the surface current $\overline{J}_s$ (A/m):

$$\begin{aligned}
\sigma_s &= \left. \hat{x} \cdot \overline{D} \right|_{x=0} = \epsilon E_0\, e^{-jkz} \quad \text{at } x = 0 \\
\sigma_s &= \left. -\hat{x} \cdot \overline{D} \right|_{x=d} = -\epsilon E_0\, e^{-jkz} \quad \text{at } x = d
\end{aligned} \tag{5.2.8}$$

and

$$\begin{aligned}
\overline{J}_s &= \left. \hat{x} \times \overline{H} \right|_{x=0} = \hat{z}\,\frac{1}{\eta}\, E_0\, e^{-jkz} \quad \text{at } x = 0 \\
\overline{J}_s &= \left. -\hat{x} \times \overline{H} \right|_{x=d} = -\hat{z}\,\frac{1}{\eta}\, E_0\, e^{-jkz} \quad \text{at } x = d
\end{aligned} \tag{5.2.9}$$

The total current flowing in the $+\hat{z}$ direction along the plate at $x = 0$ is

$$I(z) = \int_0^W \overline{J}_s(z) \cdot \hat{z}\, dy = H_y(z)\, W \tag{5.2.10}$$

and the total current in the top plate at $x = d$ is equal and opposite. The total voltage $V(z)$ between the two plates can be defined as

$$V(z) = \int_0^d \overline{E}(z) \cdot \hat{x}\, dx = E_x(z)\, d \tag{5.2.11}$$

Caution should be exercised here because the value of the integral (5.2.11) might be different for an alternate integration path, as suggested in Figure 5.2 for the paths ab and dc. Faraday's law (4.1.5) for *TEM* waves ($E_z = B_z \equiv 0$) yields

$$\oint_{abcd} \overline{E} \cdot d\overline{\ell} = -j\omega \int_A \overline{B} \cdot \hat{z}\, da = 0$$

because $\overline{B}$ points in the $\hat{y}$ direction. We may write integrals for each leg of the contour separately:

$$\int_a^b \overline{E} \cdot d\overline{\ell} + \int_b^c \overline{E} \cdot d\overline{\ell} + \int_c^d \overline{E} \cdot d\overline{\ell} + \int_d^a \overline{E} \cdot d\overline{\ell} = 0 \qquad (5.2.12)$$

But the electric field parallel to the conducting plates is zero by (5.2.1), and therefore

$$\int_b^c \overline{E} \cdot d\overline{\ell} = \int_d^a \overline{E} \cdot d\overline{\ell} = 0$$

Equation (5.2.12) therefore yields

$$V(z) = \int_a^b \overline{E} \cdot d\overline{\ell} = \int_d^c \overline{E} \cdot d\overline{\ell} \qquad (5.2.13)$$

where we note that the integration limits have been reversed to absorb a negative sign in the latter term of (5.2.13). Therefore, $V(z)$ is independent of the integration path (and hence uniquely defined) as long as the path lies in a plane perpendicular to the $\hat{z}$-axis. Note that $V(z)$ is a function of position along the guide and that in general

$$V(z_1) \neq V(z_2) \qquad (5.2.14)$$

even though perfect conductors exist at $x = 0$ and $x = d$. The reason that the conducting surface is not an equipotential in the $\hat{z}$ direction is that Faraday's law

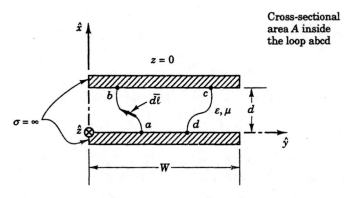

Figure 5.2   Cross-section of a parallel-plate transmission line.

integrated around a loop in the x–z plane encloses time-varying magnetic fields ($H_y \neq 0$). We should therefore expect the voltage to vary with z if the fields are not completely static (i.e., $\omega \neq 0$).

It is useful to reformulate Maxwell's curl equations for the fields (5.2.5) and (5.2.6) in terms of V and I. Faraday's and Ampere's laws (1.4.4) and (1.4.5) in the source-free region between the plates

$$\nabla \times \overline{E} = -j\omega\mu\overline{H}$$

$$\nabla \times \overline{H} = j\omega\epsilon\overline{E}$$

become, respectively,

$$\hat{y}\frac{\partial}{\partial z}E_x(z) = -j\omega\mu\,\hat{y}\,H_y(z) \tag{5.2.15}$$

$$-\hat{x}\frac{\partial}{\partial z}H_y(z) = j\omega\epsilon\,\hat{x}\,E_x(z) \tag{5.2.16}$$

Integrating (5.2.15) over x from 0 to d and (5.2.16) over y from 0 to W yields

$$\frac{\partial}{\partial z}E_x d = -j\omega\mu d H_y \tag{5.2.17}$$

$$\frac{\partial}{\partial z}H_y W = -j\omega\epsilon W E_x \tag{5.2.18}$$

Using the definitions (5.2.10) and (5.2.11) for $I(z)$ and $V(z)$, (5.2.17) and (5.2.18) become

$$\frac{dV(z)}{dz} = -j\omega L I(z) \tag{5.2.19}$$

$$\frac{dI(z)}{dz} = -j\omega C V(z) \tag{5.2.20}$$

where

$$L = \mu d / W = \text{inductance per unit length (H/m)} \tag{5.2.21}$$

and

$$C = \epsilon W / d = \text{capacitance per unit length (F/m)} \tag{5.2.22}$$

Equations (5.2.21) and (5.2.22) should be familiar expressions from quasistatics.[1] The two equations (5.2.19) and (5.2.20) are called the *TEM transmission line*

---

1. We recall that for a quasistatic parallel-plate capacitor, the electric field points from one plate to the other and has a magnitude $|\overline{E}| = V/d$, where V is the voltage between the plates and d is the plate separation distance. The total charge on each plate of length ℓ and width W is then $q = \sigma_s \ell W$, where $\sigma_s = \epsilon|\overline{E}|$. Therefore, the total capacitance of the plates is $C = q/V = \epsilon \ell W/d$, and the capacitance per unit length is $C = C/\ell = \epsilon W/d$. The inductance $\mathcal{L}$ is calculated from the ratio of the magnetic flux density $\Lambda = \int_A \overline{B} \cdot \hat{n}\,da$ and the current I, where $\Lambda = \mu|\overline{H}|\ell d$ and $I = |\overline{H}|W$. (The magnetic field points along the width W and the current along the length ℓ.) The inductance is then $\mathcal{L} = \mu\ell d/W$, which means that the inductance per unit length is $L = \mathcal{L}/\ell = \mu d/W$.

equations or *telegrapher's equations*. We shall see in Section 5.3 that **any** two-conductor *TEM* structure of arbitrary but uniform cross-section also obeys these transmission line equations.

If we differentiate (5.2.19) with respect to z and then replace dI/dz by (5.2.20), we find that

$$\frac{d^2 V(z)}{dz^2} = -\omega^2 LC V(z) \tag{5.2.23}$$

which is a homogeneous wave equation for the voltage. Differentiating (5.2.20) with respect to z and substituting (5.2.19) into the resulting expression yields a similar current wave equation:

$$\frac{d^2 I(z)}{dz^2} = -\omega^2 LC I(z) \tag{5.2.24}$$

Therefore, we find that a general solution to (5.2.23) is a plane wave

$$V = V_0 e^{\pm jkz} \tag{5.2.25}$$

and we may substitute (5.2.25) into (5.2.23) to find the dispersion relation for the parallel-plate transmission line

$$k^2 = \omega^2 LC \tag{5.2.26}$$

Therefore, the velocity of voltage and current wave propagation on the line is $v = 1/\sqrt{LC}$. Because $LC = \mu\epsilon$ from (5.2.21) and (5.2.22), the dispersion relation (5.2.26) may also be written as $k^2 = \omega^2 \mu\epsilon$. Not surprisingly, this dispersion relation is identical to that of a uniform plane wave in the unbounded medium characterized by μ and ϵ [see (1.3.10)].

The most general solution to the voltage wave equation is

$$V(z) = V_+ e^{-jkz} + V_- e^{+jkz} \tag{5.2.27}$$

where V_+ and V_- are complex constants consistent with system excitations and boundary conditions, and where we associate V_+ with a forward $(+\hat{z})$ traveling wave e^{-jkz} and V_- with the reverse $(-\hat{z})$ traveling wave e^{+jkz}. We may superpose any number of plane wave solutions having different frequencies ω and hence different wave numbers k to form an arbitrary waveform. We shall only be concerned here with **single frequency** plane wave solutions to the transmission line equations. Sections 5.5 and 6.5 discuss transient, nonsinusoidal wave behavior.

The voltage on the transmission line (5.2.27) may be differentiated with respect to z and substituted into (5.2.19) to yield

$$I(z) = \frac{k}{\omega L} \left(V_+ e^{-jkz} - V_- e^{+jkz} \right)$$

or

$$I(z) = Y_0 \left(V_+ e^{-jkz} - V_- e^{+jkz} \right) \tag{5.2.28}$$

where the *characteristic admittance* of the transmission line is

$$Y_0 = \frac{k}{\omega L} = \sqrt{\frac{C}{L}} = \frac{W}{\eta d} \qquad (5.2.29)$$

The reciprocal of (5.2.29) is called the *characteristic impedance* of the transmission line and is

$$Z_0 = \frac{1}{Y_0} = \sqrt{\frac{L}{C}} = \frac{\eta d}{W} \qquad (5.2.30)$$

The characteristic impedance of the line Z_0 is **not** to be confused with the characteristic wave impedance of the medium $\eta = \sqrt{\mu/\epsilon}$, which is **independent** of the geometry of the transmission line. The wave impedance is the ratio of electric to magnetic fields, while the characteristic impedance is a voltage to current ratio, where voltage and current depend explicitly on the *TEM* line dimensions as in (5.2.10) and (5.2.11). Therefore, $LC = \mu\epsilon$, but in general $L/C \neq \mu/\epsilon$.

Example 5.2.1

Design a strip-line geometry for printed circuits that has an approximate impedance of 10 Ω when a dielectric coating of $\epsilon = 4\epsilon_0$ is used over the ground plane. If the dielectric breaks down for field strengths greater than 40,000 V/cm, what is the thinnest strip-line structure that can propagate one kW of power?

Solution: The strip-line structure is illustrated in Figure 5.3. Because ϵ and μ are not uniform everywhere, the transmission line does not propagate pure *TEM* waves. They are *TEM* to an excellent approximation, however, because most of the stored energy resides directly between the two conductors where ϵ and μ are approximately constant. The wave impedance η of the medium is

$$\eta = \sqrt{\mu_0/4\epsilon_0} = 188 \ \Omega$$

and (5.2.30) can then be used to find the characteristic impedance of this *TEM* line:

$$Z_0 = \eta d/W = 10 \ \Omega$$

Therefore, any line with the ratio of d/W equal to 10/188 has a 10 Ω impedance.

The average power $P(W)$ in a single wave traveling on the transmission line is computed by integrating the complex Poynting vector (1.6.17) (W/m^2) over the cross-sectional area of the strip-line so that

$$P \cong \frac{1}{2} \mathrm{Re} \left\{ \int_0^d \int_0^W (\overline{E} \times \overline{H}^*) \cdot \hat{z} \, dx dy \right\} = \frac{1}{2} \mathrm{Re}\{VI^*\}$$

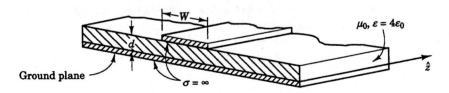

Figure 5.3   Strip-line geometry.

by virtue of (5.2.10) and (5.2.11). But the current and voltage of a single wave are related by Ohm's law, $V = Z_0 I$, so

$$P = \frac{1}{2}\operatorname{Re}\{VI^*\} = \frac{1}{2}|I|^2\operatorname{Re}\{Z_0\} = \frac{1}{2}|V|^2\operatorname{Re}\left\{\frac{1}{Z_0^*}\right\}$$

The last equality allows us to write

$$|V| = \sqrt{\frac{2P}{\operatorname{Re}\{1/Z_0^*\}}} = |E_0|\,d$$

In order to avoid dielectric breakdown, d should be chosen such that

$$d \geq \frac{\sqrt{2P/\operatorname{Re}\{1/Z_0^*\}}}{|E_{\max}|} = \frac{\sqrt{2\cdot 10^3\ \text{W}\cdot 10\ \Omega}}{40,000\ \text{V/cm}} = 35\ \mu\text{m}$$

If $d = 35\ \mu\text{m}$, then $W = 18.8d = 658\ \mu\text{m}$.

Example 5.2.2

The medium that fills a real transmission line is often lossy, and we wish to determine what effect this loss has on the behavior of *TEM* waves. A parallel-plate *TEM* transmission line has dimensions $W = 1$ cm and $d = 2$ mm and is filled with glass ($\epsilon = 4\epsilon_0$ F/m, $\sigma = 10^{-10}$ S/m). What is the characteristic impedance Z_0 of this *TEM* line? How many meters ℓ can a wave propagate along this line before it is attenuated 20 dB? If we now replace the glass with distilled water ($\epsilon = 80\epsilon_0$ F/m, $\sigma = 2 \times 10^{-4}$ S/m), what are Z_0 and ℓ? The frequency of the wave is 10 MHz. Materials that are commonly used in transmission lines generally fall somewhere in the range represented by these two materials.

Solution: The transmission line equations (5.2.19) and (5.2.20) must be modified since the medium filling the line is lossy, and therefore Faraday's and Ampere's laws (5.2.15) and (5.2.16) must be re-examined. Faraday's law remains unchanged by the addition of loss, but Ampere's law becomes $\nabla \times \overline{H} = j\omega\epsilon\overline{E} + \sigma\overline{E}$ where currents generated by Ohm's law ($\overline{J} = \sigma\overline{E}$) have been included. As in Section 3.4, we can define an *effective dielectric constant* $\epsilon_{\text{eff}} = \epsilon - j\sigma/\omega$, which means that Ampere's law for the parallel-plate line (5.2.18) becomes

$$\frac{\partial}{\partial z}H_y W = -j\omega\left(\epsilon - j\frac{\sigma}{\omega}\right)WE_x$$

The transmission line equations are thus

$$\frac{\partial V(z)}{\partial z} = -j\omega L I(z) \tag{1}$$

$$\frac{\partial I(z)}{\partial z} = -(G + j\omega C)V(z) \tag{2}$$

where $C = \epsilon W/d$ and $L = \mu d/W$ as before, and the conductance per unit length is $G = \sigma W/d$. We note that the charge relaxation time $\tau_c = \epsilon/\sigma$ is equal to C/G, a relation that is true for an arbitrary *TEM* line of uniform cross-section, as will be shown in the next section.

A voltage wave equation is derived by differentiating (1) with respect to z and substituting it into (2):

$$\frac{\partial^2 V(z)}{\partial z^2} = j\omega L(G + j\omega C)V(z)$$

This equation again has the forward-propagating solution $V(z) = V_0 e^{-jkz}$, where

$$k = \sqrt{\omega^2 LC - j\omega LG} = \omega\sqrt{LC}\sqrt{1 - j\frac{G}{\omega C}} \tag{3}$$

We note that because $LC = \mu\epsilon$ and $G/C = \sigma/\epsilon$, (3) can also be written as

$$k = \omega\sqrt{\mu\epsilon}\sqrt{1 - j\frac{\sigma}{\omega\epsilon}}$$

which is not surprisingly the same complex wave number found in (3.4.6). If the *TEM* line is only slightly lossy so the loss tangent $\sigma/\omega\epsilon = G/\omega C \ll 1$, (3) may be expanded in a Taylor series:

$$k \simeq \omega\sqrt{LC} - j\frac{G}{2}\sqrt{\frac{L}{C}} \qquad (G \ll \omega C) \tag{4}$$

The characteristic impedance is the ratio of voltage to current, and substitution of $V(z) = V_0 e^{-jkz}$ into (1) yields

$$Z_0 = \frac{\omega L}{k} = \sqrt{\frac{L}{C}}\frac{1}{\sqrt{1 - jG/\omega C}}$$

$$\simeq \sqrt{\frac{L}{C}}\left(1 + \frac{jG}{2\omega C}\right) \qquad (G \ll \omega C) \tag{5}$$

For a lossy transmission line, it is clear that both the characteristic impedance and the wave number are complex numbers. We compute the loss tangents for both glass and distilled water (dH$_2$O) at 10 MHz

$$\frac{G}{\omega C}(\text{glass}) = \frac{10^{-10}\ \text{S/m}}{2\pi \cdot 10^7\ \text{s}^{-1} \cdot 4 \times 8.85 \times 10^{-12}\ \text{F/m}} = 4.5 \times 10^{-8} \ll 1$$

$$\frac{G}{\omega C}(\text{dH}_2\text{O}) = \frac{2 \times 10^{-4}\ \text{S/m}}{2\pi \cdot 10^7\ \text{s}^{-1} \cdot 80 \times 8.85 \times 10^{-12}\ \text{F/m}} = 4.5 \times 10^{-3} \ll 1$$

so in both media the small loss approximations for k and Z_0 may be used.

For glass, $C = \epsilon W/d = 4 \times 8.85 \times 10^{-12} \times 50 = 1.77$ nF/m, $L = \mu d/W = 4\pi \times 10^{-7}/50 = 25.1$ nH/m, and $G = \sigma W/d = 10^{-10} \times 50 = 5.0$ nS/m. Therefore, substituting these values in (5) yields

$$Z_0(\text{glass}) = \sqrt{\frac{25.1\ \text{nH/m}}{1.77\ \text{nF/m}}}\left(1 + j\frac{5.0\ \text{nS/m}}{4\pi \cdot 10^7\ \text{s}^{-1} \cdot 1.77\ \text{nF/m}}\right)$$

$$= 3.77 + j8.5 \times 10^{-8}\ \Omega$$

which is very close to the lossless impedance of 3.77 Ω.

For distilled water, $C = \epsilon W/d = 80 \times 8.85 \times 10^{-12} \times 50 = 35.4$ nF/m, $L = \mu d/W = 4\pi \times 10^{-7}/50 = 25.1$ nH/m, and $G = \sigma W/d = 2 \times 10^{-4} \times 50 = 10$ mS/m. Therefore,

$$Z_0(\text{dH}_2\text{O}) = \sqrt{\frac{25.1\ \text{nH/m}}{35.4\ \text{nF/m}}}\left(1 + j\frac{10\ \text{mS/m}}{4\pi \cdot 10^7\ \text{s}^{-1} \cdot 35.4\ \text{nF/m}}\right)$$

$$= 0.84 + j0.0019\ \Omega$$

The power propagating on the line is proportional to the square of the voltage magnitude, so for the forward-propagating wave,

$$P \propto \left| e^{-jk'z - k''z} \right|^2 = e^{-2k''z} = e^{-\alpha_0 z}$$

where $\alpha_0 = 2k'' = -2\,\mathrm{Im}\{\overline{k}\} = G\sqrt{L/C}$ is the power attenuation coefficient. If the power decreases by 20 dB in ℓ m, then

$$\frac{P(z=\ell)}{P(z=0)} = e^{-\alpha_0 \ell} = 10^{-2}$$

and therefore $\ell = 2 \ln 10 / \alpha_0$.[2] For glass,

$$\ell(\text{glass}) = \frac{2 \ln 10}{5.0 \text{ nS/m} \cdot 3.77 \ \Omega} = 2.44 \times 10^8 \text{ m}$$

Therefore, the wave would have to travel 244 million meters in glass before it would decay to one percent of its initial amplitude; glass is not particularly lossy! This is one reason why glass optical fibers are so useful in communications networks. (But note that the nonideal conductors surrounding the dielectric are significantly more lossy at high frequencies—see Example 5.4.2—so the actual attenuation would be much greater.) On the other hand, the distilled water-filled conductor is attenuated 20 dB when

$$\ell(\text{dH}_2\text{O}) = \frac{2 \ln 10}{10 \text{ mS/m} \cdot 0.84 \ \Omega} = 548 \text{ m}$$

5.3* GENERAL *TEM* TRANSMISSION LINES

Many transmission line geometries can propagate *TEM* waves, an example being the generic line illustrated in Figure 5.4(a). All the transmission lines considered in this section are fabricated using two perfect conductors of an arbitrary cross-section that does not vary along the direction of wave propagation ($\hat{z}$). The medium between the conductors is isotropic, homogeneous, and characterized by μ, ϵ, and σ. Figure 5.4(b) depicts some commonly used *TEM* geometries.

Since we are considering only *TEM* solutions where $E_z = H_z = 0$, the electric and magnetic fields are purely transverse to the direction of wave propagation. We indicate that a vector component is transverse by using the subscript T. Therefore, within the *TEM* line, $\overline{E} = \overline{E}_T$ and $\overline{H} = \overline{H}_T$, where $\overline{E}_T \cdot \hat{z} = \overline{H}_T \cdot \hat{z} \equiv 0$ by definition. Of course, $\overline{E}$ and $\overline{H}$ do not point in the same direction; in general, they are orthogonal. Furthermore, we can decompose the del operator ∇ into a transverse and a longitudinal part

$$\nabla = \nabla_T + \hat{z}\frac{\partial}{\partial z} \tag{5.3.1}$$

---

2. The attenuation in *decibels* (dB) is given by

$$-10 \log_{10}\left(\frac{P(z=\ell)}{P(z=0)}\right)$$

after the wave has propagated a distance ℓ.

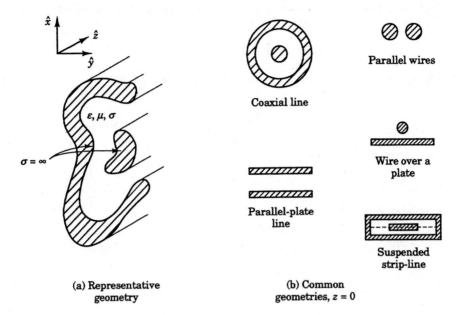

(a) Representative
geometry

(b) Common
geometries, $z = 0$

Figure 5.4   General *TEM* transmission lines.

where $\nabla_T = \hat{x}\,\partial/\partial x + \hat{y}\,\partial/\partial y$ is the transverse del operator in Cartesian coordinates. By definition, $(\nabla_T \Psi) \cdot \hat{z} \equiv 0$, where Ψ is an arbitrary scalar field. Faraday's and Ampere's laws may then be written in terms of transverse and longitudinal components: $\nabla \times \overline{E} = -j\omega\mu\overline{H}$ becomes

$$\nabla_T \times \overline{E}_T + \frac{\partial}{\partial z}(\hat{z} \times \overline{E}_T) = -j\omega\mu\overline{H}_T \tag{5.3.2}$$

and $\nabla \times \overline{H} = j\omega\epsilon\overline{E} + \overline{J}$ becomes

$$\nabla_T \times \overline{H}_T + \frac{\partial}{\partial z}(\hat{z} \times \overline{H}_T) = j\omega\epsilon\overline{E}_T + \overline{J} \tag{5.3.3}$$

If the medium is conducting, $\overline{J} = \sigma\overline{E}_T = \overline{J}_T$, and the only terms in (5.3.2) and (5.3.3) that do **not** point in a direction transverse to the direction of propagation $\hat{z}$ are $\nabla_T \times \overline{E}_T$ and $\nabla_T \times \overline{H}_T$; these two terms are $\hat{z}$-directed. The other vectors $\overline{H}_T$, $\overline{E}_T$, $\hat{z} \times \overline{H}_T$, and $\hat{z} \times \overline{E}_T$ lie in the x–y plane. Therefore, we know that

$$\nabla_T \times \overline{E}_T = 0 \tag{5.3.4}$$

$$\nabla_T \times \overline{H}_T = 0 \tag{5.3.5}$$

and

$$\frac{\partial}{\partial z}(\hat{z} \times \overline{E}_T) = -j\omega\mu\overline{H}_T \tag{5.3.6}$$

$$\frac{\partial}{\partial z}(\hat{z} \times \overline{H}_T) = j\omega\epsilon\overline{E}_T + \sigma\overline{E}_T \tag{5.3.7}$$

Furthermore, Gauss's law tells us that

$$\nabla \cdot \overline{H} = 0 = \nabla_T \cdot \overline{H}_T + \frac{\partial}{\partial z}(\hat{z} \cdot \overline{H}_T)$$

But $\hat{z} \cdot \overline{H}_T \equiv 0$, so

$$\nabla_T \cdot \overline{H}_T = 0 \qquad\qquad (5.3.8)$$

Equations (5.3.5) and (5.3.8) state that both the curl and divergence of $\overline{H}_T$ in the transverse plane are zero, which are the equations of *magnetostatics* in a medium with no current density [see also (2.1.3) and (2.1.4)]. Furthermore, $\overline{H}_T$ must obey the boundary conditions at the surface of the perfect conductors: $\hat{n} \cdot \overline{H}_T = 0$ and $\hat{n} \times \overline{H}_T = \overline{J}_s$. The solution of (5.3.5) and (5.3.8) for $\overline{H}_T$ is thus completely independent of the solution for $\overline{E}_T$, and the form of $\overline{H}_T$ is exactly what it would be for a static surface current distribution flowing on the conductors.

In the sinusoidal steady state (time-harmonic Maxwell's equations), there can be no free charge density between the conducting surfaces, so Gauss's law says that $\nabla \cdot \overline{E} = \rho/\epsilon = 0$. Because $\hat{z} \cdot \overline{E}_T = 0$, we also find that

$$\nabla_T \cdot \overline{E}_T = 0 \qquad\qquad (5.3.9)$$

Equations (5.3.4) and (5.3.9) imply that both the curl and divergence of $\overline{E}_T$ in the transverse plane are zero, which are the equations of *electrostatics* in a charge-free medium [see also (2.1.1) and (2.1.2)]. The boundary conditions at the perfectly conducting surfaces, $\hat{n} \cdot \overline{E}_T = \sigma_s/\epsilon$ and $\hat{n} \times \overline{E}_T = 0$, must also be obeyed. Again we find that the electric and magnetic fields are completely decoupled for *TEM* waves. The form of the electric field between the conducting plates is thus the same as it would be for a static charge distribution on the surface of the conductors. We note the important result that no *TEM* waves can propagate inside a perfectly conducting hollow pipe with uniform ϵ, μ because the only static $\overline{E}_T$ consistent with such an equipotential boundary is the trivial field $\overline{E}_T = 0$.

We again wish to define a voltage $V(z)$ and current $I(z)$ as we did for the parallel-plate line. If we consult Figure 5.5(a), we see that we can express the transmission line voltage as

$$V(z) = \int_a^b \overline{E}_T \cdot d\overline{\ell} \qquad\qquad (5.3.10)$$

where a and b are the endpoints of a path connecting the two conductors. As long as the path connecting the two conductors lies in a single plane perpendicular to the $\hat{z}$-axis, $V(z)$ will be a uniquely defined voltage (at a particular value of z) and thus independent of the specific path taken between the two conductors. This is easily seen by applying Stokes's theorem (4.1.4) to (5.3.4)

$$\nabla_T \times \overline{E}_T = 0 \Longrightarrow \oint_C \overline{E}_T \cdot d\overline{\ell} = 0 \qquad\qquad (5.3.11)$$

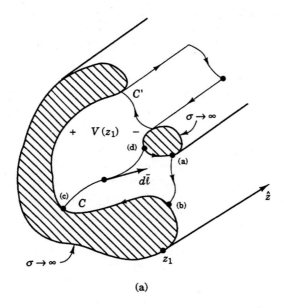

(a)

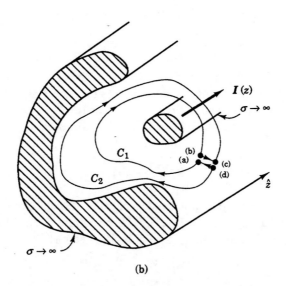

(b)

Figure 5.5 Contours, surfaces, and integration variables in a general *TEM* transmission line.

where C is a closed contour in an x–y plane at an arbitrary distance z along the transmission line. For the contour shown in Figure 5.5(a), (5.3.11) is rewritten

$$\int_a^b \overline{E}_T \cdot d\overline{\ell} + \int_b^c \overline{E}_T \cdot d\overline{\ell} + \int_c^d \overline{E}_T \cdot d\overline{\ell} + \int_d^a \overline{E}_T \cdot d\overline{\ell} = 0 \qquad (5.3.12)$$

As already discussed for the parallel-plate transmission line, the two segments of the contour along the surfaces of the conductors do not contribute to (5.3.12) because these paths are tangential to conducting surfaces and the tangential electric field at a perfect conductor is zero. Thus

$$\int_b^c \overline{E}_T \cdot d\overline{\ell} = \int_d^a \overline{E}_T \cdot d\overline{\ell} = 0$$

and so

$$\int_a^b \overline{E}_T \cdot d\overline{\ell} = \int_d^c \overline{E}_T \cdot d\overline{\ell} \equiv V(z)$$

where the limits of integration c and d have been interchanged to absorb a negative sign. Therefore, we have shown that $V(z)$ is independent of path as long as the path remains in a plane transverse to the direction of wave propagation on the line and $H_z \equiv 0$. Naturally, $V(z_1) \neq V(z_2)$ in general because the closed contour $\oint_{C'} \overline{E}_T \cdot d\overline{\ell} \neq 0$ when C' lies in the x–z plane. This contour encloses time-varying magnetic fields, and is also shown in Figure 5.5(a).

We may also define a unique current $I(z)$ from the integral form of Ampere's law. We recall from (4.1.6) that

$$\oint_C \overline{H} \cdot d\overline{\ell} = \int_A \overline{J} \cdot \hat{n} \, da + j\omega\epsilon \int_A \overline{E} \cdot \hat{n} \, da \qquad (5.3.13)$$

where $\overline{J}$ in (5.3.13) is composed of two pieces: a surface current at the conductor wall and a conduction current within the medium $\sigma \overline{E}_T$. We choose the contour C to be C_1 as shown in Figure 5.5(b); C_1 lies in a plane of constant z. The direction in which C_1 is traversed leads to the conclusion that $\hat{n} = \hat{z}$ as a result of the right-handed convention illustrated in Figure 4.1. Because $\overline{E}_T \cdot \hat{z} = 0$, only the surface current contributes to the integral over $\overline{J}$ in (5.3.13). Because $\int_A \overline{J} \cdot \hat{n} \, da = I(z)$, (5.3.13) becomes

$$\oint_{C_1} \overline{H}_T \cdot d\overline{\ell} = I(z) \qquad (5.3.14)$$

But $I(z)$ is unique only if it is independent of the choice of contour. We have already shown that $\nabla_T \times \overline{H}_T = 0$ (5.3.5) in the current-free medium between the conducting surfaces, which means that by Stoke's theorem (4.1.4),

$$\oint_{C''} \overline{H}_T \cdot d\overline{\ell} = 0$$

where C'' is any closed contour that does **not** encircle the current flowing in the conductors. Therefore, C'' given by $abcd$ in Figure 5.5(b) is an acceptable contour, and

$$\oint_{C''} \overline{H}_T \cdot d\overline{\ell} = \oint_{C_1} \overline{H}_T \cdot d\overline{\ell} + \int_b^c \overline{H}_T \cdot d\overline{\ell} - \oint_{C_2} \overline{H}_T \cdot d\overline{\ell} + \int_d^a \overline{H}_T \cdot d\overline{\ell} = 0$$

$$(5.3.15)$$

where the negative sign in (5.3.15) arises because the direction in which the contour C_2 is traced is opposite to that of C_1. Since paths bc and da are identical but reversed in direction,

$$\int_b^c \overline{H}_T \cdot d\overline{\ell} + \int_d^a \overline{H}_T \cdot d\overline{\ell} = 0$$

and we are finally left with

$$\oint_{C_1} \overline{H}_T \cdot d\overline{\ell} = \oint_{C_2} \overline{H}_T \cdot d\overline{\ell} \equiv I(z)$$

showing the uniqueness of $I(z)$ for a particular value of z.

Having uniquely defined voltage and current on a transmission line, we wish to relate $V(z)$ and $I(z)$ by simple differential equations similar to those given by (5.2.19) and (5.2.20) for a parallel-plate *TEM* line. It is first useful to find the capacitance C (F/m) and inductance L (H/m) per unit length of line.

Capacitance per unit length (F/m) is defined as the charge per unit length $q'(z)$ divided by voltage, or

$$C = \frac{q'(z)}{V(z)} \tag{5.3.16}$$

where the prime signifies a per-unit-length quantity. We can find $q'(z)$ from the integral form of Gauss's law (4.1.2)

$$\oint_A \epsilon \overline{E} \cdot \hat{n} \, da = \int_V \rho \, dv \tag{5.3.17}$$

where we choose the closed surface A to be a pillbox with identical planar faces at $z = z_1$ and $z = z_2$ enclosed by C, and an infinitesimal separation distance $\delta = z_2 - z_1$. Because $\overline{E}_T \cdot \hat{z} = 0$, the surfaces at $z = z_1$ and $z = z_2$ make no contribution to the left side of (5.3.17), so we integrate only over the ribbon-shaped surface of thickness δ. Thus,

$$\oint_A \epsilon \overline{E}_T \cdot \hat{n} \, da = \oint_C (\epsilon \overline{E}_T \cdot \hat{n} \, d\ell)\delta = q'(z)\delta \tag{5.3.18}$$

where δ, $\hat{n}$, and $d\ell$ are defined in Figure 5.6(a) and we assume that $\overline{E}_T$ does not vary much with z if δ is infinitesimal. But $\hat{n}$ is perpendicular to $d\overline{\ell}$, so we can replace $\overline{E}_T \cdot \hat{n}$ with $\left(\overline{E}_T \times \frac{d\overline{\ell}}{d\ell} \right) \cdot \hat{z}$, and (5.3.18) then becomes

$$\oint_C \epsilon (\overline{E}_T \times d\overline{\ell}) \cdot \hat{z} = q'(z) \tag{5.3.19}$$

The charge per unit length $q'(z)$ is independent of the choice of the contour C because all of the charge lies on the conductor surface inside the contour. Of course, an equal and opposite charge density $-q'(z)$ must reside on the other

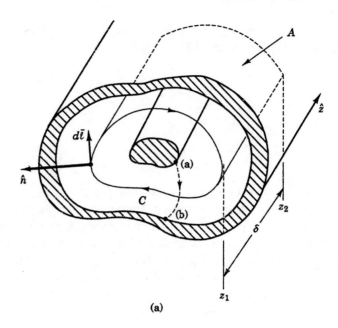

(a)

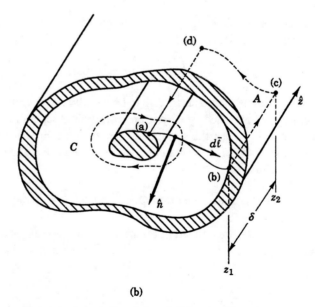

(b)

Figure 5.6   Calculation of L and C for an arbitrary *TEM* transmission line.

conductor to preserve charge balance. Combining (5.3.10), (5.3.16), and (5.3.19) yields

$$C = \frac{\oint_C \epsilon(\overline{E}_T \times d\overline{\ell}) \cdot \hat{z}}{\int_a^b \overline{E}_T \cdot d\overline{\ell}} \tag{5.3.20}$$

where the two $d\overline{\ell}$s are **not** the same in (5.3.20) because the integration paths are different. In the numerator of (5.3.20), we integrate around the central conductor along an arbitrary contour in the transverse plane, while in the denominator, we integrate from one conductor to the other, also in the transverse plane.

The inductance per unit length L (H/m) is defined as

$$L = \frac{\Lambda'(z)}{I(z)} \tag{5.3.21}$$

where $\Lambda'(z)$ is the magnetic flux per unit length linked to the current $I(z)$ flowing in one conductor. The total flux $\Lambda(z)$ linked in length $\delta = z_2 - z_1$ is given by $\Lambda(z) = \Lambda'(z)\delta$, and is

$$\Lambda(z) = \int_A \overline{B} \cdot \hat{n}\, da \tag{5.3.22}$$

where A is the surface shown in Figure 5.6(b). The perimeter of the surface A has two parallel segments (ab and cd) that connect the inner and outer conductors at $z = z_1$ and $z = z_2$, respectively, while bc and ad are $\hat{z}$-directed paths of length δ. The evaluation of (5.3.22) over this surface is thus

$$\Lambda(z) = \int_a^b \left(\mu \overline{H}_T \cdot \hat{n}\, d\ell\right)\delta \tag{5.3.23}$$

where we assume that $\overline{H}_T$ does not vary significantly with z over the surface $abcd$ if δ is infinitesimal. The quantities δ, $\hat{n}$, and $d\overline{\ell}$ are defined in Figure 5.6(b) and are **not** the same as similarly named quantities in Figure 5.6(a). Again, since $\hat{n} \cdot d\overline{\ell} = 0$, we may substitute $\overline{H}_T \cdot \hat{n} = \left(\overline{H}_T \times \frac{d\overline{\ell}}{d\ell}\right) \cdot (-\hat{z})$ into (5.3.23) and combine (5.3.14), (5.3.21), and (5.3.23) to find

$$L = \frac{\int_a^b \mu(\overline{H}_T \times d\overline{\ell}) \cdot (-\hat{z})}{\oint_C \overline{H}_T \cdot d\overline{\ell}} = \frac{\int_a^b \mu \overline{H}_T \cdot (\hat{z} \times d\overline{\ell})}{\oint_C \overline{H}_T \cdot d\overline{\ell}} \tag{5.3.24}$$

where we have used the identities $\overline{a} \cdot (\overline{B} \times \overline{C}) = \overline{B} \cdot (\overline{C} \times \overline{A})$ and $\overline{A} \times \overline{B} = -\overline{B} \times \overline{A}$ to demonstrate the last equality in (5.3.24).

Having defined the capacitance per unit length C, it is also easy to define the conductance per unit length G (S/m) by noting that $\overline{J}_T = \sigma \overline{E}_T$, where σ is the conductivity of the medium. Because the total conductance $\mathcal{G}$ (S $= \Omega^{-1}$) is defined by

$$\mathcal{G} = \frac{I(z)}{V(z)} = \frac{\int_A \overline{J} \cdot \hat{n}\, da}{V(z)} \tag{5.3.25}$$

we may use Ohm's law (3.4.1) to write

$$G = \frac{\int_A \sigma \overline{E}_T \cdot \hat{n} \, da}{V(z)} \tag{5.3.26}$$

But the surface A in (5.3.26) is the ribbon of width δ and perimeter C shown in Figure 5.6(a), and so

$$\int_A \sigma \overline{E}_T \cdot \hat{n} \, da = \oint_C \left(\sigma \overline{E}_T \cdot \hat{n} \, d\ell \right) \delta = \oint_C \left(\sigma (\overline{E}_T \times d\overline{\ell}) \cdot \hat{z} \right) \delta \tag{5.3.27}$$

by analogy with (5.3.18) and (5.3.19). We finally find, after combining (5.3.10), (5.3.26), and (5.3.27), that

$$G = G/\delta = \frac{\oint_C \sigma (\overline{E}_T \times d\overline{\ell}) \cdot \hat{z}}{\int_a^b \overline{E}_T \cdot d\overline{\ell}} = \frac{\sigma}{\epsilon} C \tag{5.3.28}$$

where the charge relaxation time ϵ/σ is now seen to be equal to C/G for an arbitrary *TEM* line with uniform ϵ and σ.

The differential equations for $V(z)$ and $I(z)$ on a general *TEM* line then follow from Maxwell's curl equations (5.3.6) and (5.3.7):

$$\frac{\partial}{\partial z} (\hat{z} \times \overline{E}_T) = -j\omega\mu \overline{H}_T \tag{5.3.29}$$

$$\frac{\partial}{\partial z} (\hat{z} \times \overline{H}_T) = (\sigma + j\omega\epsilon)\overline{E}_T \tag{5.3.30}$$

If we cross $\hat{z}$ into both sides of (5.3.29) and integrate from a to b over the path length shown in Figure 5.6(b), we find that

$$\int_a^b \frac{\partial}{\partial z} \left[(\hat{z} \times \overline{E}_T) \times \hat{z} \right] \cdot d\overline{\ell} = -j\omega\mu \int_a^b (\overline{H}_T \times \hat{z}) \cdot d\overline{\ell} \tag{5.3.31}$$

The left side of (5.3.31) may be simplified by using the vector identity $\overline{A} \times (\overline{B} \times \overline{C}) = \overline{B}(\overline{A} \cdot \overline{C}) - \overline{C}(\overline{A} \cdot \overline{B})$, so that

$$(\hat{z} \times \overline{E}_T) \times \hat{z} = \overline{E}_T(\hat{z} \cdot \hat{z}) - \hat{z}(\hat{z} \cdot \overline{E}_T) = \overline{E}_T$$

since $\hat{z} \cdot \hat{z} = 1$ and $\hat{z} \cdot \overline{E}_T = 0$. Therefore, the left side of (5.3.31) is simply

$$\int_a^b \frac{\partial}{\partial z} \overline{E}_T \cdot d\overline{\ell} = \frac{\partial V}{\partial z} \tag{5.3.32}$$

from (5.3.10). But $(\overline{H}_T \times \hat{z}) \cdot d\overline{\ell} = \overline{H}_T \cdot (\hat{z} \times d\overline{\ell})$ and therefore

$$\int_a^b (\overline{H}_T \times \hat{z}) \cdot d\overline{\ell} = \int_a^b \overline{H}_T \cdot (\hat{z} \times d\overline{\ell}) = \frac{1}{\mu} L \oint_C \overline{H}_T \cdot d\overline{\ell} = \frac{1}{\mu} L \, I(z) \tag{5.3.33}$$

where the expressions for inductance (5.3.24) and current (5.3.14) have been used. Combining (5.3.31)–(5.3.33) yields

$$\frac{\partial V(z)}{\partial z} = -j\omega L \, I(z) \tag{5.3.34}$$

which is one of the transmission line equations. We find the other transmission line equation by crossing both sides of (5.3.30) with $\hat{z}$ and integrating around the closed contour C in the transverse plane shown in Figure 5.6(a):

$$\oint_C \frac{\partial}{\partial z}\left[(\hat{z} \times \overline{H}_T) \times \hat{z}\right] \cdot d\overline{\ell} = (\sigma + j\omega\epsilon) \oint_C (\overline{E}_T \times \hat{z}) \cdot d\overline{\ell} \qquad (5.3.35)$$

By exactly the same reasoning as above and by virtue of (5.3.14), the left side of (5.3.35) is simply $\partial I/\partial z$. The integrand on the right side of (5.3.35) may be rewritten as $(\overline{E}_T \times \hat{z}) \cdot d\overline{\ell} = -(\overline{E}_T \times d\overline{\ell}) \cdot \hat{z}$, so (5.3.35) becomes

$$\frac{\partial I(z)}{\partial z} = -(G + j\omega C)V(z) \qquad (5.3.36)$$

after the expressions for capacitance (5.3.20) and conductance (5.3.28) have been used. Equations (5.3.34) and (5.3.36) are the same transmission line equations previously derived in Section 5.2 for the special case of a parallel-plate *TEM* line with $\sigma = G = 0$.

There is another way of finding these transmission line equations and it gives us a way to relate G, L, and C to σ, μ, and ϵ. If we dot $d\overline{\ell}$ into both sides of (5.3.29) and integrate around the closed, transverse contour C shown in Figure 5.6(a), we find

$$\oint_C \frac{\partial}{\partial z}(\hat{z} \times \overline{E}_T) \cdot d\overline{\ell} = -j\omega\mu \oint_C \overline{H}_T \cdot d\overline{\ell} \qquad (5.3.37)$$

But $(\hat{z} \times \overline{E}_T) \cdot d\overline{\ell}$ may be written as $(\overline{E}_T \times d\overline{\ell}) \cdot \hat{z}$ and $\oint_C \overline{H}_T \cdot d\overline{\ell} = I(z)$. Equation (5.3.37) becomes

$$\frac{1}{\epsilon} \frac{\partial}{\partial z} \oint_C \epsilon(\overline{E}_T \times d\overline{\ell}) \cdot \hat{z} = -j\omega\mu I(z) \qquad (5.3.38)$$

Substituting the expression for capacitance (5.3.20) into this equation and recalling that $\int_a^b \overline{E}_T \cdot d\overline{\ell} = V(z)$ from (5.3.10) yields

$$\frac{\partial V(z)}{\partial z} = \frac{-j\omega\mu\epsilon}{C} I(z) \qquad (5.3.39)$$

A comparison of (5.3.34) and (5.3.39) establishes the important result that

$$LC = \mu\epsilon \qquad (5.3.40)$$

for **any** *TEM* line, a result we had already shown for the special case of a parallel-plate *TEM* line.

A similar procedure whereby $d\overline{\ell}$ is dotted into both sides of (5.3.30) and integrated from a to b over the transverse path shown in Figure 5.6(b) yields

$$\frac{\partial I(z)}{\partial z} = -\frac{\mu}{L}(\sigma + j\omega\epsilon)V(z) \qquad (5.3.41)$$

after (5.3.10), (5.3.14), and (5.3.24) are invoked. We again may compare (5.3.36) and (5.3.41) to find that $LC = \mu\epsilon$ but also that

$$G = \frac{\mu\sigma}{L} \tag{5.3.42}$$

which is consistent with $G = \sigma C / \epsilon$.

The telegrapher's equations (5.3.34) and (5.3.36) can be combined to form a wave equation in $V(z)$ by differentiating (5.3.34) with respect to z and substituting in (5.3.36):

$$\frac{\partial^2 V(z)}{\partial z^2} = j\omega L(G + j\omega C)V(z) \tag{5.3.43}$$

For practical transmission lines, the conductors are slightly lossy and introduce a finite resistance per unit length R (Ω/m) in series with the inductance $j\omega L$. Because the $\hat{z}$-directed surface current now flows through nonideal conductors, a $\hat{z}$-directed electric field component is necessary to ensure continuity of the tangential electric field at the conducting walls. The waves are therefore no longer purely *TEM*. If $R \ll \omega L$, the energy stored in this $\hat{z}$-directed field will be very small, and the wave behavior will be approximately the same as that of the pure *TEM* waves that result when $R = 0$. That is, we may ignore any non-*TEM* field components if R is sufficiently small. If R is small but nonzero, (5.3.43) becomes

$$\frac{\partial^2 V(z)}{\partial z^2} = (R + j\omega L)(G + j\omega C)V(z) \tag{5.3.44}$$

The solution to the wave equation (5.3.44) has the same form as that of (5.2.27):

$$V(z) = V_+ e^{-jkz} + V_- e^{+jkz} \tag{5.3.45}$$

where

$$k^2 = -(R + j\omega L)(G + j\omega C) \tag{5.3.46}$$

In general, k is a complex wave vector, but for $\sigma = G = R = 0$, the wave propagates with the usual lossless dispersion relation $k = \omega\sqrt{LC} = \omega\sqrt{\mu\epsilon}$ for an isotropic medium.

A *TEM* wave in a general transmission line propagates at phase velocity $v_p = \omega/k$ from (3.5.1). Thus, use of (5.3.46) for $G = R = 0$ yields

$$v_p = 1/\sqrt{LC} = 1/\sqrt{\mu\epsilon} \text{ (m/s)} \tag{5.3.47}$$

If we include R, (5.3.34) becomes

$$\frac{\partial V(z)}{\partial z} = -(R + j\omega L)I(z) \tag{5.3.48}$$

which, when combined with (5.3.45) leads to

$$I(z) = \frac{jk}{R + j\omega L} \left(V_+ e^{-jkz} - V_- e^{+jkz} \right)$$

$$= Y_0(V_+ e^{-jkz} - V_- e^{+jkz}) \tag{5.3.49}$$

where

$$Y_0 = \frac{jk}{R + j\omega L} = \sqrt{\frac{G + j\omega C}{R + j\omega L}}$$

$$= \sqrt{C/L} \;\; \text{if} \; G = R = 0 \tag{5.3.50}$$

The characteristic admittance Y_0 of a *TEM* line with nonzero, positive conductance G and/or resistance R is therefore complex, and waves on this line will decay exponentially as they propagate. A line with negative conductance ($G < 0$) can support exponential growth or amplification. Note that the expressions for voltage and current given by (5.3.45) and (5.3.49) are exact for $R = 0$ but only approximate if $R \neq 0$. True solutions to the $R \neq 0$ transmission line require non-*TEM* wave components to satisfy the boundary conditions at the nonideal conducting surfaces. If R is small enough ($R \ll \omega L$), however, (5.3.45) and (5.3.49) are quite reasonable approximations. Lossy lines are discussed further in Section 5.6.

Because every *TEM* transmission line is completely described by the transmission line equations (5.3.36) and (5.3.48) for voltage and current, we need only know the parameters L, C, R, and G of a line to know everything about it. In the next section, we discuss methods for finding *TEM* line parameters and discuss several examples of *TEM* lines.

5.4 THE COAXIAL CABLE AND OTHER TRANSMISSION LINE EXAMPLES

We now specialize to a particular *TEM* line configuration—the lossless coaxial cable, illustrated in Figure 5.7. A coaxial cable has an inner conductor of radius a surrounded by a concentric conducting sheath of radius b, and the region between the conductors is filled with a dielectric having material properties ϵ, μ, and σ. Coaxial transmission lines are widely used to provide power to antennas, waveguides, and resonators, and are an integral part of cable television and network serial data lines.

We know from (5.3.4) and (5.3.9) that the electric field $\overline{E}_T$ in a single plane of constant z must have zero curl and divergence within that plane, and therefore must satisfy the equations of electrostatics in a charge-free medium. We also know that $\overline{E}_T$ must have only a radial component at the surfaces of the two conductors because the tangential electric field at the surface of a perfect conductor is zero from (5.2.1), and thus $E_\theta = 0$ at $r = a, b$. We thus expect that $\overline{E}_T$ will be radially directed and proportional to $1/r$ since there must be no divergence of the electric field between the plates: $\nabla \cdot \overline{E}_T = 0$ in a charge-free medium. Since the surface area of the cylinder through which the radially directed electric field passes is proportional to $2\pi r$, the field must fall off as $1/r$ for the total flux $\oint_A \overline{E} \cdot \hat{n} \, da$ to remain constant. Therefore, we suspect that

$$\overline{E}_T = \hat{r} \frac{V_0}{r} e^{-jkz} \tag{5.4.1}$$

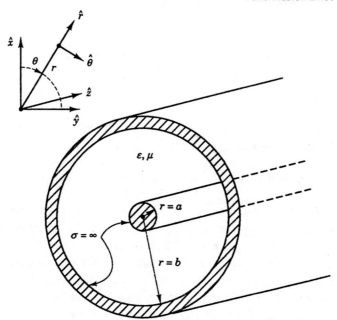

Figure 5.7   Coaxial cable transmission line.

is the form of the forward-propagating electric field. We integrate (5.4.1) from
$r = a$ to $r = b$ to find the voltage:

$$V(z) = \int_a^b \overline{E}_T \cdot d\overline{\ell} = \int_a^b \frac{V_0}{r} e^{-jkz} \, dr = V_0 \ln (b/a) \, e^{-jkz} \qquad (5.4.2)$$

Similarly, $\overline{H}_T$ must satisfy the laws of magnetostatics because the transverse
curl and divergence of $\overline{H}_T$ are zero by (5.3.5) and (5.3.8). Again, a divergence-free
$\overline{H}_T$ suggests that $\overline{H}_T \propto 1/r$ and we expect that $\overline{H}_T$ will be $\hat{\theta}$-directed because
the magnetic field must circulate around the $\hat{z}$-directed surface current. Also, since
$\overline{E} \times \overline{H}^*$ must point in the direction of propagation $\hat{z}$ for a simple isotropic medium
and $\overline{E}$ is $\hat{r}$-directed, $\overline{H}$ must point in the $\hat{\theta}$ direction if the wave is to be *TEM*.
Thus,

$$\overline{H}_T = \hat{\theta} \, \frac{I_0}{r} e^{-jkz} \qquad (5.4.3)$$

and we integrate $\overline{H}_T$ around a closed circular path of radius r ($a < r < b$) to find
the current on the inner conductor:

$$I(z) = \oint_C \overline{H}_T \cdot d\overline{\ell} = \int_0^{2\pi} \frac{I_0}{r} e^{-jkz} \, (r d\theta) = 2\pi I_0 e^{-jkz} \qquad (5.4.4)$$

The electric and magnetic fields are related by Faraday's law

$$\nabla \times \overline{E}_T = -j\omega\mu\overline{H}_T$$

$$\hat{\theta}\,\frac{\partial E_r}{\partial z} = -\hat{\theta}\,\frac{jkV_0}{r}\,e^{-jkz} = -\hat{\theta}\,\frac{j\omega\mu I_0}{r}\,e^{-jkz}$$

which means that the ratio of electric to magnetic field $|\overline{E}_T|/|\overline{H}_T| = V_0/I_0 = \omega\mu/k = \sqrt{\mu/\epsilon} = \eta$, the wave impedance of the medium between the two conductors.

We can find the capacitance for this coaxial cable by finding the surface charge on the inner conductor:

$$\sigma_s = \hat{n}\cdot\epsilon\overline{E}\Big|_{r=a} = \hat{r}\cdot\left(\hat{r}\,\frac{\epsilon V_0}{r}\,e^{-jkz}\right)\Big|_{r=a} = \frac{\epsilon V_0}{a}\,e^{-jkz} \qquad (5.4.5)$$

The capacitance per unit length is then

$$C = \frac{q'(z)}{V(z)} = \frac{2\pi a\sigma_s}{V(z)} = \frac{2\pi\epsilon}{\ln(b/a)} \quad \text{(F/m)} \qquad (5.4.6)$$

where (5.3.16), (5.4.2) and (5.4.5) have been used. The magnetic flux per unit length $\Lambda'(z)$ of the coax is given by

$$\Lambda'(z) = \int_a^b \overline{B}\cdot\hat{\theta}\,dr = \int_a^b \mu\frac{I_0}{r}\,e^{-jkz}\,dr = \mu I_0 \ln(b/a)\,e^{-jkz} \qquad (5.4.7)$$

and therefore the inductance per unit length of the cable is

$$L = \frac{\Lambda'(z)}{I(z)} = \frac{\mu}{2\pi}\ln(b/a) \quad \text{(H/m)} \qquad (5.4.8)$$

where (5.3.21), (5.4.4) and (5.4.7) have been used. We can check (5.4.6) and (5.4.8) by multiplying them together and verifying that $LC = \mu\epsilon$ (5.3.40). The characteristic impedance of this transmission line is

$$Z_0 = \sqrt{\frac{L}{C}} = \frac{1}{2\pi}\sqrt{\frac{\mu}{\epsilon}}\ln(b/a) = \frac{\eta}{2\pi}\ln(b/a) \quad \Omega \qquad (5.4.9)$$

which is **not** the same as the wave impedance η. Once we have found L and C (or Z_0 and v, where v is the velocity of wave propagation on the line, and $L = Z_0/v$, $C = 1/Z_0 v$), then the physical structure of the *TEM* transmission line is no longer important when discussing the behavior of waves on it. The transmission line equations (5.3.34) and (5.3.36) are **always** valid for the *TEM* solutions to a transmission line of uniform cross-section, supporting wave propagation in the $\pm\hat{z}$ directions.

We now discuss another way to determine the characteristic impedance $Z_0 = 1/Y_0$ of an arbitrary *TEM* line, which we shall illustrate with the coaxial cable. We notice first that when the electric and magnetic fields are of the form

$$\overline{E}(z) = \overline{E}_T(z) = \overline{E}_{T0}\, e^{-jkz}$$
$$\overline{H}(z) = \overline{H}_T(z) = \overline{H}_{T0}\, e^{-jkz}$$
(5.4.10)

we can substitute these transverse fields into Faraday's law (1.4.4) to find

$$\nabla \times \overline{E}_T = -jk\hat{z} \times \overline{E}_{T0}\, e^{-jkz} = -j\omega\mu\overline{H}_{T0}\, e^{-jkz}$$

or

$$|\overline{E}_{T0}|/|\overline{H}_{T0}| = \frac{\omega\mu}{k} = \eta$$

as was previously illustrated for both the parallel-plate *TEM* line and coaxial cable. Therefore, $\overline{E}_T$ and $\overline{H}_T$ for a single traveling wave are in phase, orthogonal, and the ratio of their magnitudes is the wave impedance of the medium. Locally, the electric and magnetic fields resemble uniform plane waves inside an infinitesimal parallel-plate transmission line.

We use this concept to break up an arbitrary *TEM* line into an infinite number of parallel-plate transmission lines of thickness d and width W, as shown in Figure 5.8(a). The conducting walls of these hypothetical, infinitesimal *TEM* lines, each possessing a characteristic impedance $\eta d/W$, must be placed along the equipotential surfaces generated by the real, two-conductor transmission line so that they do not disturb the *TEM* fields of that line. (Because the electric field is perpendicular to the surface of a perfect conductor, we can place a thin metal plate in the field without disturbing it, as long as the plate surface is everywhere perpendicular to the electric field lines.) For a coaxial cable, these infinitesimal conductors are oriented along surfaces of constant radius, as shown in Figure 5.8(b). We place n infinitesimal plates at a given radius r from the center of the coax ($a \le r \le b$) and use m layers of these hypothetical plates to fill the space between the coax walls. The distance between adjacent layers of the plates is d, and the width of each plate is W, where by simple geometric considerations,

$$n = \frac{2\pi r}{W}$$
(5.4.11)

If the plate layers are (unevenly) spaced such that the same voltage $v = E_T d$ is observed between each pair of adjacent layers, then $V = mE_T d$, where $V = V_0 \ln(b/a)\, e^{-jkz}$ from (5.4.2) is the total voltage between the coax conductors at $r = a$ and $r = b$. But since $E_T = (V_0/r)\, e^{-jkz}$ from (5.4.1), we see that

$$m = \frac{r}{d} \ln(b/a)$$
(5.4.12)

If m and n are to be constants, it is clear from (5.4.11) and (5.4.12) that W and d must both be proportional to r. This means that the impedance of each infinitesimal parallel-plate *TEM* line $\eta d/W$ is **independent** of r. The impedance of the coaxial cable can then easily be calculated from the equivalent circuit model pictured in Figure 5.8(c), because the cable is composed of a stack of m infinitesimal parallel-plate transmission lines (each of impedance $\eta d/W$) connected in

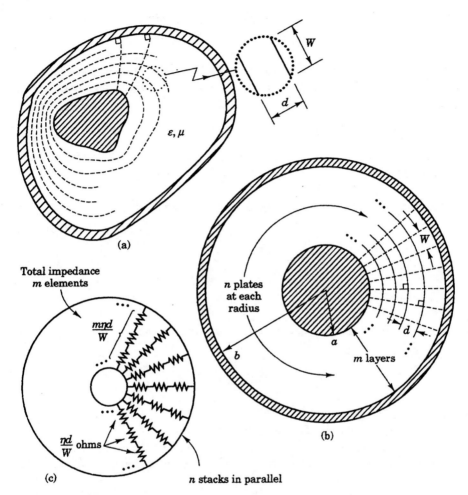

Figure 5.8   General *TEM* transmission line decomposed into infinitesimal parallel-plate transmission lines.

series; n such stacks are connected in parallel. The impedance of a stack of m identical lines connected in series is $m\eta d/W$ and n such impedances connected in parallel yield the total impedance

$$Z_0 = \frac{m}{n}\left(\frac{\eta d}{W}\right) = \frac{\eta}{2\pi}\ln(b/a)$$

after combining (5.4.11) and (5.4.12), which is exactly the same result given by (5.4.9). Any two-conductor transmission line may thus be decomposed into infinitesimal parallel-plate *TEM* lines that are oriented such that their conducting plates lie along equipotential surfaces of the two-conductor system.

Example 5.4.1

Express the instantaneous energy stored per unit length of a coaxial cable in terms of L, C, $I(z, t)$, and $V(z, t)$.

Solution: The electric and magnetic energy densities $W_e(z, t)$ and $W_m(z, t)$ are $\frac{1}{2}\epsilon |\overline{E}(z, t)|^2$ and $\frac{1}{2}\mu |\overline{H}(z, t)|^2$, respectively, where both W_e and W_m are expressed in J/m^3. Combining (5.4.1) and (5.4.2) gives an expression for $\overline{E}_T$ in terms of the voltage:

$$\overline{E}_T(z, t) = \hat{r}\, \frac{V(z, t)}{r \ln(b/a)} \tag{1}$$

Integrating (1) over the cross-section of the *TEM* line yields the instantaneous electric energy per unit length

$$w_e(z, t) = \frac{\epsilon}{2} \int_a^b \int_0^{2\pi} [V(z, t)/r \ln(b/a)]^2\, rd\theta dr$$

$$= \frac{\pi \epsilon V^2(z, t)}{\ln^2(b/a)} \int_a^b \frac{dr}{r} = \pi \epsilon V^2(z, t)/\ln(b/a)$$

$$= \frac{1}{2} CV^2(z, t) \quad (J/m)$$

where (5.4.6) gives us the expression for C. Similarly,

$$w_m(z, t) = \frac{\mu}{2} \int_a^b \int_0^{2\pi} [I(z, t)/2\pi r]^2\, rd\theta dr$$

$$= \frac{\mu}{4\pi} I^2(z, t) \int_a^b \frac{dr}{r} = \frac{\mu}{4\pi} I^2(z, t) \ln(b/a)$$

$$= \frac{1}{2} LI^2(z, t) \quad (J/m)$$

where the expression for L (5.4.8) has been used.

Although this derivation was performed for coaxial cables, the same simple results apply to *TEM* structures of any uniform cross-section. Equivalent expressions for the complex variables $V(z)$ and $I(z)$ are

$$\langle w_e \rangle = \frac{1}{4} C |V(z)|^2 \quad (J/m)$$

$$\langle w_m \rangle = \frac{1}{4} L |I(z)|^2 \quad (J/m)$$

It is often easiest to find C and L for a *TEM* line of arbitrary shape by integrating $W_e = \frac{1}{2}\epsilon |\overline{E}|^2$ and $W_m = \frac{1}{2}\mu |\overline{H}|^2$ over the cross-sectional area of the line and then equating these results to $\frac{1}{2}C|V|^2$ and $\frac{1}{2}L|I|^2$, respectively. This enables us to find C without needing to solve for the charge per unit length q' on one of the conductors, and to find L without having to evaluate the magnetic flux per unit length Λ'.

Example 5.4.2

A coaxial cable is to be used for a community cable television system for frequencies up to 400 MHz. If $\epsilon = 3\epsilon_0$, $\mu = \mu_0$, $a = 0.5$ mm, $b = 2$ mm, and copper ($\sigma \cong 5 \times 10^7$ S/m) is used for the conductors, then approximately how far apart must the repeater amplifiers be placed if their gain is to be 40 dB at 400 MHz ?

Solution: First, we must find the power attenuation of the transmission line.   The average power P traveling in the line is given by

$$P = \frac{1}{2} \operatorname{Re}\{VI^*\} = \frac{1}{2} |I|^2 \operatorname{Re}\{Z_0\} \tag{1}$$

as was shown in Example 5.2.1, where $\operatorname{Re}\{Z_0\} = Z_0$ for a lossless transmission line. The characteristic impedance of the coaxial cable is given by (5.4.9):

$$Z_0 = \eta \ln(b/a)/2\pi = \left[(377/\sqrt{3}) \ln 4 \right] \Big/ 2\pi = 48 \ \Omega$$

The power dissipated is basically lost within a skin depth $\delta = \sqrt{2/\omega\mu\sigma} \approx 3.5 \ \mu m \ll a = 0.5$ mm of the surfaces of the two conductors. Therefore, the current $I(z)$ flows through an area of approximately $2\pi a\delta$ m^2 adjacent to the inner conductor at $r = a$ and $2\pi b\delta$ m^2 at $r = b$. The resistance per unit length R of the cable is thus

$$R \simeq \frac{1}{2\pi a\delta\sigma} + \frac{1}{2\pi b\delta\sigma}$$

The average power dissipated per meter P_d is

$$P_d \cong \frac{1}{2} |I|^2 R \tag{2}$$

and is related to the average power flow by power conservation. If the total power at $z = z_0$ is $P(z_0)$ and the total power at $z = z_0 + \delta z$ is $P(z_0 + \delta z)$, then the amount of power dissipated per unit length is $[P(z_0) - P(z_0 + \delta z)]/\delta z$, which becomes $-dP/dz$ as $\delta z \to 0$. We have already seen that for a wave propagating in the $\hat{z}$ direction in a lossy medium, $\overline{E} \propto e^{-k''z}$. Because power is proportional to $|\overline{E}|^2$, it follows that $P = P_0 e^{-2k''z} = P_0 e^{-\alpha_0 z}$ where $\alpha_0 = 2k''$ is the power attenuation coefficient and P_0 is the total power propagating in the line at $z = 0$. Therefore,

$$P_d = -\frac{dP}{dz} = -\frac{d}{dz}\left(P_0 e^{-\alpha_0 z} \right) = \alpha_0 P \quad (W/m)$$

But $P_d/P = R/Z_0$ from (1) and (2), which means that

$$\alpha_0 = \frac{P_d}{P} = \frac{R}{Z_0} = \frac{1}{Z_0} \left(\frac{1}{2\pi a\delta\sigma} + \frac{1}{2\pi b\delta\sigma} \right)$$

for a coaxial cable. The formula $\alpha_0 = R/Z_0$ for slightly lossy *TEM* lines of arbitrary cross-section may be useful to remember.[3]   The skin depth of the coaxial cable $\delta = \sqrt{2/\omega\mu\sigma}$ is 3.5 μm at 400 MHz, so

$$\alpha_0 \cong \frac{1}{48 \times 5 \times 10^7 \times 3.5 \times 10^{-6} \times 2\pi} \left(\frac{1}{0.5 \times 10^{-3}} + \frac{1}{2 \times 10^{-3}} \right)$$

$$\cong 0.047 \ \text{m}^{-1}$$

An amplifier gain of 40 dB is necessary when the power attenuation between amplifiers spaced ℓ m apart is

$$e^{-\alpha_0 \ell} = 10^{-4}$$

---

3. We compare the power attenuation coefficient $\alpha_0 = R/Z_0 = RY_0$ for nonideal conductors filled with lossless media to $\alpha_0 = G\sqrt{L/C} \simeq GZ_0$ for ideal conductors filled with lossy media (see Example 5.2.2). If the *TEM* line has **both** a lossy medium and nonideal conductors, then $\alpha_0 \simeq RY_0 + GZ_0$.

and therefore $\ell = 196$ m. Therefore, a few repeater amplifiers per kilometer of cable are typically required, and this number increases as the maximum frequency and R increase. In practice, the dielectric losses (G) are normally smaller than the resistive losses for such RF cables, as was shown in Example 5.2.2.

Example 5.4.3

Find the characteristic impedance of a *TEM* line consisting of a long, thin cylindrical conductor of radius R placed a distance $a \gg R$ over a conducting strip in free space, shown in Figure 5.9.

Solution: We first need to find the capacitance per unit length of this *TEM* line, a purely electrostatic problem. We assume that the conducting strip is wide enough to be approximated as an infinite ground plane. If the radius of the conducting cylinder is much less than the distance from the cylinder to the conducting plane, then we can also approximate the cylinder as a line of charge located at a distance a from the plane and use the method of images to find the electric potential everywhere. We can find the electric field from an infinite line of charge (q' C/m) located at the origin by using Gauss's law:

$$\oint_A \epsilon_0 \overline{E} \cdot \hat{n}\, da = \int_V \rho\, dv$$

$$\epsilon_0 E_r 2\pi r \ell = q' \ell$$

$$\overline{E} = \hat{r}\, \frac{q'}{2\pi \epsilon_0 r}$$

Thus, the *electric scalar potential* is

$$\Phi = -\int_{r_0}^r E_r\, dr = -\frac{q'}{2\pi \epsilon_0} \ln\left(\frac{r}{r_0}\right)$$

where r_0 is an arbitrary reference point (integration constant).

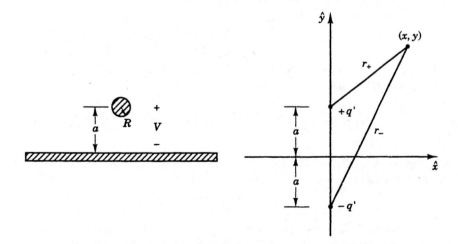

Figure 5.9   Cylinder over a plane *TEM* transmission line.

If we have two line charges, $+q'$ at $y = a$ and $-q'$ at $y = -a$, then superposition leads to the total potential

$$\Phi = -\frac{q'}{2\pi\epsilon_0} \ln\left(\frac{r_+}{r_0}\right) + \frac{q'}{2\pi\epsilon_0} \ln\left(\frac{r_-}{r_0}\right)$$

$$= \frac{q'}{2\pi\epsilon_0} \ln\left(\frac{r_-}{r_+}\right)$$

$$= \frac{q'}{2\pi\epsilon_0} \ln\sqrt{\frac{x^2 + (y+a)^2}{x^2 + (y-a)^2}}$$

where $r_\pm = \sqrt{x^2 + (y \mp a)^2}$ is the distance from the line charge at $\left\{\begin{smallmatrix} y=+a \\ y=-a \end{smallmatrix}\right\}$ to an observer at (x, y), as shown in Figure 5.9. We note that if the observation point is on the conducting plane at $y = 0$, $r_+ = r_-$ and therefore $\Phi = 0$, which demonstrates that the ground plane is the expected equipotential surface. If the observation point is located at any point on the surface of the small cylinder (i.e., at $x = 0$, $y = a \pm R$ or $x = \pm R$, $y = a$), it should also be on an equipotential surface of V volts. For $a \gg R$, we find

$$V = \Phi(\text{cylinder}) \simeq \frac{q'}{2\pi\epsilon_0} \ln(2a/R)$$

and thus

$$C = q'/V \simeq \frac{2\pi\epsilon_0}{\ln(2a/R)}$$

There is no need to calculate L independently, as we already know from (5.3.40) that $LC = \mu\epsilon$ for a *TEM* line. So

$$L = \frac{\mu_0}{2\pi} \ln(2a/R)$$

and therefore

$$Z_0 = \sqrt{\frac{L}{C}} = \frac{\eta_0}{2\pi} \ln(2a/R)$$

is the characteristic impedance of this *TEM* line.

5.5 TIME-DOMAIN ANALYSIS OF TRANSMISSION LINES

The characterization of transmission line behavior in the sinusoidal steady state is desirable because of its simplicity and is described further in Sections 6.1–6.4. Any sinusoidal signal of frequency ω introduced into a possibly lossy or dispersive linear system remains sinusoidal with the same frequency ω as it propagates through that system. Because only the amplitude and phase of the signal may vary, the mathematical analysis of the sinusoidal steady state is particularly tractable. But not all signals are sinusoids. In general, an arbitrary input waveform may be constructed from a possibly infinite set of sinusoids using Fourier or Laplace transform methods. Because the *TEM* line is linear, each input sinusoid may be analyzed separately and the resulting output sinusoids recombined to form a total output signal.

Alternatively, an entire nondispersive linear system may be analyzed directly in the time domain, and in many cases such an approach is much simpler and more useful.

To begin our discussion of time-domain wave behavior, we return to the basic differential equations for lossless *TEM* transmission lines. The time-dependent transmission line equations (5.3.34) and (5.3.36) are

$$\frac{\partial V(z, t)}{\partial z} = -L \frac{\partial I(z, t)}{\partial t} \tag{5.5.1}$$

$$\frac{\partial I(z, t)}{\partial z} = -C \frac{\partial V(z, t)}{\partial t} \tag{5.5.2}$$

where the replacement $j\omega \rightarrow \partial/\partial t$ has been made. The voltage wave equation follows by differentiating (5.5.1) with respect to z and (5.5.2) with respect to t:

$$\frac{\partial^2 V(z, t)}{\partial z^2} = LC \frac{\partial^2 V(z, t)}{\partial t^2} \tag{5.5.3}$$

A similar equation is found for the current $I(z, t)$:

$$\frac{\partial^2 I(z, t)}{\partial z^2} = LC \frac{\partial^2 I(z, t)}{\partial t^2} \tag{5.5.4}$$

Therefore, both the voltage and the current satisfy the homogeneous wave equation discussed in Section 1.3.

We recall that the most general solution to (5.5.3) is the superposition of forward- and backward-traveling voltage waves

$$V(z, t) = V_+ \left(t - \frac{z}{v} \right) + V_- \left(t + \frac{z}{v} \right) \tag{5.5.5}$$

where $V_{\{\pm\}}$ signifies a wave traveling in the $\left\{ \begin{smallmatrix} +\hat{z} \\ -\hat{z} \end{smallmatrix} \right\}$ direction.[4] We substitute (5.5.5) into the voltage wave equation (5.5.3) to find

$$\left(-\frac{1}{v} \right)^2 V_+'' \left(t - \frac{z}{v} \right) + \left(\frac{1}{v} \right)^2 V_-'' \left(t + \frac{z}{v} \right)$$
$$= LC \left[V_+'' \left(t - \frac{z}{v} \right) + V_-'' \left(t + \frac{z}{v} \right) \right] \tag{5.5.6}$$

where the double prime represents differentiation with respect to the **entire** argument in parentheses. Equation (5.5.6) is seen to be satisfied when

$$v = 1/\sqrt{LC} \quad \text{(m/s)} \tag{5.5.7}$$

---

4. The waveform $V_+(t - z/v)$ is fixed in both space and time when $t - z/v = \text{constant}$. Differentiating this expression with respect to time yields $dz/dt = +v$, meaning that the wave propagates in the $+\hat{z}$ direction with velocity v. Similarly, we can follow a point on the $V_-(t + z/v)$ waveform by requiring that $t + z/v = \text{constant}$, in which case $dz/dt = -v$. This waveform travels at velocity v in the $-\hat{z}$ direction.

where v is the velocity of wave propagation. Substituting (5.5.5) into (5.5.2) yields

$$\frac{\partial I(z,t)}{\partial z} = -C\left[V'_+\left(t - \frac{z}{v}\right) + V'_-\left(t + \frac{z}{v}\right)\right]$$

which is integrated to give

$$I(z,t) = Cv\left[V_+\left(t - \frac{z}{v}\right) - V_-\left(t + \frac{z}{v}\right)\right]$$

$$= Y_0\left[V_+\left(t - \frac{z}{v}\right) - V_-\left(t + \frac{z}{v}\right)\right] \tag{5.5.8}$$

where $Y_0 = Cv = \sqrt{C/L} = 1/Z_0$ is the same characteristic admittance as found before in (5.3.50). Note that it would be equally valid to write (5.5.5) as

$$V(z,t) = \hat{V}_+(z - vt) + \hat{V}_-(z + vt) \tag{5.5.9}$$

where $\hat{V}_+$ and $\hat{V}_-$ resemble V_+ and V_- but have arguments that are reversed and scaled. Whether we use (5.5.5) or (5.5.9) depends on personal preference and on the nature of the boundary conditions in space and time. It is usually easier to use (5.5.5) if the boundary condition is generated by a source $V(t)$ at a fixed position z, and to use (5.5.9) if the initial conditions $V(z, t = 0)$ and $I(z, t = 0)$ are specified. A more complete explanation of the time-domain behavior of transmission lines is found in Section 6.5, where we explore the reflection and transmission of waves across multiply connected *TEM* line systems and lines terminated with frequency-dependent impedances.

5.6 PERIODIC TRANSMISSION LINES WITH LUMPED ELEMENTS

Lossless *TEM* transmission lines can be characterized by their inductance L and capacitance C per unit length, where the governing differential equations (5.3.36) and (5.3.48) are

$$\frac{dV(z)}{dz} = -j\omega L\, I(z) \tag{5.6.1}$$

$$\frac{dI(z)}{dz} = -j\omega C\, V(z) \tag{5.6.2}$$

and $R = G = 0$. This suggests that the circuit model in Figure 5.10(a) may be equivalent as $\Delta z \to 0$.

Difference equations for $V(z)$ and $I(z)$ in this lumped-element circuit are

$$V(z + \Delta z) - V(z) = \Delta V(z) = -j\omega L \Delta z I(z) \tag{5.6.3}$$

$$I(z + \Delta z) - I(z) = \Delta I(z) = -j\omega C \Delta z V(z) \tag{5.6.4}$$

Dividing (5.6.3) and (5.6.4) by Δz and letting $\Delta z \to 0$ yields the differential equations (5.6.1) and (5.6.2), respectively.

Because *TEM* transmission lines are typically fabricated with conductors having large but finite conductivity and dielectric insulators with small but nonzero

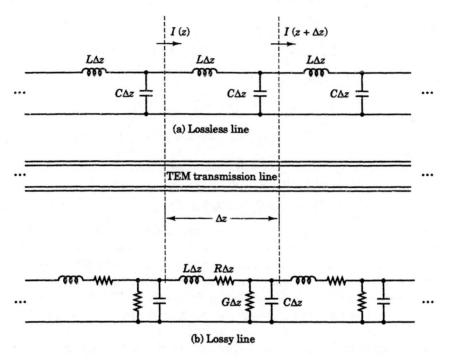

Figure 5.10   Incremental approximate equivalent circuits for *TEM* transmission lines.

conductivity, the circuit illustrated in Figure 5.10(b) is more accurate due to its incorporation of a series resistance R and shunt conductance G per unit length. The limit $\Delta z \to 0$ then yields

$$\frac{dV(z)}{dz} = -(R + j\omega L)I(z) \tag{5.6.5}$$

$$\frac{dI(z)}{dz} = -(G + j\omega C)V(z) \tag{5.6.6}$$

If we assume solutions to (5.6.5) and (5.6.6) are of the form $V = V_{+}e^{-jkz}$, then the voltage wave equation (5.3.44) found by combining (5.6.5) and (5.6.6) yields the dispersion relation

$$k = \sqrt{-(R + j\omega L)(G + j\omega C)} = \omega \sqrt{L\left(1 - \frac{jR}{\omega L}\right)C\left(1 - \frac{jG}{\omega C}\right)} \tag{5.6.7}$$

$$= k' - jk'' \cong \omega\sqrt{LC}\left(1 - \frac{jR}{2\omega L} - \frac{jG}{2\omega C}\right) \text{ if } R \ll \omega L, \ G \ll \omega C \tag{5.6.8}$$

For small values of $R/\omega L$ and $G/\omega C$, we see that the phase velocity $v_p = \omega/k'$ and attenuation k'' are approximately frequency independent, and so a pulse propagating modest distances is not strongly distorted by small amounts of loss. Exponential attenuation of the wavefront is a much more significant effect. For large loss, (5.6.7) suggests that both k' and k'' have frequency dependencies that could strongly distort pulses, with one important exception. In the case where $R/L = G/C$, (5.6.7) becomes

$$k = \omega\sqrt{LC}(1 - jR/\omega L) = \omega\sqrt{LC} - jR\sqrt{C/L} \qquad (5.6.9)$$

Since $k'' = R\sqrt{C/L}$, the attenuation is independent of frequency and since $k' = \omega\sqrt{LC}$, the line is nondispersive. Such a structure is called a *distortionless transmission line*.

If we assume $V(z) = V_+ e^{-jkz}$ and use (5.6.7) for k, then (5.6.5) yields

$$I(z) = \sqrt{\frac{G + j\omega C}{R + j\omega L}}\, V_+ e^{-jkz} = Y_0 V_+ e^{-jkz} \qquad (5.6.10)$$

The characteristic admittance Y_0 of such a lossy line is purely real only if the distortionless condition holds: $R/L = G/C$.

If we approximate a *TEM* transmission line by discrete (lumped) elements, the approximation is good only if $\Delta z \ll \lambda$, as we can see from studying the lossless *lumped-element transmission line* in Figure 5.11. To find the basic difference equations, we apply Kirchhoff's current and voltage laws to the nth and $n + 1^{st}$ nodes of the discrete *TEM* line to obtain

$$V_n - V_{n+1} = j\omega L_0 I_{n+1} \qquad (5.6.11)$$

$$I_n - I_{n+1} = j\omega C_0 V_n \qquad (5.6.12)$$

which can be combined, leading to the discrete voltage wave equation

$$V_{n+1} - 2V_n + V_{n-1} = -\omega^2 L_0 C_0 V_n \qquad (5.6.13)$$

If the excitation is purely from one direction, we guess that the relation between any two adjacent nodes is $V_{n+1}/V_n = $ a complex constant (independent of n) $= e^{\pm jk\ell}$ since the wave cannot distinguish between individual nodes and it travels a distance

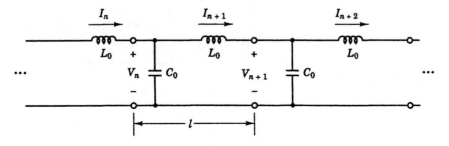

Figure 5.11   Lumped-element transmission line.

ℓ between them. Therefore, the most general voltage V_n at the nth node is a superposition of forward and backward waves

$$V_n = V_+ e^{-jn\theta} + V_- e^{jn\theta} \qquad (5.6.14)$$

where the dimensionless parameter $\theta = k\ell$ may be complex even for a lossless lumped-element line. Likewise, the current I_n at the nth node can be written

$$I_n = I_+ e^{-jn\theta} + I_- e^{jn\theta} \qquad (5.6.15)$$

If we substitute the forward-propagating voltage wave (5.6.14) (with $V_- = 0$) into the discrete wave equation (5.6.13), we find

$$e^{-j\theta} - 2 + e^{j\theta} = -\omega^2 L_0 C_0$$

and an identical equation results from the backward-propagating wave. After noting that $e^{j\theta} + e^{-j\theta} = 2\cos\theta$ and $1 - \cos\theta = 2\sin^2(\theta/2)$, we emerge with the discrete dispersion relation

$$\sin^2(\theta/2) = \frac{\omega^2 L_0 C_0}{4} = \frac{\omega^2}{\omega_0^2} \qquad (5.6.16)$$

The frequency $\omega_0 = 2/\sqrt{L_0 C_0}$ is the *cut-off frequency* for this discrete line, for if $\omega > \omega_0$, θ must be imaginary and the waves evanescent.

The characteristic impedance of the line may be found by substituting (5.6.14) and (5.6.15) into (5.6.11) to yield

$$\left(V_+ e^{-jn\theta} + V_- e^{jn\theta}\right) - \left(V_+ e^{-j(n+1)\theta} + V_- e^{j(n+1)\theta}\right)$$
$$= j\omega L_0 \left(I_+ e^{-j(n+1)\theta} + I_- e^{j(n+1)\theta}\right) \qquad (5.6.17)$$

If we equate the coefficients in front of the linearly independent functions $e^{jn\theta}$ and $e^{-jn\theta}$ in (5.6.17), we find that

$$V_+ \left(1 - e^{-j\theta}\right) = j\omega L_0 I_+ e^{-j\theta}$$
$$V_- \left(1 - e^{j\theta}\right) = j\omega L_0 I_- e^{j\theta} \qquad (5.6.18)$$

Because $1 - e^{\pm j\theta}$ can be written as $\mp 2j\, e^{\pm j\theta/2} \sin(\theta/2)$, we find that

$$I_+ = \frac{2V_+}{\omega L_0} e^{j\theta/2} \sin(\theta/2)$$
$$I_- = -\frac{2V_-}{\omega L_0} e^{-j\theta/2} \sin(\theta/2) \qquad (5.6.19)$$

so the characteristic impedance of the forward-going wave is

$$Z_0 = \frac{V_+}{I_+} = \sqrt{\frac{L_0}{C_0}}\, e^{-j\theta/2} \qquad (5.6.20)$$

in view of (5.6.19) and the dispersion relation (5.6.16).

The dispersion relation (5.6.16) is plotted in Figure 5.12. Note that the points a and a' at $\theta = \theta_a$ and $\theta_a + 2\pi$ are identical because

$$e^{-jn\theta_a} = e^{-jn(\theta_a + 2\pi)}$$

and therefore only those points θ within a band 2π wide are unique. The *principal band* where $-\pi < \theta \le \pi$ is usually chosen, and because both forward and backward waves are explicitly represented by (5.6.14) and (5.6.15), we restrict θ to the positive values $0 < \theta \le \pi$.

In the limit $\omega \ll \omega_0$, the dispersion relation (5.6.16) becomes

$$\sin^2 \frac{\theta}{2} \simeq \frac{\theta^2}{4} = \frac{\omega^2 L_0 C_0}{4} \qquad (5.6.21)$$

which means that

$$k \simeq \pm\omega\sqrt{(L_0/\ell)(C_0/\ell)} \qquad (5.6.22)$$

Equation (5.6.22) is the dispersion relation of a continuous, lossless *TEM* line. Because ω/k is approximately constant for $\omega \ll \omega_0$, the phase and group velocities are the same in this regime. Therefore, $v_p = v_g \simeq \ell/\sqrt{L_0 C_0}$ as suggested by the

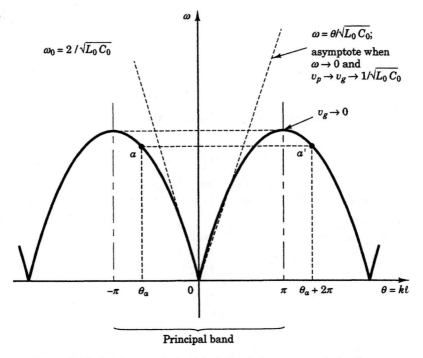

Figure 5.12 Dispersion relation for a lumped-element transmission line.

plot of ω versus $k\ell$ in Figure 5.12. Note that as ω approaches the cut-off frequency ω_0, $\theta = \pi$ and the group velocity $v_g = \partial\omega/\partial k$ goes to zero.

The principal practical significance of lumped-element circuits is that the product $L_0 C_0$ can be made almost arbitrarily large compared to $\mu\epsilon$, and therefore the effective phase velocity

$$v_p = \omega/k = \ell/\sqrt{L_0 C_0} \tag{5.6.23}$$

can be made very slow.

When the frequency ω exceeds the cut-off frequency ω_0, no real value of θ can satisfy the dispersion relation (5.6.16), so

$$\theta = \theta_r - j\alpha \tag{5.6.24}$$

where θ_r and α are real. The dispersion relation (5.6.16) then becomes

$$\sin\left(\frac{\theta_r - j\alpha}{2}\right) = \frac{\omega}{\omega_0}$$

where the positive square root of (5.6.16) is taken corresponding to propagation in the direction of increasing node number n. Using a well-known trigonometric identity, the dispersion relation is

$$\sin(\theta_r/2)\cosh(\alpha/2) - j\cos(\theta_r/2)\sinh(\alpha/2) = \omega/\omega_0$$

Since ω/ω_0 is real,

$$\cos(\theta_r/2)\sinh(\alpha/2) = 0 \tag{5.6.25}$$

so $\alpha = 0$ or $\theta_r = \pi$. If $\alpha = 0$, then

$$\sin(\theta_r/2) = \omega/\omega_0 \tag{5.6.26}$$

which is the propagating case we studied previously for $\omega < \omega_0$. If $\theta_r = \pi$, then $\omega > \omega_0$ and

$$\cosh(\alpha/2) = \omega/\omega_0 \tag{5.6.27}$$

which corresponds to an evanescent wave. Thus, we expect a phase shift of π and incremental attenuation of $e^{-\alpha}$ each time we move to an adjacent cell along the line when $\omega > \omega_0$.

In the limit $\omega \gg \omega_0$, (5.6.27) becomes

$$\omega/\omega_0 = \frac{e^{\alpha/2} + e^{-\alpha/2}}{2} \cong \frac{1}{2}e^{\alpha/2} \tag{5.6.28}$$

and the voltage on the nth node becomes

$$V = V_+ e^{-jn\pi - n\alpha} = V_+ \left(-e^{-\alpha}\right)^n \tag{5.6.29}$$

But (5.6.16) and (5.6.28) yield an expression for e^α

$$e^\alpha \cong 4\omega^2/\omega_0^2 = \omega^2 L_0 C_0$$

so that (5.6.29) becomes

$$V \cong V_+ \left(-\frac{1}{\omega^2 L_0 C_0}\right)^n \tag{5.6.30}$$

This result can also be obtained from an examination of the circuit illustrated in Figure 5.11 in the high frequency limit ($\omega \gg 2/\sqrt{L_0 C_0}$). The circuit then approximates a series of $L_0 C_0$ voltage dividers, each of which strongly reduces the voltage by a factor

$$V_{n+1}/V_n \cong \frac{1/j\omega C_0}{j\omega L_0 + 1/\omega C_0} \approx -1/\omega^2 L_0 C_0 \ll 1$$

which is equivalent to (5.6.30). Strong attenuation results because L_0 is nearly an open circuit, and C_0 nearly a short circuit when $\omega \gg \omega_0$. The signal becomes weaker as n increases due to this reactive low-pass filter, not because of dissipative losses.

Another interesting circuit is obtained if we interchange the inductors and capacitors in Figure 5.11. The dispersion relation is easily found by replacing $j\omega L_0$ with $1/j\omega C_0$ and $1/j\omega C_0$ with $j\omega L_0$ in (5.6.11) and (5.6.12). The dispersion relation $\sin^2(\theta/2) = \omega^2 L_0 C_0/4 = -(j\omega L_0)(j\omega C_0)/4$ then becomes

$$\sin^2 \frac{\theta}{2} = -\frac{1}{4}\left(\frac{1}{j\omega C_0}\right)\left(\frac{1}{j\omega L_0}\right) = \frac{1}{4\omega^2 L_0 C_0} = \frac{\omega_0^2}{\omega^2} \tag{5.6.31}$$

where now $\omega_0 = 1/2\sqrt{L_0 C_0}$. This circuit and dispersion relation are illustrated in Figure 5.13. This line has the very interesting property that when the group velocity is positive, corresponding to an energy packet moving in the $+n$ direction, the phase velocity is negative. This is called a *backward-wave transmission line*. Note that real values of θ exist only for $\omega > \omega_0$, and so this structure acts as a high-pass filter. When $\omega < \omega_0$, this line supports only evanescent waves.

With this more complete model for lumped-element lines, we consider the consequences of seeking distortionless behavior by altering L_0. The series resistance losses R dominate the conductance losses G in most telephone lines, so that $R/L \gg G/C$, causing serious distortion for long lines. In 1900, Pupin of Columbia University showed experimentally that adding loading coils L_p periodically to telephone lines reduced speech distortion. We assume that these Pupin coils are spaced much less that a wavelength apart and that the intervening transmission line can be accurately modeled by the lumped-element circuit in Figure 5.14.

To find the dispersion relation, we replace $j\omega L_0$ in (5.6.16) by $j\omega(L_p + L_0) = j\omega L_s$ and include resistive losses in the *TEM* line conductors by adding series resistance R_0 as in (5.6.5):

$$\sin(\theta/2) = \frac{1}{2}\sqrt{-(j\omega C_0)(R_0 + j\omega L_s)}$$

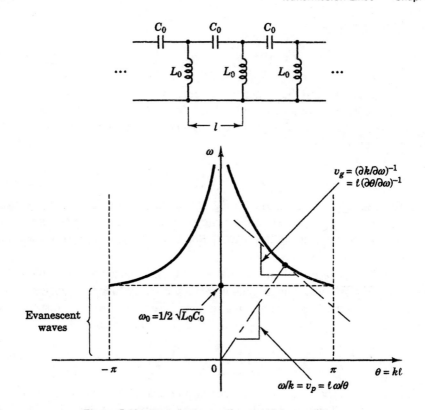

Figure 5.13   Backward-wave line and high-pass filter.

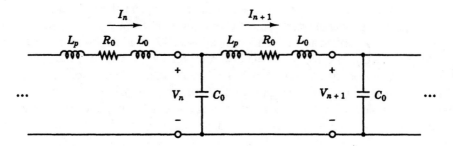

Figure 5.14   Lumped-element transmission line model with Pupin coils.

The conductance G_0 has been neglected. For $\theta \ll \pi$, $\sin(\theta/2) \approx \theta/2$ and we find

$$\theta \cong \omega\sqrt{L_s C_0}\sqrt{1 - j\frac{R}{\omega L_s}}$$

$$\cong \omega\sqrt{L_s C_0}\left(1 + \frac{R_0}{8\omega^2 L_s^2}\right) - j\frac{R_0}{2}\sqrt{\frac{C_0}{L_s}}$$

Thus, by increasing L_p and therefore L_s, both the attenuation and distortion are reduced. They would approach zero as $L_s \to \infty$ except for the fact that such coils also have a small series resistance that must be considered for large values of L_p. The waves also slow down as the Pupin inductance increases, since $v_p \simeq v_g \simeq 1/\sqrt{L_s C_0}$.

Example 5.6.1

Design a 10^{-5} s lumped-element delay line with a characteristic impedance of $\sim 100 \ \Omega$ and a delay time that varies less than 10^{-6} s over the bandwidth $0 - 1$ MHz. Ignore resistive and conductive losses in this line.

Solution: The dispersion relation for the discrete-element delay line is

$$\sin(\theta/2) = \frac{\omega\sqrt{L_0 C_0}}{2} \tag{1}$$

The group velocity of this line may be calculated by differentiating both sides of (1) with respect to k ($\theta = k\ell$), so that

$$\frac{\ell}{2}\cos(k\ell/2) = \frac{1}{2}\sqrt{L_0 C_0}\frac{\partial\omega}{\partial k}$$

which means that

$$v_g = \frac{\partial\omega}{\partial k} = \frac{\ell}{\sqrt{L_0 C_0}}\cos(k\ell/2) = \frac{N\ell}{T} \tag{2}$$

where N is the number of delay line sections and T is the time it takes a signal to propagate from one end of the line to the other; (2) defines the delay time T. In order for T, defined to be 10^{-5} s at 0 Hz, to vary less than 10^{-6} s over the bandwidth 0–1 MHz,

$$\left|\frac{T(\omega = 2\pi \cdot 10^6) - T(\omega = 0)}{T(\omega = 0)}\right| \leq \frac{10^{-6}}{10^{-5}} = 0.1$$

so that

$$1 \leq \frac{T(\omega = 2\pi \cdot 10^6)}{T(\omega = 0)} \leq 1.1$$

Since $T(\omega = 0)/T(\omega = 2\pi \cdot 10^6) = \cos(k\ell/2)$, we find that $k\ell/2 \leq 0.43$ radians. The dispersion relation (1) can then be used determine a value for the $L_0 C_0$ product

$$\sqrt{L_0 C_0} = \frac{2}{\omega}\left|\sin\left(\frac{k\ell}{2}\right)\right|_{\omega = 2\pi \cdot 10^6} \leq 0.133 \ \mu s$$

so the number of delay line sections necessary is given by (2):

$$N \geq \frac{T}{\sqrt{L_0 C_0}} \cos\left(\frac{k\ell}{2}\right)\Bigg|_{\omega=0} = \frac{10^{-5}}{1.33 \times 10^{-7}} = 75.4$$

If the characteristic impedance of the line is $Z_0 = \sqrt{L_0/C_0} = 100 \ \Omega$, and $\sqrt{L_0 C_0} = 0.133 \ \mu s$, then

$$L_0 = 100 \times 1.33 \times 10^{-7} = 13.3 \ \mu H$$

$$C_0 = 1.33 \times 10^{-7}/100 = 1.33 \ nF$$

5.7 SUMMARY

Transverse electromagnetic (TEM) waves propagate in the $\pm \hat{z}$ direction with z dependence given by $e^{\mp jkz}$, where $E_z = H_z = 0$. These *TEM* waves can propagate not only in homogeneous, unbounded media but also in structures with cross-sections independent of z, such as *parallel plates, coaxial lines, parallel wires*, and so forth. On each of these *transmission lines*, a unique voltage $V(z)$ and current $I(z)$ can be defined, satisfying the *telegrapher's equations*

$$\frac{\partial V(z)}{\partial z} = -j\omega L I(z)$$

$$\frac{\partial I(z)}{\partial z} = -(G + j\omega C)V(z)$$

where L, C, and G are *inductance, capacitance*, and *conductance* per unit length, respectively. If the conductors from which the transmission line structure is fabricated are lossy, then pure *TEM* waves no longer propagate. However, if the conductors are only slightly lossy, with *resistance* per unit length $R \ll \omega L$, then the telegrapher's equations are still approximately valid and $j\omega L$ can be replaced with $R + j\omega L$.

The most general waves propagating on *TEM* lines thus have voltage and current given by

$$V(z) = V_+ e^{-jkz} + V_- e^{+jkz}$$

$$I(z) = Y_0 V_+ e^{-jkz} - Y_0 V_- e^{+jkz}$$

where

$$Y_0 = 1/Z_0 = \sqrt{\frac{G + j\omega C}{R + j\omega L}}$$

and

$$k = \sqrt{-(R + j\omega L)(G + j\omega C)}$$

For lossless *TEM* lines ($R = G = 0$), $Y_0 = \sqrt{C/L}$ and $k = \omega\sqrt{LC}$ where Y_0 and k are real. If $R/L = G/C$, then $Y_0 = \sqrt{C/L}$ and $k = \omega\sqrt{LC} - jRY_0$ and the line is *distortionless*. For structures that are truly *TEM* $(E_z = H_z \equiv 0)$, all frequencies can propagate whether or not the line is lossless.

Signals possessing many frequencies may best be treated in the time domain, provided that the *characteristic impedance* Z_0 of the line is not frequency-dependent. The total voltage and current are superpositions of separate forward- and backward-traveling waves:

$$V(z, t) = V_+\left(t - \frac{z}{v}\right) + V_-\left(t + \frac{z}{v}\right)$$

$$I(z, t) = Y_0\left[V_+\left(t - \frac{z}{v}\right) - V_-\left(t + \frac{z}{v}\right)\right]$$

where v is the *velocity* of wave propagation.

It can be useful to model *TEM* lines as *periodic lumped-element circuits* containing inductors, capacitors, and resistors characterized by difference equations rather than the differential telegrapher's equations. The solutions to these difference equations yield *dispersion relations* similar to those of continuous structures, but often with *cut-off frequencies* above or below which only *evanescent waves* may occur. Often, lumped-element structures have desirable features, making them superior to (or more economically feasible than) continuous structures.

5.8 PROBLEMS

5.2.1 Neglecting fringing fields, what is the approximate characteristic impedance Z_0 of a *TEM* line formed by a printed-circuit wire 0.2 mm wide on a plastic sheet ($\epsilon = 9\epsilon_0$) that is 20 μm thick and bonded to a ground plane? What are the associated inductance L (H/m) and capacitance C (F/m) of this line? What is the velocity of *TEM* waves on this line?

5.2.2 One kilowatt of time-averaged power is propagating in the $+\hat{z}$ direction along an air-filled 100-Ω parallel-plate *TEM* line with 1-cm plate separation in the $\hat{x}$ direction at 100 MHz.

(a) Find numerical expressions for $\overline{E}(x, y, z, t)$ and $\overline{H}(x, y, z, t)$ for the case of no reflection. At $z = 0$ and $t = 0$, $\overline{E} = 0$ and $\partial\overline{E}/\partial t > 0$.

(b) Numerically evaluate the surface charges density $\sigma_s(y, z, t)$ and surface current density $\overline{J}_s(y, z, t)$ for part (a).

(c) Repeat parts (a) and (b) for the case where $\epsilon = 4\epsilon_0$ and $\mu = 9\mu_0$.

5.2.3 A 50-Ω microstrip *TEM* line has dimensions 10 μm $\times$ 400 μm and is filled with a low-loss dielectric ($\mu = \mu_0$, $\sigma = 3 \times 10^{-4}$ S/m).

(a) What is the capacitance C (F/m) and conductance G (S/m) of this line?

(b) After how many meters will the power drop by 20 dB in this line?

5.2.4 An air-filled, 0.1-GHz parallel-plate *TEM* line is 1 cm thick and 50 cm wide. The maximum possible electric field on this line is 3×10^7 V/m and the maximum magnetic field is 10 T.

(a) What is the maximum power carried by this transmission line (watts)?

(b) If the conducting walls are metal with $\sigma = 5 \times 10^5$ S/m, what is the power dissipation in the walls (W/m)?

(c) What is the power attenuation coefficient α (m^{-1}) for this *TEM* line? How many meters does it take for the power to fall to $1/e$ of its original value?

5.3.1* Prove for a *TEM* line of arbitrary cross-section that

$$V\, I^* = \int_A \left(\overline{E} \times \overline{H}^*\right) \cdot \hat{z}\, da$$

where $\hat{z}$ is the unit vector in the direction of propagation and A is the cross-sectional area of the line.

5.3.2* Show that the time-averaged electric and magnetic energy densities of a *TEM* waveguide of arbitrary cross-section are given by

$$\langle w_e'\rangle = \frac{1}{4}\,C\,|V|^2 = \frac{1}{4}\int_A \epsilon |\overline{E}|^2\, da \quad \text{(J/m)}$$

$$\langle w_m'\rangle = \frac{1}{4}\,L\,|I|^2 = \frac{1}{4}\int_A \mu |\overline{H}|^2\, da \quad \text{(J/m)}$$

where A is the cross-sectional area of the *TEM* line.

5.4.1 Large amounts of 1-GHz RF power are to be fed to the magnetically confined plasma of a nuclear fusion reactor. Assume an air-filled coaxial cable is to be used, made of copper with $\sigma = 5 \times 10^7$ S/m.

(a) If the outer radius b of the cable is fixed, approximately what value for the inner radius a (in terms of b) permits maximum power transfer to a matched load such that the instantaneous breakdown voltage 50,000 V/cm is never exceeded? Note that the peak field strength occurs at the conductors.

(b) What minimum dimensions a and b are needed to handle 1-GW time-averaged power?

(c) How much heat (watts) must be dissipated by the cooling system per meter of coaxial length if this cable were to carry 1 GW steadily? Assume the cable walls are thick compared to the skin depth.

5.4.2* A pair of parallel wires in free space, each of diameter 0.1 mm and separated by 1 cm, acts as a *TEM* line.

(a) What is the approximate characteristic impedance of this line?

(b) What are the inductance and capacitance L (H/m) and C (F/m) of this line?

(c) What is the total energy per meter stored in this line if it carries 100 W?

5.4.3 Using the graphical technique of Figure 5.8 in Section 5.4, roughly estimate the impedance of a *TEM* line composed of two parallel metal strips 0.1 mm thick, 1 cm wide (each), and separated by 1 mm at their edges; they lie in the same plane. Assume $\epsilon = \epsilon_0$ and $\mu = \mu_0$ everywhere, and note that the total cross-section of the line is $10^{-4} \times 0.021$ m^2.

5.5.1 A 10-MHz, 1-W wave is propagating in a 100-Ω air-filled *TEM* line. The electric field is zero and $\partial \overline{E}/\partial t$ is greater than zero at $t = 0$ and $z = 0$.

(a) What is $V_+(t - z/c)$?

(b) What is $\bar{V}_+(z - ct)$?

5.5.2 An air-filled 100-Ω *TEM* transmission line 0.5 m long is excited by a step voltage $2u(t)$ in series with the resistor Z_0 at $z = 0$. A resistor R_L is placed across the line at $z = 0.5$ m. We observe $V(t)$ at $z = 0$ as a steady 1 V until 10 ns, when it drops to 1/3 volt and remains at that level for all time.

(a) What is the inductance per meter L of the *TEM* line?

(b) What is the load resistance R_L?

(c) Sketch and dimension $V(z)$ at $t = 8$ ns.

(d) Sketch and dimension $I(z)$ at $t = 8$ ns.

(e) Sketch $V(t)$ for the case where R_L is replaced by a 30-pF capacitor.

5.6.1 Design a 10^{-4} s lumped-element delay line with a characteristic impedance of 100 Ω and a delay time that varies less than 10^{-5} s over the band 0-100 kHz.

(a) What values did you choose for L_0, C_0 and how did you configure them?

(b) How many sections did you use and why? (Use the minimum number generally consistent with the specification for delay-time variability.)

5.6.2 A certain twisted pair of insulated wires approximates a 1-MHz *TEM* line with $Z_0 \cong$ 50 Ω and a propagation velocity of 10^8 m/s. If the shunt conductance G is 10^{-5} S/m and the series resistance of each is 0.1 Ω/m, then

(a) What are L (H/m) and C (F/m) for this line? (Give quantitative answers.)

(b) What periodic lossless lumped reactances would you add to produce a distortion-less line? (That is, give a numerical value for X and determine the appropriate spacing of the elements.) Assume the line is used below 1 MHz for data communication inside buildings.

(c) If the distortionless line of part (b) were short-circuited at one end, find Γ and the complex impedance Z of the shorted line as seen from a point 112.5 meters away from the short.

5.6.3 A lumped-element transmission line is constructed as shown in Figure 5.13 (see Section 5.4).

(a) Determine a set of first-order difference equations relating I_{n+1}, and V_{n+1} to I_n and V_n. Each cell has length ℓ.

(b) Find the propagation constant k as a function of frequency ω using these equations.

(c) What is the pass-band of this structure?

(d) Repeat parts (a)–(c) for the case where each capacitor C_0 of Figure 5.13 is in parallel with an inductor $L_0/10$.

6

TRANSMISSION LINE SYSTEMS

6.1 IMPEDANCE, REFLECTION, AND TRANSMISSION

Because the behavior of a transmission line can be characterized by the complex, spatially varying impedance of the line $Z(z) = V(z)/I(z)$, transmission lines are able to transform one impedance into almost any other. *TEM* lines are thus extremely flexible and useful elements in many electrical systems. Chapter 6 focuses on the design and analysis of transmission line systems, in which transmission lines are connected to other transmission lines or to passive lumped elements. Both time- and frequency-domain analysis is presented, using Smith chart and *ABCD* matrix techniques for the latter. *TEM* lines are also used to model the transmission of electromagnetic waves at planar boundaries, and many parallels can be drawn between these analogous systems.

We begin by describing the generic *TEM* transmission line illustrated in Figure 6.1, which possesses a characteristic impedance Z_0, wave number k, and is terminated with a load impedance Z_L. This *TEM* line may be a parallel-plate line, a coaxial cable, or any other arbitrary two-conductor configuration of uniform cross-section. As the previous sections have shown, the transmission line equations (5.3.36) and (5.3.48) are obeyed for all such lines. It is often mathematically convenient to place the origin $z = 0$ at the load; the start of the line is then at $z = -\ell$ for a *TEM* line of length ℓ. We wish to determine the relationship between forward- and backward-propagating *TEM* waves at this load, and we therefore rewrite (5.3.45) and (5.3.49) in terms of the *reflection coefficient at the load* Γ_L:

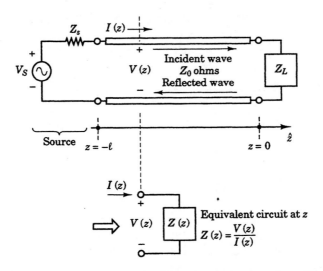

Figure 6.1   Generic transmission line and equivalent impedance $Z(z)$.

$$V(z) = V_+ \left(e^{-jkz} + \Gamma_L e^{+jkz} \right) \tag{6.1.1}$$

$$I(z) = Y_0 V_+ \left(e^{-jkz} - \Gamma_L e^{+jkz} \right) \tag{6.1.2}$$

where $\Gamma_L \equiv V_-/V_+$. The ratio of the total voltage $V(z)$ to the total current $I(z)$ is defined as the complex impedance $Z(z) = V(z)/I(z)$, so $Z(z)$ is a function of position z along the *TEM* line. We interject a reminder at this point as to how the complex impedance is defined. In general, $Z(z)$ is written

$$Z(z) = R(z) + jX(z)$$

where the real quantities R and X are called *resistance* and *reactance*, respectively. The resistance $R(z)$ should not be confused with R, the series resistance of a *TEM* line (Ω/m) introduced in (5.3.44) to account for the dissipative loss in the two conductors. Instead, $R(z)$ is the real part of $Z(z)$ and depends primarily on Z_0 (which may include effects of R), line length ℓ, and load impedance Z_L. For a passive system, $R(z) \geq 0$ for all z. The reactance $X(z)$ can be either positive or negative. Since the impedance of an inductor is $Z_L = j\omega L = jX_L$, we associate **positive** values of X with **inductive** behavior; conversely, **negative** values of X suggest **capacitive** behavior because $Z_C = 1/j\omega C = -jX_C$. The reciprocal of impedance is also a useful quantity, and it too has both a real and an imaginary part, as

$$Y(z) = G(z) + jB(z)$$

where $Y(z) = 1/Z(z)$ is called *admittance*, G is *conductance*, and B is *susceptance*; both G and B are real numbers ($G \geq 0$ if $R \geq 0$). All six quantities Z, R,

X, Y, G, and B are functions of position, and whether we utilize impedance or admittance is generally a matter of convenience.

The ratio of (6.1.1) to (6.1.2) evaluated at $z = 0$ yields

$$Z(z = 0) = Z_L = \frac{V(z = 0)}{I(z = 0)} = Z_0 \left(\frac{1 + \Gamma_L}{1 - \Gamma_L} \right) \tag{6.1.3}$$

and (6.1.3) may be solved for Γ_L to give

$$\Gamma_L = \frac{Z_L - Z_0}{Z_L + Z_0} \tag{6.1.4}$$

In certain cases, Γ_L assumes simple values:

(1) If $Z_L = 0$, then $\Gamma_L = -1$ and $V_- = -V_+$. The incident wave is reflected with full amplitude but $180°$ out of phase so the total voltage at $z = 0$ is zero. This is consistent with a short-circuit load, and we note that no power is dissipated in the load because the wave magnitude suffers no attenuation.

(2) If $Z_L = \infty$, $\Gamma_L = +1$ and $V_- = V_+$. The wave is fully reflected from an open circuit with no change of sign, satisfying the open-circuit boundary condition $I(z = 0) = 0$. Again, no power is dissipated in the load because no current flows through it.

(3) If $Z_L = Z_0$, $\Gamma_L = 0$ and $V_- = 0$. No wave is reflected because the termination impedance is in the same ratio $V(z)/I(z)$ as an infinite extension of the transmission line. Such a termination is called a *matched load*. Because no reflected wave is observed, **all** of the incident power is dissipated in the load.

We have already mentioned that the impedance $Z(z)$ generally varies with position. Dividing (6.1.1) by (6.1.2) and substituting (6.1.4) for Γ_L yields

$$\begin{aligned} Z(z) = \frac{V(z)}{I(z)} &= Z_0 \left(\frac{(Z_L + Z_0) e^{-jkz} + (Z_L - Z_0) e^{+jkz}}{(Z_L + Z_0) e^{-jkz} - (Z_L - Z_0) e^{+jkz}} \right) \\ &= Z_0 \left(\frac{Z_L \cos kz - j Z_0 \sin kz}{-j Z_L \sin kz + Z_0 \cos kz} \right) \\ &= Z_0 \left(\frac{Z_L - j Z_0 \tan kz}{Z_0 - j Z_L \tan kz} \right) \end{aligned} \tag{6.1.5}$$

At the source end of the transmission line of length ℓ, we thus find that

$$Z_{\text{in}} = Z(z = -\ell) = Z_0 \left(\frac{Z_L + j Z_0 \tan k\ell}{Z_0 + j Z_L \tan k\ell} \right) \tag{6.1.6}$$

which means that the input impedance of the line varies periodically with period $\ell = \pi/k = \lambda/2$ as we move away from the load.

Consider the special case of a short-circuit load, $Z_L = 0$. In this case, (6.1.6) reduces to

$$Z_{in} = Z(z = -\ell, Z_L = 0) = jZ_0 \tan k\ell = jX \qquad (6.1.7)$$

where the reactance X is plotted in Figure 6.2. At $k\ell = 0$, π, 2π, 3π, ..., where $\ell = 0$, $\lambda/2$, λ, $3\lambda/2$, ..., we find that $Z_{in} = 0$. Therefore, looking into a shorted *TEM* line an integral number of half wavelengths long is equivalent to looking into a short circuit. At $k\ell = \pi/2$, $3\pi/2$, ..., or $\ell = \lambda/4$, $3\lambda/4$, ..., the reactance becomes infinite and the input impedance of the *TEM* line appears to be an open circuit. For other values of ℓ not equal to integer multiples of $\lambda/4$, the input impedance at $z = -\ell$ is finite and can be either inductive or capacitive. The resemblance of a *TEM* line to a capacitor or an inductor refers only to the sign of X at a single frequency, since the impedance looking into a *TEM* line is generally not directly or inversely proportional to frequency, even over a very narrow bandwidth. The fact that an open- or short-circuited *TEM* line of the proper length may be used to simulate **any** capacitance or inductance (at a given frequency) is of great practical importance.

We proceed to make several observations about the input impedance of a *TEM* line of characteristic impedance Z_0 and length ℓ:

(1) As we vary the length of a line terminated by a short circuit, the equivalent impedance Z_{in} varies from a short circuit ($Z_{in} = 0$) to an open circuit ($Z_{in} \to \pm\infty$). Therefore, the termination impedance Z_L is not sufficient to determine the input impedance of the line; line length is important too.

(2) The input impedance is a periodic function of ℓ with period $\ell = \lambda/2$.

(3) If the line is lossless (Z_0 is purely real) and the termination Z_L is purely reactive ($Z_L = jX_L$), then the input impedance $Z_{in}(z = -\ell)$ is purely reactive

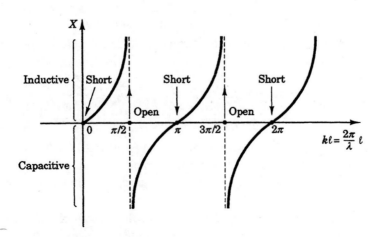

Figure 6.2 Impedance $Z(z = -\ell) = jX$ on a short-circuited *TEM* transmission line.

as well. This may be shown by substituting $Z_L = jX_L$ into (6.1.6) and observing that $\text{Re}\{Z_{\text{in}}\} = 0$. No power can be dissipated anywhere on a line if the termination and the line itself are both lossless.

(4) If $Z_L = Z_0$ then the load is matched to the *TEM* line. In this case, (6.1.6) becomes

$$Z_{\text{in}} = Z(z = -\ell, Z_L = Z_0) = Z_0$$

Because we have already found that no reflected wave is generated at a matched load, it is unsurprising that the voltage-to-current ratio is simply the characteristic impedance Z_0 everywhere on the line.

We are used to thinking of a conducting surface as being an electrostatic equipotential. Clearly, this is not the case with wave propagation along *TEM* lines because $V(z)$, the voltage across the two plates, has already been shown to vary with position. However, if the line is made short enough (much less than one wavelength), we should recover quasi- static behavior. If $Z_L = 0$ and $k\ell \ll 1$, (6.1.6) may be rewritten as

$$Z_{\text{in}}(Z_L = 0) = jZ_0 \tan k\ell \simeq jZ_0 k\ell = j\sqrt{\frac{L}{C}} \, \omega\sqrt{LC} \, \ell$$

$$= j\omega L\ell = j\omega\mathcal{L} \tag{6.1.8}$$

where expressions (5.3.46) and (5.3.50) for the wave number and characteristic impedance as functions of L and C have been used. (The *TEM* line is assumed to be lossless here, so $R = G = 0$.) The total inductance in this transmission line is $\mathcal{L} = L\ell$ (H), so the line impedance right next to this short circuit is just that of an inductor. Very close to a short circuit, currents are much larger than voltages, so that the average stored magnetic energy greatly exceeds the average stored electric energy, and the equivalent circuit near a short-circuited load is thus inductive. As ℓ increases, the stored electric energy becomes more important, and the exact expression for input impedance (6.1.6) must be used.

If the load impedance is infinite, corresponding to an open circuit, and $k\ell \ll 1$, (6.1.6) reduces to

$$Z_{\text{in}}(Z_L \to \pm\infty) = -jZ_0 \cot k\ell \simeq \frac{-jZ_0}{k\ell} = -j\sqrt{\frac{L}{C}} \frac{1}{\omega\sqrt{LC} \, \ell}$$

$$= -j/\omega C\ell = 1/j\omega\mathcal{C} \tag{6.1.9}$$

where again (5.3.46) and (5.3.50) have been used (with $R = G = 0$), and $\mathcal{C}$ is the total capacitance, equal to $C\ell$ (F). Near an open circuit, voltages are much larger than currents, and so the electric energy dominates and the impedance appears to be capacitive.

For many calculations using *TEM* lines, it would be useful to replace the *TEM* line with simple circuit elements. Because transmission lines are linear, circuit theory allows this substitution; any combination of linear circuit elements may be replaced by a *Thevenin* or *Norton equivalent circuit*. In the Thevenin case, we can replace everything to the left of the load at $z = 0$ with a series combination of a voltage source V_{Th} and a complex impedance Z_{Th}, where both V_{Th} and Z_{Th} depend on the length and characteristic impedance of the *TEM* line. No matter what load we place across the terminals of this equivalent circuit, the voltage and current at $z = 0$ must be exactly the same as would be found if an identical load were used to terminate the transmission line. The Norton equivalent circuit is a parallel combination of a current source I_N and impedance Z_N, which also has the same terminal voltage and current as the *TEM* line for any load placed across it. A generic transmission line, along with its Thevenin and Norton equivalent circuits, is pictured in Figure 6.3.

We can calculate the Thevenin equivalent voltage source by placing convenient loads at $z = 0$ in both circuits in Figure 6.3. Initially, we imagine that the load impedance of both the Thevenin circuit and the *TEM* line is infinite. In this case, the terminal voltage of the Thevenin equivalent circuit is just $V(z = 0) = V_{Th}$. With an open-circuited load, the load reflection coefficient of the *TEM* line Γ_L is $+1$, so the load voltage of the *TEM* line is $V(z = 0) = 2V_+$ by (6.1.1). We therefore conclude that

$$V_{Th} = 2V_+ \tag{6.1.10}$$

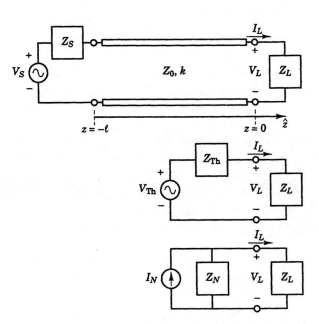

Figure 6.3   Thevenin and Norton equivalent circuits for a *TEM* line.

Since V_{Th} does not depend on the load (by definition), (6.1.10) is true in general, as will also be demonstrated for time-domain signals in Section 6.5. We can calculate V_+ by finding the voltage at $z = -\ell$ from (6.1.1):

$$V(z = -\ell, \Gamma_L = +1) = V_+(e^{jk\ell} + e^{-jk\ell}) = 2V_+ \cos k\ell \qquad (6.1.11)$$

But $V(z = -\ell)$ is also related to the source voltage V_S and source impedance $Z(z = -\ell)$ by the voltage-divider equation

$$V(z = -\ell) = V_S \left(\frac{Z(z = -\ell)}{Z(z = -\ell) + Z_S} \right) \qquad (6.1.12)$$

because the impedance looking into the *TEM* line at $z = -\ell$ is in series with the source impedance Z_S. For an open-circuited line ($Z_L \to \infty$), $Z(z = -\ell)$ is just $-jZ_0 \cot k\ell$ [see (6.1.9)], so we can combine (6.1.10)–(6.1.12) to find

$$V_{\text{Th}} = 2V_+ = \left(\frac{-jZ_0 \csc k\ell}{Z_S - jZ_0 \cot k\ell} \right) V_S \qquad (6.1.13)$$

The Thevenin impedance Z_{Th} may be found in at least two ways. By far the simplest is to remove the load resistors from both circuits in Figure 6.3 and then to short-circuit V_S [and hence V_{Th}, which is directly proportional to V_S in (6.1.13)]. We then calculate the impedance looking back into each circuit from $z = 0$. In the Thevenin circuit, this impedance is just Z_{Th}. For the *TEM* line, the impedance is that of a *TEM* line of length ℓ and characteristic impedance Z_0 terminated by an impedance Z_S, so

$$Z_{\text{Th}} = Z_0 \left(\frac{Z_S + jZ_0 \tan k\ell}{Z_0 + jZ_S \tan k\ell} \right) \qquad (6.1.14)$$

where we have used (6.1.6) with Z_S taking the role of a load impedance.

Alternatively, the Thevenin impedance may be found by replacing the load impedance in both the *TEM* line and equivalent circuit with a short circuit: $Z_L = 0$ and $\Gamma_L = -1$. In this case, $I(z = 0) = V_{\text{Th}}/Z_{\text{Th}}$ for the Thevenin equivalent circuit, and $I(z = 0)$ may be calculated from (6.1.2) for the short-circuited *TEM* line: $I(z = 0) = 2\hat{V}_+/Z_0$. A hat is placed over V_+ to remind us that $\hat{V}_+$ is different from V_+ defined in (6.1.10) and (6.1.11) because the amplitude of the forward-propagating voltage wave is dependent on the load impedance. We can now relate $\hat{V}_+$ to $V(z = -\ell)$ using (6.1.1):

$$V(z = -\ell, \Gamma_L = -1) = \hat{V}_+(e^{jk\ell} - e^{-jk\ell}) = 2j\hat{V}_+ \sin k\ell \qquad (6.1.15)$$

Again, $V(z = -\ell)$ and V_S are linked by (6.1.10), where this time the short-circuited *TEM* line input impedance $Z(z = -\ell)$ is $jZ_0 \tan k\ell$ from (6.1.6) with $Z_L = 0$. Combining (6.1.10) and (6.1.15) thus yields

$$I(z = 0) = \frac{2\hat{V}_+}{Z_0} = \frac{V(z = -\ell, \Gamma_L = -1)}{jZ_0 \sin k\ell} = \left(\frac{\sec k\ell}{Z_S + jZ_0 \tan k\ell} \right) V_S \qquad (6.1.16)$$

which is also equal to V_{Th}/Z_{Th} in the Thevenin equivalent circuit. Therefore, (6.1.13) and (6.1.16) result in the identical expression for Thevenin impedance (6.1.4) derived more intuitively above.

The Norton equivalent circuit is easily found by comparison with the Thevenin circuit. For an open-circuit load, we find that $V(z = 0) = V_{Th} = Z_N I_N$, and for a short-circuit load, $I(z = 0) = V_{Th}/Z_{Th} = I_N$. Therefore,

$$I_N = V_{Th}/Z_{Th}$$
$$Z_N = Z_{Th}$$
(6.1.17)

for this alternate circuit; (6.1.17) is true in general. Note that we could have picked any two test load impedances to find V_{Th} and Z_{Th}—the open- and short-circuit cases were just particularly simple to analyze.

To compute the total time-averaged power P_d absorbed by a complex load at the end of a lossless *TEM* line, it is usually simplest to compute the real power at the input to the line. That is, we can find $Z(z = -\ell)$ using (6.1.6), and then compute

$$P_d = \frac{1}{2}\, \mathrm{Re}\left\{V I^*\right\}\big|_{z=-\ell}$$
$$= \frac{1}{2}\, \mathrm{Re}\left\{|V|^2/Z^*\right\}\big|_{z=-\ell}$$
(6.1.18)

where $V(z = -\ell)$ is given by (6.1.12). We can also compute P_d by considering the power absorbed in the load:

$$P_d = \frac{1}{2}\, \mathrm{Re}\left\{V I^*\right\}\big|_{z=0}$$
$$= \frac{1}{2Z_0}|V_+|^2(1 - |\Gamma_L|^2)$$
(6.1.19)

Equations (6.1.18) and (6.1.19) should result in identical expressions for the time-averaged dissipated power.

Example 6.1.1

A voltage source of $V_S(t) = V\cos\omega t$ $(\omega = 2\pi \cdot 10^6 \text{ rad/s})$ drives a transmission line $(Z_0 = 100\ \Omega,\ \ell = \lambda/8)$ terminated with a resistor $R = 100\ \Omega$ in parallel with an inductor $L = 16\ \mu\text{H}$, as illustrated in Figure 6.4. What are the input impedance Z_{in}, the input voltage $V_{in}(t)$, and the output (load) voltage $V_L(t)$? What power is dissipated in the resistor R?

Solution: The impedance of the inductor at 1 MHz is $j\omega L = j(2\pi \cdot 10^6)(16 \times 10^{-6}) \simeq j100\ \Omega$, so the parallel combination of the inductor and resistor gives a total load impedance of

$$Z_L \simeq \frac{100(j100)}{100 + j100} = 50(1 + j)$$
(1)

The input impedance of the *TEM* line may be found from (6.1.6), so

$$Z_{in} = Z_0 \left(\frac{Z_L + jZ_0 \tan k\ell}{Z_0 + jZ_L \tan k\ell} \right) = 100 \left(\frac{50(1+j) + j100 \tan k\ell}{100 + j50(1+j) \tan k\ell} \right) \qquad (2)$$

where

$$\tan k\ell = \tan\left(\frac{2\pi}{\lambda}\right)\left(\frac{\lambda}{8}\right) = \tan\left(\frac{\pi}{4}\right) = 1$$

Therefore,

$$Z_{in} = 100\frac{1+j3}{1+j} = 100(2+j)$$

The voltage at $z = -\lambda/8$ is computed via the voltage-divider equation (6.1.12), such that

$$V_{in} = 1 \cdot \frac{100(2+j)}{100(2+j) + 100} \simeq 0.7 + j0.1 = 0.71\, e^{j0.14}$$

where $V_S(t) = \cos \omega t = \mathrm{Re}\{e^{j\omega t}\}$ becomes the phasor $V_S = 1$ and the angle of the phasor is expressed in radians. We can convert the phasor V_{in} back to a time-dependent form:

$$V_{in}(t) = \mathrm{Re}\{0.71\, e^{j0.14}\, e^{j\omega t}\} = 0.71\, \cos(\omega t + 0.14) \qquad (3)$$

The load reflection coefficient Γ_L is found from (6.1.4):

$$\Gamma_L = \frac{Z_L - Z_0}{Z_L + Z_0} = \frac{50(1+j) - 100}{50(1+j) + 100} = -0.2 + j0.4$$

To find $V_L = V(z = 0)$, we evaluate (6.1.1) at both $z = 0$ and $z = -\ell = -\lambda/8$, resulting in

$$V_L = V_+(1 + \Gamma_L)$$
$$V_{in} = V_+(e^{jk\ell} + \Gamma_L\, e^{-jk\ell})$$

which means that

$$V_L = \frac{V_{in}(1 + \Gamma_L)}{e^{jk\ell} + \Gamma_L\, e^{-jk\ell}}$$

$$\simeq \frac{(0.7 + j0.1)(0.8 + j0.4)}{(1+j)/\sqrt{2} + (-0.2 + j0.4)(1-j)/\sqrt{2}}$$

$$= 0.42 - j0.14 = 0.45\, e^{-j0.32} \qquad (4)$$

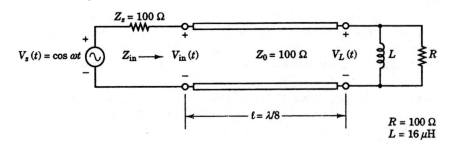

Figure 6.4 Transmission line circuit for Example 6.1.1.

where $e^{-jk\ell} = e^{-j\pi/4} = (1-j)/\sqrt{2}$. The time-dependent load voltage $V_L(t)$ is then

$$V_L(t) = 0.45\cos(\omega t - 0.32)$$

The average power P_d dissipated in R can be calculated in at least two ways. From (2) and (3), we find

$$P_d = \frac{1}{2}\text{Re}\left\{\frac{|V_{in}|^2}{Z_{in}^*}\right\} = \frac{1}{2}(0.71)^2\text{Re}\left\{\frac{1}{100(2-j)}\right\}$$
$$= 10^{-3}\text{ W}$$

Alternatively, we get the same answer by calculating the power dissipated in the load using (1) and (4):

$$P_d = \frac{1}{2}\text{Re}\left\{\frac{|V_L|^2}{Z_L^*}\right\} = \frac{1}{2}(0.45)^2\text{Re}\left\{\frac{1}{50(1-j)}\right\}$$
$$= 10^{-3}\text{ W}$$

Note that use of the first approach for finding P_d avoids the necessity of solving for V_L.

Example 6.1.2
    What are the input impedance Z_{in} and the total power P supplied by the source V_S ($Z_S = 10$ Ω) for the circuit illustrated in Figure 6.5, consisting of two quarter-wavelength TEM lines of characteristic impedance 25 Ω and 100 Ω terminated by a capacitor with $\omega C = 0.01$?

    Solution: We begin at the load and work backward toward the generator.

$$Z_L = 1/j\omega C = -j100$$

$$k\ell = \frac{2\pi}{\lambda}\left(\frac{\lambda}{4}\right) = \frac{\pi}{2}$$

$$Z_A = Z_0\left(\frac{Z_L + jZ_0\tan k\ell}{Z_0 + jZ_L\tan k\ell}\right) = Z_0^2/Z_L = +j100$$

Because the transmission line closest to the source is also one-quarter wavelength long,

$$Z_{in} = (Z_0')^2/Z_A = (25)^2/(j100) = -j6.25$$

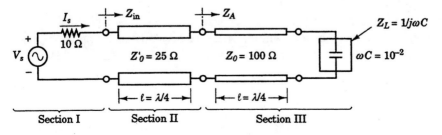

Figure 6.5   Transmission line circuit for Example 6.1.2.

and therefore the total power supplied by the source is

$$P = \frac{1}{2} \text{Re}\{V_S I_S^*\} = \frac{1}{2} \text{Re}\{|V_S|^2 / (10 + Z_{\text{in}})^*\} = 0.036 \, |V_S|^2 \quad \text{(W)}$$

6.2 ABCD MATRICES

If more than one transmission line and/or separate lumped elements are cascaded together, the mathematics for analyzing wave propagation quickly becomes rather cumbersome. It is useful in this case to introduce the concept of *ABCD matrices*, which relate the voltage and current at one point in a *TEM* circuit to the voltage and current at another point in this circuit. Thus,

$$\begin{bmatrix} V_2 \\ I_2 \end{bmatrix} = \begin{bmatrix} A & B \\ C & D \end{bmatrix} \begin{bmatrix} V_1 \\ I_1 \end{bmatrix} \tag{6.2.1}$$

where A, B, C, and D are complex numbers. A and D are dimensionless, while B has units of impedance and C has units of admittance. The utility of *ABCD* matrices lies in their ability to be easily cascaded. For example, if we know how to relate the voltage and current at position z_2 (V_2, I_2) to that at z_1 (V_1, I_1) by the coefficients A_1, B_1, C_1, and D_1 in a matrix equation like (6.2.1), and we also can relate V_2, I_2 to V_3, I_3 via A_2, B_2, C_2, and D_2, then we find

$$\begin{bmatrix} V_3 \\ I_3 \end{bmatrix} = \begin{bmatrix} A_2 & B_2 \\ C_2 & D_2 \end{bmatrix} \begin{bmatrix} V_2 \\ I_2 \end{bmatrix} = \begin{bmatrix} A_2 & B_2 \\ C_2 & D_2 \end{bmatrix} \begin{bmatrix} A_1 & B_1 \\ C_1 & D_1 \end{bmatrix} \begin{bmatrix} V_1 \\ I_1 \end{bmatrix} \tag{6.2.2}$$

which means that the matrix

$$\begin{bmatrix} A_3 & B_3 \\ C_3 & D_3 \end{bmatrix} = \begin{bmatrix} A_2 & B_2 \\ C_2 & D_2 \end{bmatrix} \begin{bmatrix} A_1 & B_1 \\ C_1 & D_1 \end{bmatrix} \tag{6.2.3}$$

relates V_3, I_3 to V_1, I_1 directly. We note that the **order** of matrix multiplication is important; in general, matrices do not commute. This concept allows us to break up a complicated network into smaller two-port elements, find the *ABCD* matrices for each element, and then multiply the individual-element matrices back together to find the overall *ABCD* matrix for the system. Figure 6.6 illustrates how the *ABCD* matrices for two different two-port lumped-element networks may be cascaded to synthesize a larger network.

The transmission line of characteristic impedance Z_0 and length ℓ also has an *ABCD* matrix associated with it, which we may calculate from the expressions (6.1.1) and (6.1.2) for voltage and current on a *TEM* line. We recall that

$$V(z) = V_+(e^{-jkz} + \Gamma_L e^{+jkz})$$

and

$$I(z) = Y_0 V_+(e^{-jkz} - \Gamma_L e^{+jkz})$$

and choose $z = 0$ to be position 1 and $z = -\ell$ to be position 2. If we open-circuit the *TEM* line at $z = 0$, then $\Gamma_L = +1$ in order that $I(z = 0) = I_1 = 0$. In this case,

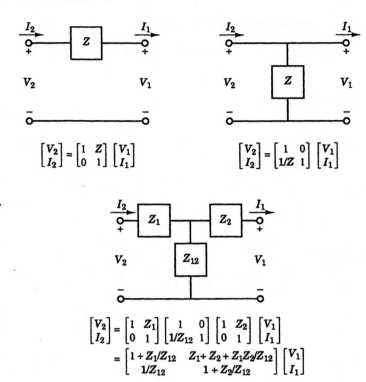

$$\begin{bmatrix} V_2 \\ I_2 \end{bmatrix} = \begin{bmatrix} 1 & Z \\ 0 & 1 \end{bmatrix} \begin{bmatrix} V_1 \\ I_1 \end{bmatrix}$$

$$\begin{bmatrix} V_2 \\ I_2 \end{bmatrix} = \begin{bmatrix} 1 & 0 \\ 1/Z & 1 \end{bmatrix} \begin{bmatrix} V_1 \\ I_1 \end{bmatrix}$$

$$\begin{bmatrix} V_2 \\ I_2 \end{bmatrix} = \begin{bmatrix} 1 & Z_1 \\ 0 & 1 \end{bmatrix} \begin{bmatrix} 1 & 0 \\ 1/Z_{12} & 1 \end{bmatrix} \begin{bmatrix} 1 & Z_2 \\ 0 & 1 \end{bmatrix} \begin{bmatrix} V_1 \\ I_1 \end{bmatrix}$$

$$= \begin{bmatrix} 1 + Z_1/Z_{12} & Z_1 + Z_2 + Z_1 Z_2/Z_{12} \\ 1/Z_{12} & 1 + Z_2/Z_{12} \end{bmatrix} \begin{bmatrix} V_1 \\ I_1 \end{bmatrix}$$

Figure 6.6 *ABCD* matrices for two-port lumped-element networks.

$$I_1 = I(z = 0) = 0$$
$$V_1 = V(z = 0) = 2V_+$$
$$I_2 = I(z = -\ell) = Y_0 V_+(e^{jk\ell} - e^{-jk\ell}) = jY_0 V_1 \sin k\ell$$
$$V_2 = V(z = -\ell) = V_+(e^{jk\ell} + e^{-jk\ell}) = V_1 \cos k\ell$$

If we set $I_1 = 0$ in (6.2.1), we find that

$$A = \cos k\ell$$
$$C = jY_0 \sin k\ell$$
(6.2.4)

Conversely, we may short-circuit the *TEM* line at $z = 0$ so that $V(z = 0) = V_1 = 0$; the load reflection coefficient is $\Gamma_L = -1$. In this case,

$$V_1 = V(z = 0) = 0$$
$$I_1 = I(z = 0) = 2Y_0 \hat{V}_+$$
$$V_2 = V(z = -\ell) = \hat{V}_+(e^{jk\ell} - e^{-jk\ell}) = jZ_0 I_1 \sin k\ell$$
$$I_2 = I(z = -\ell) = Y_0 \hat{V}_+(e^{jk\ell} + e^{-jk\ell}) = I_1 \cos k\ell$$

If we set $V_1 = 0$ in (6.2.1), we then find

$$B = jZ_0 \sin k\ell$$
$$D = \cos k\ell$$

(6.2.5)

and we can finally use (6.2.4) and (6.2.5) to write the *ABCD* matrix equation for any *TEM* transmission line:

$$\begin{bmatrix} V_2(z = -\ell) \\ I_2(z = -\ell) \end{bmatrix} = \begin{bmatrix} \cos k\ell & jZ_0 \sin k\ell \\ jY_0 \sin k\ell & \cos k\ell \end{bmatrix} \begin{bmatrix} V_1(z = 0) \\ I_1(z = 0) \end{bmatrix}$$

(6.2.6)

The *ABCD* matrix in (6.2.6) transforms the variables voltage and current from their values at $z = 0$ to those at $z = -\ell$. Note that because $z = 0$ is arbitrary, the *ABCD* matrix in (6.2.6) actually transforms from position $z = z_0$ to $z = z_0 - \ell$, where z_0 can take on any value. If we wanted instead to transform from $z = -\ell$ to $z = 0$, it would be the same transformation as that from $z = 0$ to $z = \ell$ by virtue of the arbitrary location of the origin, and this transformation could be made by simply reversing the sign of ℓ in (6.2.6):

$$\begin{bmatrix} V_1(z = 0) \\ I_1(z = 0) \end{bmatrix} = \begin{bmatrix} \cos k\ell & -jZ_0 \sin k\ell \\ -jY_0 \sin k\ell & \cos k\ell \end{bmatrix} \begin{bmatrix} V_2(z = -\ell) \\ I_2(z = -\ell) \end{bmatrix}$$

(6.2.7)

Notice that the two matrices in (6.2.6) and (6.2.7) are inverses of each other because one transforms from $z = 0$ to $z = -\ell$ and the other undoes the transformation by moving from $z = -\ell$ back to $z = 0$. Thus the net effect of the two transformations is to leave the voltage and current values unchanged; hence

$$\begin{bmatrix} \cos k\ell & -jZ_0 \sin k\ell \\ -jY_0 \sin k\ell & \cos k\ell \end{bmatrix} \begin{bmatrix} \cos k\ell & jZ_0 \sin k\ell \\ jY_0 \sin k\ell & \cos k\ell \end{bmatrix} = \begin{bmatrix} 1 & 0 \\ 0 & 1 \end{bmatrix}$$

as may be easily verified.

ABCD matrices become especially useful when computing voltages and currents in large cascades of transmission lines. Consider the case of N *TEM* lines attached in series with a voltage source V_S at the left end and a load Z_L at the right end of the line, as shown in Figure 6.7. The ith line has a characteristic impedance Z_{0i}, a length l_i, and a wave number k_i. The *ABCD* matrix for the ith line is thus

$$\overline{\overline{T}}_i = \begin{bmatrix} \cos k_i\ell_i & jZ_{0i} \sin k_i\ell_i \\ jY_{0i} \sin k_i\ell_i & \cos k_i\ell_i \end{bmatrix}$$

(6.2.8)

The voltage and current at the load termination are related by Ohm's law

$$I_L = Y_L V_L$$

and so we can write the source voltage and current as the product of individual *ABCD* matrices

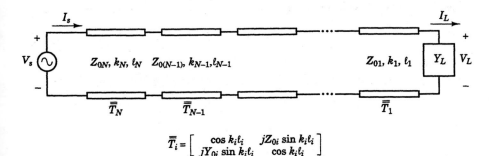

$$\overline{\overline{T}}_i = \begin{bmatrix} \cos k_i \ell_i & j Z_{0i} \sin k_i \ell_i \\ j Y_{0i} \sin k_i \ell_i & \cos k_i \ell_i \end{bmatrix}$$

Figure 6.7   Cascade of N *TEM* transmission lines in series.

$$\begin{bmatrix} V_S \\ I_S \end{bmatrix} = \overline{\overline{T}}_N \cdot \overline{\overline{T}}_{N-1} \cdots \overline{\overline{T}}_2 \cdot \overline{\overline{T}}_1 \begin{bmatrix} V_L \\ Y_L V_L \end{bmatrix}$$

$$= V_L \begin{bmatrix} A_N & B_N \\ C_N & D_N \end{bmatrix} \begin{bmatrix} 1 \\ Y_L \end{bmatrix} \tag{6.2.9}$$

where

$$\overline{\overline{T}}_N \cdot \overline{\overline{T}}_{N-1} \cdots \overline{\overline{T}}_2 \cdot \overline{\overline{T}}_1 = \begin{bmatrix} A_N & B_N \\ C_N & D_N \end{bmatrix}$$

We can multiply out the first line of (6.2.9) to find

$$V_S = V_L(A_N + Y_L B_N)$$

in which case

$$V_L = V_S / (A_N + Y_L B_N) \tag{6.2.10}$$

and as soon as V_L is known, all of the other voltages and currents follow easily. Once the coefficients A_N and B_N in (6.2.9) are initially calculated, it is a simple one-step process to recompute V_L via (6.2.10) each time the load admittance Y_L is changed.

Example 6.2.1

For the circuit illustrated in Figure 6.5, use *ABCD* matrices to find the input impedance Z_{in} and the total power supplied by V_S.

Solution: The circuit is broken up into the three segments shown in Figure 6.5. The *ABCD* matrix for Section I is found from (6.2.6) with $k\ell = \pi/2$ for a quarter-wavelength line and $Z_0 = 100 \; \Omega$:

$$\begin{bmatrix} A & B \\ C & D \end{bmatrix}_I = \begin{bmatrix} \cos k\ell & j Z_0 \sin k\ell \\ j Y_0 \sin k\ell & \cos k\ell \end{bmatrix} = \begin{bmatrix} 0 & j100 \\ j0.01 & 0 \end{bmatrix}$$

For Section II, we also have a line where $k\ell = \pi/2$, and because $Z_0' = 25 \; \Omega$,

$$\begin{bmatrix} A & B \\ C & D \end{bmatrix}_{II} = \begin{bmatrix} 0 & j25 \\ j0.04 & 0 \end{bmatrix}$$

Section III simply contains the source resistor, and we have already shown in Figure 6.6 that

$$\begin{bmatrix} A & B \\ C & D \end{bmatrix}_{III} = \begin{bmatrix} 1 & Z_S \\ 0 & 1 \end{bmatrix} = \begin{bmatrix} 1 & 10 \\ 0 & 1 \end{bmatrix}$$

If we now cascade these transfer matrices as described by (6.2.9), we find that

$$\begin{bmatrix} V_S \\ I_S \end{bmatrix} = \begin{bmatrix} 1 & 10 \\ 0 & 1 \end{bmatrix} \begin{bmatrix} 0 & j25 \\ j0.04 & 0 \end{bmatrix} \begin{bmatrix} 0 & j100 \\ j0.01 & 0 \end{bmatrix} \begin{bmatrix} V_L \\ Y_L V_L \end{bmatrix}$$

$$= \begin{bmatrix} -0.25 & -40 \\ 0 & -4 \end{bmatrix} \begin{bmatrix} V_L \\ Y_L V_L \end{bmatrix}$$

which means that

$$V_S = (-0.25 - 40Y_L)V_L = (-0.25 - j0.4)V_L$$

and

$$I_S = -4Y_L V_L = -j0.04V_L$$

for $Y_L = 1/Z_L = 0.01j$. The input impedance Z_{in} is related to V_S, I_S, and Z_S by $I_S = V_S/(Z_{in} + Z_S)$, or

$$Z_{in} = \frac{V_S}{I_S} - Z_S = \frac{-0.25 - j0.4}{-j0.04} - 10 = -j6.25$$

as was also found in Example 6.1.2. The power supplied by the source is again

$$P = \frac{1}{2}\text{Re}\{V_S I_S^*\} = \frac{1}{2}\text{Re}\left\{ V_S \left(\frac{V_S}{10 - j6.25} \right)^* \right\}$$

$$= 0.036|V_S|^2 \text{ (W)}$$

6.3 GAMMA PLANE AND SMITH CHART ANALYSIS METHODS

Although the methods discussed in Sections 6.1 and 6.2 are adequate to solve for $V(z)$ and $I(z)$ everywhere on a network composed of lumped elements and *TEM* lines, other analysis techniques can also prove helpful. In particular, we shall consider the complex gamma plane $\Gamma(z)$ and its mapping into complex impedance $Z(z)$, which is carried out graphically using a Smith chart. The Smith chart, while not so useful a computational method in this age of computers, still provides physical insight into how impedances transform on a *TEM* line. It is also a useful way to quickly estimate impedances that could be exactly (but more tediously!) computed analytically.

We first recall (5.3.45) and (5.3.49):

$$V(z) = V_+ e^{-jkz} + V_- e^{+jkz} \tag{6.3.1}$$

$$I(z) = Y_0 \left[V_+ e^{-jkz} - V_- e^{+jkz} \right] \tag{6.3.2}$$

These equations become

$$V(z) = V_+ e^{-jkz}[1 + \Gamma(z)] \tag{6.3.3}$$

$$I(z) = Y_0 V_+ e^{-jkz}[1 - \Gamma(z)] \tag{6.3.4}$$

where

$$\Gamma(z) \equiv \frac{V_- e^{jkz}}{V_+ e^{-jkz}} = \Gamma_L e^{2jkz} \qquad (6.3.5)$$

and $\Gamma_L = \Gamma(z=0) = V_-/V_+ = (Z_L - Z_0)/(Z_L + Z_0)$ from (6.1.4). It follows that

$$Z(z) = \frac{V(z)}{I(z)} = Z_0 \left(\frac{1+\Gamma(z)}{1-\Gamma(z)} \right) \qquad (6.3.6)$$

where $\Gamma(z)$ and hence $Z(z)$ are periodic functions of position z. Equivalently,

$$\Gamma(z) = \frac{Z_n(z) - 1}{Z_n(z) + 1} \qquad (6.3.7)$$

where the *normalized impedance* $Z_n(z)$ is defined as $Z(z)/Z_0$.

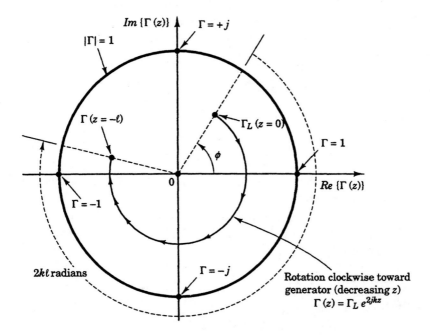

Figure 6.8   Complex gamma plane for $\Gamma(z)$.

Before considering the one-to-one mapping [given by (6.3.6) and (6.3.7)] between $\Gamma(z)$ and $Z_n(z)$ that motivated invention of the Smith chart, we first establish the character of $\Gamma(z)$ and the relationship between $\Gamma(z)$ and the total line voltage $V(z)$. The complex *gamma plane* is illustrated in Figure 6.8, having axes Re$\{\Gamma\}$ and Im$\{\Gamma\}$. From (6.3.5), we note immediately that $|\Gamma(z)| = |\Gamma_L|$.[1] If we use (6.3.5) to plot $\Gamma(z)$ in the gamma plane, we find that as we move in the negative

---

1. In this analysis, we assume that the characteristic impedance Z_0 is a real number so that k is also real.

z direction away from the load Z_L, the angle of $\Gamma(z)$ **decreases** with negative z. Therefore, $\Gamma(z)$ rotates **clockwise** along a circle of radius $|\Gamma_L|$ as z moves away from the load, one complete revolution occurring each time that $2k\Delta z$ equals 2π, or $\Delta z = \lambda/2$.

The voltage magnitude is found from (6.3.3) to be

$$|V(z)| = |V_+| \, |1 + \Gamma(z)| \tag{6.3.8}$$

which is represented graphically in Figure 6.9(a), where $1 + \Gamma(z)$ is the complex vector connecting the point -1 and $\Gamma(z)$ in the gamma plane. For lossless passive media, we have already mentioned that $|\Gamma(z)| \le 1$, because the reflected power must not exceed the incident power at a junction.[2]

This geometric construction shows how $|V(z)|$ varies between $V_{max} = |V_+|(1 + |\Gamma|)$ and $V_{min} = |V_+|(1 - |\Gamma|)$ with a period $\Delta z = \lambda/2$. The pattern $|V(z)|$ is generally **not** sinusoidal in z. The voltage magnitude is plotted as a function of z in Figure 6.9(b) for the normalized load impedance $Z_{Ln} = 0.8 - j0.6$. We define the *voltage standing wave ratio* or *VSWR* as

$$\text{VSWR} \equiv \frac{V_{max}}{V_{min}} = \frac{1 + |\Gamma(z)|}{1 - |\Gamma(z)|} = \frac{1 + |\Gamma_L|}{1 - |\Gamma_L|} \tag{6.3.9}$$

Therefore,

$$|\Gamma(z)| = |\Gamma_L| = \frac{\text{VSWR} - 1}{\text{VSWR} + 1} \tag{6.3.10}$$

and so measurements of the VSWR, which usually can be made easily, yield $|\Gamma_L|$ and the power reflectivity of a load $|\Gamma_L|^2$.

Consider, for example, the special case of a purely reactive load $Z_L = jX$. In this case, no power is dissipated and so $|\Gamma(z)|^2 = |\Gamma_L|^2 = 1$. This result also follows from (6.1.4):

---

2. Note that lossy lines with $k = k' - jk''$ can exhibit interesting behavior in the $\Gamma(z)$ plane. For example, (6.3.5) becomes

$$\Gamma(z = -\ell) = \Gamma_L \, e^{-2jk'\ell - 2k''\ell}$$

and $\Gamma(z = -\ell)$ spirals exponentially inward toward $\Gamma = 0$ as $\ell \to \infty$. Thus the impedance at $z = -\ell$ approaches Z_0 as $\ell \to \infty$. It is also interesting to note that $|\Gamma|$ can exceed unity for a lossy line because $Z_0 = \sqrt{(R + j\omega L)/(G + j\omega C)}$ is in general complex, and Z_0 can have phase angle ϕ where $-\pi/4 < \phi < \pi/4$. Thus, $Z_n = Z_L/Z_0$ can have an angle $\alpha = \beta - \phi$ up to $\pm 3\pi/4$. The angle β of Z_L is limited to $-\pi/2 \le \beta \le \pi/2$ since the real part of Z_L is always positive for passive media. It can easily be seen that

$$|\Gamma| = \frac{|Z_n - 1|}{|Z_n + 1|} = \frac{||Z_n| \, e^{j\alpha} - 1|}{||Z_n| \, e^{j\alpha} + 1|}$$

is largest when $\alpha = \pm 3\pi/4$ and $\text{Re}\{|Z_n| \, e^{j\alpha}\}$ is negative. The maximum possible value of $|\Gamma|$ is $1 + \sqrt{2}$. Note than when $|\Gamma| > 1$ on a lossy line, it does not mean that there is power amplification. Here, $|\Gamma| > 1$ is the result of the reactive elements, and power in the forward or reverse directions is no longer simply equal to $|V_+|^2/2Z_0$ or $|V_-|^2/2Z_0$ because Z_0 is complex.

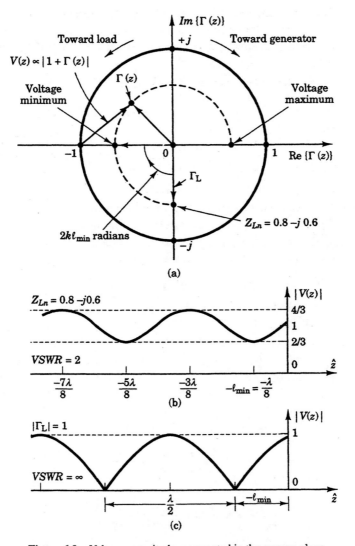

Figure 6.9   Voltage magnitude represented in the gamma plane.

$$|\Gamma_L| = \left|\frac{Z_L - Z_0}{Z_L + Z_0}\right| = \left|\frac{jX - Z_0}{jX + Z_0}\right| = 1$$

For a purely reactive load, the VSWR defined by (6.3.9) is infinite, and $|V(z)|$ is illustrated in Figure 6.9(c). We can also calculate $|V(z)|$ analytically in this special case by noting that $\Gamma_L = e^{j\phi}$ where ϕ is determined from X and Z_0: $\phi = \tan^{-1}[2XZ_0/(X^2 - Z_0^2)]$. Equation (6.3.8) then yields

$$|V(z)| = |V_+||1 + e^{j\phi}\,e^{2jkz}|$$
$$= 2|V_+||\cos(kz - \phi/2)|$$

which is exactly what is shown in Figure 6.9(c).

Note that measurements of both the VSWR and the positions of the minima (or maxima) in $|V(z)|$ allow us to find the magnitude and phase of $\Gamma_L = |\Gamma_L| e^{j\phi}$. The magnitude of Γ_L is simply given by (6.3.10) if the VSWR is known. We define the distance from the load to the first voltage minimum to be ℓ_{min}. As may be seen in Figure 6.9(a), $\Gamma(z)$ is located on the negative real axis at a voltage minimum. Therefore, we rotate clockwise along a circle of radius Γ_L from the angle of the load ϕ to π as we move from $z = 0$ to $z = -\ell_{min}$; $\phi = 2k\ell_{min} - \pi$. The load reflection coefficient is then

$$\Gamma_L = -\left(\frac{\text{VSWR} - 1}{\text{VSWR} + 1}\right) e^{2jk\ell_{min}} \qquad (6.3.11)$$

Measurements of $|I(z)|$ can also provide the same information about Γ_L.

Now we return to (6.3.6) and (6.3.7), which define the one-to-one relationship between $Z(z)$ and $\Gamma(z)$. This relationship permits us to overlay the complex gamma plane $\Gamma(z)$ with contours corresponding to normalized resistance R_n and normalized reactance X_n, where the normalized impedance Z_n is

$$Z_n(z) = Z(z)/Z_0 = R_n + jX_n$$

It can be shown that the contours of constant R_n and constant X_n are all circles, as illustrated in Figure 6.10.[3] We call this chart of $Z_n(z)$ superimposed on the gamma plane a *Smith chart* after its inventor.

The initial utility of the Smith chart nomogram is that we can immediately determine the input impedance at $z = -\ell$ of a *TEM* line of length ℓ terminated by a load impedance Z_L. The basic procedure for using the Smith chart is as follows:

---

3. These contours of constant R_n and X_n are determined by finding the real and imaginary parts of Γ in terms of R_n and X_n

$$\Gamma = \frac{Z_n - 1}{Z_n + 1} = \frac{R_n + jX_n - 1}{R_n + jX_n + 1}$$

which means that

$$x \equiv \text{Re}\{\Gamma\} = \frac{R_n^2 + X_n^2 - 1}{(R_n + 1)^2 + X_n^2}$$

and

$$y \equiv \text{Im}\{\Gamma\} = \frac{-2X_n}{(R_n + 1)^2 + X_n^2}$$

If these two equations are solved for x and y as a function of R_n alone, then after some algebra we find that

$$\left(x - \frac{R_n}{R_n + 1}\right)^2 + y^2 = \left(\frac{1}{R_n + 1}\right)^2$$

which is the equation for a circle centered at $(R_n/(R_n + 1), 0)$ of radius $1/(R_n + 1)$. We note that this circle always intersects the point $(1,0)$ in the gamma plane. These are the contours of constant R_n.

Conversely, we may solve for x and y in terms of X_n alone, giving

$$(x - 1)^2 + \left(y + \frac{1}{X_n}\right)^2 = \left(\frac{1}{X_n}\right)^2$$

which is also the equation of a circle centered at $(1, -1/X_n)$ with radius $1/X_n$. The portions of these

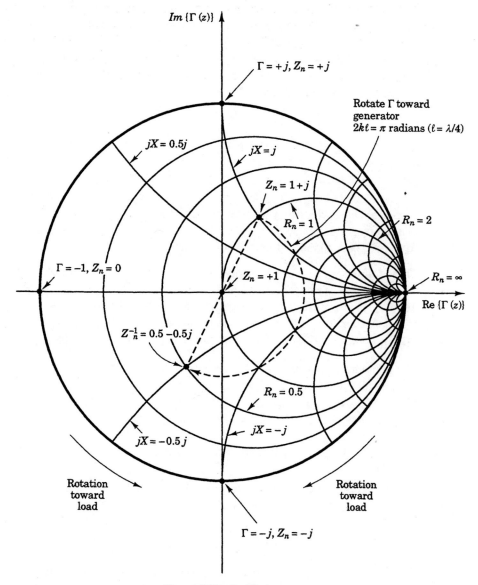

Figure 6.10   Smith chart.

(1) Calculate the normalized load impedance by dividing Z_L by the characteristic impedance of the transmission line: $Z_{Ln} = Z_L/Z_0$.

(2) Find the point $Z_{Ln} = R_{Ln} + jX_{Ln}$ on the Smith chart by looking for the

---

circles that lie within the unit circle on the gamma plane ($|\Gamma| < 1$) are the contours of constant X_n, and these circles also intersect the point $(1, 0)$ in the Γ-plane.

intersection of the R_{Ln} and X_{Ln} contours. This point is also the location of Γ_L on our usual gamma plane, i.e., the distance from the point Z_{Ln} on the Smith chart to the origin is $|\Gamma_L|$ and the angle of the vector to Γ_L and the Re$\{\Gamma\}$-axis is the angle ϕ since $\Gamma_L = |\Gamma_L| e^{j\phi}$.

(3) If we wish to find the impedance at $z = -\ell$ (away from the load, toward the generator), we refer to (6.3.5):

$$\Gamma(z = -\ell) = \Gamma_L e^{2jkz}\big|_{z=-\ell} = \Gamma_L e^{-2jk\ell}$$

This corresponds to rotating the point Γ_L by an additional $2k\ell$ radians in the clockwise direction (decreasing angle) at the same radius $|\Gamma_L|$, which we can perform graphically by using a compass on the Smith chart.

(4) The new point at $\Gamma_L e^{-2jk\ell}$ also corresponds to a normalized impedance on the Smith chart, which may be determined by finding the R_n and X_n contours that intersect this point. We convert the normalized input impedance $Z_n = R_n + jX_n$ back to units of impedance by multiplying Z_n by Z_0.

These four steps represent the general methodology of Smith chart manipulation. An illustration of how a 100-Ω *TEM* line terminated by $Z_L = 100(1 - j)$ Ω has a $100(2 + j)$-Ω impedance three-eighths of a wavelength from load is given in Figure 6.11. We could also find this answer analytically by using (6.1.6).

The more important properties of the Smith chart are summarized below:

(1) The contours of constant R_n are perpendicular to the contours of constant X_n, and all such contours are circles.

(2) When the impedance is purely real ($X_n = 0$), the reflection coefficient lies on the real axis of the Γ-plane because Γ has no imaginary part:

$$\text{Im}\{\Gamma\} = \text{Im}\left\{\frac{R_n - 1}{R_n + 1}\right\} = 0$$

(3) When the impedance is purely imaginary ($R_n = 0$), the reflection coefficient lies on the circle $|\Gamma| = 1$ in the gamma plane, because

$$|\Gamma| = \left|\frac{jX_n - 1}{jX_n + 1}\right| = 1$$

(4) The origin of the gamma plane ($\Gamma = 0$) corresponds to the normalized impedance $Z_n = 1$, which is Z_0 in un-normalized form. This is the matched-load impedance case, where no reflections are observed.

(5) If a lossless line is terminated with the conjugate impedance Z_L^*, then the reflection coefficient must be complex conjugated also:

$$\Gamma^* = \frac{Z_n^* - 1}{Z_n^* + 1}$$

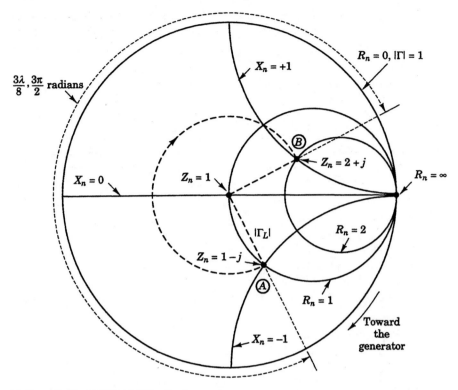

(1) $Z_{Ln} = Z_L/Z_o = 100(1-j)/100 = 1-j$
(2) Find Z_{Ln} on the Smith Chart at (A), the intersection of the $R_n = 1$ and $X_n = -1$ contours,
(3) Rotate (A) along a circle of radius $|\Gamma_L|$ clockwise toward the generator $3\lambda/8$ or $3\pi/2$ radians to (B).
(4) (B) lies at the intersection of $R_n = 2$, $X_n = 1$; $Z_n = 2+j$. Therefore $Z = Z_n Z_o = 100(2+j)$.

Figure 6.11   General procedure for Smith chart manipulation.

(6) If the *TEM* line is terminated with an inverse normalized impedance ($Z_n \to Z_n' = 1/Z_n$), then the reflection coefficient changes sign (is rotated $180°$ in the gamma plane).

$$\Gamma \to \Gamma' = \frac{Z_n' - 1}{Z_n + 1} = \frac{1/Z_n - 1}{1/Z_n + 1} = -\frac{Z_n - 1}{Z_n + 1} = -\Gamma$$

Therefore, the normalized admittance $Y_n(z)$ is found by rotating $Z_n(z) = 1/Y_n(z)$ by $180°$ on the Smith chart, as illustrated for three specific cases in Figure 6.12. The Smith chart can thus be used equally well to represent impedances or admittances.

(7) A change Δz in position along the transmission line is represented by rotating Γ by an angle $2k\Delta z$ radians; Γ_L at $z = 0$ is transformed into $\Gamma(z)$ at position z by

$$\Gamma(z) = \Gamma_L e^{2jkz}$$

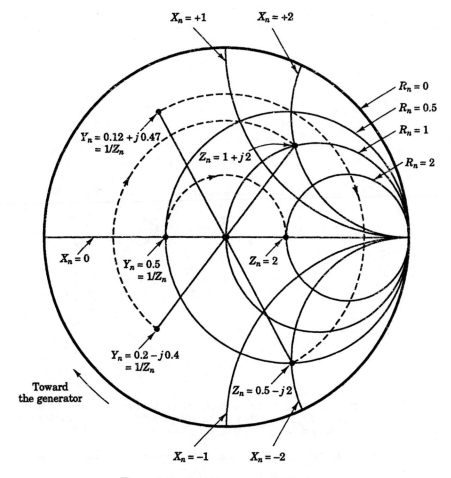

Figure 6.12 Admittances on a Smith chart.

That is, we draw a circle of radius $|\Gamma_L|$ and rotate clockwise from Γ_L to determine the impedance at positions closer to the generator, and counterclockwise to determine the impedance at positions closer to the load, where the load is in the direction of increasing z. As noted earlier, a complete rotation occurs when $2k\Delta z = 2\pi$, or when $\Delta z = \lambda/2$.

(8) The VSWR may be obtained directly from the Smith chart by comparing it with the definition of normalized impedance (6.3.9):

$$\text{VSWR} = \frac{1 + |\Gamma|}{1 - |\Gamma|}$$

versus

$$Z_n(z) = \frac{1 + \Gamma(z)}{1 - \Gamma(z)}$$

We notice that the normalized impedance is equal to the VSWR when $\Gamma(z)$ is real and positive, i.e., at the point in the gamma plane where the circle of radius $|\Gamma(z)| = |\Gamma_L|$ intersects the positive real axis. (The point where $\Gamma(z)$ intersects the negative real axis would give 1/VSWR .)

As an example, we consider a line with $Z_0 = 50\ \Omega$, terminated by $Z_L = 100 - j150$. Then,

$$Z_{Ln} = \frac{100 - j150}{50} = 2 - j3$$

If we plot this normalized load impedance on the Smith chart in Figure 6.13, we find that $|\Gamma_L| = 0.745$. We then draw a circle of radius 0.745 (compared to the unit-radius circle that defines the boundary of the Smith chart) centered at the origin of the Smith chart, and note that the circle intersects the positive real axis at $Z_n = 6.85 = $ VSWR . To find Z_{in} at a distance 0.2λ toward the generator from this load, we rotate 0.2λ or $4\pi/5$ radians clockwise

$$Z_{in,n} = 0.147 - j0.081$$

and therefore $Z_{in} = Z_0 Z_{in,n} = 7.34 - j4.04\ \Omega$.

We could also use the Smith chart to transform admittances. Suppose $Y_0 = 100$, $Y_L = 15.4 + j23.1$, and we wish to find Y_{in} at a position 0.2λ toward the generator. Then,

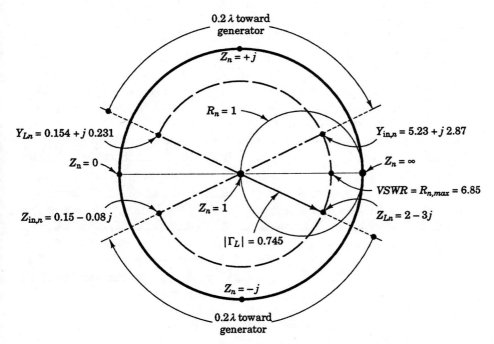

Figure 6.13   Impedance and admittance transformations using a Smith chart.

$$Y_{Ln} = Y_L/Y_0 = 0.154 + j0.231$$

which is plotted in Figure 6.13. The corresponding Z_{Ln} is directly opposite on the chart, as discussed in observation (6), and is given by $Z_{Ln} = 2 - j3$. We rotate this 0.2λ clockwise to find $Z_{in,n} = 0.147 - j0.081$ as before. The normalized admittance is then directly opposite $Z_{in,n}$, and is equal to $Y_{in,n} = 5.23 + j2.87$. The total input admittance is thus

$$Y_{in} = (5.23 + j2.87)100 = 523 + j287 \ \Omega^{-1}$$

But we could have also utilized admittances directly:

(1) $Y_{Ln} = Y_L/Y_0 = (15.4 + j23.1)/100 = 0.154 + j0.231$
(2) Rotate Y_{Ln} clockwise (toward the generator) 0.2λ to get $Y_{in,n} = 5.23 + j2.87$.
(3) $Y_{in} = Y_{in,n}Y_0 = 523 + j287$.

The Smith chart can also facilitate finding Z_L from VSWR measurements. Suppose we measure the voltage maximum to be $\ell_{max} = 0.2\lambda$ from the unknown load Z_L, and the VSWR is 5. Then we know that Z_{Ln} must lie on the Γ circle intercepting the point $R_n = 5$, $jX_n = 0$, where Γ is real and positive. Since voltage maxima occur when Γ is real and positive, we know that $Z_{in,n} = 5$ at this point. To find Z_{Ln}, we move 0.2λ **counterclockwise** toward the load, yielding $Z_{Ln} = 0.220 + j0.311$ and $Z_L = Z_{Ln}Z_0$, illustrated in Figure 6.14.

For a quarter wavelength of transmission line ($\ell = \lambda/4$), the Smith chart makes it clear that Z_{Ln} is transformed into $Y_{Ln} = 1/Z_{Ln}$. This can be quite useful in certain applications, one of which is the *quarter-wave transformer*. Suppose we have an impedance $Z_L = 400 \ \Omega$ to which we wish to deliver maximum power from a transmission line of characteristic impedance $Z_0 = 100 \ \Omega$. If all of the power is delivered to the 400-Ω resistor, then no reflections should be found on the 100-Ω line; i.e., the equivalent circuit for the 100-Ω line should be terminated with a 100-Ω load.

We connect a transmission line of characteristic impedance Z_0' and length $\lambda/4$ between the $Z_0 = 100 \ \Omega$ line and the $Z_L = 400 \ \Omega$ load, as shown in Figure 6.15. The normalized load impedance Z_{Ln} is thus $Z_{Ln} = Z_L/Z_0'$, which transforms to $Z_{An} = 1/Z_{Ln} = Z_0'/Z_L$ at the junction between the two *TEM* lines, because we have rotated $\lambda/4$ or π radians counterclockwise around the Smith chart ($\Gamma \rightarrow -\Gamma \Rightarrow Z_n \rightarrow 1/Z_n$). But $Z_A = Z_{An}Z_0' = (Z_0')^2/Z_L$, and this impedance Z_A is set equal to Z_0 to match the $Z_0 = 100 \ \Omega$ line. Therefore,

$$Z_0' = \sqrt{Z_0 Z_L} = \sqrt{100 \cdot 400} = 200 \ \Omega$$

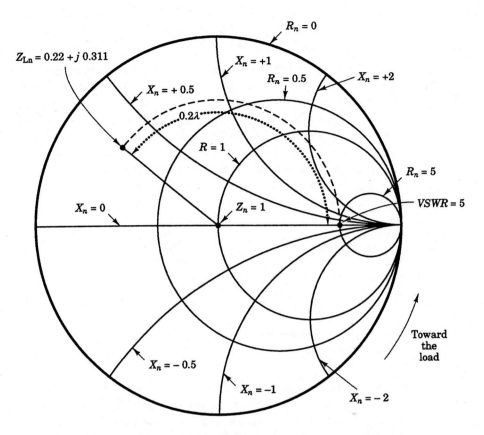

Figure 6.14   Load determination for Z_{Ln} using a Smith chart.

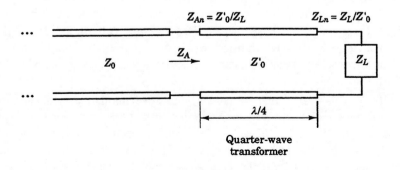

Figure 6.15   Transmission line quarter-wave transformer.

which means that a quarter-wave transformer with characteristic impedance equal to the geometric mean of Z_0 and Z_L can perfectly match Z_0 to Z_L. We note that reflections are certainly present on the Z_0' section of transmission line because the load Z_L is not matched to Z_0' nor is Z_0' matched to Z_0, but these reflections are confined to the Z_0' *TEM* line. If that line is lossless, then all of the power must eventually be dissipated in the load.

One of the principal uses of a Smith chart is in design. For example, it is often desirable to couple power from a source to a load with maximum efficiency, as was illustrated by the quarter-wave transformer when the load is purely resistive. We find this maximum efficiency condition by considering a voltage source V_S connected to a source impedance Z_S in series with a load impedance Z_L. The power dissipated in the load is given by

$$P_d = \frac{1}{2}\operatorname{Re}\{V_L I_L^*\} = \frac{1}{2}|I_L|^2\operatorname{Re}\{Z_L\} = \frac{1}{2}\left|\frac{V_S}{Z_S + Z_L}\right|^2 \operatorname{Re}\{Z_L\}$$

$$= \frac{|V_S|^2}{2}\frac{R_L}{(R_S + R_L)^2 + (X_S + X_L)^2}$$

(6.3.12)

If we now maximize P_d with respect to R_L and X_L by setting $\partial P_d/\partial R_L = \partial P_d/\partial X_L = 0$, we find that $R_L = R_S$ and $X_L = -X_S$; these are the *matching conditions*.

Consider a load of normalized admittance Y_{Ln} connected to a *TEM* line. If we move back toward the generator along the line, we trace out a clockwise circle of radius $|\Gamma_L|$ in the gamma plane, where Γ_L is found from Y_{Ln}. This circle intersects the contour $R_n = 1$ at two places, as shown in Figure 6.16, so at two positions within a half wavelength of the load, the admittance is $Y_n = 1 \pm jB_n$. The distance between the load and the stub is found by computing the number of wavelengths necessary to rotate the normalized load admittance around to the $R_n = 1$ contour. In choosing a position on the line where the real part of Y_n is unity, we have matched the resistive part of the impedance as we did in the quarter-wave transformer. Now we need add only a susceptive element to cancel the imaginary part of Y_n to completely match the *TEM* line to the load.

If $Y_n = 1 - jB_n$, then we add a capacitive normalized stub admittance $Y_{sn} = jB_n$ in parallel across the transmission line. (Admittances add in parallel, impedances in series.) If $Y_n = 1 + jB_n$, then we add an inductive normalized stub $Y_{sn} = -jB_n$, also in parallel across the *TEM* line. These admittances can be lumped susceptive (reactive) elements or they can be formed using short sections of transmission line that are losslessly terminated. A short-circuit stub termination is most common because it typically radiates and dissipates the least amount

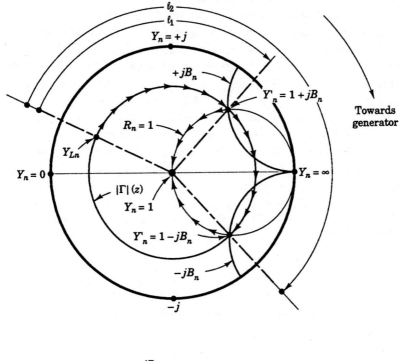

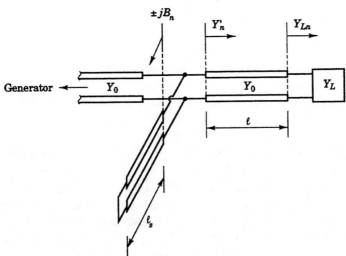

Figure 6.16   Single-stub matching using a Smith chart.

of energy. The length of the short-circuit stub is found by noting the number of wavelengths needed to rotate the admittance $Y_{Ln} = \infty$ clockwise to $Y_{sn} = \pm j B_n$.

This type of matching system, consisting of a single *TEM* tuning element, is called a *single-stub tuner*. The short-circuit load can typically be moved up and down along the *TEM* stub to "tune" the circuit in case the load impedance is not known very accurately. Such stubs can alternatively be added in series. In this case, the load impedance $Z_{Ln} = R_n + j X_n$ is transformed over some distance ℓ on the transmission line so that $Z'_{Ln} = 1 \pm j X'_n$. Again, a lossless impedance of $\mp j X'_n$ can be added in series so as to produce a matched load.

Sometimes it is not practical to fabricate a stub tuner for which the distance ℓ between the load and the stub is adjustable. In this case more than one stub tuner can be used. Suppose two stubs are added in parallel to a *TEM* line of characteristic admittance Y_0, as illustrated in Figure 6.17. If one stub (normalized admittance Y_{1n}) is placed just before the load, then the total normalized admittance at the end of the *TEM* line is

$$Y'_n = Y_{Ln} + Y_{1n}$$

where the admittances are defined in Figure 6.12, and we use admittance instead of impedance because the stubs are connected in parallel. We suppose that the second stub is located at $\ell = 3\lambda/8$ from the load, and the impedance looking into the line at that point is Y''_n. Thus,

$$Y'''_n = Y''_n + Y_{2n}$$

where Y_{2n} is the normalized admittance of the second stub and Y'''_n is the equivalent admittance that terminates the *TEM* line. In order that there be no reflected waves to the left of the stub closest to the generator, we insist that $Y'''_n = 1$ (matched-load condition). We may then use a Smith chart to determine how the stubs should be designed, keeping in mind that each stub only contributes a susceptance B because the stubs are losslessly terminated bits of *TEM* line. (After all, we don't want to waste any power by dissipating it in the stubs.)

We can work backward from $Y'''_n = G'''_n + j B'''_n = 1$. Since $Y'''_n = Y''_n + Y_{2n}$ and Y_{2n} is purely imaginary, the real part of Y''_n must be 1. Therefore, Y''_n may lie anywhere on the $G_n = 1$ circle on the Smith chart. If we rotate Y''_n toward the load (counterclockwise) by $3\lambda/8$, we find that each possible value of Y''_n on the $G_n = 1$ circle is rotated counterclockwise by $3\pi/2$ radians, leading to another circle on which possible values of Y'_n must lie. (This second circle is not a contour of constant G_n or B_n, however.) This circle of possible Y'_n values intersects each of the contours of constant G_n **except** those circles that are smaller than the circle of constant G_n tangent to the Y'_n circle. Geometric considerations can be used to show that this is the $G_n = 2$ circle. Therefore, the normalized load admittance

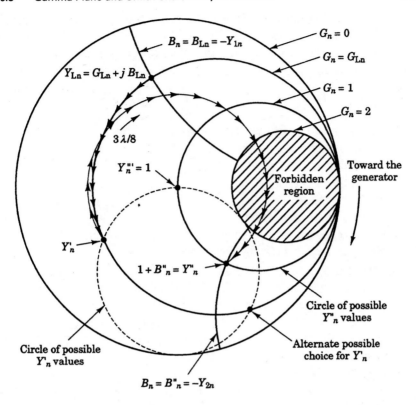

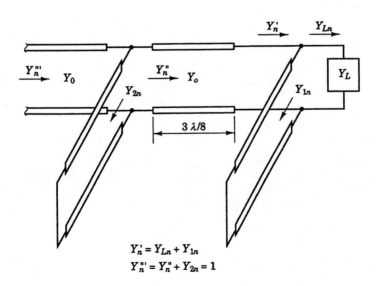

Figure 6.17   Double-stub tuner.

must not have a real part that exceeds $G_n = 2$ if it is to be matched using only two lossless stubs separated by $3\lambda/8$, and this "forbidden region" is shown in Figure 6.17. In order to find the lengths of the two stub tuners, we carry out the following procedure:

(1) Plot $Y_{Ln} = Y_L/Y_0$ on the Smith chart. If $\text{Re}\{Y_{Ln}\} > 2$, the two-stub tuner described here with distance $\ell = 3\lambda/8$ separating the stubs will not work. In this case, it will be necessary to add a quarter-wave section between the load and the first stub (which sends $Y_{Ln} \rightarrow 1/Y_{Ln}$), to change the spacing between the stubs, or to insert a third stub.

(2) Rotate along the contour of constant $\text{Re}\{Y_{Ln}\}$ until it intersects the circle of possible Y'_n values, shown in Figure 6.17. The two points of intersection then give the actual values of Y'_n, and the difference between Y'_n and Y_{Ln} is Y_{1n}, a pure susceptance. We may choose either of the two intersections, but the first is less reactive and therefore preferable.

(3) If the stub Y_{1n} is made from a shorted *TEM* line of admittance Y_0, then we may find ℓ_1 by simply noting how many radians (or wavelengths) it takes to rotate a short circuit (infinite admittance) to the point Y_{1n}.

(4) The normalized admittance Y'_n is now rotated $3\lambda/8$ toward the generator (clockwise) to find Y''_n, which should satisfy $\text{Re}\{Y''_n\} = 1$. Therefore, $Y_{2n} = 1 - Y''_n$ where Y_{2n} is then also a pure susceptance. Step (3) may be used with Y_{2n} to determine the length ℓ_2 of the shorted stub. The normalized admittance looking into the line just in front of the stubs is then $Y'''_n = 1$, and the line is matched.

Notice that every step in this calculation involved **normalized** admittances, which were easily manipulated because the line and both stubs had the same characteristic admittance Y_0. Were all the characteristic admittances different, it would be necessary to "un-normalize" admittances associated with one transmission line and then to "renormalize" using the characteristic admittance of the next *TEM* line in the sequence.

Example 6.3.1

Measurements of $|V(z)|$ on a 100 Ω *TEM* line reveal VSWR $= 3$ and a minimum of $V(z)$ located $\lambda/8$ from the load Z_L at the end of the line. What is the reflection coefficient Γ and what is Z_L?

Solution: The reflection coefficient is

$$|\Gamma| = (\text{VSWR} - 1)/(\text{VSWR} + 1) = 1/2$$

from (6.3.10). At $z = -\lambda/8$, the voltage is a minimum, which means that Γ is negative and real ($\Gamma(z = -\lambda/8) = -1/2$). This value of Γ is then rotated counterclockwise $\lambda/8$ or $\pi/2$ radians to yield Γ_L. From (6.3.5),

$$\Gamma_L = \Gamma(z = 0) = \Gamma(z = -\lambda/8) \cdot e^{2jk\lambda/8} = -0.5\, e^{j\pi/2} = -j0.5$$

Once Γ_L is known, Z_{Ln} may be read off the Smith chart or evaluated from (6.3.6):

$$Z_{Ln} = \frac{1 + \Gamma_L}{1 - \Gamma_L} = \frac{1 - j0.5}{1 + j0.5} = 0.6 - j0.8$$

$$Z_L = Z_0 Z_{Ln} = 60 - j80$$

Example 6.3.2

The measured VSWR is 4 and 3 at distances of 1 km and 2 km from a load Z_L. What is $|\Gamma_L|$ at the end of this lossy line?

Solution: The reflection coefficient is

$$|\Gamma| = \frac{\text{VSWR} - 1}{\text{VSWR} + 1}$$

so $|\Gamma_1| = 3/5$ and $|\Gamma_2| = 1/2$ at 1 km and 2 km, respectively. But

$$\Gamma(z) = \Gamma_L e^{2jkz} = \Gamma_L e^{2jk'z - 2k''z}$$

Therefore,

$$|\Gamma_2| = |\Gamma_1|\, e^{-2k'' \cdot 10^3} = |\Gamma_L|\, e^{-2k'' \cdot 2 \cdot 10^3}$$

which means that

$$|\Gamma_L| = |\Gamma_1|\, e^{+2k'' \cdot 10^3} = |\Gamma_1| \left(\frac{|\Gamma_1|}{|\Gamma_2|} \right) = \frac{18}{25}$$

Example 6.3.3

Design a single-stub tuner to match a load $Z_L = 60 - j80\ \Omega$ on a 100-Ω *TEM* transmission line. Place the stub as close to the load as possible, and then make it as short as possible.

Solution: The normalized load impedance $Z_{Ln} = 0.6 - j0.8$ and the corresponding normalized admittance $Y_{Ln} = 1/Z_{Ln} = 0.6 + j0.8$ are both plotted on the Smith chart in Figure 6.18. It is clear that rotating from the normalized load admittance $0.6 + j0.8$ clockwise (toward the generator) to the $G_n = 1$ circle is a shorter distance than rotating clockwise from the impedance $Z_{Ln} = 0.6 - j0.8$ to the $R_n = G_n = 1$ circle. Therefore, we choose to use admittances rather than impedances and must place the stub in parallel with the *TEM* line rather than in series. The rotation on the Smith chart from $0.6 + j0.8$ to the unit admittance circle is seen to be 0.042λ. At this intersection, the normalized line admittance **without** the stub is $Y_n(z = -0.042\lambda) = 1 + j1.16$, and with a stub $Y_{sn} = -j1.16$, it is

$$Y_n = Y_{sn} + Y_n(z = -0.042\lambda) = 1$$

which is the match we desire. The minimum length of a short-circuited stub ($Y_{sLn} = \infty$) with input admittance $Y_{sn} = -j1.16$ is seen from the Smith chart to be 0.112λ, whereas the minimum length of a corresponding open-circuited stub would be $\lambda/4$ longer. We could also add an inductor in parallel instead of the stub, such that $\omega L = 1/1.16$.

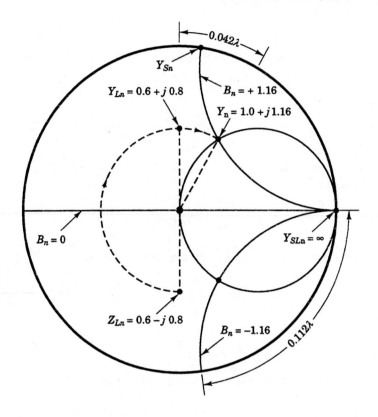

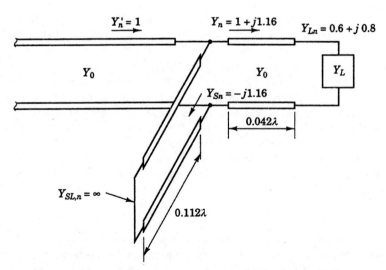

Figure 6.18   Solution to single-stub tuner design in Example 6.3.3.

6.4 *TEM* WAVES AT PLANAR BOUNDARIES

The formalism of Sections 6.1–6.3 can be applied to *TEM* waves at a multiplicity of parallel planar boundaries. We begin by considering examples involving reflection and transmission at normal incidence. Consider, first, the similarity between the basic equations for uniform plane waves normally incident upon a planar boundary and waves traveling across the junction between two *TEM* transmission lines. For $\hat{x}$-polarized waves, we recall from (4.2.1)–(4.2.4) that

$$\overline{E}(z < 0) = \hat{x}\, E_0 \left(e^{-jk_iz} + \Gamma\, e^{+jk_iz}\right) \tag{6.4.1}$$

$$\overline{H}(z < 0) = \hat{y}\, \frac{E_0}{\eta_i} \left(e^{-jk_iz} - \Gamma\, e^{jk_iz}\right) \tag{6.4.2}$$

$$\overline{E}(z > 0) = \hat{x}\, T E_0\, e^{-jk_tz} \tag{6.4.3}$$

$$\overline{H}(z > 0) = \hat{y}\, T \frac{E_0}{\eta_t}\, e^{-jk_tz} \tag{6.4.4}$$

where ϵ_i, μ_i and ϵ_t, μ_t characterize the two media, $k_i = \omega\sqrt{\mu_i\epsilon_i}$, $k_t = \omega\sqrt{\mu_t\epsilon_t}$, $\eta_i = \sqrt{\mu_i/\epsilon_i}$, $\eta_t = \sqrt{\mu_t/\epsilon_t}$, and Γ and T are the complex reflection and transmission coefficients.

For two *TEM* transmission lines connected at $z = 0$, the corresponding equations are

$$V(z < 0) = V_+ \left(e^{-jk_iz} + \Gamma\, e^{+jk_iz}\right) \tag{6.4.5}$$

$$I(z < 0) = \frac{V_+}{Z_{0i}} \left(e^{-jk_iz} - \Gamma\, e^{+jk_iz}\right) \tag{6.4.6}$$

$$V(z > 0) = V_+ T\, e^{-jk_tz} \tag{6.4.7}$$

$$I(z > 0) = \frac{V_+}{Z_{0t}} T\, e^{-jk_tz} \tag{6.4.8}$$

The equations (6.4.1)–(6.4.4) are identical to (6.4.5)–(6.4.8), respectively, if we observe some equivalencies between variables:

$$V(z) \leftrightarrow \overline{E}(z) \cdot \hat{x} \tag{6.4.9}$$

$$I(z) \leftrightarrow \overline{H}(z) \cdot \hat{y} \tag{6.4.10}$$

$$Z_{0i} = \sqrt{L_i/C_i} \leftrightarrow \eta_i = \sqrt{\mu_i/\epsilon_i}$$

$$Z_{0t} = \sqrt{L_t/C_t} \leftrightarrow \eta_t = \sqrt{\mu_t/\epsilon_t} \tag{6.4.11}$$

$$k_i = \omega\sqrt{L_iC_i} \leftrightarrow k_i = \omega\sqrt{\mu_i\epsilon_i}$$

$$k_t = \omega\sqrt{L_tC_t} \leftrightarrow k_t = \omega\sqrt{\mu_t\epsilon_t} \tag{6.4.12}$$

For the equivalence to be complete, not only must the equations for the media be equivalent, but so must the boundary conditions. Consider an example from optics in which a thin dielectric coating on a lens is designed to act as a quarter-wave transformer and thereby significantly reduce unwanted reflections. The equivalent transmission line model of this coating is shown in Figure 6.19.

The boundary conditions at the junctions of the *TEM* transmission lines specify that both the voltage $V(z)$ and current $I(z)$ be continuous across each junction.

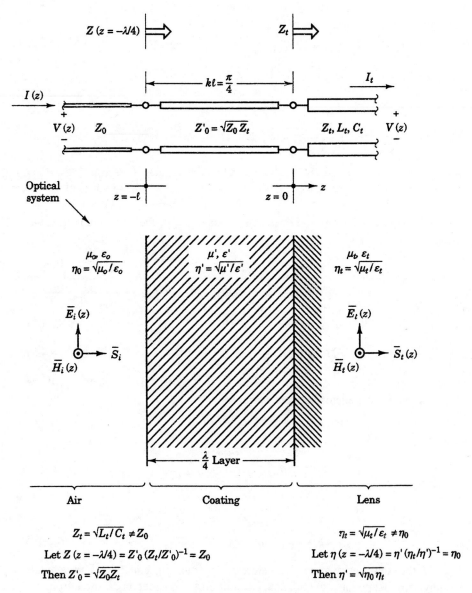

Figure 6.19   Quarter-wave transformer in coated optics.

The boundary conditions for the plane wave interfaces force the tangential components of both $\overline{E}$ and $\overline{H}$ to be continuous, and since $\overline{E} \cdot \hat{x} \leftrightarrow V$ and $\overline{H} \cdot \hat{y} \leftrightarrow I$, these two sets of boundary conditions are also equivalent. Thus, the quarter-wave transmission line transformer shown in Figure 6.15 and used for matching two *TEM* lines of different impedances is completely analogous to an optical quarter-wave transformer illustrated in Figure 6.19 of impedance $\eta_0' = \sqrt{\eta_0 \eta_t}$ and $\ell = \lambda'/4$ (λ' is the wavelength inside the transformer). Camera lenses are often coated with a dielectric layer $\mu' = \mu_0$, $\epsilon' = \sqrt{\epsilon_t \epsilon_0}$, where ϵ_t is the permittivity of the lens. The wavelength for which $\ell = \lambda'/4$ in optical material is usually chosen to be in the center of the visible band (i.e., yellow). Because the reflectivity is nonzero for wavelengths at the band edges, lens coatings often appear to be blue or purple.

The transmission line equivalent circuit for plane waves at non-normal incidence can be derived by first identifying the corresponding equivalent variables and boundary conditions. For a *TE* wave incident upon a planar interface, illustrated in Figure 6.20(a), the logical correspondence is

$$V(z) \leftrightarrow \overline{E}(z) \cdot \hat{y} = E_0 \tag{6.4.13}$$

$$I(z) \leftrightarrow \overline{H}(z) \cdot \hat{x} = -H_0 \cos \theta \tag{6.4.14}$$

$$Z_0^{TE} = \left| \frac{V(z)}{I(z)} \right| \leftrightarrow \left| \frac{E_y}{H_x} \right| = \frac{\eta}{\cos \theta} \tag{6.4.15}$$

We recall that both the tangential electric and magnetic fields must be continuous at a planar interface, and so it is these parallel fields that we relate to voltage and current, also continuous at a *TEM* transmission line junction.

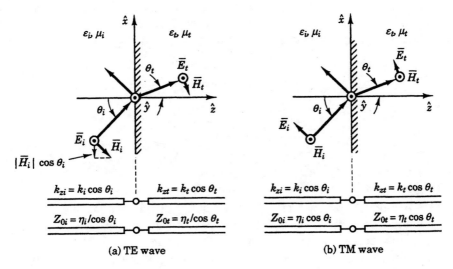

Figure 6.20   Transmission line equivalent circuits for waves at non-normal incidence.

Because waves on a *TEM* line travel in the $\pm \hat{z}$ direction, the $\hat{z}$ component of $\overline{k}$ corresponds to the wavenumber k on the *TEM* line (for both *TE* and *TM* waves):

$$k_i \leftrightarrow k_{iz} = k_i \cos \theta_i = \omega \sqrt{\mu_i \epsilon_i} \cos \theta_i = 2\pi / \lambda_{zi} \qquad (6.4.16)$$

$$k_t \leftrightarrow k_{tz} = k_t \cos \theta_t = \omega \sqrt{\mu_t \epsilon_t} \cos \theta_t = 2\pi / \lambda_{zt} \qquad (6.4.17)$$

A slightly different transmission line equivalent circuit results for *TM* waves, illustrated in Figure 6.20(b), where again the tangential components of $\overline{E}$ and $\overline{H}$ correspond to the voltage and current:

$$V(z) \leftrightarrow \overline{E} \cdot \hat{x} = E_0 \cos \theta \qquad (6.4.18)$$

$$I(z) \leftrightarrow \overline{H} \cdot \hat{y} = H_0 \qquad (6.4.19)$$

$$Z_0^{TM} = \left| \frac{V(z)}{I(z)} \right| \leftrightarrow \frac{E_x}{H_y} = \eta \cos \theta \qquad (6.4.20)$$

Comparison of these equivalent circuits clarifies why *TM* waves incident upon dielectrics with $\mu = \mu_0$ everywhere can have a Brewster's angle at which all of the incident light is transmitted, while *TE* waves cannot. For convenience, we choose $\epsilon_t > \epsilon_i$, but this choice does not affect the following discussion. At the junction between the two media, the normalized impedance for *TE*-wave equivalent transmission lines is

$$Z_n^{TE} = \frac{Z_{0t}}{Z_{0i}} = \frac{\eta_t \cos \theta_i}{\eta_i \cos \theta_t} \qquad (6.4.21)$$

But if $\epsilon_t > \epsilon_i$, then for $\mu_t = \mu_i$ it follows that $\sqrt{\mu_t / \epsilon_t} = \eta_t < \eta_i = \sqrt{\mu_i / \epsilon_i}$. From Snell's law, we note that $\cos \theta_i < \cos \theta_t$, and so $Z_n^{TE} < 1$. Thus Z_n^{TE} is less than unity for any θ_i, and there is no Brewster's angle for a *TE* wave with $\mu_i = \mu_t$. For *TM* waves with $\mu_t = \mu_i$ and $\epsilon_t > \epsilon_i$, we have

$$Z_n^{TM} = \frac{\eta_t \cos \theta_t}{\eta_i \cos \theta_i} \qquad (6.4.22)$$

where

$$0 < \eta_t < \eta_i < \infty$$

$$\cos \theta_t > \cos \theta_i \geq 0$$

Therefore, a Brewster's angle θ_B exists for which $Z_n^{TM} = 1$.

A more interesting example of the utility of transmission line equivalent circuits predicts the power transfer between two media separated by a gap in which only evanescent waves exist, as illustrated in Figure 6.21. If $\theta_i > \theta_c = \sin^{-1} \sqrt{\epsilon_g / \epsilon_i}$, only evanescent waves can exist in the gap between the two dielectrics, and k_g and Z_g for the transmission line equivalent circuit model of the

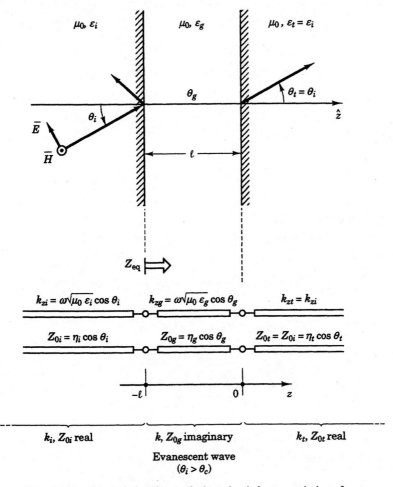

Figure 6.21 Transmission line equivalent circuit for transmission of power through an evanescent-wave zone.

gap are both imaginary. As an example, we let $\epsilon_i = \epsilon_t = 9\epsilon_0$, $\epsilon_g = \epsilon_0$ and $\theta_i = \pi/4$ for a *TM* wave. In this specific case, we find that

$$Z_i^{TM} = Z_t^{TM} = \sqrt{\frac{\mu_0}{9\epsilon_0}}\cos\theta_i = \frac{1}{3\sqrt{2}}\eta_0$$

$$Z_g^{TM} = \eta_0\cos\theta_g$$

and

$$k_{iz} = k_{tz} = 3k_0\cos\theta_i$$

$$k_{gz} = k_0\cos\theta_g$$

where we have used the equivalencies (6.4.16), (6.4.17) and (6.4.20) for *TM* waves and $k_0 = \omega/c$. In order to find θ_g, we use Snell's law (4.3.11) at either interface:

$$\sin \theta_g = 3/\sqrt{2}$$

But since $3/\sqrt{2} > 1$, θ_g is not a real angle and must therefore be complex: $\cos \theta_g = \sqrt{1 - \sin^2 \theta_g} = \pm j\sqrt{7/2}$. We choose the sign of $\cos \theta_g$ by noting that a wave in the evanescent region must have z-dependence given by $e^{-jk_{gz}z} = e^{-\alpha z}$ for $\alpha > 0$. If the source of incident radiation is on the left of the gap, this means that

$$\alpha = jk_{gz} = jk_g \cos \theta_g = jk_0 \left(-j\sqrt{\frac{7}{2}} \right) = \sqrt{\frac{7}{2}} k_0 > 0$$

and therefore $\cos \theta_g = -j\sqrt{7/2}$.

We summarize the characteristic impedances and wave numbers of the equivalent circuit:

$$Z_i = Z_t = \frac{1}{3\sqrt{2}} \eta_0 \qquad (z < -\ell, z > 0)$$

$$Z_g = -j\sqrt{\frac{7}{2}} \eta_0 \qquad (-\ell < z < 0)$$

$$k_{iz} = k_{tz} = \frac{3}{\sqrt{2}} k_0 \qquad (z < -\ell, z > 0)$$

$$k_{gz} = -j\sqrt{\frac{7}{2}} k_0 \qquad (-\ell < z < 0)$$

We can calculate the equivalent impedance looking into the circuit at $z = -\ell$ by using (6.1.6). The load impedance is simply Z_t, and

$$Z_{\text{in}}(z = -\ell) = Z_g \left(\frac{Z_t + jZ_g \tan k_{gz}\ell}{Z_g + jZ_t \tan k_{gz}\ell} \right)$$

where $\tan k_{gz}\ell = \tan(-jk_0\ell\sqrt{7/2}) = -j\tanh(k_0\ell\sqrt{7/2})$. We note that while Z_{in} generally has both a real and an imaginary part, corresponding to partial transmission across the gap, there are two special cases of interest:

(1) ℓ is very small. In this case, $Z_{\text{in}} \approx \eta_0/3\sqrt{2} - j32\eta_0 k_0\ell/9$ and therefore the *TEM* line to the left of $z = -\ell$ is terminated with close to a matched impedance. In this case, only $|\Gamma|^2 \approx 512(k_0\ell)^2/9$ of the incident power is reflected. If the gap vanishes entirely, then $|\Gamma| = 0$ and all of the incident power is transmitted.

(2) $\ell \to \infty$. In this case, $\tanh(k_0\ell\sqrt{7/2}) \to 1$ and so $Z_{\text{in}} \to Z_g = -j\eta_0\sqrt{7/2}$, corresponding to a purely capacitive equivalent impedance; all of the input power is reflected. This should again come as no surprise, since evanescent

wave amplitudes decay exponentially to zero infinitely far from a boundary, and thus do not survive to propagate in the transmitting medium.

Once transmission line equivalent circuits are constructed for a given problem, all of the insights they can provide become available.

Example 6.4.1

A thin planar sheet of resistance η_s ohms per square is to be placed parallel to a perfectly conducting planar surface a distance ℓ away so as to eliminate all reflections at wavelength λ. What are η_s and ℓ as a function of the incidence angle for *TE* and *TM* waves?

Solution: The *TEM* equivalent circuits (from Figure 6.20) are shown in Figure 6.22. The equivalent circuit for the conducting sheet follows from the definition of equivalent variables: $V(z)_{TE} \equiv E_y(z)$, $I(z)_{TE} = -H_x(z)$; $V(z)_{TM} \equiv E_x(z)$, $I(z)_{TM} \equiv H_y(z)$. We therefore find that

$$V(z = -\ell)/[I(z = -\ell - \delta) - I(z = -\ell + \delta)] = Z_s$$

corresponds to

$$E_y(z = -\ell)/[H_x(z = -\ell + \delta) - H_x(z = -\ell - \delta)] = \eta_s$$

and that for *TE* and *TM* waves, $Z_s \leftrightarrow \eta_s$, where Z_s is the equivalent impedance of the conducting sheet.

One method of eliminating reflections in the *TE* case is to choose $\ell = \lambda_z/4 = \lambda/(4 \cos \theta)$ and $\eta_s = \eta_0/\cos \theta$; for *TM* waves, $\ell = \lambda/(4 \cos \theta)$ and $\eta_s = \eta_0 \cos \theta$. In each case, the short circuit at $z = 0$ (equivalent to the perfectly conducting planar surface) becomes an open circuit at $z = -\ell$ by virtue of the fact that ℓ is a quarter wavelength long. The open circuit in parallel with η_s is equivalent to a terminating impedance η_s, and we therefore set η_s equal to the line impedance $\eta_0/\cos \theta$ for *TE* waves or $\eta_0 \cos \theta$ for *TM* waves to yield the desired match.

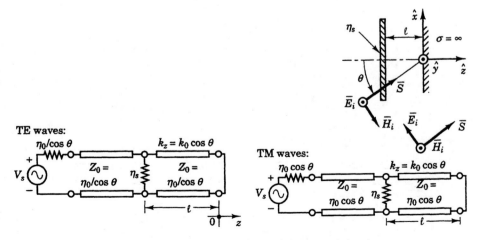

Figure 6.22   Equivalent circuits for an absorbing resistive sheet.

Example 6.4.2

A narrow-band multilayer dielectric filter is desired to reduce optical radiation of frequency f by 50 dB. The filter consists of m thin layers of dielectric ϵ_a alternating with thin layers of ϵ_b, all embedded in material of dielectric ϵ_b; $\mu = \mu_0$ everywhere. How thick should the layers be, and what m is required? Assume normal incidence.

Solution: The *TEM* equivalent circuit is illustrated in Figure 6.23. The thinnest transformer that converts one real impedance to another is a quarter wavelength thick. So ϵ_a should be $\lambda_a/4$ thick and ϵ_b should be $\lambda_b/4$ thick, where $\lambda_a = 1/f\sqrt{\epsilon_a\mu_0}$, $\lambda_b = 1/f\sqrt{\epsilon_b\mu_0}$. The normalized load impedance is

$$Z_{n1} = \eta_b/\eta_a$$

which we then transform to its reciprocal after a length $\lambda_a/4$ of line η_a. The un-normalized impedance Z_2 is thus $Z_2 = \eta_a^2/\eta_b$. We can then call Z_2 the load impedance for the next segment of transmission line and normalize it to the η_b impedance of this segment:

$$Z_{n2} = \eta_a^2/\eta_b^2$$

Thus,

$$Z_{n3} = (\eta_b/\eta_a)^3$$
$$Z_{n4} = (\eta_a/\eta_b)^4$$

for the second layer of η_a, and by extension

$$Z_{n(2m)} = (\eta_a/\eta_b)^{2m}$$

The reflection coefficient for the $2m^{th}$ layer is then just

$$\Gamma_{2m} = \left(\frac{Z_{n(2m)} - 1}{Z_{n(2m)} + 1}\right) = \left(\frac{\eta_a^{2m} - \eta_b^{2m}}{\eta_a^{2m} + \eta_b^{2m}}\right)$$

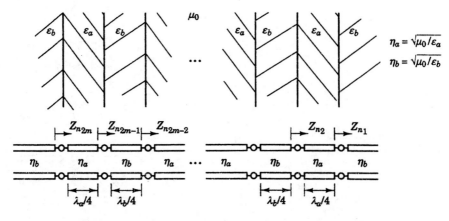

Figure 6.23   Equivalent circuit for a multilayer dielectric filter.

For the case $\epsilon_b = 4\epsilon_a$, we have $\eta_a/\eta_b = \sqrt{\epsilon_b/\epsilon_a} = 2$ and

$$\Gamma_{2m} = \left[1 - (\eta_b/\eta_a)^{2m}\right]/\left[1 + (\eta_b/\eta_a)^{2m}\right]$$
$$\cong 1 - 2(\eta_b/\eta_a)^{2m}$$
$$|\Gamma_{2m}|^2 \cong 1 - 4(\eta_b/\eta_a)^{2m}$$

But the total amount of power reflected from the multilayer filter is just $1 - |\Gamma_{2m}|^2$ which we choose to be less than 10^{-5} so that the necessary 50 dB power reduction is obtained. Therefore, $10^{-5} \le 4(\eta_b/\eta_a)^{2m}$ and $m = 9.3$ layers, so ten layers must be used to ensure adequate reduction of the optical radiation.

6.5 TIME-DOMAIN ANALYSIS OF TRANSMISSION LINES

Any specific problem imposes boundary conditions on $V(z, t)$ and/ or $I(z, t)$ in time and space, where these conditions uniquely determine the solution to a well-posed problem. Our approach to solving time-domain problems is the same as previously described for any electromagnetic *boundary value problem*. That is, we:

(1) Find general expressions for the waves in the separate media, where the wave expressions are in terms of certain unknowns;

(2) Impose boundary conditions, which fix certain variables or functions of variables at given times and places;

(3) Solve the resulting equations for the unknowns; and

(4) Test the resulting solutions with those of Maxwell's equations that have not already been imposed.

Consider the simple case of an infinite transmission line of characteristic impedance Z_0 driven by a *unit step* voltage source $u(t)$ ($u(t) = 1$, $t \ge 0$; $u(t) = 0$, $t < 0$). The line has no stored energy for $t < 0$. Let $z = 0$ at the junction between the voltage source and the transmission line, and let z increase to infinity along the line. Because there is no energy on the line for $t < 0$ and the source at $z = 0$ only generates a wave traveling in the $+\hat{z}$ direction, there is no $-\hat{z}$-propagating wave for $t \ge 0$. The backward wave $V_-(t + z/v)$ in (5.5.5) is therefore zero. Thus, the voltage is given by

$$V(z, t) = V_+\left(t - \frac{z}{v}\right) \tag{6.5.1}$$

which must satisfy the boundary condition at $z = 0$:

$$V(z = 0, t) = V_+(t) = u(t) \tag{6.5.2}$$

This means that

$$V(z, t) = V_+\left(t - \frac{z}{v}\right) = u\left(t - \frac{z}{v}\right) \tag{6.5.3}$$

because we can operate identically on all the arguments of an equality and still preserve that equality. That is, if $f(x) = g(y)$, then $f[T(x)] = g[T(y)]$ where

T is any arbitrary transformation of the argument. The current corresponding to (6.5.3) is found from (5.5.8):

$$I(z, t) = Y_0 u\left(t - \frac{z}{v}\right) \tag{6.5.4}$$

Both the voltage and current are sketched in Figure 6.24(a).

If the voltage source were connected to the infinite *TEM* line via a series source resistor Z_S, then the magnitudes of the voltage and current in (6.5.3) and (6.5.4) would have to be modified. Because there is only a forward-propagating wave on the infinite line, the voltage-to-current ratio is still just Z_0, and he equivalent impedance for the infinite line is just Z_0. The source voltage must therefore be divided between the resistor Z_S and the equivalent impedance Z_0, and so $V(z, t)$ and $I(z, t)$ are scaled by $Z_0/(Z_0 + Z_S)$

$$V(z, t) = \frac{Z_0}{Z_0 + Z_S} u\left(t - \frac{z}{v}\right) \tag{6.5.5}$$

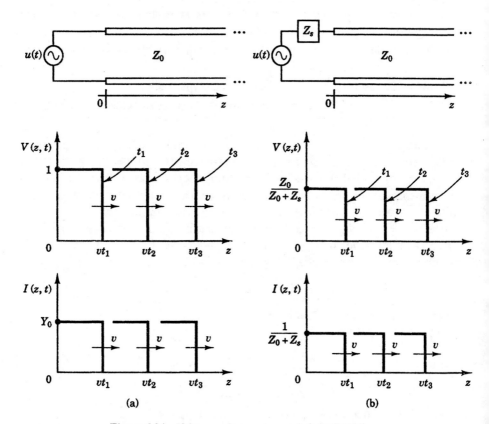

Figure 6.24   Voltage and current on an infinite *TEM* line.

$$I(z, t) = \frac{1}{Z_0 + Z_S} u\left(t - \frac{z}{v}\right) \tag{6.5.6}$$

as shown in Figure 6.24(b).

If the infinite *TEM* circuit described above (including Z_S) is now terminated with a frequency-independent load impedance Z_L, then a new boundary condition is imposed at $z = \ell$ (ℓ is the line length), which only becomes relevant **after** the signal generated at $z = 0$ has had a chance to reach $z = \ell$, i.e., for $t > \ell/v$. The voltage and current waves given by (6.5.5) and (6.5.6) are **still** valid for $0 < t < \ell/v$, but if $Z_L \neq Z_0$, then a reflected wave must be generated at $z = \ell$ in order to satisfy the boundary condition $V(z = \ell)/I(z = \ell) = Z_L$. (If $Z_L = Z_0$, then the boundary condition is automatically satisfied and no reflected wave is necessary. This is the same reflectionless phenomenon observed in the sinusoidal steady state for a matched load.) This terminated *TEM* line is shown in Figure 6.25 for $Z_L = 3Z_0$ and $Z_S = Z_0/2$.

With the addition of a reflected wave, we then have a voltage and current at $z = \ell$ given by

$$V(z = \ell, t) = \frac{Z_0}{Z_0 + Z_S}\left[u\left(t - \frac{\ell}{v}\right) + V_-^{(1)}\left(t + \frac{\ell}{v}\right)\right]$$

$$I(z = \ell, t) = \frac{1}{Z_0 + Z_S}\left[u\left(t - \frac{\ell}{v}\right) - V_-^{(1)}\left(t + \frac{\ell}{v}\right)\right] \tag{6.5.7}$$

where $V_-^{(1)}$ is the (first) reflected, $-\hat{z}$-propagating wave. Since the ratio of voltage to current at $z = \ell$ must be Z_L, we find

$$Z_L = Z_0\left(\frac{u(t - \frac{\ell}{v}) + V_-^{(1)}(t + \frac{\ell}{v})}{u(t - \frac{\ell}{v}) - V_-^{(1)}(t + \frac{\ell}{v})}\right)$$

$$= Z_0\left(\frac{1 + \Gamma_L}{1 - \Gamma_L}\right) \tag{6.5.8}$$

where $\Gamma_L = V_-^{(1)}(t + \frac{\ell}{v})/u(t - \frac{\ell}{v}) = (Z_L - Z_0)/(Z_L + Z_0)$.

The reflection coefficient at the load Γ_L is just a frequency-independent number if Z_L is a resistance, in which case the reflected wave generated at $z = \ell$ is

$$V_-^{(1)}\left(t + \frac{\ell}{v}\right) = \Gamma_L u\left(t - \frac{\ell}{v}\right) \tag{6.5.9}$$

If we allow $V_-^{(1)}$ to have z-dependence, we find that

$$V_-^{(1)}\left(t + \frac{z}{v}\right) = \Gamma_L u\left(t + \frac{z}{v} - \frac{2\ell}{v}\right) \tag{6.5.10}$$

where the argument of the right side of (6.5.10) must have $t + z/v$ dependence so that the reflected wave will propagate in the $-\hat{z}$ direction with velocity v. The $2\ell/v$ term is subtracted from the same argument so that (6.5.10) reduces to

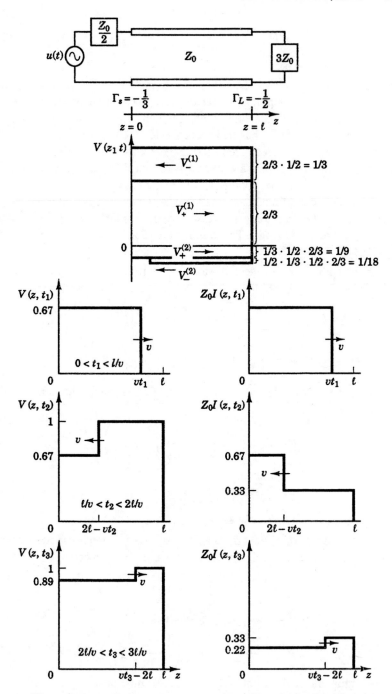

Figure 6.25   Unit step voltage source driving a nondispersive *TEM* line.

(6.5.9) at $z = \ell$. The argument $t + z/v - 2\ell/v$ may also be determined by adding $z/v - \ell/v$ to the arguments of **both** sides of (6.5.9). Because we have transformed the arguments in the same way, the equality in (6.5.9) is unaffected. Therefore, we can rewrite (6.5.7) as

$$V(z, t) = \frac{Z_0}{Z_0 + Z_S}\left[u\left(t - \frac{z}{v}\right) + \Gamma_L u\left(t + \frac{z}{v} - \frac{2\ell}{v}\right)\right]$$

$$I(z, t) = \frac{1}{Z_0 + Z_S}\left[u\left(t - \frac{z}{v}\right) - \Gamma_L u\left(t + \frac{z}{v} - \frac{2\ell}{v}\right)\right]$$

(6.5.11)

where (6.5.11) is valid for $0 < t < 2\ell/v$, i.e., before the reflected waveform has had a chance to reach the $z = 0$ boundary.

The $-\hat{z}$-propagating wave $V_-^{(1)}$ generated at $z = \ell$ now in turn acts as the source of a new, $+\hat{z}$-propagating wave $V_+^{(2)}$, generated as a result of an impedance mismatch at $z = 0$. As $V_-^{(1)}$ travels toward $z = 0$, the source end of the transmission line appears to be terminated with just the impedance Z_S. No voltage appears across the voltage source for the $-\hat{z}$-traveling wave because the first $+\hat{z}$-propagating wave already satisfies the boundary condition on the voltage at $z = 0$. For these backward waves, the voltage source looks like a short circuit.[4] Because the *TEM* line is terminated with the impedance Z_S for the $-\hat{z}$-propagating wave, we find that

$$V_+^{(2)}(z = 0, t) = \Gamma_S V_-^{(1)}(z = 0, t) = \frac{Z_0}{Z_0 + Z_S}\Gamma_S\Gamma_L u\left(t - \frac{2\ell}{v}\right) \qquad (6.5.12)$$

where $\Gamma_S = (Z_S - Z_0)/(Z_S + Z_0)$ in exact analogy with the discussion surrounding (6.5.7) and (6.5.8) relating $V_-^{(1)}$ to $V_+^{(1)}$ at the $z = \ell$ termination.

This new source term (6.5.12) produces a response propagating in the $+\hat{z}$ direction with velocity v

$$V_+^{(2)}(z, t) = \frac{Z_0}{Z_0 + Z_S}\left[\Gamma_S\Gamma_L u\left(t - \frac{z}{v} - \frac{2\ell}{v}\right)\right] \qquad (6.5.13)$$

which reduces to (6.5.12) at $z = 0$. Therefore, the voltage $V(z, t)$ and current $I(z, t)$ become

---

4. Were the source at $z = 0$ a current source instead of a voltage source, the initial forward-propagating wave would satisfy an initial current condition, and therefore any subsequent $-\hat{z}$-propagating waves generated at $z = \ell$ would see zero current looking back into the current source. For these backward-propagating waves, the current source would appear to be an open circuit.

$$V(z, t) = \frac{Z_0}{Z_0 + Z_S} \left[u\left(t - \frac{z}{v}\right) + \Gamma_L u\left(t + \frac{z}{v} - \frac{2\ell}{v}\right) \right.$$

$$\left. + \Gamma_S \Gamma_L u\left(t - \frac{z}{v} - \frac{2\ell}{v}\right) \right]$$

(6.5.14)

$$I(z, t) = \frac{1}{Z_0 + Z_S} \left[u\left(t - \frac{z}{v}\right) - \Gamma_L u\left(t + \frac{z}{v} - \frac{2\ell}{v}\right) \right.$$

$$\left. + \Gamma_S \Gamma_L u\left(t - \frac{z}{v} - \frac{2\ell}{v}\right) \right]$$

where (6.5.14) is now valid for $0 < t < 3\ell/v$, i.e., until the (second) reflected wave $V_+^{(2)}$ generated at $z = 0$ has a chance to reach $z = \ell$. We notice that because the third term in (6.5.12) does not "turn on" until $t \geq 2\ell/v$, (6.5.14) reduces to (6.5.11) for $0 < t < 2\ell/v$ and thus is consistent with earlier expressions for voltage and current.

A general description of wave propagation on a frequency-independent *TEM* line terminated with frequency-independent impedances is summarized below:

(1) The source at $z = 0$ generates a $+\hat{z}$-propagating wave $V_+^{(1)}$, so initially, the line termination is unimportant. This solution is valid for $0 < t < \ell/v$, where ℓ is the line length and v the wave velocity. If a source impedance connects the *TEM* line to a voltage source, then the magnitude of each *partial wave* is scaled down by $Z_0/(Z_0 + Z_S)$ because the equivalent circuit seen by this initial wave is just the *TEM* impedance Z_0 in series with Z_S.

(2) At $t = \ell/v$, the $+\hat{z}$-propagating wave $V_+^{(1)}$ reaches the load termination, and a reflected wave $V_-^{(1)}$ is consequently generated so that the voltage-to-current ratio at $z = \ell$ is Z_L (where Z_L is assumed to be frequency-independent). The magnitude of this reflected wave is Γ_L times the incident wave, where $\Gamma_L = (Z_L - Z_0)/(Z_L + Z_0)$.

(3) To find the total voltage, the $-\hat{z}$-propagating wave $V_-^{(1)}$ is added to the $+\hat{z}$-propagating wave $V_+^{(1)}$. To find the total current, the $V_-^{(1)}$ is subtracted from $V_+^{(1)}$ and the result multiplied by Y_0, as described by (5.5.8). This solution is valid for $0 < t < 2\ell/v$.

(4) At $t = 2\ell/v$, the wave has made a round trip on the line and reached the $z = 0$ source termination. Because the boundary condition at $z = 0$ has already been met with the first $+\hat{z}$-propagating wave $V_+^{(1)}$, a second $+\hat{z}$-propagating wave $V_+^{(2)}$ must be added to cancel the effects of $V_-^{(1)}$. In effect, the voltage source appears to be short-circuited and the source end of the *TEM* line is thus terminated with Z_S. The ratio of $V_+^{(2)}$ to $V_-^{(1)}$ at $z = 0$ is then just the reflection coefficient $\Gamma_S = (Z_S - Z_0)/(Z_S + Z_0)$. This new forward-propagating wave $V_+^{(2)}$ is added to the previous voltage and also scaled by Y_0

and added to the previous current. The total voltage and current solution are valid for $0 < t < 3\ell/v$. If a current source were driving the line instead, it would be open-circuited for this computation.

(5) At $t = 3\ell/v$, the second $+\hat{z}$-propagating wave $V_+^{(2)}$ creates a second $-\hat{z}$-propagating wave $\Gamma_L V_-^{(2)}$ times smaller because of further impedance mismatch at $z = \ell$.

This process continues, with each newly generated (partial) wave creating in turn its own reflected partial wave propagating in the opposite direction, scaled either by Γ_L or Γ_S depending on the end at which the reflection takes place. These partial waves are illustrated in Figure 6.25.

We can finally write voltage and current expressions for all time and space for this transmission line:

$$V(z, t) = \frac{Z_0}{Z_0 + Z_S} \left[u\left(t - \frac{z}{v}\right) + \Gamma_L u\left(t + \frac{z}{v} - \frac{2\ell}{v}\right) \right.$$
$$\left. + \Gamma_S \Gamma_L u\left(t - \frac{z}{v} - \frac{2\ell}{v}\right) + \Gamma_L \Gamma_S \Gamma_L u\left(t + \frac{z}{v} - \frac{4\ell}{v}\right) \right.$$
$$\left. + \Gamma_S \Gamma_L \Gamma_S \Gamma_L u\left(t - \frac{z}{v} - \frac{4\ell}{v}\right) + \cdots \right] \qquad (6.5.15)$$

$$I(z, t) = \frac{1}{Z_0 + Z_S} \left[u\left(t - \frac{z}{v}\right) - \Gamma_L u\left(t + \frac{z}{v} - \frac{2\ell}{v}\right) \right.$$
$$\left. + \Gamma_S \Gamma_L u\left(t - \frac{z}{v} - \frac{2\ell}{v}\right) - \Gamma_L \Gamma_S \Gamma_L u\left(t + \frac{z}{v} - \frac{4\ell}{v}\right) \right.$$
$$\left. + \Gamma_S \Gamma_L \Gamma_S \Gamma_L u\left(t - \frac{z}{v} - \frac{4\ell}{v}\right) + \cdots \right] \qquad (6.5.16)$$

Note that the total voltage is simply the sum of all the partial reflected waves, while the total current is found by adding all of the $+\hat{z}$-traveling partial waves and subtracting all of the $-\hat{z}$-propagating waves, as prescribed by (5.5.5) and (5.5.8).

Because $|\Gamma_L|$, $|\Gamma_S| \leq 1$, the series in (6.5.15) and (6.5.16) converge with increasing time. In fact, as steady state ($t \to \infty$) is reached, we find that

$$V(z, t \to \infty) = \frac{Z_0}{Z_0 + Z_S} [1 + \Gamma_L + \Gamma_S \Gamma_L + \Gamma_L \Gamma_S \Gamma_L + \Gamma_S \Gamma_L \Gamma_S \Gamma_L + \cdots]$$
$$= \frac{Z_0}{Z_0 + Z_S} (1 + \Gamma_L) \sum_{n=0}^{\infty} (\Gamma_S \Gamma_L)^n$$
$$= \frac{Z_0}{Z_0 + Z_S} \left(\frac{1 + \Gamma_L}{1 - \Gamma_S \Gamma_L}\right) = \frac{Z_L}{Z_L + Z_S} \qquad (6.5.17)$$

and

$$I(z, t \to \infty) = \frac{1}{Z_0 + Z_S} [1 - \Gamma_L + \Gamma_S \Gamma_L - \Gamma_L \Gamma_S \Gamma_L + \Gamma_S \Gamma_L \Gamma_S \Gamma_L - \cdots]$$

$$= \frac{1}{Z_0 + Z_S} (1 - \Gamma_L) \sum_{n=0}^{\infty} (\Gamma_S \Gamma_L)^n$$

$$= \frac{1}{Z_0 + Z_S} \left(\frac{1 - \Gamma_L}{1 - \Gamma_S \Gamma_L} \right) = \frac{1}{Z_L + Z_S} \qquad (6.5.18)$$

where the geometric series $\sum_{n=0}^{\infty} a^n = (1 - a)^{-1}$ for $|a| < 1$ has been used. The voltage and current thus become nearly constant on the *TEM* line if we wait long enough, and

$$\frac{V(z, t \to \infty)}{I(z, t \to \infty)} = Z_L \qquad (6.5.19)$$

which is just what we expect. The forward- and backward-traveling waves ultimately adjust so that the boundary condition at $z = \ell$ is met.

We could also find the DC (steady state) limit without having to go through the trouble of analyzing the partial waves on the *TEM* line. For $\omega = 0$, the lossless transmission line equations (5.3.36) and (5.3.48) (with $R = G = 0$) become

$$\frac{\partial V}{\partial z} = \frac{\partial I}{\partial z} = 0$$

which means that I and V are constant along the length of the *TEM* line. Therefore, in the static limit, the *TEM* line may be thought of as nothing but a pair of circuit wires with no wave properties, as illustrated in Figure 6.26. The voltage on the line is then simply $Z_L/(Z_L + Z_S)$ V, because the source $u(t)$ is 1 V for $t > 0$ and the line voltage in the static limit is divided between the impedances Z_L and Z_S. The current is simply $1/(Z_L + Z_S)$, which is again consistent with the non-wavelike behavior of the *TEM* line. Initially, the source drives current through the *TEM* line as if it were a Z_0-Ω resistor (without regard to the load impedance), but only the load impedance Z_L is ultimately important as $t \to \infty$, because the wave-like properties of the *TEM* line are nonexistent in the static limit. The characteristic impedance Z_0 should not appear in any voltage or current expressions as $\omega \to 0$.

So far, we have been concerned only with a frequency-independent load impedance, for which the pulse shape is maintained (though possibly scaled) upon reflection. If the termination impedance is a capacitor or an inductor that reflects each frequency component of the incident waveform with a slightly different magnitude, then the waveform shape will be distorted upon reflection and the methods described above are insufficient to characterize the resulting wave behavior. The easiest way to analyze a *TEM* system terminated by a dispersive load is to reduce the *TEM* line to a lumped-element equivalent circuit valid for some period of time, and then to solve the lumped-element system in the usual fashion.

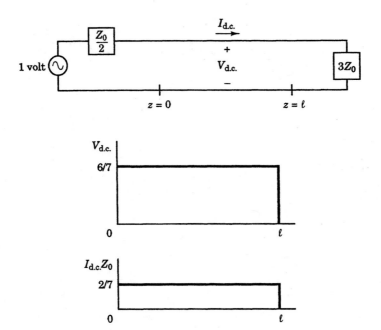

Figure 6.26   Static limit of wave propagation on a *TEM* line.

We may use a Thevenin or Norton equivalent circuit with equal ease. Consider the Thevenin equivalent circuit for the *TEM* line in Figure 6.27 with characteristic impedance Z_0, length ℓ, and velocity v terminated by a capacitor C. The source voltage $u(t)$ is connected to the transmission line by a source impedance Z_0. For $0 < t < \ell/v$, the waveform does not yet see the termination impedance, and the transmission line has an equivalent impedance Z_0 (the impedance of an infinite length *TEM* line). Therefore, the voltage is divided equally between the series resistor Z_0 and the *TEM* line equivalent impedance Z_0, yielding the voltage and current

$$V(z, t) = \frac{1}{2}u\left(t - \frac{z}{v}\right)$$
$$I(z, t) = \frac{Y_0}{2}u\left(t - \frac{z}{v}\right)$$

(6.5.20)

for $0 < t < \ell/v$. The Thevenin circuit for the output of the *TEM* line in Figure 6.28 is a voltage source V_{Th} connected in series with a Thevenin impedance Z_{Th}, as was also described Section 6.1 for the sinusoidal steady state. The Thevenin equivalent circuit must produce the same voltage and current at the load ($z = \ell$) as the *TEM* circuit, no matter what load is chosen. If we terminate the Thevenin circuit with an open circuit, then the voltage V_L at the load is equal to the Thevenin voltage V_{Th}. If we then open-circuit the *TEM* line on which the voltage wave $\frac{1}{2}u(t - z/v)$

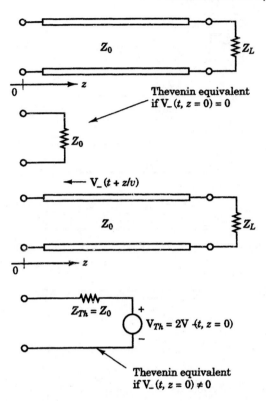

Figure 6.27 Thevenin equivalent circuits for time-domain signals.

is propagating, we find a total output voltage at $z = \ell$, which is **twice** the forward-propagating voltage there because the reflection coefficient at an open circuit is $\Gamma_L = +1$. Therefore,

$$V_L = V_{Th} = 2 \cdot \frac{1}{2} u \left(t - \frac{\ell}{v} \right) \tag{6.5.21}$$

If we terminate the Thevenin circuit with $Z_L = 0$ instead, then $I_L = V_{Th}/Z_{Th}$. If we then short-circuit the transmission line, the total current at $z = \ell$ is twice the forward-propagating current at $z = \ell$ because $\Gamma_L = -1$. Thus,

$$I_L = \frac{V_{Th}}{Z_{Th}} = 2 \cdot \frac{Y_0}{2} u \left(t - \frac{\ell}{v} \right) \tag{6.5.22}$$

or, combining (6.5.21) and (6.5.22),

$$Z_{Th} = Z_0 \tag{6.5.23}$$

Therefore, we must solve the lumped-element circuit in Figure 6.28 to find the voltage at $z = \ell$, which by traditional circuit methods is

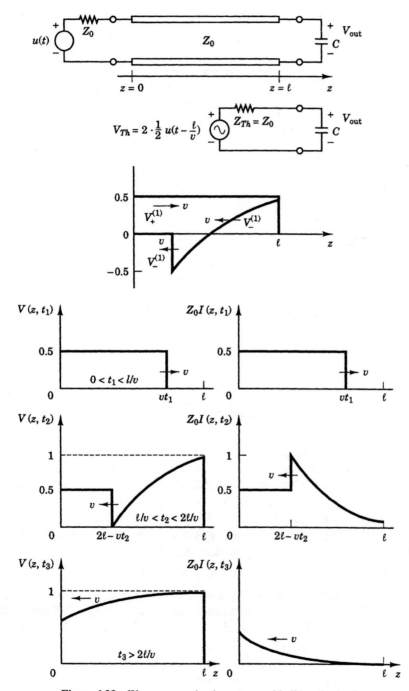

Figure 6.28   Wave propagation in systems with dispersive loads.

$$V_L = \left(1 - e^{-(t-\ell/v)/Z_0 C}\right) u\left(t - \frac{\ell}{v}\right)$$

But V_L is just the superposition of the forward-traveling wave $\frac{1}{2}u(t - \ell/v)$ at $z = \ell$ and a reflected wave $V_-(z = \ell, t)$:

$$V_L = \frac{1}{2} u\left(t - \frac{\ell}{v}\right) + V_-(z = \ell, t)$$

$$= \left(1 - e^{-(t-\ell/v)/Z_0 C}\right) u\left(t - \frac{\ell}{v}\right) \tag{6.5.24}$$

Solving (6.5.24) for $V_-(z = \ell, t)$ yields

$$V_-(z = \ell, t) = \left(\frac{1}{2} - e^{-(t-\ell/v)/Z_0 C}\right) u\left(t - \frac{\ell}{v}\right) \tag{6.5.25}$$

Because $V_-(z, t)$ must have an argument that includes the dependence $t + z/v$ to ensure propagation with velocity v in the $-\hat{z}$ direction,

$$V_-(z, t) = \left(\frac{1}{2} - e^{-(t+z/v-2\ell/v)/Z_0 C}\right) u\left(t + \frac{z}{v} - \frac{2\ell}{v}\right) \tag{6.5.26}$$

valid for $0 < t < 2\ell/v$, and therefore,

$$V(z, t) = \frac{1}{2} u\left(t - \frac{z}{v}\right) + \left(\frac{1}{2} - e^{-(t+z/v-2\ell/v)/Z_0 C}\right) u\left(t + \frac{z}{v} - \frac{2\ell}{v}\right)$$

$$I(z, t) = \frac{Y_0}{2} u\left(t - \frac{z}{v}\right) - Y_0\left(\frac{1}{2} - e^{-(t+z/v-2\ell/v)/Z_0 C}\right) u\left(t + \frac{z}{v} - \frac{2\ell}{v}\right) \tag{6.5.27}$$

where (5.5.5) and (5.5.8) have been used to combine the forward- and backward-propagating waves. The current and voltage expressions given by (6.5.27) are valid for $0 < t < 2\ell/v$, but it happens that (6.5.27) is valid for all t as well. Because the source resistance Z_S is matched to the line impedance Z_0, there are no further reflected waves generated when the $-\hat{z}$ traveling wave reaches $z = 0$. We recall that $\Gamma_S = (Z_S - Z_0)/(Z_S + Z_0) = 0$ in this case. Thus, (6.5.27) needs no further modifications and these waveforms are illustrated in Figure 6.28.

We can verify that (6.5.27) is correct in several limits:

(1) Before the initial wave reaches $z = \ell$, the termination impedance is irrelevant. Thus, the voltage and current for $0 < t < \ell/v$ are not affected by the capacitor C and have the ratio Z_0.

(2) If we wait long enough, the transients generated by $u(t)$ decay and the capacitor becomes an open circuit ($Z_C = 1/j\omega C \to \infty$ as $\omega \to 0$). Therefore, as $t \to \infty$, the transmission line looks like a 1-V voltage source terminated by infinite impedance, and $V(z, t \to \infty) = 1$, $I(z, t \to \infty) = 0$, as can be verified by (6.5.27).

(3) At $z = \ell$ and $t = \ell/v$ (just as the step function of 0.5 V magnitude reaches the end of the line), the high frequency part of the step function effectively short-circuits the capacitor. Therefore, $V(z = \ell, t = \ell/v) = 0$ and $I(z = \ell, t = \ell/v) = 2 \cdot \frac{1}{2}Y_0 = Y_0$ because the total current doubles in magnitude. This is easily verified from (6.5.27).

If the source resistor were not so conveniently matched to the transmission line impedance, we would have to find new Thevenin equivalent circuits for $t > 2\ell/v$. The new Thevenin voltage for the transmission line as seen by the source at $z = 0$ would be twice the first reflected wave evaluated at $z = 0$

$$V'_{\text{Th}} = 2 \left(\frac{1}{2} - e^{-(t-2\ell/v)/Z_0 C} \right) u \left(t - \frac{2\ell}{v} \right) \tag{6.5.28}$$

while the Thevenin impedance would again be Z_0, and the source impedance Z_S would provide the load for this circuit. In general, each forward-propagating wave $V_+^{(i)}$ gives rise to a negative-propagating wave $V_-^{(i)}$, and each $V_-^{(i)}$ in turn produces a $V_+^{(i+1)}$. Each separate signal $V_{\pm}^{(i)}$ can be represented by its own Thevenin (or Norton) equivalent circuit and generates its own separate response. All such responses are superposed to yield the total voltage and current on the *TEM* line.

This same procedure can also be used to analyze the interactions between transients on transmission lines and lumped nonlinear systems, except that the Thevenin (or Norton) equivalent source cannot be formed for each partial wave $V_{\pm}^{(i)}$ independently. At each time t **all** of the relevant partial waves must be summed to form the total voltage from which a corresponding Thevenin equivalent may be constructed. Superposition does not apply when nonlinearities are present.

We have considered boundary conditions in space for transient problems, and we now consider boundary conditions in time. Suppose $V(z)$ and $I(z)$ are specified on a transmission line at $t = 0$, as suggested in Figure 6.29. The total voltage and current on the line at $t = 0$ may be thought of as the superposition of a forward-traveling wave $V_+(z - vt)$ and a backward-traveling wave $V_-(z + vt)$. We can solve for these waves by manipulating the general expressions

$$V(z, t) = V_+(z - vt) + V_-(z + vt) \tag{6.5.29}$$

$$I(z, t) = Y_0[V_+(z - vt) - V_-(z + vt)] \tag{6.5.30}$$

to yield

$$V_+(z - vt) = [V(z, t) + Z_0 I(z, t)]/2 \tag{6.5.31}$$

$$V_-(z + vt) = [V(z, t) - Z_0 I(z, t)]/2 \tag{6.5.32}$$

Once these separate waves V_+ and V_- are known as functions of both space and time, their destinies can be followed separately, and $V(z, t)$ and $I(z, t)$ can be found by superposition via (6.5.29) and (6.5.30).

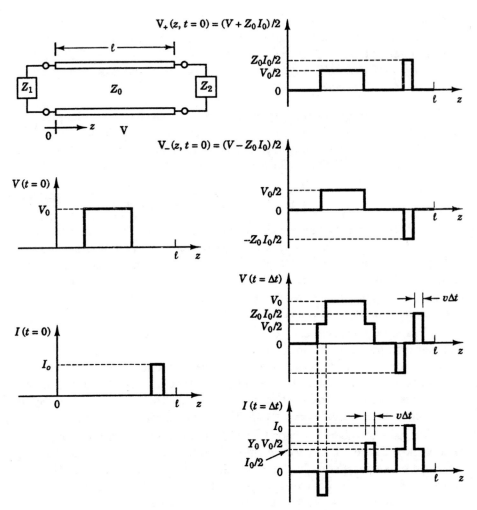

Figure 6.29   Initial conditions for transient signals on a *TEM* transmission line.

For example, the line in Figure 6.29 has one section that is initially charged to V_0 volts and zero current. On this section, $V_+ = V_- = V_0/2$ at $t = 0$ using (6.5.31) and (6.5.32). Another section has I_0 amperes and zero voltage; there, $V_+ = -V_- = Z_0 I_0/2$ at $t = 0$. Everywhere else on the line, the voltage and current are zero ($V_+ = V_- = 0$ at $t = 0$). If we wait a short time Δt, the V_+ pulse moves a distance $v\Delta t$ to the right ($+\hat{z}$ direction) while the V_- pulse moves $v\Delta t$ to the left ($-\hat{z}$ direction). Each forward and backward pulse may be tracked independently, and then superposed using (6.5.29) and (6.5.30) any time the total voltage and total current are desired. Reflections at the ends of the transmission line are treated just as before, with the $+\hat{z}$- and $-\hat{z}$-traveling waves considered independently. Figure 6.29 illustrates the wave behavior at two instants of time.

Example 6.5.1

A unit step current source of $u(t)$ A drives an air-filled 100-Ω *TEM* line ℓ m long and open-circuited at the end. What is the input voltage $v(t)$?

Solution: For $0 < t < 2\ell/c$, the voltage $v(t) = i(t) \cdot Z_0 = 100$ V, because no reflections from the $z = \ell$ termination have been received. At $z = \ell$, $\Gamma_L = +1$, so that $V_-^{(1)} = V_+^{(1)} = 100$ V, as illustrated in Figure 6.30.

When the $-\hat{z}$-traveling wave reaches the current source, it is reflected with $\Gamma_S = +1$ because the boundary condition $I(z = 0, t) = u(t)$ has already been met with $V_+^{(1)}$. Each subsequent wave must now reflect so as to contribute zero current at $z = 0$, and the source therefore appears to be an open circuit. Therefore $V_+^{(2)} = V_-^{(1)}$, as illustrated. The output voltage $v(t) = V_+^{(1)} + V_-^{(1)} + V_+^{(2)} = 300$ V for $2\ell/c < t < 4\ell/c$. At $t = 4\ell/c$, the voltage $v(t)$ again jumps by another 200 V. Only voltage breakdown and other departures from the

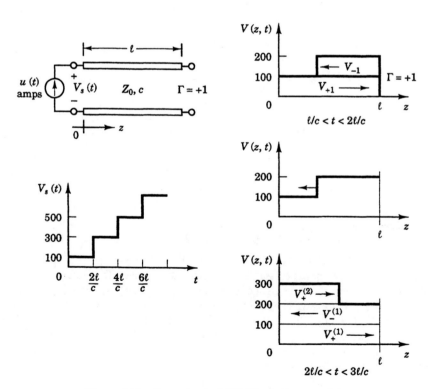

Figure 6.30   Open-circuited *TEM* line in Example 6.5.1.

idealized circuit element models would prevent the line voltage from eventually increasing to infinity.

The rate of increase of voltage with respect to time is approximately given by

$$\frac{\partial V}{\partial t} \approx \frac{v(t)}{t} = \frac{Z_0 I_0}{\ell/c} = I_0 \sqrt{\frac{L}{C}} \frac{1}{\ell \sqrt{LC}}$$

$$= \frac{I_0}{C\ell} = \frac{I_0}{C}$$

which is the differential equation for the capacitor $C = C\ell$ charged by a constant current source I_0. Therefore, a very short open-circuited *TEM* line is an approximate model of a capacitor, as was also shown in Section 6.1.

Example 6.5.2

An air-filled *TEM* line is 2 km long and has a capacitance per unit length $C = 33$ pF/m. The line is connected to a 10-kV DC power supply at one end ($z = 0$) and is terminated at the other end by a 50-Ω load impedance, as shown in Figure 6.31. At $t = 0$, lightning strikes the transmission line at $z = 1$ km, shorting the line for $\Delta t = 1$ μs. Describe the voltage and current waves propagating on the line at $t = 2$ μs.

Solution: We first need to find the characteristic impedance Z_0 of this transmission line. If $Z_0 = \sqrt{L/C}$ and $c = 1/\sqrt{LC} = 3 \times 10^8$ m/s in air, then

$$Z_0 = \frac{1}{cC} = \frac{1}{3 \times 10^8 \text{ m/s} \times 33 \times 10^{-12} \text{ F/m}} = 100 \ \Omega$$

We now find the initial current and voltage on the line, assuming that no transients exist for $t < 0$. In this case, the *TEM* line has no wave properties, and we find that

$$V(z, t = 0^-) = 10 \text{ kV} \equiv V_0 = V_+ + V_-$$

$$I(z, t = 0^-) = \frac{10 \text{ kV}}{50 \ \Omega} = \frac{V_0}{Z_0/2} = \frac{1}{Z_0}(V_+ - V_-)$$

which means that $V_+ = 3V_0/2$ and $V_- = -V_0/2$ for $t < 0$. At $t = 0^+$, the line is shorted in the middle; $V(z = \ell/2, t = 0^+) = 0$. This obviously creates a disturbance in the V_+ and V_- component waves, but the $V_{(\pm)}$ waveform may **only** be affected to the $\left\{ \begin{array}{c} \text{right} \\ \text{left} \end{array} \right\}$ of $z = \ell/2$ because the disturbance must necessarily propagate with the wave. Therefore, V_+ to the left of the 1-km point and V_- to the right of the 1-km point are unaffected by the short circuit. At the end of 1 μs, the wave has been able to propagate $c\Delta t = 3 \times 10^8 \times 10^{-6} = 300$ m. The short circuit has thus affected V_+ in the region from 1000 to 1300 m, and in that region $V_+ = V_0/2$ because $V_- = -V_0/2$ to the right of the 1-km point. The two voltages must therefore sum to zero. Likewise, the short circuit has affected V_- in the region from 700 to 1000 m at $t = 1$ μs, and $V_- = -3V_0/2$ because $V_+ = 3V_0/2$ there. Once the independent forward- and backward-propagating waves are known for an initial time $t = 1$ μs, then each wave may be treated independently. Figure 6.31 shows the voltage and current profiles at 2 μs.

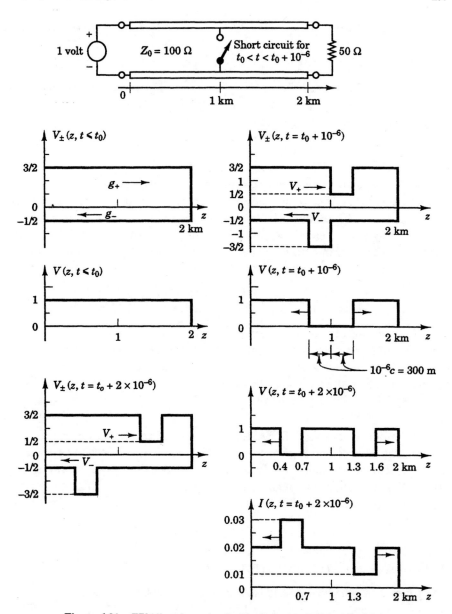

Figure 6.31 *TEM* line short-circuited by lightning in Example 6.5.2.

6.6 SUMMARY

The complex, spatially varying *impedance* of a transmission line results from interference between the forward and backward waves. If the *TEM* line is terminated by a load impedance Z_L, the impedance is

$$Z(z) = \frac{V(z)}{I(z)} = Z_0 \left(\frac{1 + \Gamma(z)}{1 - \Gamma(z)} \right)$$

where $\Gamma(z) = \Gamma_L e^{2jkz}$ and $\Gamma_L = (Z_L - Z_0)/(Z_L + Z_0)$ is the complex *reflection coefficient* of the load. The *voltage standing wave ratio* (VSWR) is the ratio of maximum to minimum voltage on the line:

$$\text{VSWR} = \frac{1 + |\Gamma(z)|}{1 - |\Gamma(z)|} = \frac{1 + |\Gamma_L|}{1 - |\Gamma_L|}$$

Although transmission line problems can be solved analytically, it is often useful to describe *TEM* behavior using a *Smith chart*, on which impedance transformations are made pictorially. *ABCD matrices* are also convenient for representing a cascade of many different lines, with lumped impedances added to some of the junctions. The relation between voltage and current at two points separated by a distance ℓ on a *TEM* line is given by

$$\begin{bmatrix} V_2(z = -\ell) \\ I_2(z = -\ell) \end{bmatrix} = \begin{bmatrix} \cos k\ell & jZ_0 \sin k\ell \\ jY_0 \sin k\ell & \cos k\ell \end{bmatrix} \begin{bmatrix} V_1(z = 0) \\ I_1(z = 0) \end{bmatrix}$$

TEM models can also be constructed for many propagation problems that are not truly *TEM*, provided that the forward and backward waves are proportional to $e^{\pm jkz}$, do not couple in a distributed fashion, and obey boundary conditions that require two different wave parameters to be continuous across the boundary, such as E_x and H_y. These parameters can be related to $V(z)$ and $I(z)$ in a *TEM equivalent circuit*.

Time-domain analysis of *TEM* lines may also be carried out for transmission line systems by matching *boundary* and *initial conditions*. Reflection coefficients, defined as the ratio of individual forward- and backward-propagating waves, may also be defined in the time domain for nondispersive loads. The most general expression for waves on a *TEM* line is a superposition of forward- and backward-propagating waves:

$$V(z, t) = V_+(z - vt) + V_-(z + vt)$$

$$I(z, t) = Y_0 \left[V_+(z - vt) - V_-(z + vt) \right]$$

which means that the forward and backward waves can be constructed from

$$V_+(z - vt) = \frac{1}{2} [V(z, t) + Z_0 I(z, t)]$$

$$V_-(z + vt) = \frac{1}{2} [V(z, t) - Z_0 I(z, t)]$$

Once these separate waves are established at a particular time, their motion may then be independently tracked and $V(z, t)$ and $I(z, t)$ found at a later time by superposing the two waves.

6.7 PROBLEMS

6.1.1 What are the Thevenin equivalent circuits for the following *TEM* lines at 1 GHz? The *TEM* line in each case has a characteristic impedance of 50 Ω and is air-filled and lossless.

    (a) Open-circuited *TEM* line $7\lambda/8$ long.

    (b) Short-circuited *TEM* line $7\lambda/8$ long.

    (c) 100-Ω resistor at the end of a line $\lambda/4$ long.

    (d) 100-Ω resistor at the end of a line $\lambda/8$ long.

    (e) Capacitor of $C = (\pi \times 10^7)^{-1}$ farads at the end of a line $3\lambda/4$ long.

    (f) Same as part (e) for a line $3\lambda/8$ long.

6.1.2 What are the Thevenin equivalent circuits for a 100-Ω *TEM* line $\lambda/4$ long driven at the far end by:

    (a) A voltage source V_0.

    (b) A current source I_0 in parallel with a 100-Ω resistor.

    (c) A current source I_0 in parallel with a 400-Ω resistor.

6.1.3 Measurements of $|V(z)|$ on an air-filled 50-Ω *TEM* transmission line driven by a generator at the left end of the cable yield voltage minima of 10 V 30 cm apart. The maxima are 30 V and the first minimum is 10 cm from the load at the right end of the line.

    (a) What is the frequency of the generator?

    (b) What is $|\Gamma|$ on the line?

    (c) What is the load impedance Z_L at $z = 0$?

    (d) What fraction of the power incident upon the load Z_L is reflected from it?

    (e) What power (watts) does the load absorb?

6.1.4 A certain lossy *TEM* line at 10 MHz can be incrementally modeled as an inductance $L = 2 \times 10^{-6}$ H/m in series with $R = 10$ Ω/m, and a shunt of $C = 5 \times 10^{-11}$ F/m in parallel with $G = 2 \times 10^{-4}$ S/m.

    (a) What approximately is the complex characteristic impedance Z_0?

    (b) What is the wavelength λ?

    (c) What is the approximate input impedance of a short circuit seen through this transmission line if it has length (i) 10 m, (ii) 12 m, and (iii) 100 m?

6.1.5 A lossless *TEM* line is connected to a source voltage V_S and terminated by a load impedance Z_L. Show that power dissipation in this circuit computed by $P_d = \frac{1}{2} \operatorname{Re}\{V \, I^*\}$ is the same whether the voltage and current are evaluated at the load ($z = 0$) or the source ($z = -\ell$).

6.2.1 A voltage source of $2 \cos \omega t$ V is connected to a quarter-wave 50-Ω *TEM* line, which is in turn connected to an eighth-wave 100-Ω *TEM* line connected to a 50-Ω load.

    (a) What are the *ABCD* matrices $\overline{\overline{T}}$ for each *TEM* line? Give numerical values for all coefficients.

    (b) What is the overall transfer matrix for the pair of *TEM* lines?

    (c) Find a numerical value for the load voltage V_L.

    (d) Repeat parts (a)–(c) for a voltage source of $\cos 2\omega t$ V where the rest of the problem is unchanged.

6.2.2 Show that the *ABCD* matrix for a *TEM* line of length L is equal to the product of the *ABCD* matrices corresponding to lines of length a and b, where $a + b = L$. All the lines have the same characteristic impedance Z_0.

6.2.3 Show that the *ABCD* matrices given for each two-port network in Figure 6.6 are correct.

6.3.1 Design a quarter-wave transformer to match a 100-Ω line to a 1000-Ω resistive load. Assume glass-filled *TEM* lines are used with index of refraction $n = 2$ at 100 MHz.

6.3.2 The equivalent impedance of a very slightly lossy 100-Ω *TEM* line of length ℓ terminated with a 300-Ω resistor can be (i) 100 Ω, (ii) 300 Ω, (iii) 33 Ω, and (iv) $60 - j80$ Ω under certain circumstances. What are the lengths of line for each of these cases?

6.3.3* A microwave radiometer of impedance Z_0 can measure thermal radiation at microwave wavelengths, the intensity being directly proportional to the average physical temperature of any matched load. A transmission line of impedance Z_0 connects the radiometer and a mismatched load Z_L. Assume $\omega = 10^9$ rad/s, $Z_0 = 100$ Ω, and $Z_L = 200 + j100$.

    (a) A single capacitor C is to be placed in series with the transmission line to match the load Z_L. What is the closest distance to the load the capacitor should be placed, and what value of C (if any) should be used to produce a match?

    (b) Repeat part (a) for a capacitor C placed in parallel with the line.

    (c) Repeat part (a) for an inductor L placed in series with the line.

    (d) Does the VSWR between the load Z_L and the tuning element (L or C) depend on the value of that L or C? What is the VSWR for the matching system of part (a)?

6.3.4 A *TEM* coaxial line has a characteristic impedance of 100 Ω, $\mu = \mu_0$, and a characteristic velocity of 2×10^8 m/s. Its copper walls have conductivity $\sigma \cong 5 \times 10^7$ S/m, $\epsilon = \epsilon_0$, and $\mu = \mu_0$.

    (a) Let the total series resistance R_s (Ω/m) $= R_{s0} + R_{si}$, where R_{si} and R_{s0} are the series resistances of the inner and outer conductors, respectively ($R_s \cong 0.484$ Ω/m). In terms of R_s and other parameters, what is Γ 100 m from a short circuit at 100 MHz?

    (b) At 100 MHz what is the skin depth in the copper?

    (c) What is the approximate equivalent series resistance R_{si} (Ω/m) for current flowing on the inner conductor at 100 MHz? The radius of the inner conductor is 1 mm.

6.3.5 Whiz Electronics, Inc., is developing a computer using logic with a rise time of 10^{-10} s. If we Fourier transform the pulse trains propagating through this machine, we have significant frequency components as high as ~ 1 GHz.

    (a) A single line of the printed circuit can be modeled as a parallel-plate *TEM* line of width 0.1 mm, thickness 0.04 mm, and $\epsilon = 4\epsilon_0$. What is Z_0 for this *TEM* line if we neglect fringing fields?

    (b) Particular *TEM* wire runs 15 cm. The input impedance of the connected transistor circuit is 1000 Ω. Sketch the approximate complex input impedance seen at the other end of the wire that drives this transistor as a function of frequency, from zero to 1 GHz. Indicate quantitatively any maxima or minima in the impedance function.

(c*) Assume the driven end of this wire can be modeled as a Thevenin equivalent circuit having a source impedance of 10 Ω and a square-wave voltage source up to 150 MHz. In this case, will the circuit architects for the machine need to understand electromagnetic waves in order to perfect their product? Explain. What rule-of-thumb design criteria would you suggest to them?

6.3.6 A certain RF amplifier is built using a tunnel diode (TD) biased so as to produce a small-signal load impedance of -15 Ω. Amplification occurs when a signal on a 100-Ω *TEM* line is reflected from this negative impedance load.

(a) What are Γ and the power gain $|\Gamma|^2$ for this amplifier? Assume that the RF input signal effectively sees a load impedance of -15 Ω and does not exceed the dynamic range of the diode.

(b) At a distance $\lambda/8$ from the diode, what is the equivalent impedance Z seen by the source?

(c) What is the VSWR on this line?

6.3.7 The current magnitude on an air-filled 50 Ω line has been measured. It has a 1-m period between minima, and the first minimum is 0.25 m from the load. The maxima and minima are 3 mA and 1 mA, respectively.

(a) What is the frequency?

(b) What are the VSWR and $|\Gamma|^2$?

(c) What is the complex load impedance Z_L?

6.3.8 A 300-Ω twin-lead (*TEM*) cable is delivering RF power to a heating device with impedance $150(1 - j)$ Ω. A single-stub tuner is to be designed to provide a perfect match to this load. The tuner is shorted at one end and has length L. The characteristic impedance is also 300 Ω.

(a) If the stub tuner were connected in series with the 300-Ω line, what is the closest distance to the load the stub could be placed to yield a match?

(b) Repeat part (a) for the case where the stub is connected in parallel.

(c) For part (b), what should be the length of the short-circuited stub?

6.4.1 We wish to determine what fraction of light in glass incident from the left upon an air gap $D = 1$ μm wide is transmitted across the gap into an identical piece of glass. Assume the incident waves are $\hat{y}$-polarized *TE* waves, $\theta_i = 60°$, $\epsilon_{glass} = 2\epsilon_0$, and λ_0 (free space) $= 0.5$ μm. Note that the waves in the gap are evanescent, and therefore the equivalent *TEM* line for the gap has purely imaginary values for Z_0 and k_0. Nonetheless, a *TEM* transmission line equivalent circuit is still valid, as are the equations that characterize how such lines transform impedances.

(a) Derive the parameters for the three equivalent *TEM* transmission lines (each of the two glass regions plus the gap).

(b) Calculate the equivalent impedance seen looking toward the gap from the right end of the left-most transmission line.

(c) Calculate Γ for E_y, and $|\Gamma|^2$ at this point.

(d) What fraction of the incident power is transmitted?

6.4.2 Use the parameters of Problem 6.4.1, except let $\theta_i = 45°$ and $D = 0.1 \ \mu m$.

    (a) Find the parameters Z_{0g} for glass and Z_{0a} for air, and k_{zg}, k_{za} for the equivalent *TEM* transmission lines.

    (b) Calculate the equivalent impedance of the air gap seen from the right end of the left-most transmission line.

    (c) What fraction of the incident power is transmitted?

6.4.3 Typical glass has a dielectric constant $\epsilon = 2.5\epsilon_0$.

    (a) For light at normal incidence, what fraction of the power is reflected by the front surface of a glass camera lens?

    (b) A quarter-wave transformer coating of dielectric is applied to the front surface of this lens to reduce reflections to zero at the wavelength of yellow light. In vacuum, $\lambda_{yellow} \simeq 0.56 \ \mu m$. What percentage of incident blue-green light ($\lambda = 0.49 \ \mu m$) would be reflected by this coating?

    (c) Determine the following parameters for a simple *TEM* transmission line circuit equivalent to this coated lens when it operates at $0.49 \ \mu m$ free-space wavelength with 45° incident angle and *TE* polarization: Z_{01}, Z_{02}, Z_{03}, k_{z1}, k_{z2}, k_{z3} and ℓ. The characteristic impedances Z_{01}, Z_{02}, and Z_{03} correspond to the lens surface in the air, coating, and glass, respectively, where ℓ is the equivalent length of the line Z_{02}.

6.5.1 For circuits (a) and (b) in Figure 6.32 sketch and dimension both $V(z)$ and $I(z)$ at $t = 10^{-8}$ s, 3×10^{-8} s, and 5×10^{-8} s (6 sketches per circuit). The characteristic impedance is $Z_0 = 100 \ \Omega$ and $\ell = 6$ m in both circuits.

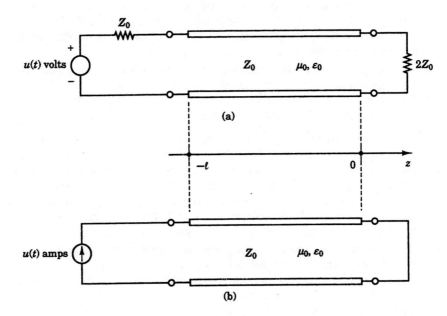

Figure 6.32   Circuits for Problem 6.5.1.

6.5.2 The signals on a *TEM* transmission line can be represented as

$$V(z, t) = f_+ \left(t - \frac{z}{v} \right) + f_- \left(t + \frac{z}{v} \right)$$

Prove that the total power flowing in the $+\hat{z}$ direction is equal to the sum of the separate powers associated with the positive- and negative-propagating waves, where we associate negatively directed power with the $-\hat{z}$-directed wave. Note that powers do not generally add in most systems unless the two signals are in some sense orthogonal.

6.5.3 A high-speed computer has a transistor line driver that triggers a flip-flop at the other end of a 20-cm *TEM* line of characteristic impedance 200 Ω. The dielectric in the line has $\epsilon = 4\epsilon_0$. The line driver is equivalent to a 10-V unit step in series with 50 Ω.

(a) Sketch and dimension $V(z)$ and $I(z)$ on the transmission line at $t = 1$ ns.
(b) Repeat (a) for $t = 2$ ns.
(c) Sketch quantitatively the load voltage $V_L(t)$ as a function of time until that time t when the flip-flop is triggered; it flips when $V_L > 4$ V.
(d) What is the asymptotic value of $V_L(t)$ as $t \to \infty$?
(e) Note that triggering is excessively delayed by this transmission line. If Z_0 were 10 Ω instead of 200 Ω, would there still be a long delay? Suppose $Z_0 = 50$ Ω. What is the delay now?

6.5.4 A power supply is connected to a particular system by an air-filled 100-Ω *TEM* cable 10 m long. The impedances of the source and load are 50 and 200 ohms, respectively, and the Thevenin source voltage is 100 V DC. The system is operating well until $t = 0$, when a short circuit at $z = 5$ m occurs. Sketch and dimension $V(z)$ and $I(z)$ at $t = 10^{-8}$ s and $t = 2 \times 10^{-8}$ s.

6.5.5 A test circuit is used to characterize unknown loads at the end of a *TEM* line of length D with Z_0, $4\mu_0$, and ϵ_0. The source is a step function $V_s u(t)$ V in series with a resistor Z_0. When the unknown load is connected, we measure $V(t)$ at the source end of the line as a 1-V step lasting 1 ms before the voltage steps a second time to 1.5 V total. We measure $I(t)$ at the same point to be a step of 1 mA, also followed later by a second discontinuity.

(a) What is the line length D (meters)?
(b) What is Z_0 (ohms)?
(c) Find a load (quantitatively) that is consistent with the given data.
(d) Sketch and dimension $I(t)$ for $0 < t < 3$ ms.

6.5.6 A break (open circuit) in a high-voltage DC power line matched to its load occurs at time $t = 0$. The line was carrying a DC voltage V_0 and DC current I_0 before the break occurred in the middle of the circuit at $z = 0$.

(a) Sketch V and I on the line at some time t after the break, but before any reflections from the source and load ends have occurred. The characteristic impedance of the line is Z_0.
(b) Consider a 600-kV two-wire line carrying 1 GW of power with a characteristic impedance of 500 Ω. What is the peak voltage on the line after the break occurs?

6.5.7* A common method for generating short, high-voltage pulses is to use a charged transmission line with a fast switch. The basic idea of such devices is illustrated in this problem.

(a) Consider the system shown in Figure 6.33(a), where the line has been charged to a voltage V_0 by a DC high voltage source with internal resistance $R \gg Z_0$. (You can consider R to be an open circuit in analyzing the "fast" transients.) Sketch and dimension the voltage across the load $V_L(t)$ if the switch closes at $t = 0$ with the line fully charged to $V = V_0$.

(b) If the transmission line is an air-filled coaxial cable with $Z_0 = 50$ Ω, how long should the line be to give a voltage pulse of 0.1 μs duration? How big must the DC supply voltage be to deliver one joule of total pulse energy to the load R_L?

(c) To see if this is physically reasonable, calculate the smallest possible inner radius of the (air-filled) 50-Ω coaxial line with 100-kV charging voltage, if the breakdown electric field is taken as 3×10^6 V/m.

(d) A modified form of the above scheme is the Blumlein line, shown in Figure 6.33(b). Both lines are of length ℓ. Sketch and dimension $V_L(t)$. What is the advantage of this scheme?

6.5.8 A matched source drives an air-filled 50-Ω *TEM* line 10 m long. The voltage source is a step function $5\,u(t)$ V. Sketch and quantitatively dimension $V(z, t)$ and $I(z, t)$ on the line at $t = 5 \times 10^{-8}$ s for the case where the load is (i) a 100-Ω resistor, (ii) an

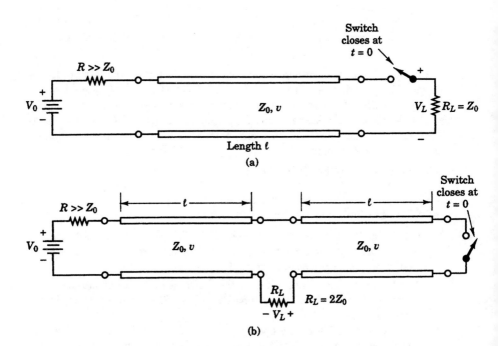

Figure 6.33   Transmission lines for high-speed switching.

inductor $L = 10^{-6}$ H, (iii) a capacitor $C = 10^{-10}$ F, and (iv) a diode oriented to pass positive current but back-biased with a 2-V battery.

6.5.9 Repeat Problem 6.5.8 for the case where the matched source has a Thevenin equivalent voltage which is a zero-mean 1-GHz square wave of 10 V peak-to-peak and the load is (i) a 10-Ω resistor, (ii) an inductor $L = 1$ μH, (iii) a 50-Ω resistor in series with a 20-pF capacitor, and (iv) a back-biased diode like that in part (iv) of Problem 6.5.8.

6.5.10 An 0.5 μH inductor terminates a 3-m air-filled *TEM* line with 100-Ω characteristic impedance. The *TEM* line is connected to a current source in parallel with a 100-Ω resistor. If the current source puts out a 5-ns square pulse of 10 mA, sketch the voltage and current on the line at 5 ns, 15 ns, and 25 ns.

7

WAVEGUIDES

7.1 PARALLEL-PLATE TRANSMISSION LINES

Transmission lines are structures that propagate electromagnetic waves from one region of space to another, such that both $\overline{E}$ and $\overline{H}$ are proportional to $e^{\pm jkz}$ ($\mp \hat{z}$ is the direction of propagation). In Chapter 5, we considered *TEM* wave propagation in such structures, where $E_z = H_z = 0$. We now relax this constraint on the $\hat{z}$ components of the electric and magnetic fields, and observe that many other kinds of waves (modes) can exist on transmission lines. In particular, we shall discuss transverse electric (*TE*) waves ($E_z = 0$ but $H_z \neq 0$) and transverse magnetic (*TM*) waves ($H_z = 0$, $E_z \neq 0$). Any arbitrary wave may be constructed from a superposition of *TE* and *TM* modes.

Perhaps the simplest type of waveguide consists of two perfectly conducting parallel plates between which electromagnetic waves are trapped and guided. In order to find the equations describing the *TE* modes of a parallel-plate guide filled with an ϵ, μ medium, we first recall the behavior of *TE* waves incident upon a single perfectly conducting plate at $x = 0$. From Section 4.3, we know that a *TE* wave propagating at an angle of incidence θ_i from the $\hat{x}$-axis toward the conducting plate at $x = 0$ has an electric field

$$\overline{E}_i = \hat{y} E_0 e^{jk_x x - jk_z z} \tag{7.1.1}$$

where $k_x = k_0 \cos \theta_i$, $k_z = k_0 \sin \theta_i$, and $k_0 = \omega\sqrt{\mu\epsilon}$, as illustrated in Figure 7.1. Because the tangential electric field E_y must be zero at the perfectly conducting boundary at $x = 0$, a reflected wave is necessarily generated, having the electric field

$$\overline{E}_r = -\hat{y}\, E_0\, e^{-jk_x x - jk_z z} \tag{7.1.2}$$

Both the incident and reflected waves obey the same dispersion relation

$$k_x^2 + k_z^2 = k_0^2 = \omega^2 \mu\epsilon \tag{7.1.3}$$

and we recall that the k_z components of both waves must be the same because of the phase-matching condition (4.3.8) at $x = 0$. Therefore, both waves have the same $|k_x|$ and the angle of incidence $\theta_i = \tan^{-1} |k_z / k_x|$ is the same as the angle of reflection. The $\left\{ \begin{smallmatrix} \text{incident} \\ \text{reflected} \end{smallmatrix} \right\}$ wave propagates in the direction $\left\{ \begin{smallmatrix} -\hat{x}\, k_x + \hat{z}\, k_z \\ \hat{x}\, k_x + \hat{z}\, k_z \end{smallmatrix} \right\}$, as shown in Figure 7.1.

The total electric field for the *TE* wave is just the superposition of the incident and reflected waves (7.1.1) and (7.1.2):

$$\overline{E} = \overline{E}_i + \overline{E}_r = \hat{y}\, 2j E_0 \sin k_x x\, e^{-jk_z z} \tag{7.1.4}$$

The magnetic field results from Faraday's law

$$\overline{H} = -\frac{\nabla \times \overline{E}}{j\omega\mu} = \frac{1}{j\omega\mu}\left(\hat{x}\frac{\partial}{\partial z} - \hat{z}\frac{\partial}{\partial x} \right) E_y$$
$$= \frac{2E_0}{\eta}\left(\hat{x}\,\frac{-jk_z}{k_0} \sin k_x x - \hat{z}\,\frac{k_x}{k_0}\cos k_x x \right) e^{-jk_z z} \tag{7.1.5}$$

where $\eta = \sqrt{\mu/\epsilon}$.

As expected, both the tangential electric field E_y and normal magnetic field H_x are zero at $x = 0$, satisfying boundary conditions (4.1.19) and (4.1.22) for waves at a perfectly conducting surface. But we also make the very important observation that these two boundary conditions are also satisfied at **any** value of

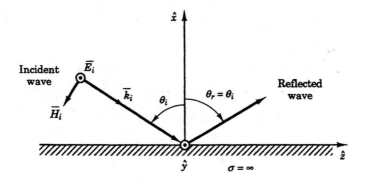

Figure 7.1   *TE* wave reflected from a perfect conductor.

$x = d$ such that $\sin k_x d = 0$, i.e., for which $k_x d = m\pi$ where m is an integer. This means that we can slip a second perfect conductor into the field at these special values of $x = d = m(\lambda_x/2)$ **without disturbing the existing field**. Electromagnetic waves will then exist in the region between the plates $(0 \leq x \leq d)$, but $\overline{E}$ and $\overline{H}$ will be zero for $x < 0$ and $x > d$. Two parallel conducting plates confine and guide the *TE* wave as it bounces between them. Such a structure is called a *parallel-plate waveguide*.

Notice that although the incident and reflected waves (7.1.1) and (7.1.2) are each separately pure traveling waves, their superposition forms a standing wave pattern in the $\hat{x}$ direction that propagates along the guide in the $\hat{z}$ direction. This is also consistent with the expression for the complex Poynting vector. We recall that $\overline{S} = \overline{E} \times \overline{H}^*$ from (1.6.17)

$$\overline{S} = \frac{|2E_0|^2}{\eta} \left(\hat{z} \frac{k_z}{k_0} \sin^2 k_x x - \hat{x} \frac{jk_x}{k_0} \sin k_x x \cos k_x x \right) \tag{7.1.6}$$

from which we can deduce that the time-averaged power $\langle \overline{S} \rangle = \frac{1}{2}\mathrm{Re}\{\overline{S}\}$ is nonzero only in the $\hat{z}$ direction.

Because we have shown that $k_x = m\pi/d$ for a *TE* wave to exist within the guide, this wave is designated as the TE_m mode. Excluding the nulls ($m/2$ $\hat{x}$-projected wavelengths λ_x) at $x = 0$ and $x = d$, the TE_m electric field pattern has $m - 1$ nulls between the two conducting walls of the guide. The total $\overline{E}$ and $\overline{H}$ fields for the TE_1 mode are pictured in Figure 7.2(a) at $t = 0$, where $\overline{E}(t) = \mathrm{Re}\left\{\overline{E}\,e^{j\omega t}\right\}$ and $\overline{H}(t) = \mathrm{Re}\left\{\overline{H}\,e^{j\omega t}\right\}$. Note that the TE_1 mode is the first nontrivial *TE* mode; $\overline{E} = \overline{H} = 0$ for the TE_0 mode.

But parallel-plate transmission lines also support *TM* modes. Returning to reflection from a single boundary, we consider a *TM* wave incident on the perfectly conducting plate at $x = 0$. The magnetic field in this case is

$$\overline{H}_i = \hat{y}\, H_0\, e^{jk_x x - jk_z z} \tag{7.1.7}$$

which in turn generates a reflected magnetic field with reflection coefficient equal to $+1$:

$$\overline{H}_r = \hat{y}\, H_0\, e^{-jk_x x - jk_z z} \tag{7.1.8}$$

The total magnetic field is then the sum of (7.1.7) and (7.1.8)

$$\overline{H} = \overline{H}_i + \overline{H}_r = \hat{y}\, 2H_0 \cos k_x x\, e^{-jk_z z} \tag{7.1.9}$$

which gives rise to an electric field by way of Ampere's law:

$$\overline{E} = \frac{\nabla \times \overline{H}}{j\omega\epsilon} = \frac{1}{j\omega\epsilon} \left(-\hat{x}\frac{\partial}{\partial z} + \hat{z}\frac{\partial}{\partial x} \right) H_y$$

$$= 2\eta H_0 \left(\hat{x} \frac{k_z}{k_0} \cos k_x x + \hat{z} \frac{jk_x}{k_0} \sin k_x x \right) e^{-jk_z z} \tag{7.1.10}$$

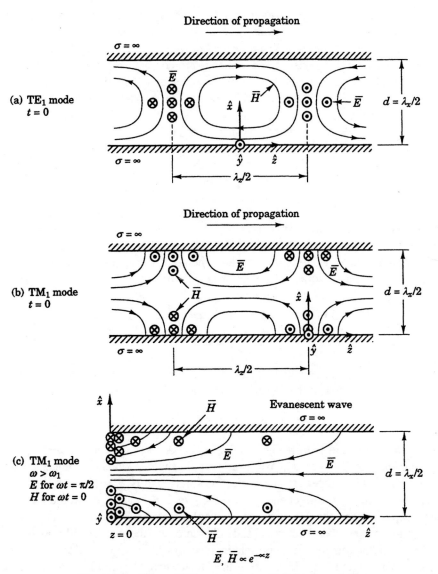

Figure 7.2 TE_1 and TM_1 modes in parallel-plate waveguide.

These electric and magnetic fields again satisfy the boundary conditions that the tangential $\overline{E}$ and normal $\overline{H}$ fields be zero at the conducting plates; i.e., $E_y = E_z = H_x = 0$ at $x = 0$.[1]

---

1. Note that duality may not be used to find the *TM* solutions from the *TE* solutions because the boundary conditions here are not dual. The tangential electric field and the normal magnetic field

Because the fields are periodic in x, E_z and H_x are also zero at any value of x such that $\sin k_x x = 0$. Therefore, we can again insert a conducting plate into the field at these special values of $x = d = m\pi/k_x$ (m is an integer) without disturbing the field. As in the TE case, the TM modes are confined between the two plates that guide the radiation. Each of the infinite set of TM modes is designated by TM_m, where $k_x = m\pi/d$ for the mth mode. The $\overline{E}$ and $\overline{H}$ fields for the TM_1 mode are sketched in Figure 7.2(b) at $t = 0$. Note that the TM_0 mode **does** exist; when $k_x = 0$, $k_z = k_0$ from the dispersion relation (7.1.3) and

$$\begin{aligned}
\overline{E} &= \hat{x} \, 2\eta H_0 \, e^{-jk_z z} \\
\overline{H} &= \hat{y} \, 2H_0 \, e^{-jk_z z}
\end{aligned} \tag{7.1.11}$$

which are the fields for a TEM wave. Therefore, the TEM wave discussed extensively in Chapter 5 is simply one of an infinite series of modes (the TM_0 mode) that can exist in the waveguide. In order to determine which modes will actually propagate in a waveguide, we need to know how the guide is excited (i.e., the frequency, orientation, and distribution of the source). Excitation of modes is addressed in Section 7.3.

For both TE and TM waves, the *guidance condition* was found to be

$$k_x d = m\pi \tag{7.1.12}$$

where $m = 0, 1, 2, \dots$ for the TM modes and $m = 1, 2, \dots$ for the TE modes. Combining (7.1.12) with the dispersion relation (7.1.3) yields an expression for k_z:

$$k_z = \sqrt{\omega^2 \mu\epsilon - (m\pi/d)^2} \tag{7.1.13}$$

In Figure 7.2(a) and (b), the TE_1 and TM_1 modes are sketched, where k_z is a real number ($\omega\sqrt{\mu\epsilon} > \pi/d$). We recall that the angle at which the modes bounce in the guide with respect to the surface normals $\pm\hat{x}$ is $\theta_i = \theta_r = \tan^{-1}|k_z/k_x|$. Were we to lower the frequency of the TE_1 or TM_1 mode, k_z would decrease while k_x remained the same; hence θ_i would decrease. When ω was equal to $\pi/d\sqrt{\mu\epsilon}$ so that $k_z = 0$, θ_i would also be zero and the wave would bounce up and down in the same place (at normal incidence) without moving along the guide, as shown in Figure 7.3. If we were then to lower ω beyond this threshold, k_z would become imaginary and the waves in the guide would become evanescent ($\overline{E}$, $\overline{H} \propto e^{-\alpha z}$ where $\alpha > 0$). The frequency at which propagation becomes evanescent for the TE_m or TM_m modes is called the *cut-off frequency* ω_m. For parallel-plate waveguides, ω_m is

---

must be zero at a perfectly conducting surface for **both** the TE and TM modes, and these boundary conditions are not interchangeable when the roles of $\overline{E}$ and $\overline{H}$ are exchanged ($\overline{E} \to \overline{H}$, $\overline{H} \to -\overline{E}$). For transmission and reflection at a nonconducting dielectric interface, the tangential electric and tangential magnetic fields are continuous and these boundary conditions **do** obey duality.

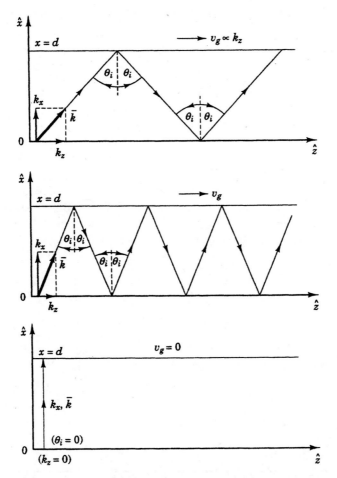

Figure 7.3 Propagation of waves in a parallel-plate waveguide: effect of decreasing θ_i.

$$\omega_m = \frac{m\pi}{d\sqrt{\mu\epsilon}} \tag{7.1.14}$$

which may be found by setting $k_z = 0$ in (7.1.13). Because the cut-off frequency for the *TEM* mode is $\omega_0 = 0$, *TEM* modes are able to propagate at any frequency. If we choose the signal bandwidth such that $\omega < \pi/d\sqrt{\mu\epsilon}$, then all of the higher modes ($m \geq 1$) are cut off (evanescent) and **only** the *TEM* wave can propagate, which is a useful way of exciting only a single mode in the waveguide.

The *waveguide wavelength* λ_z is defined as

$$\lambda_z = \frac{2\pi}{k_z} = \frac{2\pi}{\sqrt{\omega^2\mu\epsilon - (m\pi/d)^2}} \tag{7.1.15}$$

which becomes infinite at the cut-off frequency $\omega = \omega_m$. The wavelength in the $\hat{x}$ direction is

$$\lambda_x = \frac{2\pi}{k_x} = \frac{2d}{m} \tag{7.1.16}$$

and from the dispersion relation we note that

$$\left(\frac{1}{\lambda_x}\right)^2 + \left(\frac{1}{\lambda_z}\right)^2 = \left(\frac{1}{\lambda_0}\right)^2$$

where $\lambda_0 = 2\pi/k_0 = 2\pi/\omega\sqrt{\mu\epsilon}$. The relationship between λ_x, λ_z, λ_0, k_x, k_z, k_0, θ_i, and m is suggested in Figure 7.4.

The phase and group velocities of these waves can be determined from the expression for k_z as a function of ω, given by (7.1.13). Since waves within the guide propagate in the $+\hat{z}$ direction as $e^{-jk_z z}$, it follows from (3.5.1) and (3.5.2) that the phase and group velocities are

$$v_p = \frac{\omega}{k_z} = v \Big/ \sqrt{1 - \left(\frac{m\pi v}{\omega d}\right)^2} \tag{7.1.17}$$

$$v_g = \left(\frac{\partial k_z}{\partial \omega}\right)^{-1} = v\sqrt{1 - \left(\frac{m\pi v}{\omega d}\right)^2} \tag{7.1.18}$$

where $v = 1/\sqrt{\mu\epsilon}$ is the velocity of the electromagnetic waves within the guide ($v = c$ for vacuum). Note that both v_p and v_g approach v as ω approaches

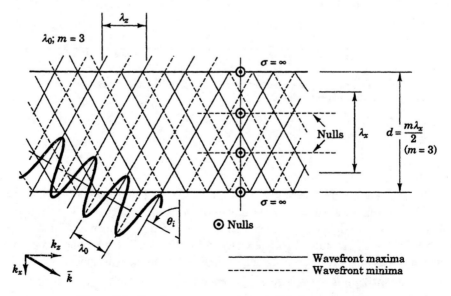

Figure 7.4 Interference patterns in a parallel-plate waveguide.

infinity, and that the product $v_p v_g$ is always v^2 for waveguide propagation, where $v_p \geq v$, $v_g \leq v$. (This latter equality is true for any quadratic dispersion relation of the form $\omega^2 = ak_z^2 + b$.) The phase velocity is the rate at which the traveling wave patterns of Figure 7.2 translate down the waveguide, whereas the group velocity is the rate at which narrow-band packets of energy or information move.

Lastly, we consider the form of the field equations for a mode excited below its cut-off frequency. The expressions (7.1.9) and (7.1.10) for the TM_m mode become

$$\overline{H} = \hat{y}\, 2 H_0 \cos\left(\frac{m\pi x}{d}\right) e^{-\alpha z} \tag{7.1.19}$$

where

$$\alpha = j k_z = j\sqrt{\omega^2 \mu\epsilon - k_x^2} = \sqrt{(m\pi/d)^2 - \omega^2 \mu\epsilon} \tag{7.1.20}$$

and

$$\overline{E} = 2\eta H_0 \left(-\hat{x}\, \frac{j\alpha}{k_0} \cos\left(\frac{m\pi x}{d}\right) + \hat{z}\, \frac{jm\pi}{k_0 d} \sin\left(\frac{m\pi x}{d}\right) \right) e^{-\alpha z} \tag{7.1.21}$$

The electric field is sketched at $\omega t = \pi/2$ in Figure 7.2(c) for the TM_1 mode, at which time $\overline{H} = 0$. The magnetic field is sketched at $\omega t = 0$, when $\overline{E} = 0$. Because $\overline{E}$ and $\overline{H}$ are 90° out of phase, no time-averaged power propagates, which is the characteristic of an evanescent wave: $\langle \overline{S} \rangle = \frac{1}{2}\mathrm{Re}\{\overline{S}\} = 0$.

Example 7.1.1

An air-filled parallel-plate waveguide has a plate separation of 1 cm. At what frequencies can the modes TE_0, TE_1, TE_2, TM_0, TM_1, and TM_2 propagate? What are the phase and group velocities for the TE_1 mode at 20 GHz? Near 15 GHz?

Solution: The TE_0 mode does not exist because it would require a nonzero electric field parallel to the conducting plates, while the TM_0 mode is equivalent to the TEM mode and can propagate at all frequencies. The cut-off frequency for the TE_m or TM_m mode is given by (7.1.14)

$$\omega_m = \frac{1}{\sqrt{\mu_0 \epsilon_0}} \left(\frac{m\pi}{d}\right) = \frac{m\pi c}{d}$$

so that

$$f_m = \frac{\omega_m}{2\pi} = \frac{mc}{2d} = m\left(\frac{3 \times 10^8}{2 \times 10^{-2}}\right) = m \cdot 15 \text{ GHz}$$

Therefore the cut-off frequency for the TE_1 and TM_1 modes is $f_1 = 15$ GHz, and for the TE_2 and TM_2 modes is 30 GHz.

The phase and group velocities are $v_p = \omega/k$ and $v_g = \left(\dfrac{\partial k}{\partial \omega}\right)^{-1}$, respectively, and are given by (7.1.17) and (7.1.18). For the TE_1 mode at 20 GHz, we find

$$v_p = \frac{\omega}{k} = \frac{c}{\sqrt{1 - (m\pi c/\omega d)^2}}$$

$$= 3 \times 10^8 \left(1 - \left(\frac{1 \cdot \pi \cdot 3 \times 10^8}{2\pi \cdot 20 \times 10^9 \cdot 10^{-2}}\right)^2\right)^{-1/2}$$

$$\cong 4.5 \times 10^8 \text{ m/s}$$

$$v_g = \left(\frac{dk_z}{d\omega}\right)^{-1} = \frac{c^2}{v_p}$$

$$\cong 2 \times 10^8 \text{ m/s}$$

As expected, $v_p > c$ and $v_g < c$.

At $f = 15$ GHz, the TE_1 mode is cut off and $k_z = 0$. This means that $v_p \to \infty$ and $v_g = 0$. Near the cut-off frequency $\omega_1 = 2\pi f_1$, we can find a more analytic expression for the phase and group velocities. If $\omega = \omega_1 + \delta\omega$, the phase velocity is

$$v_p = c\left(1 - \left(\frac{\omega_1}{\omega_1 + \delta\omega}\right)^2\right)^{-1/2} = c\left(1 - \left(\frac{1}{1 + \delta\omega/\omega_1}\right)^2\right)^{-1/2}$$

$$\simeq c\left(1 - \left(1 - \frac{\delta\omega}{\omega_1}\right)^2\right)^{-1/2} \simeq c\sqrt{\frac{\omega_1}{2\,\delta\omega}}$$

where terms on the order of $(\delta\omega/\omega_1)^2$ have been neglected. The group velocity is therefore

$$v_g = \frac{c^2}{v_p} \simeq c\sqrt{\frac{2\,\delta\omega}{\omega_1}}$$

which goes to zero as $\sqrt{\delta\omega}$.

7.2 DIELECTRIC-SLAB TRANSMISSION LINES

A transmission line filled with dielectric material but without conducting walls is another structure that may be used to guide electromagnetic waves, a notable example being an optical fiber. The cylindrical geometry of these fibers leads to a more complicated mathematical analysis than is needed to gain physical insight into the behavior of dielectric guides, so for simplicity, a rectangular slab surrounded by vacuum (ϵ_0, μ_0) is considered instead. The slab has permittivity ϵ, permeability μ, thickness $2d$ in the $\hat{x}$ direction, and infinite width ($W \gg 2d$) in the $\hat{y}$ direction so that $\partial/\partial y = 0$. The slab also has a uniform cross-section in the $\hat{z}$ direction, and is shown in Figure 7.5. Note that we choose $x = 0$ in the center of the slab to take advantage of symmetry.

If we allow waves to propagate inside the slab, all of the issues addressed in Chapter 4 concerning reflection and transmission at planar boundaries between dielectric regions will apply at $x = \pm d$. Because the purpose of the slab is to localize and direct electromagnetic waves, we do not want these waves to "leak"

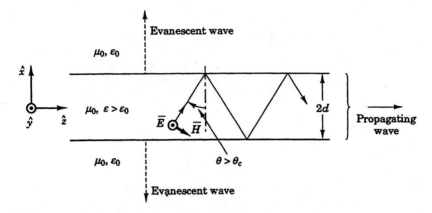

Figure 7.5   *TE* wave in a dielectric-slab waveguide.

power outside the guide. In other words, we want the fields outside the slab to be evanescent. We recall that for total internal reflection to occur, the index of refraction of the slab must be greater than that of the vacuum ($\mu\epsilon > \mu_0\epsilon_0$), and the angle of incidence of propagation inside the slab must exceed the critical angle $\theta_c = \sin^{-1}\sqrt{\mu_0\epsilon_0/\mu\epsilon}$ given by (4.3.13).

The analysis of the parallel-plate transmission line presented in Section 7.1 leads us to expect that a series of *TE* and *TM* modes will propagate in the dielectric slab as well. The major difference in the two cases is the set of boundary conditions at $x = \pm d$. Initially, we shall discuss the behavior of the *TE* modes and later use the concept of electric and magnetic field duality to construct the *TM* modes. A plot of the electric field intensity for the first few *TE* modes is given in Figure 7.6(a). From the symmetry of the slab/vacuum system about the $\hat{x}$-axis, we note that these modes will either be symmetric (TE_1, TE_3, TE_5, ...) or antisymmetric (TE_2, TE_4, TE_6, ...). The subscripts refer to the number of zero-crossings in the field pattern (i.e., the TE_m mode has $m - 1$ nodes between $+d$ and $-d$, just as the TE_m mode for the parallel-plate transmission line had $m - 1$ nodes between the conducting plates). Instead of the electric field being forced to go to zero at the conducting walls of the parallel-plate guide, the field of the dielectric waveguide has an exponentially decaying (evanescent) "tail" outside the slab. These simple field sketches enable us to write the form of the *TE* solutions with ease, where again $E_z = 0$ for the *TE* wave:

$$\overline{E} = \hat{y}\, E_0 \begin{Bmatrix} \sin k_x x \\ \cos k_x x \end{Bmatrix} e^{-jk_z z} \qquad |x| \le d \qquad (7.2.1)$$

$$\overline{E} = \hat{y}\, E_1 e^{-\alpha x - jk_z z} \qquad x \ge d$$

$$\overline{E} = \hat{y} \begin{Bmatrix} - \\ + \end{Bmatrix} E_1 e^{+\alpha x - jk_z z} \qquad x \le -d \qquad (7.2.2)$$

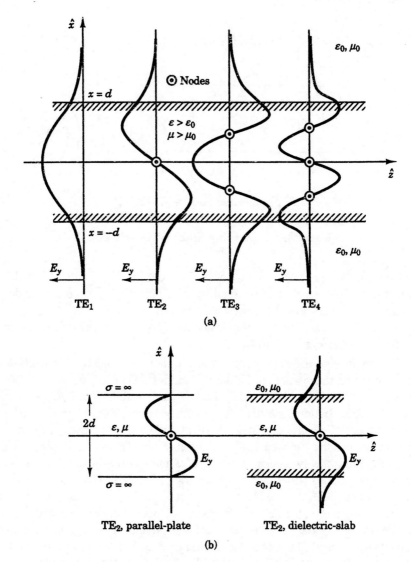

(a)

(b)

TE$_2$, parallel-plate   TE$_2$, dielectric-slab

Figure 7.6 Electric field for the *TE* modes in a dielectric-slab waveguide.

The quantities k_x, k_z, and α are real, while E_0 and E_1 are in general complex. The top and bottom lines in braces refer to the antisymmetric and symmetric *TE* modes, respectively.

Equation (7.2.1) describes a standing wave in the $\hat{x}$ direction inside the guide (made up of propagating incident and reflected waves), while outside the guide the waves are evanescent as indicated in (7.2.2). Both inside and outside the slab, the field patterns propagate in the $+\hat{z}$-direction as they did also for the parallel-plate guide. To be physically reasonable, α must be positive so that the fields attenuate away from the guide.

The corresponding magnetic fields may be found by using Faraday's law with (7.2.1) and (7.2.2). Inside the slab, the magnetic field is found from (7.2.1):

$$\overline{H} = \frac{E_0}{\omega\mu} \left(-\hat{x}\, k_z \begin{Bmatrix} \sin k_x x \\ \cos k_x x \end{Bmatrix} - \hat{z}\, jk_x \begin{Bmatrix} -\cos k_x x \\ \sin k_x x \end{Bmatrix} \right) e^{-jk_z x}, \quad |x| \le d \quad (7.2.3)$$

Outside the slab, the magnetic field is found from (7.2.2):

$$\overline{H} = \frac{E_1}{\omega\mu_0} \left(-\hat{x}\, k_z - \hat{z}\, j\alpha \right) e^{-\alpha x - jk_z z}, \quad x \ge d$$

$$\overline{H} = \begin{Bmatrix} - \\ + \end{Bmatrix} \frac{E_1}{\omega\mu_0} \left(-\hat{x}\, k_z + \hat{z}\, j\alpha \right) e^{+\alpha x - jk_z z}, \quad x \le -d$$

$$(7.2.4)$$

We can easily compare the dielectric slab wave solutions (7.2.1) and (7.2.2) to the fields for a parallel-plate transmission line with perfectly conducting walls at $x = \pm d$. In both cases, (7.2.1) is an appropriate description of the field inside the guide, because for both systems symmetric and antisymmetric TE modes exist. (The symmetries that exist in the parallel-plate guide are not as apparent in Section 7.1 because of the asymmetrical choice of coordinate system.) The only difference between the two types of guides is the restrictions placed on $k_x d$ to ensure propagation. Figure 7.6(b) compares the TE_2 modes for the dielectric slab and parallel-plate transmission lines. The conducting walls of the parallel-plate guide force the tangential electric field at $x = \pm d$ to be zero, so there is no electric field outside the guide. The dielectric slab has evanescent fields outside the slab, because the boundary conditions only force continuity of the tangential electric and magnetic fields.

Phase matching at the planar boundaries ($x = \pm d$) as described in Section 4.3 implies right away that the $\hat{z}$-components of the wave vectors inside and outside the slab are the same. Because the electric fields given by (7.2.1) and (7.2.2) must satisfy the wave equation (1.3.7), the following two dispersion relations result:

$$k_x^2 + k_z^2 = \omega^2 \mu\epsilon \qquad |x| \le d \qquad (7.2.5)$$

$$-\alpha^2 + k_z^2 = \omega^2 \mu_0\epsilon_0 \qquad |x| \ge d \qquad (7.2.6)$$

The boundary conditions at the slab/vacuum interfaces ($x = \pm d$) are that both tangential electric and magnetic fields be continuous.[2] (The tangential elec-

---

2. If $\mu = \mu_0$, continuity of tangential $\overline{H}$ is equivalent to continuity of the normal derivative of $\overline{E}$ by Faraday's law. For the TE modes, this means that E_y and $\partial E_y/\partial x$ are both continuous at $x = \pm d$, as illustrated in Figure 7.6. A similar statement may be made about TM modes when $\epsilon = \epsilon_0$; H_y and $\partial H_y/\partial x$ will then both be continuous at $x = \pm d$.

tric field is always continuous, and the tangential magnetic field is discontinuous only if there is a surface current at the boundary, certainly not the case in a non-conducting dielectric medium.) For simplicity, we first consider only the symmetric ($\cos k_x x$) TE solutions, which are the odd-numbered TE modes. Equating the lower line of (7.2.1), the symmetric tangential electric field inside the dielectric, with (7.2.2), the tangential electric field in the vacuum, at $x = d$ yields

$$E_0 \cos k_x d\, e^{-jk_z z} = E_1 e^{-\alpha d}\, e^{-jk_z z} \tag{7.2.7}$$

for the symmetric TE modes. Continuity of the symmetric tangential magnetic fields may be established by setting the $\hat{z}$ component of the lower line of (7.2.3) equal to the $\hat{z}$ component of (7.2.4) at $x = d$

$$-\frac{jk_x E_0}{\omega\mu} \sin k_x d\, e^{-jk_z z} = -\frac{j\alpha E_1}{\omega\mu_0} e^{-\alpha d}\, e^{-jk_z z} \tag{7.2.8}$$

Notice that because of the symmetry of the problem, applying the boundary conditions at either $x = d$ or $x = -d$ gives the same two equations (7.2.7) and (7.2.8). Dividing (7.2.8) by (7.2.7) yields the guidance condition for the symmetric TE modes:

$$k_x d \tan k_x d = \frac{\mu}{\mu_0} \alpha d \quad \text{(symmetric TE modes)} \tag{7.2.9}$$

The dispersion relations (7.2.5) and (7.2.6) may be combined to give another equation containing k_x and α:

$$k_x^2 + \alpha^2 = \omega^2(\mu\epsilon - \mu_0\epsilon_0) \tag{7.2.10}$$

Substitution of (7.2.10) into (7.2.9) gives a transcendental equation in $k_x d$:

$$\tan k_x d = \frac{\mu}{\mu_0} \sqrt{\frac{\omega^2(\mu\epsilon - \mu_0\epsilon_0)d^2}{(k_x d)^2} - 1} \tag{7.2.11}$$

which may be solved graphically by simultaneously plotting both sides of (7.2.11) as a function of $k_x d$. The values of $k_x d$ at which intersection of these two plots occurs are the values of $k_x d$ for which propagation in the slab is possible. Figure 7.7 is the result of such a graphical procedure.

The following observations may be readily made by looking at Figure 7.8, the pictorial representation of the guidance condition:

(1) Intersection points that lie in the range $m\pi \le k_x d < (m + \frac{1}{2})\pi$ correspond to TE_{2m+1} modes ($m = 0, 1, 2, \dots$). We recall that the guidance condition that we have derived is applicable only to the symmetric (odd-numbered) TE modes. An equivalent guidance condition is given below for the antisymmetric (even-numbered) TE modes.

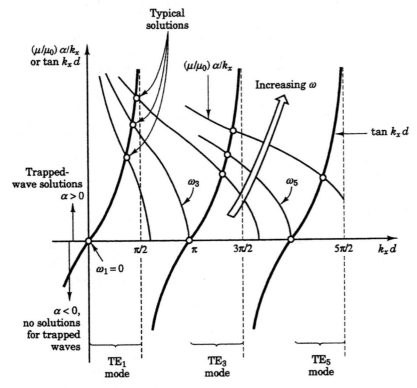

Figure 7.7   Graphical solution for the symmetric TE_{2m+1} modes in a dielectric-slab waveguide.

(2) The cut-off frequency for the $2m + 1^{st}$ mode is that frequency below which propagation of the TE_{2m+1} mode cannot occur. Pictorially, the TE_{2m+1} mode can propagate when the right side of (7.2.11) is real-valued within the range $m\pi \le k_x d < (m + \frac{1}{2})\pi$. Cut-off occurs when $k_x d = m\pi$, and $\tan k_x d = 0$. In this case, the right side of (7.2.11) must be zero as well, and $\omega_{2m+1} = k_x / \sqrt{\mu\epsilon - \mu_0\epsilon_0} = m\pi/d\sqrt{\mu\epsilon - \mu_0\epsilon_0}$. But this is just another way of saying that $\alpha = 0$ at cut-off, which is the point where the waves outside the guide just begin to propagate. At cut-off, we find from (7.2.6) that $k_z = \omega_{2m+1}\sqrt{\mu_0\epsilon_0}$, and we know that $k_x = m\pi/d$, so we can calculate the angle at which the waves bounce inside the guide:

$$\theta_i = \tan^{-1}\frac{k_z}{k_x} = \sin^{-1}\frac{k_z}{\sqrt{k_x^2 + k_z^2}} = \sin^{-1}\sqrt{\frac{\mu_0\epsilon_0}{\mu\epsilon}} = \theta_c$$

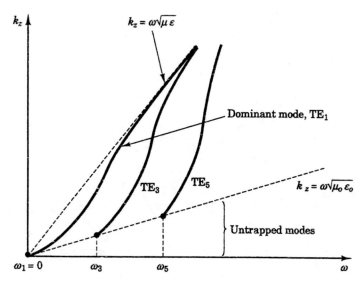

Figure 7.8 Dispersion diagram for the symmetric TE_{2m+1} modes of a dielectric-slab waveguide.

We recognize θ_c as the critical angle! Therefore, for angles of bounce greater than θ_c, waves can propagate **outside** the slab and the integrity of the mode is destroyed. Because $\omega_1 = 0$, the TE_1 mode will be able to propagate at all frequencies.

(3) As the frequency ω is changed, the wave vector k_x adjusts accordingly, in contrast to the parallel-plate waveguide of width d in which $k_x = m\pi/d$ independent of frequency.

(4) As the frequency is increased, the intersection points of the graphical solution shift closer to the $k_x d = (m + \frac{1}{2})\pi$ asymptote of the TE_{2m+1} mode. Therefore, at high frequencies, the guidance condition $k_x(2d) = (2m + 1)\pi$ for the TE_{2m+1} mode of the dielectric slab waveguide is the same as the guidance condition $k_x(2d) = m\pi$ for the odd-numbered TE_m modes of a parallel-plate transmission line of thickness $2d$. [Recall that $k_x d = m\pi$ (7.1.12) was derived for a plate separation distance d.] The reason that the high frequency behavior of the dielectric slab approaches that of a parallel-plate waveguide is that for any given mode TE_{2m+1}, $k_x d$ is confined to the range $[m\pi, (m + \frac{1}{2})\pi]$. From (7.2.10), we see that as $\omega \to \infty$, α also goes to infinity, which means that the mode is heavily attenuated outside of the guide. [This can also be observed from the graphical solution in Figure 7.8; $\tan k_x d$ evaluated at $(m + \frac{1}{2})\pi$ is infinity, and therefore $\alpha\mu/k_x\mu_0$ is infinite as well.]

With very little field present for $|x| > d$, the $\overline{E}$ field is increasingly pinned close to zero at the dielectric boundaries $x = \pm d$, which is also the boundary condition for the parallel-plate transmission line.

Additional insight into the properties of the dielectric slab may be gained by considering the relation between k_z and ω for the different modes in the guide. As $\omega \to \infty$, we have just demonstrated that $k_x d \to (m + \frac{1}{2})\pi$ for the odd-numbered TE modes. Therefore, $k_z = \sqrt{\omega^2 \mu \epsilon - ((m + \frac{1}{2})\pi/d)^2}$ from (7.2.5), which means that $k_z \cong \omega\sqrt{\mu\epsilon}$ for sufficiently high frequencies. This is entirely consistent with the dielectric slab acting like a parallel-plate transmission line at high frequencies because as the frequency increases, we have just seen that the attenuation of the fields outside the slab becomes more severe. This means that the energy in the system is increasingly confined to the region of the slab, and that the dispersion relation is dominated by the permittivity and permeability of the slab. Wave propagation has little dependence on ϵ_0 or μ_0 at high frequencies. At the other end of the spectrum, the cut-off frequency for the TE_{2m+1} mode occurs when $k_x d = m\pi$ as discussed above. But in that limit, $\alpha = 0$ and so $k_z = \omega\sqrt{\mu_0 \epsilon_0}$ from (7.2.6) for each TE mode at cut-off. In this regime, most of the power in the electromagnetic fields is being transported **outside** the guide, since there is essentially no attenuation of the evanescent waves. Therefore, it makes physical sense that the dispersion relation is dominated by the parameters of the vacuum, and that ϵ and μ for the slab are irrelevant at cut-off. The dispersion relation between ω and k_z for the symmetric TE modes of the slab guide is given in Figure 7.8. Note that since k_x, k_z, and α are real, (7.2.5) and (7.2.6) allow us to confine k_z to the region $\omega\sqrt{\mu_0 \epsilon_0} \leq k_z \leq \omega\sqrt{\mu\epsilon}$.

The above discussion has only addressed the properties of the symmetric (odd-numbered) TE modes, but the guidance condition for the antisymmetric (even-numbered) TE modes is readily found by matching the antisymmetric fields (7.2.1)–(7.2.4) at $x = d$:

$$-k_x d \cot k_x d = \frac{\mu}{\mu_0}\alpha d \quad \text{(antisymmetric TE modes)} \qquad (7.2.12)$$

Equation (7.2.12) can be combined with the dispersion relations (7.2.5) and (7.2.6) to form a transcendental equation in $k_x d$. The graphical solution to this equation is similar to that plotted in Figure 7.7 for the symmetric TE modes. We note that intersection points that are in the range $(m - \frac{1}{2})\pi < k_x d < m\pi$ correspond to TE_{2m} modes ($m = 1, 2, \ldots$). The cut-off frequency for the TE_{2m} mode is $\omega_{2m} = (m - \frac{1}{2})\pi/d\sqrt{\mu\epsilon - \mu_0\epsilon_0}$ because $k_x d = (m - \frac{1}{2})\pi$ and $\alpha = 0$ at cut-off. At high frequencies, $k_x(2d) = 2m\pi$, which is also the guidance condition for the even-numbered TE modes of a parallel-plate waveguide of thickness $2d$.

Because the electric field of each antisymmetric *TE* mode has a null in the plane $x = 0$, a perfectly conducting sheet could be placed at $x = 0$ without changing the field distributions for $x \geq 0$. Therefore, we immediately find the electric and magnetic fields for a dielectric slab of thickness d bounded by a perfect conductor on one side. These fields are the antisymmetric ($\overline{E} \propto \sin k_x x$) expressions given by the upper braces of (7.2.1) and (7.2.3) for $0 \leq x \leq d$, and (7.2.2) and (7.2.4) for $x > d$. The electric and magnetic fields are of course zero for $x < 0$. Only the antisymmetric modes exist in such a one-sided dielectric slab; the symmetric modes do not obey the boundary condition $E_y(x = 0) = 0$.

TM mode solutions to the dielectric slab problem may be written by inspection using duality of the electric and magnetic fields. The boundary conditions at $x = \pm d$ are the same for both $\overline{E}$ and $\overline{H}$; i.e., both tangential fields are continuous across the dielectric interfaces. Letting $\epsilon \rightarrow \mu$, $\mu \rightarrow \epsilon$, $\overline{E} \rightarrow \overline{H}$, and $\overline{H} \rightarrow -\overline{E}$, we have

$$\overline{H} = \hat{y} H_0 \begin{Bmatrix} \sin k_x x \\ \cos k_x x \end{Bmatrix} e^{-jk_z z}, \qquad |x| < d \tag{7.2.13}$$

$$\overline{H} = \hat{y} H_1 e^{-\alpha x - jk_z z}, \qquad x > d$$

$$\overline{H} = \hat{y} \begin{Bmatrix} - \\ + \end{Bmatrix} H_1 e^{+\alpha x - jk_z z}, \qquad x < -d \tag{7.2.14}$$

$$\overline{E} = \frac{H_0}{\omega \epsilon} \left(\hat{x} k_z \begin{Bmatrix} \sin k_x x \\ \cos k_x x \end{Bmatrix} + \hat{z} j k_x \begin{Bmatrix} -\cos k_x x \\ \sin k_x x \end{Bmatrix} \right) e^{-jk_z z}, |x| < d \tag{7.2.15}$$

$$\overline{E} = \frac{H_1}{\omega \epsilon_0} (\hat{x} k_z + \hat{z} j\alpha) e^{-\alpha x - jk_z z}, \qquad x > d$$

$$\overline{E} = \begin{Bmatrix} - \\ + \end{Bmatrix} \frac{H_1}{\omega \epsilon_0} (\hat{x} k_z - \hat{z} j\alpha) e^{+\alpha x - jk_z z}, \qquad x < -d \tag{7.2.16}$$

where the top and bottom lines in the braces refer to the antisymmetric and symmetric *TM* modes, respectively. The guidance conditions for the *TM* modes are given by

$$k_x d \tan k_x d = \frac{\epsilon}{\epsilon_0} \alpha d \qquad \text{(symmetric *TM* modes)}$$
$$-k_x d \cot k_x d = \frac{\epsilon}{\epsilon_0} \alpha d \qquad \text{(antisymmetric *TM* modes)} \tag{7.2.17}$$

and are found by applying the boundary conditions that H_y and E_z are continuous at $x = \pm d$, or by interchanging $\epsilon \leftrightarrow \mu$ and $\epsilon_0 \leftrightarrow \mu_0$ directly in (7.2.9) and (7.2.12). Plots of the electric field intensity for the TE_1 and TE_2 modes are given in Figure 7.9, showing how the field distribution changes as ω increases from the cut-off frequency.

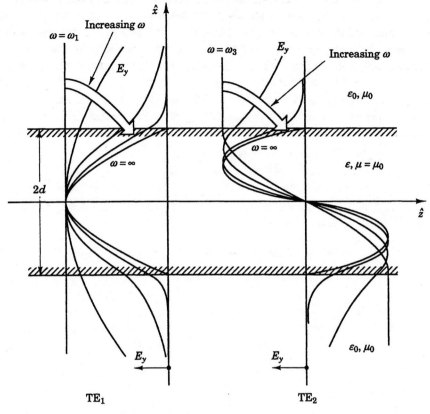

Figure 7.9 Electric field intensity for the TE_1 and TE_2 modes as a function of frequency.

7.3* EXCITATION AND ORTHOGONALITY OF MODES

Waveguides in which only one mode can propagate respond to excitations in a straightforward manner because only the amplitude and phase of that mode can vary within the guide. Chapter 5 discussed such single-mode propagation along transmission lines where the source frequency was sufficiently low so that only the *TEM* mode could propagate. In a general waveguide, however, many modes can propagate simultaneously, and energy may be exchanged between modes. What appears to be a much more complicated task of evaluating the amplitude and phase of each excited mode separately in fact turns out to be relatively simple because the set of modes that exist for any arbitrary waveguide are *orthogonal*. The concept of orthogonal modes is best illustrated by example, and we therefore return to the description of wave propagation in a parallel-plate transmission line.

Consider an infinitely long parallel-plate waveguide excited at $z = 0$ by a current sheet $\overline{J}_s = \hat{y} \, J_s(x)$ (A/m), as illustrated in Figure 7.10. Radiation by currents in free space was discussed earlier in Section 4.5, and similar methods may be applied here. In general, we might expect any of the possible TE_m and TM_m modes to be excited by an arbitrary source. We shall show presently, however, that no TM_m modes can be excited by this $\hat{y}$-directed current source, so we shall initially focus on the TE_m modes. We allow for the possibility that each of the infinite number of TE modes is excited by representing the actual TE electric field for $z > 0$ as an infinite (Fourier) superposition of TE modes

$$\overline{E}(z > 0) = \sum_{m=1}^{\infty} \hat{y} \, E_{+m} \sin k_{xm} x \, e^{-jk_{zm}z} \tag{7.3.1}$$

where E_{+m} is an arbitrary complex coefficient representing the magnitude and phase of the TE_m mode, and

$$k_{xm} = m\pi/d \tag{7.3.2}$$

$$k_{zm} = \sqrt{\omega^2\mu\epsilon - k_{xm}^2} = \sqrt{\omega^2\mu\epsilon - (m\pi/d)^2} \tag{7.3.3}$$

Equations (7.3.2) and (7.3.3) are just the guidance condition (7.1.12) and dispersion relation (7.1.13) for the parallel-plate waveguide, necessary so that (7.3.1) satisfies the wave equation (1.4.9) and also the boundary condition $E_y(x = 0) = E_y(x =$

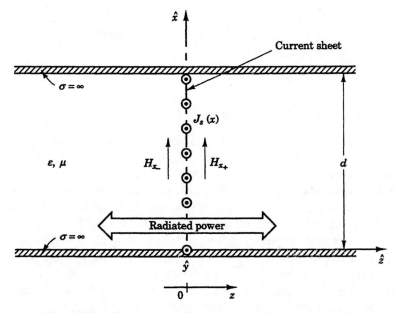

Figure 7.10   Current-sheet excitation of a parallel-plate waveguide.

$d) = 0$. From Faraday's law, we can find the magnetic field inside the guide

$$\overline{H}(z > 0) = -\sum_{m=1}^{\infty} \frac{E_{+m}}{\eta} \left(\hat{x} \frac{k_{zm}}{k_0} \sin k_{xm}x - \hat{z} \frac{jk_{xm}}{k_0} \cos k_{xm}x \right) e^{-jk_{zm}z} \quad (7.3.4)$$

where $\eta = \sqrt{\mu/\epsilon}$ and $k_0 = \omega\sqrt{\mu\epsilon}$. Because the parallel-plate waveguide is infinitely long in the $-\hat{z}$ direction, we also have an infinite series of modes propagating to the left (as $e^{+jk_{zm}z}$), with electric and magnetic fields given by

$$\overline{E}(z < 0) = \sum_{m=1}^{\infty} \hat{y} \, E_{-m} \sin k_{xm}x \, e^{jk_{zm}z} \quad (7.3.5)$$

$$\overline{H}(z < 0) = -\sum_{m=1}^{\infty} \frac{E_{-m}}{\eta} \left(-\hat{x} \frac{k_{zm}}{k_0} \sin k_{xm}x - \hat{z} \frac{jk_{xm}}{k_0} \cos k_{xm}x \right) e^{jk_{zm}z} \quad (7.3.6)$$

where (7.3.5) and (7.3.6) may be derived directly from (7.3.1) and (7.3.4) by using symmetry and by making the substitutions $E_{+m} \to E_{-m}$ and $k_{zm} \to -k_{zm}$. The guidance condition (7.3.2) and dispersion relation (7.3.3) are of course the same for the $-\hat{z}$-propagating waves.

As with all the other boundary value problems we have discussed in previous chapters, we now impose boundary conditions on these general solutions and solve for the unknown coefficients. At $z = 0$, the boundary conditions are given by (4.1.19) and (4.1.21). The tangential electric field is continuous, while the imposed surface current $\overline{J}_s$ gives rise to a discontinuity in the tangential magnetic field. Mathematically, this leads to

$$\hat{z} \times [\overline{E}(z > 0) - \overline{E}(z < 0)]\big|_{z=0} = 0$$

$$E_y(z = 0^+) - E_y(z = 0^-) = 0$$

$$\sum_{m=1}^{\infty} (E_{+m} - E_{-m}) \sin \left(\frac{m\pi x}{d} \right) = 0 \quad (7.3.7)$$

for the electric field, and

$$\hat{z} \times [\overline{H}(z > 0) - \overline{H}(z < 0)]\big|_{z=0} = \overline{J}_s = \hat{y} \, J_s(x)$$

$$H_x(z = 0^+) - H_x(z = 0^-) = J_s(x)$$

$$-\sum_{m=1}^{\infty} \frac{k_{zm}}{\eta k_0} (E_{+m} + E_{-m}) \sin \left(\frac{m\pi x}{d} \right) = J_s(x) \quad (7.3.8)$$

for the magnetic field, where we have used (7.3.1)–(7.3.6). Before we can make any further progress evaluating (7.3.7) and (7.3.8), we must first discuss mode orthogonality.

A set of functions $\{f_m(x)\}$ is said to be *orthogonal* over the interval $[a, b]$ if the relation

$$\int_a^b w(x) \, f_m(x) \, f_n(x) \, dx = \delta_{mn} \equiv \begin{cases} 1 & m = n \\ 0 & m \neq n \end{cases} \quad (7.3.9)$$

is true, where δ_{mn} is the Kronecker delta function and $w(x)$ is a weighting function.[3]

The orthogonality relation for the set of functions $\{f_m(x)\} = \{\sin(m\pi x/d)\}$ is given by

$$\int_0^d \frac{2}{d} \sin\left(\frac{m\pi x}{d}\right) \sin\left(\frac{n\pi x}{d}\right) dx = \delta_{mn} \qquad (7.3.10)$$

where $w(x) = 2/d$ and (7.3.10) may be easily verified using integration by parts. If we now multiply each term in (7.3.7) by $(2/d)\sin(n\pi x/d)$ and integrate over the interval $[0, d]$, we find that

$$\sum_{m=1}^{\infty} (E_{+m} - E_{-m}) \int_0^d \frac{2}{d} \sin\left(\frac{m\pi x}{d}\right) \sin\left(\frac{n\pi x}{d}\right) dx = 0$$

$$\sum_{m=1}^{\infty} (E_{+m} - E_{-m})\delta_{mn} = 0$$

$$E_{+n} - E_{-n} = 0 \qquad (7.3.11)$$

The boundary condition on the tangential electric field thus implies that the $+\hat{z}$- and $-\hat{z}$-propagating waves have identical coefficients $E_{+m} = E_{-m}$ multiplying every one of the TE_m modes. Substituting (7.3.11) into (7.3.8), multiplying both sides of (7.3.8) by $(2/d)\sin(n\pi x/d)$ and integrating from $x = 0$ to $x = d$ yields

$$\sum_{m=1}^{\infty} \frac{2E_{+m}k_{zm}}{\eta k_0} \int_0^d \frac{2}{d} \sin\left(\frac{m\pi x}{d}\right) \sin\left(\frac{n\pi x}{d}\right) dx = -\int_0^d J_s(x)\frac{2}{d} \sin\left(\frac{n\pi x}{d}\right) dx$$

$$\sum_{m=1}^{\infty} \frac{2E_{+m}k_{zm}}{\eta k_0} \delta_{mn} = -\frac{2}{d} \int_0^d J_s(x) \sin\left(\frac{n\pi x}{d}\right) dx$$

$$\frac{2E_{+n}k_{zn}}{\eta k_0} = -\frac{2}{d} \int_0^d J_s(x) \sin\left(\frac{n\pi x}{d}\right) dx$$

or, finally

$$E_{+n} = -\frac{\eta k_0}{k_{zn}d} \int_0^d J_s(x) \sin\left(\frac{n\pi x}{d}\right) dx \qquad (7.3.12)$$

---

3. The expression (7.3.9) for the orthogonal set of functions $\{f_m(x)\}$ is completely analogous to the vector orthogonality relations

$$\hat{x} \cdot \hat{x} = \hat{y} \cdot \hat{y} = \hat{z} \cdot \hat{z} = 1$$

$$\hat{x} \cdot \hat{y} = \hat{x} \cdot \hat{z} = \hat{y} \cdot \hat{z} = 0$$

for the Cartesian coordinate system. Just as any vector may be expressed as a linear combination of the orthogonal basis vectors $\{\hat{x}, \hat{y}, \hat{z}\}$, (i.e., $\overline{A} = \hat{x}A_x + \hat{y}A_y + \hat{z}A_z$), any function $f(x)$ may be expressed as a linear combination of the orthogonal basis functions $\{f_m(x)\}$. Thus,

$$f(x) = \sum_{m=1}^{\infty} C_m f_m(x)$$

where C_m is a complex coefficient weighting the mth orthogonal function.

The orthogonality relation thus permits us to select a single mode from the infinite sum of modes comprising the total wave, and to find the complex coefficient multiplying that mode.[4] Since n is just a dummy variable, we could equally well substitute m for n in (7.3.12). The coefficients of each of the TE_m modes are then

$$E_{+m} = E_{-m} = -\frac{\omega\mu}{d\sqrt{\omega^2\mu\epsilon - (m\pi/d)^2}} \int_0^d J_s(x) \sin\left(\frac{m\pi x}{d}\right) dx \qquad (7.3.13)$$

Note that if $J_s(x)$ is orthogonal to $\sin(m\pi x/d)$, then the TE_m mode is not excited by this current source. If the TE_m mode is driven at a frequency below cut-off for that mode ($\omega^2\mu\epsilon < (m\pi/d)^2$), then the TE_m electric field and current $J_s(x)$ are 90° out of phase. In this case, no average power propagates in the TE_m mode; the mode is evanescent. Once the current distribution $J_s(x)$ is specified, a numerical value for each of the TE_m mode coefficients is found by performing the integration in (7.3.13).

But what of the TM modes? We find the general form of the electric and magnetic fields by superposing an infinite number of the TM_m modes given by (7.1.9) and (7.1.10)

$$\overline{H}(z > 0) = \sum_{m=0}^{\infty} \hat{y} H_{+m} \cos k_{xm}x \, e^{-jk_{zm}z} \qquad (7.3.14)$$

$$\overline{E}(z > 0) = \sum_{m=0}^{\infty} \eta H_{+m}(\hat{x} \frac{k_{zm}}{k_0} \cos k_{xm}x + \hat{z} \frac{jk_{xm}}{k_0} \sin k_{xm}x) \, e^{-jk_{zm}z}$$

$$(7.3.15)$$

$$\overline{H}(z < 0) = \sum_{m=0}^{\infty} \hat{y} H_{-m} \cos k_{xm}x \, e^{jk_{zm}z} \qquad (7.3.16)$$

$$\overline{E}(z < 0) = \sum_{m=0}^{\infty} \eta H_{-m}(-\hat{x} \frac{k_{zm}}{k_0} \cos k_{xm}x + \hat{z} \frac{jk_{xm}}{k_0} \sin k_{xm}x) \, e^{jk_{zm}z}$$

$$(7.3.17)$$

where (7.3.2) and (7.3.3) are still valid.

---

4. To continue our analogy with vectors, we consider the arbitrary vector $\overline{A}$, which we may express in Cartesian coordinates as

$$\overline{A} = \hat{x} A_x + \hat{y} A_y + \hat{z} A_z$$

If we wish to find the coefficient of the $\hat{x}$ vector, all we need do is to take the dot product of $\overline{A}$ and $\hat{x}$

$$\overline{A} \cdot \hat{x} = (\hat{x} \cdot \hat{x})A_x + (\hat{y} \cdot \hat{x})A_y + (\hat{z} \cdot \hat{x})A_z = A_x$$

and the proper coefficient will be selected because the vector has been expanded in an orthogonal basis. Therefore, if we wish to find the coefficient of the mth mode, we multiply the entire wave by $\sin(m\pi x/d)$ and integrate over the appropriate interval—only the mth mode survives this integration.

The boundary conditions are again given by (4.1.19) and (4.1.21). Continuity of the tangential electric fields (7.3.15) and (7.3.17) yields

$$\hat{z} \times \left[\overline{E}(z > 0) - \overline{E}(z < 0)\right]\Big|_{z=0} = 0$$

$$\sum_{m=1}^{\infty} \hat{x} \frac{\eta k_{zm}}{k_0}(H_{+m} + H_{-m}) \cos\left(\frac{m\pi x}{d}\right) = 0 \qquad (7.3.18)$$

But since

$$\int_0^d \frac{2}{d} \cos\left(\frac{m\pi x}{d}\right) \cos\left(\frac{n\pi x}{d}\right) dx = \delta_{mn} \qquad (7.3.19)$$

is also a valid orthogonality relation,[5] (7.3.18) implies that

$$H_{+m} = -H_{-m} \qquad (7.3.20)$$

The boundary condition on the tangential magnetic field is

$$\hat{z} \times \left[\overline{H}(z > 0) - \overline{H}(z < 0)\right]\Big|_{z=0} = \hat{y} J_s(x) \qquad (7.3.21)$$

Because the TM modes have $\hat{y}$-directed magnetic fields, the left side of (7.3.21) is $\hat{x}$-directed so (7.3.14), (7.3.16), and (7.3.19) imply that

$$H_{+m} = H_{-m} \qquad (7.3.22)$$

The only way for (7.3.20) and (7.3.22) to be true simultaneously is if $H_{+m} = H_{-m} = 0$. We have thus conclusively shown that no TM modes are excited by this current source, which isn't surprising; a $\hat{y}$-directed surface current cannot possibly excite TM modes. Were the current source $\hat{x}$-directed instead, TM rather than TE modes would be excited in this parallel-plate waveguide.

If we wish to know only whether a given mode is excited, we need not go to the trouble of finding the exact solutions for the electric and magnetic fields. Instead, we can compute the power supplied by the source to the specific mode of interest. From (4.5.11), we recall that

$$P = -\frac{1}{2} \int_V \overline{E} \cdot \overline{J}^* \, dv \qquad (7.3.23)$$

is the total power supplied to the waveguide by the current source $\overline{J}$. Because the total electric field may be written as an infinite superposition of modes, we may also express (7.3.23) as

$$P = -\frac{1}{2} \int_V \sum_{m=0}^{\infty} \overline{E}_m \cdot \overline{J}^* \, dv = \sum_{m=0}^{\infty} \left(-\frac{1}{2} \int_V \overline{E}_m \cdot \overline{J}^* \, dv\right) = \sum_{m=0}^{\infty} P_m$$

where $P_m = -\frac{1}{2} \int_V \overline{E}_m \cdot \overline{J}^* \, dv$ is the power delivered to the mth mode. The total power can be simply written as a sum of the individual mode powers because all of

---

5. In the case where $m = 0$, the weighting function is reduced by a factor of 2 to $1/d$.

the modes are orthogonal. Therefore, to determine that a given mode is excited, we simply demonstrate that P_m is nonzero for that mode.

From this type of analysis, we see that the $\hat{y}$-directed current source $\overline{J} = \hat{y} J_s(x)\delta(z)$ generates a *TE* electric field $\overline{E}_m$ such that $\hat{y} \cdot \overline{E}_m \neq 0$ if $\overline{E}_m \cdot \overline{J}^*$ is to be nonzero. For *TM* modes, the magnetic field is $\hat{y}$-directed and therefore $\overline{E}_m \cdot \overline{J}^* = 0$; *TM* modes are not excited by this current source. On the other hand, the electric field is $\hat{y}$-directed for the *TE* modes, so excitation of *TE* modes is quite possible. We also note that if a mode is excited below its cut-off frequency, it has an imaginary value for k_{zm}. For $+\hat{z}$-traveling waves, $\text{Im}\{k_{zm}\} > 0$ so these modes will exponentially attenuate in the $+\hat{z}$ direction. From (7.3.13), $E_{+m} = E_{-m}$ is proportional to $+j\overline{J}_s(x)$ when excited below cut-off, so the power in the mth mode has a positive imaginary part. These *TE* evanescent waves thus store excess magnetic energy.

Symmetry is often a useful tool for predicting whether modes will be excited. For example, if we happen to know that $J_s(x)$ is symmetric about the waveguide axis $x = d/2$ [i.e., $J_s(x) = J_s(d - x)$], then we will also find that no even-numbered *TE* modes are excited by this current source. The TE_{2m} modes are antisymmetric with respect to the $x = d/2$ axis and thus the integral of $\overline{E}_m \cdot \overline{J}^*$ over the interval $[0, d]$ is zero for these modes.

Example 7.3.1

Design an attenuator that would affect the TM_1 mode but not the TE_1 mode. Also design an attenuator that would do the reverse.

Solution: In order that a conductive structure not attenuate a mode, it must be composed of thin sheets or wires that are everywhere perpendicular to the electric field of that mode, so that the boundary conditions for that mode will not be affected. Thin conducting sheets perpendicular to E_y for the TE_1 mode would thus have no effect on the mode because J_y would equal zero, even in the limit that $\sigma \to \infty$. The TM_1 mode, on the other hand, **would** be attenuated by conducting sheets parallel to the x–z plane because the conductor would absorb power from the mode and $\frac{1}{2} \int_V \overline{E} \cdot \overline{J}^* \, dv$ would not be zero in this case. The TM_1 attenuator is pictured in Figure 7.11(a).

To avoid attenuating the TM_1 mode, the conductivity must be nonzero only in the $\hat{y}$ direction, which implies $\hat{y}$-directed wires. These wires can attenuate the TE_1 mode if they are connected such that induced currents flow through a resistive element, as illustrated in Figure 7.11(b). In each case, the loss can be estimated using perturbation techniques provided that the currents induced in the resistive elements radiate fields which are much smaller than the unperturbed fields. This happens when the power dissipated per waveguide wavelength is small compared to the unperturbed power in the wave.

Example 7.3.2

An air-filled parallel-plate waveguide with a plate separation distance of 3 mm is excited by a current source $\overline{J}(\overline{r}, t) = \hat{y} I_0 \delta(x - a)\delta(z) \cos \omega t$ (A/m), where $f = \omega/2\pi = 155$ GHz. Which modes can propagate if $a = d$? $a = d/2$? $a = d/3$?

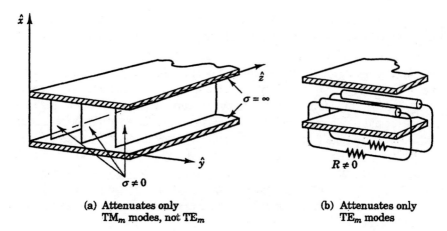

(a) Attenuates only   (b) Attenuates only
    TM_m modes, not TE_m TE_m modes

Figure 7.11   Attenuator in Example 7.3.1.

Solution: From (7.1.14) we know that the cut-off frequency for a parallel-plate waveguide is

$$f_m = \frac{\omega_m}{2\pi} = \frac{mc}{2d} = m \cdot 50 \text{ GHz}$$

If $f = 155$ GHz, then only the seven modes TE_1, TE_2, TE_3, TM_0, TM_1, TM_2, and TM_3 may possibly propagate. For $m \geq 4$, both the TE_m and TM_m modes are evanescent at this frequency. Because the current source is $\hat{y}$-directed and $E_y = 0$ for the TM modes, the integral

$$P_m = -\frac{1}{2} \int_V \overline{E}_m \cdot \overline{J}^* \, dv$$

is zero for the TM modes; no TM modes are therefore excited. Only the TE_1, TE_2, and TE_3 modes are still candidates for possibly excitable modes, and the electric field for each TE mode is proportional to $\sin(m\pi x/d)$. But because the current source is nonzero only at $x = a$, only the value of the electric field at $x = a$ is relevant:

$$|E_m| \propto \sin\left(\frac{m\pi a}{d}\right)$$

If $a = d$, then $\overline{E}_m = 0$ for all m and neither TE nor TM modes are excited. If $a = d/2$, then the TE_1 and TE_3 modes can propagate, but the TE_2 mode cannot; TE_2 has a null at $a = d/2$. Likewise, if $a = d/3$, only the TE_1 and TE_2 modes are excited because TE_3 is zero at the only position where the current is nonzero.

7.4 RECTANGULAR WAVEGUIDES

A *rectangular waveguide* is a hollow rectangular conducting structure of width a and height b where $a \geq b$, as illustrated in Figure 7.12. Because we have already studied solutions to Maxwell's equations in free space for the parallel-plate waveguide, we need only constrain these solutions to match the boundary

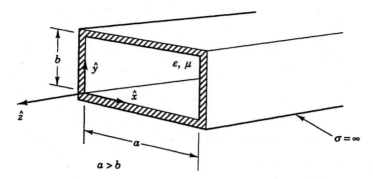

Figure 7.12 Coordinates and dimensions of a rectangular waveguide.

conditions imposed by the conducting walls at $x = 0$, a and $y = 0$, b. The boundary conditions on the electric field at the conducting walls are

$$E_y = E_z = 0 \qquad \text{at } x = 0, \ a \tag{7.4.1}$$

$$E_x = E_z = 0 \qquad \text{at } y = 0, \ b \tag{7.4.2}$$

As in the case of the parallel-plate waveguide, it is convenient to study the *TE* modes ($E_z = 0$) and *TM* modes ($H_z = 0$) separately. Both must satisfy the free-space wave equation (1.4.9)

$$\left(\nabla^2 + \omega^2 \mu \epsilon \right) \left\{ \begin{matrix} \overline{E} \\ \overline{H} \end{matrix} \right\} = 0 \tag{7.4.3}$$

where $\nabla^2 = \dfrac{\partial^2}{\partial x^2} + \dfrac{\partial^2}{\partial y^2} + \dfrac{\partial^2}{\partial z^2}$. Thus $\overline{E}$ and $\overline{H}$ are both composed of products of sines and cosines. For the *TE* modes, we try

$$E_y = E_0 \sin k_x x \left(A \sin k_y y + B \cos k_y y \right) e^{-jk_z z} \tag{7.4.4}$$

which satisfies the boundary condition $E_y = 0$ at $x = 0$ (7.4.1), exhibits standing wave behavior in the $\hat{x}$ and $\hat{y}$ directions, and travels in the $\hat{z}$ direction. To satisfy (7.4.1) at $x = a$, the guidance condition

$$k_x a = m\pi, \qquad m = 0, 1, 2, \ldots \tag{7.4.5}$$

must apply. Furthermore, substitution of (7.4.4) in (7.4.3) yields the familiar dispersion relation

$$k_x^2 + k_y^2 + k_z^2 = \omega^2 \mu \epsilon \tag{7.4.6}$$

In the same fashion, we can write

$$E_x = E_0 \sin k_y y \left(C \sin k_x x + D \cos k_x x \right) e^{-jk_z z} \tag{7.4.7}$$

where

$$k_y b = n\pi, \qquad n = 0, 1, 2, \ldots \tag{7.4.8}$$

in order to satisfy (7.4.2) at $y = b$. An expression for the wave number k_z follows from (7.4.5), (7.4.6), and (7.4.8):

$$k_z = \sqrt{\omega^2 \mu\epsilon - (m\pi/a)^2 - (n\pi/b)^2} \tag{7.4.9}$$

To find the constants A, B, C, and D we need another equation. One possible choice is Gauss's law in a charge-free medium

$$\nabla \cdot \overline{E} = \frac{\partial E_x}{\partial x} + \frac{\partial E_y}{\partial y} + \frac{\partial E_z}{\partial z} = 0$$

$$= k_x E_0 \sin k_y y \left(C \cos k_x x - D \sin k_x x \right) e^{-jk_z z}$$

$$+ k_y E_0 \sin k_x x \left(A \cos k_y y - B \sin k_y y \right) e^{-jk_z z} \tag{7.4.10}$$

where $E_z = 0$; E_x and E_y are given by (7.4.7) and (7.4.4). In order for (7.4.10) to be zero for all values of x, y, and z, it follows for nonzero k_x and k_y that

$$A = C = 0 \tag{7.4.11}$$

$$-k_x E_0 D - k_y E_0 B = 0 \tag{7.4.12}$$

Substitution of (7.4.11) and (7.4.12) into (7.4.4) and (7.4.7) finally yields equations for the *TE* modes

$$E_x = \frac{k_y}{k_0} E_0 \cos k_x x \sin k_y y \, e^{-jk_z z} \tag{7.4.13}$$

$$E_y = -\frac{k_x}{k_0} E_0 \sin k_x x \cos k_y y \, e^{-jk_z z} \tag{7.4.14}$$

where $k_0 = \omega\sqrt{\mu\epsilon}$ is included as a factor in (7.4.13) and (7.4.14) so that the arbitrary constant E_0 has the usual dimensions of V/m. The corresponding magnetic field may be found from Faraday's law for the *TE* modes

$$\overline{H} = -\frac{\nabla \times \overline{E}}{j\omega\mu} = \frac{E_0}{\eta} \left[\hat{x} \frac{k_x k_z}{k_0^2} \sin k_x x \cos k_y y + \hat{y} \frac{k_y k_z}{k_0^2} \cos k_x x \sin k_y y \right.$$

$$\left. - j\hat{z} \frac{(k_x^2 + k_y^2)}{k_0^2} \cos k_x x \cos k_y y \right] e^{-jk_z z} \tag{7.4.15}$$

where $\eta = \sqrt{\mu/\epsilon}$. Notice that $\nabla \cdot \overline{H} = 0$ is satisfied by (7.4.15).

The same procedure for *TM* modes yields

$$H_x = \frac{k_y}{k_0} H_0 \sin k_x x \cos k_y y \, e^{-jk_z z} \tag{7.4.16}$$

$$H_y = -\frac{k_x}{k_0} H_0 \cos k_x x \sin k_y y \, e^{-jk_z z} \tag{7.4.17}$$

$$\overline{E} = \frac{\nabla \times \overline{H}}{j\omega\epsilon} = -\eta H_0 \left[\hat{x} \frac{k_x k_z}{k_0^2} \cos k_x x \sin k_y y + \hat{y} \frac{k_y k_z}{k_0^2} \sin k_x x \cos k_y y \right.$$

$$\left. + j\hat{z} \frac{(k_x^2 + k_y^2)}{k_0^2} \sin k_x x \sin k_y y \right] e^{-jk_z z} \qquad (7.4.18)$$

where again

$$k_x a = m\pi$$

$$k_y b = n\pi \qquad\qquad (7.4.19)$$

$$k_z = \sqrt{\omega^2 \mu\epsilon - (m\pi/a)^2 - (n\pi/b)^2}$$

The *TM* electric field thus obeys the same boundary conditions (7.4.1) and (7.4.2) as the *TE* modes, and $\nabla \cdot \overline{E} = \nabla \cdot \overline{H} = 0$.

Because there are two guidance conditions in (7.4.19) that must be met for each *TE* or *TM* mode, the modes are designated as TE_{mn} or TM_{mn} where m and n are both integers. By convention, we associate the first index m with the length a of the broad wall of the waveguide and the second index n with the shorter wall. The field patterns for the TE_{10} mode are sketched in Figure 7.13(a), and those for the TE_{11} and TM_{11} modes are sketched in Figure 7.13(b) and (c), respectively. Note that in order to match the boundary conditions, the projected wavelengths λ_x and λ_y must be such that $m\lambda_x/2 = a$ and $n\lambda_y/2 = b$.

Not all combinations of m and n correspond to solutions. Examination of (7.4.13)–(7.4.18) reveals that the TE_{00}, TM_{00}, TM_{m0}, and TM_{0n} modes do not exist; i.e., only with zero field strength could these modes satisfy the boundary conditions for a rectangular waveguide. We thus arrive at the very important conclusion that *TEM* modes cannot exist in rectangular waveguides, a conclusion that was also reached in Section 5.3 by different reasoning.

Each TE_{mn} and TM_{mn} mode can only propagate for frequencies above the cut-off frequency ω_{mn} for that mode. For rectangular waveguide, the cut-off frequency is found by solving for the frequency at which k_z first becomes imaginary. From (7.4.9) or (7.4.19), we have

$$k_z = \sqrt{\omega_{mn}^2 \mu\epsilon - (m\pi/a)^2 - (n\pi/b)^2} = 0 \qquad (7.4.20)$$

at the cut-off frequency. Therefore, ω_{mn} for the TE_{mn} or TM_{mn} mode is

$$\omega_{mn} = \frac{1}{\sqrt{\mu\epsilon}} \sqrt{\left(\frac{m\pi}{a}\right)^2 + \left(\frac{n\pi}{b}\right)^2} \qquad (7.4.21)$$

Since the TE_{00}, TM_{00}, TM_{m0}, and TM_{0n} modes do not exist, and by convention $a \geq b$, the lowest cut-off frequency belongs to the TE_{10} mode. For this reason, The TE_{10} mode is called the *dominant mode* of a rectangular waveguide. In the

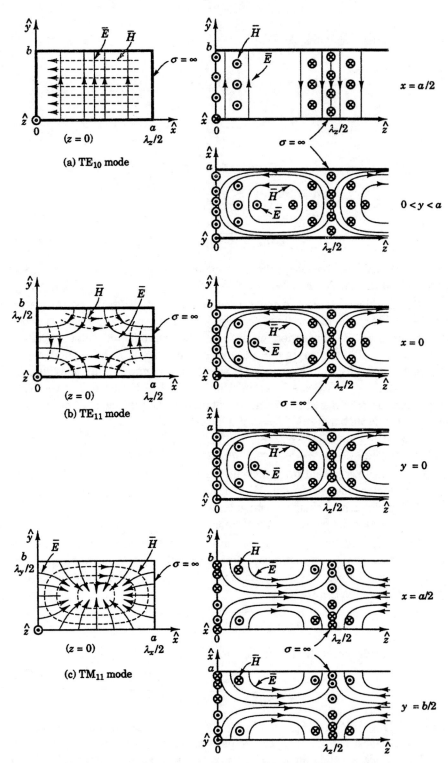

Figure 7.13 TE_{10}, TE_{11} and TM_{11} modes of a rectangular waveguide.

TABLE 7.1   Cut-Off Frequencies of a Rectangular Waveguide.

Mode	Cut-off Frequencies	Cut-off Wavelengths
TE_{10}	$c\pi/a$	$2a$
TE_{01}	$c\pi/b$	$2b$
TE_{11}, TM_{11}	$c\pi\sqrt{(1/a)^2 + (1/b)^2}$	$2/\sqrt{(1/a)^2 + (1/b)^2}$
TE_{mn}, TM_{mn}	$c\pi\sqrt{(m/a)^2 + (n/b)^2}$	$2/\sqrt{(m/a)^2 + (n/b)^2}$

special case $a = b$, we find that $\omega_{10} = \omega_{01}$. Table 7.1 lists the cut-off frequencies for the modes of a rectangular waveguide.

The dispersion relation (ω versus k_z) for several rectangular waveguide modes is sketched in Figure 7.14. Note that the phase velocity $v_p = \omega/k_z$ exceeds the speed of light for each mode and that the group velocity v_g is less than or equal to c. The relation $v_p v_g = c^2$ is obeyed by each mode because the dispersion relation is of the form $\omega^2 = c^2 k_z^2 + b$. For any given frequency ω, an infinite number of modes exist. Only some of these modes will have cut-off frequencies below ω (and can therefore propagate); the rest have an imaginary value of k_z and are evanescent. The degree to which any of these modes, including the evanescent ones, is excited depends upon the form of the source currents as was discussed in Section 7.3.

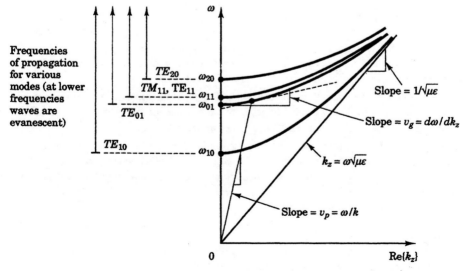

Figure 7.14   Dispersion relations for representative modes of a rectangular waveguide.

Example 7.4.1

What is the cut-off frequency in a 1×2 cm air-filled rectangular waveguide for the TE_{22} mode? Sketch $\overline{E}$ and $\overline{H}$ for the TE_{22} mode at $t = 0$ for a wave propagating in the $+\hat{z}$ direction. Sketch $\overline{E}$ in the plane $z = 0$ and $\overline{H}$ in the plane $x = a/2$.

Solution: Using equation (7.4.21), we find

$$
f_{22} = \frac{\omega_{22}}{2\pi} = c\sqrt{\left(\frac{1}{a}\right)^2 + \left(\frac{1}{b}\right)^2}
$$

$$
= 3 \times 10^8 \sqrt{\left(1/10^{-2}\right)^2 + \left(1/2 \times 10^{-2}\right)^2}
$$

$$
= 33.5 \text{ GHz}
$$

where $m = n = 2$.

The electric field $\overline{E}$ for TE_{mn} modes is given by (7.4.13) and (7.4.14), where $k_x = m\pi/a = 2\pi/a$ (7.4.5), and $k_y = n\pi/b = 2\pi/b$ (7.4.8). The time-dependent electric field is $\overline{E}(t) = \text{Re}\{\overline{E}\,e^{j\omega t}\} = \text{Re}\{\overline{E}\}$ at $t = 0$; $\overline{H}(t = 0) = \text{Re}\{\overline{H}\}$ as well, where $\overline{H}$ is given by (7.4.15). Therefore, (7.4.13)–(7.4.15) can be written as

$$
E_x(t = 0, z = 0) = \frac{2\pi}{b} E_0 \cos\left(\frac{2\pi x}{a}\right) \sin\left(\frac{2\pi y}{b}\right)
$$

$$
E_y(t = 0, z = 0) = -\frac{2\pi}{a} E_0 \sin\left(\frac{2\pi x}{a}\right) \cos\left(\frac{2\pi y}{b}\right)
$$

$$
H_x(t = 0, x = a/2) = 0
$$

$$
H_y(t = 0, x = a/2) = -\frac{2\pi E_0 k_z}{\omega\mu b} \sin\left(\frac{2\pi y}{b}\right) \cos k_z z
$$

$$
H_z(t = 0, x = a/2) = \frac{E_0}{\omega\mu}\left[\left(\frac{2\pi}{a}\right)^2 + \left(\frac{2\pi}{b}\right)^2\right] \cos\left(\frac{2\pi y}{b}\right) \sin k_z z
$$

At $z = 0$, $|E_x|$ is a maximum at $y = b/4$ and $3b/4$ when $x = 0$, $a/2$, or a. Similarly, $|E_y|$ is maximized at $y = 0$, $b/2$, and b when $x = a/4$ or $3a/4$. These locations of maximum $|\overline{E}|$ are sketched for $z = 0$ in Figure 7.15(a).

Knowing that the strength of the electric field is sinusoidal in x and y separately, and that $\nabla \cdot \overline{E} = 0$ (i.e., no electric field lines in the x–y plane terminate except on the boundaries), the rest of $\overline{E}(x, y)$ can easily be sketched as shown.

Likewise, the magnetic field $|H_y|$ has maxima at $z = 0$, π/k_z, and $2\pi/k_z$ when $y = b/4$ or $3b/4$, and $|H_x|$ has maxima at $z = \pi/2k_z$ and $3\pi/2k_z$ when $y = 0$, $b/2$, or b. Since $\nabla \cdot \overline{H} = 0$, the magnetic field lines circulate, and the full magnetic field is easily sketched in Figure 7.15(b). Note that $(\overline{E} \times \overline{H}) \cdot \hat{z}$ is always positive here, corresponding to $+\hat{z}$-directed power flow as specified.

Example 7.4.2

Design a rectangular air-filled waveguide with the largest possible bandwidth over which only the dominant mode TE_{10} can propagate, and that carries maximum power. The cut-off frequency for the TE_{10} mode is $\omega_{10}/2\pi = 10$ GHz.

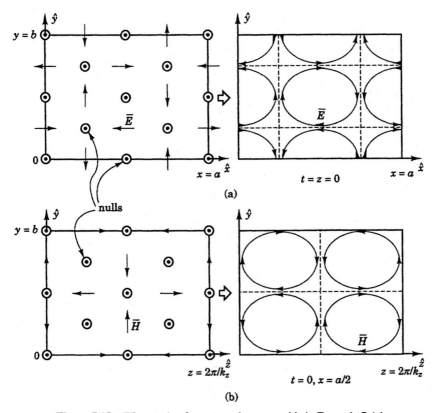

Figure 7.15 TE_{22} mode of a rectangular waveguide in Example 7.4.1.

Solution: Equation (7.4.21) implies that $\omega_{20} = 2\omega_{10}$ for any value of ω_{10}, and therefore the bandwidth over which TE_{10} alone can propagate is not more than $\omega_{10}/2\pi$, or 10 GHz. In order to maximize this bandwidth, we want all other mode cut-off frequencies ω_{mn} to be greater than ω_{20}, where

$$\omega_{mn} = c\pi \sqrt{(m/a)^2 + (n/b)^2}$$

Additionally, ω_{20} must be less than ω_{01} to ensure that the frequency of the (20) mode lies below all the $(0n)$ modes, so

$$\frac{\pi c}{b} \geq \frac{2\pi c}{a}$$

We thus design the waveguide such that $b \leq a/2$. Although it may be intuitively clear that the largest possible b $(b = a/2)$ is desired to maximize power propagation, this can be shown by noting that the forms of $\overline{E}(x, z)$ and $\overline{H}(x, z)$ are independent of y, and therefore the power P in the waveguide for the TE_{10} mode

$$P_{10} = \frac{1}{2} \, \text{Re} \left\{ \int_0^b \int_0^a (\overline{E} \times \overline{H}^*) \cdot \hat{z} \, dx \, dy \right\}$$

is simply proportional to b if the maximum field strength is fixed. High-power waveguides are often filled with special high-pressure gases so the breakdown electric field strength is as large as possible.

Example 7.4.3

Suppose the four perfectly conducting walls of a rectangular waveguide are now made slightly lossy ($\sigma \gg \omega\epsilon$). What is the attenuation coefficient of the TE_{10} mode (ratio of power dissipated per unit length to power flow in the guide)?

Solution: The TE_{10} mode has $m = 1$ and $n = 0$, or $k_x = \pi/a$ and $k_y = 0$ from (7.4.5) and (7.4.8). Equations (7.4.13)–(7.4.15) give the electric and magnetic fields for the TE_{10} mode

$$E_y = -\frac{\pi}{a} E_0 \sin\left(\frac{\pi x}{a}\right) e^{-jk_z z}$$

$$H_x = \frac{\pi}{a} \left(\frac{k_z E_0}{\omega\mu}\right) \sin\left(\frac{\pi x}{a}\right) e^{-jk_z z}$$

$$H_z = -j \left(\frac{\pi}{a}\right)^2 \left(\frac{E_0}{\omega\mu}\right) \cos\left(\frac{\pi x}{a}\right) e^{-jk_z z}$$

where $k_z = \sqrt{\omega^2 \mu\epsilon - (\pi/a)^2}$ and $E_x = E_z = H_y = 0$. We can easily compute the time-averaged power per unit area flowing in this mode from the complex Poynting vector

$$\langle \overline{S} \rangle = \frac{1}{2} \text{Re}\left\{\overline{E} \times \overline{H}^*\right\} = -\hat{z}\, \frac{1}{2} E_y H_x^*$$

$$= \hat{z}\, \frac{k_z |E_0|^2}{2\omega\mu} \left(\frac{\pi}{a}\right)^2 \sin^2\left(\frac{\pi x}{a}\right) \quad (\text{W/m}^2)$$

which means that the total time-averaged power in the TE_{10} mode is

$$P_0 = \int_0^b \int_0^a \langle \overline{S} \rangle \cdot \hat{z}\, dx\, dy$$

$$= \frac{k_z |E_0|^2}{2\omega\mu} \left(\frac{\pi}{a}\right)^2 \frac{ab}{2} \quad (\text{W})$$

The power dissipated per unit length in the $x = 0$ wall of the waveguide is

$$P_d(x = 0) = \frac{1}{2} \int_0^b \int_0^\delta \overline{E} \cdot \overline{J}^* dx\, dy = \frac{1}{2} \int_0^b \overline{E} \cdot \overline{J}_s^* \bigg|_{x=0} dy \quad (\text{W/m})$$

since $\overline{J} = \overline{J}_s \delta$ if most of the power is dissipated within a skin depth $\delta = \sqrt{2/\omega\mu\sigma}$. But $\overline{E} = \overline{J}/\sigma \simeq \overline{J}_s/\delta\sigma$ by Ohm's law, so

$$P_d(x = 0) \simeq \frac{1}{2\sigma\delta} \int_0^b |\overline{J}_s|^2 \big|_{x=0} dy \quad (\text{W/m})$$

A very similar derivation is given for the power dissipated in a slightly lossy reflecting surface in Section 4.2.

We have already shown that $\overline{J}_s = \hat{n} \times \overline{H}$ at the surface of a perfect conductor (4.1.21), an equation that is also approximately true at an almost-perfect conductor (see Example 4.2.2). For the wall at $x = 0$, $\hat{n} = \hat{x}$ and so $\overline{J}_s = -\hat{y} H_z$. Therefore,

$$P_d(x=0) \simeq \frac{1}{2\sigma\delta} \int_0^b |H_z|^2 \big|_{x=0} \, dy = \frac{1}{2\sigma\delta} \left(\frac{\pi}{a}\right)^4 \left(\frac{E_0}{\omega\mu}\right)^2 b \quad \text{(W/m)}$$

and an identical expression is computed for $P_d(x=a)$. A similar derivation for the imperfectly conducting walls at $y = 0$ and $y = b$ leads to an expression for power dissipation per unit length

$$P_d(y=0,b) \simeq \frac{1}{2\sigma\delta} \int_0^a \left(|H_x|^2 + |H_z|^2\right)\big|_{y=0,b} \, dx$$

$$= \frac{1}{2\sigma\delta} \left(\frac{\pi}{a}\right)^2 \left(\frac{E_0}{\omega\mu}\right)^2 \left[k_z^2 + \left(\frac{\pi}{a}\right)^2\right] \frac{a}{2} \quad \text{(W/m)}$$

where $k_z = \sqrt{\omega^2\mu\epsilon - (\pi/a)^2}$ The total power dissipated per unit length in all four walls is the sum of the power dissipated in each individual wall:

$$P_d \simeq \frac{1}{2\sigma\delta} \left(\frac{\pi}{a}\right)^2 \left(\frac{E_0}{\omega\mu}\right)^2 \left[(2b+a)\left(\frac{\pi}{a}\right)^2 + ak_z^2\right] \quad \text{(W/m)}$$

Therefore, the attenuation coefficient α for the TE_{10} mode is

$$\alpha = \frac{P_d}{P_0} = \frac{\delta}{k_z ab} \left[(2b+a)\left(\frac{\pi}{a}\right)^2 + ak_z^2\right] \quad \text{(m}^{-1})$$

which goes to zero as $\delta \to 0$ (no loss). The expression $\alpha = P_d/P_0$ is derived in Example 5.4.2 and will not be repeated here.

7.5* SEPARATION OF VARIABLES

The derivation of waveguide fields in Sections 7.1, 7.2 and 7.4 was straightforward since we began with known solutions to the wave equation in rectangular coordinates and then imposed boundary conditions to solve for all unknown constant coefficients in the solutions. For waveguides of arbitrary cross-section, we shall use a similar approach where we shall need to determine waveguide solutions in non-Cartesian coordinate systems.

We wish to find the electric and magnetic fields for a perfectly conducting hollow tube of arbitrary uniform cross-section (independent of z) filled with a homogeneous medium ϵ, μ. The ratio between the fields at z and those at the slightly different location $z + \delta z$ must be independent of z because the shape of the tube is independent of z, and this ratio must also approach unity as $\delta z \to 0$. The simplest equation obeying these constraints is

$$\frac{\overline{E}(z+\delta z)}{\overline{E}(z)} = \frac{\overline{H}(z+\delta z)}{\overline{H}(z)} = 1 - jk_z \delta z \tag{7.5.1}$$

where $k_z \, \delta z \ll 1$. The electric and magnetic fields thus have the general form

$$\overline{E}(\overline{r}) = \overline{E}(x, y) \, e^{-jk_z z} \tag{7.5.2}$$

$$\overline{H}(\overline{r}) = \overline{H}(x, y) \, e^{-jk_z z} \tag{7.5.3}$$

where

$$\overline{E}(x, y) = \overline{E}_T(x, y) + \hat{z} \, E_z(x, y) \tag{7.5.4}$$

$$\overline{H}(x, y) = \overline{H}_T(x, y) + \hat{z} \, H_z(x, y) \tag{7.5.5}$$

and $\overline{E}_T$ and $\overline{H}_T$ are the transverse (x, y) field components.

The homogeneous wave equation (1.4.9) is rewritten below as

$$\left(\nabla^2 + \omega^2 \mu \epsilon\right) \left\{ \frac{\overline{E}}{\overline{H}} \right\} = 0 \tag{7.5.6}$$

where the Laplacian can be separated into transverse (x, y) and $\hat{z}$-directed components:

$$\nabla^2 = \frac{\partial^2}{\partial x^2} + \frac{\partial^2}{\partial y^2} + \frac{\partial^2}{\partial z^2} = \nabla_T^2 + \frac{\partial^2}{\partial z^2} \tag{7.5.7}$$

The transverse del operator ∇_T is $\hat{x} \, \partial/\partial x + \hat{y} \, \partial/\partial y$ in Cartesian coordinates When (7.5.4) or (7.5.5) is substituted into (7.5.6), the $\hat{z}$ component of the resulting equation is

$$\left(\nabla_T^2 - k_z^2 + \omega^2 \mu \epsilon\right) \left\{ \begin{array}{c} E_z(x, y) \\ H_z(x, y) \end{array} \right\} = 0 \tag{7.5.8}$$

We have thus reduced the problem of solving for the fields of an arbitrarily shaped tubular waveguide to that of solving (7.5.8) for $E_z(x, y)$ and $H_z(x, y)$. Later we will derive all the transverse field components from just these $\hat{z}$-directed fields. Consider *TM* solutions to (7.5.8) first, for which $H_z = 0$, $E_z \neq 0$.

The wave equation (7.5.8) can be simplified by using a technique called *separation of variables*. The coordinates (variables) used in the transverse plane are generally those that conform most simply to the cross-sectional shape of the structure. Consider first the case of a rectangular waveguide, for which x, y coordinates would be most natural. Here we assume that the electric field may be written as a product of functions, one depending only on x and the other depending only on y:

$$E_z(x, y) = X(x) \cdot Y(y) \tag{7.5.9}$$

We then substitute (7.5.9) into the wave equation (7.5.8) to yield

$$X''Y + XY'' = \left(k_z^2 - \omega^2 \mu \epsilon\right) XY \tag{7.5.10}$$

where $X''(x) = \dfrac{\partial^2 X}{\partial x^2}$ and $Y''(y) = \dfrac{\partial^2 Y}{\partial y^2}$. Dividing both sides of (7.5.10) by XY yields

$$\frac{X''}{X} + \frac{Y''}{Y} = k_z^2 - \omega^2 \mu\epsilon \qquad (7.5.11)$$

Because X''/X is a function only of x, and Y''/Y is a function only of y, it follows that $X''/X + Y''/Y$ can be equal to a constant for all values of x and y **only** if X''/X and Y''/Y are themselves constants. We thus define $X''/X = -k_x^2$ and $Y''/Y = -k_y^2$ so that (7.5.11) will yield the usual dispersion relation $k_x^2 + k_y^2 + k_z^2 = \omega^2 \mu\epsilon$.

These two differential equations

$$X'' = -k_x^2 X \qquad (7.5.12)$$

$$Y'' = -k_y^2 Y \qquad (7.5.13)$$

have general solutions of the form

$$X(x) = A \sin k_x x + B \cos k_x x \qquad (7.5.14)$$

$$Y(y) = C \sin k_y y + D \cos k_y y \qquad (7.5.15)$$

In a rectangular waveguide, the boundary conditions require that $E_z = 0$ at $x = 0$, $x = a$, $y = 0$, and $y = b$, and therefore a nontrivial solution for E_z requires that $B = D = 0$ and $k_x a = m\pi$, $m = 1, 2, \ldots$, and $k_y n = n\pi$, $n = 1, 2, \ldots$, where we recall that the TM_{00}, TM_{m0}, and TM_{0n} modes do not exist. The $\hat{z}$ component of the electric fields for the TM modes are thus

$$E_z(x, y) \propto \sin\left(\frac{m\pi x}{a}\right) \sin\left(\frac{n\pi y}{b}\right)$$
$$H_z(x, y) = 0 \qquad (7.5.16)$$

as was also shown in (7.4.18).

For the TE modes, $E_z = 0$ and the $\hat{z}$ component of the magnetic field may be written as a product of functions in x and y:

$$H_z = X_H(x) \cdot Y_H(y) \qquad (7.5.17)$$

Expressions for $X_H(x)$ and $Y_H(y)$ similar to (7.5.14) and (7.5.15) can be simplified by noting that the normal magnetic field derivative $\partial H_z/\partial n$ is required to be zero at the four conducting walls because $\hat{n} \times \overline{E} \propto \hat{n} \times (\nabla \times \overline{H}) = \partial H_z/\partial n = 0$ at these boundaries, where $\hat{n}$ is a vector normal to the conducting walls of the waveguide. Therefore, the TE fields are

$$H_z(x, y) \propto \cos\left(\frac{m\pi x}{a}\right) \cos\left(\frac{n\pi y}{b}\right)$$
$$E_z(x, y) = 0 \qquad (7.5.18)$$

Once E_z and H_z have been found for either the *TE* or *TM* modes, the other field components can be computed in a straightforward manner. We start by writing Ampere's and Faraday's laws as

$$\left(\nabla_T - jk_z\hat{z}\right) \times \left(\overline{E}_T + \hat{z}\,E_z\right) = -j\omega\mu\left(\overline{H}_T + \hat{z}\,H_z\right) \qquad (7.5.19)$$

$$\left(\nabla_T - jk_z\hat{z}\right) \times \left(\overline{H}_T + \hat{z}\,H_z\right) = j\omega\epsilon\left(\overline{E}_T + \hat{z}\,E_z\right) \qquad (7.5.20)$$

where the subscript T refers to components in the transverse (x, y) plane. The transverse components of these two equations are

$$\nabla_T \times \hat{z}\,E_z - jk_z\hat{z} \times \overline{E}_T = -j\omega\mu\overline{H}_T \qquad (7.5.21)$$

$$\nabla_T \times \hat{z}\,H_z - jk_z\hat{z} \times \overline{H}_T = j\omega\epsilon\overline{E}_T \qquad (7.5.22)$$

Eliminating $\overline{H}_T$ from (7.5.22) by utilizing (7.5.21) yields

$$-j\omega\mu\nabla_T \times \hat{z}\,H_z - jk_z\hat{z} \times \left(\nabla_T \times \hat{z}\,E_z - jk_z\hat{z} \times \overline{E}_T\right) = \omega^2\mu\epsilon\overline{E}_T$$

But $\hat{z} \times (\nabla_T \times \hat{z}\,E_z) = \nabla_T E_z$ and $\hat{z} \times (\hat{z} \times \overline{E}_T) = -\overline{E}_T$, so the transverse electric field as a function of E_z and H_z is

$$\overline{E}_T = \frac{-j}{\omega^2\mu\epsilon - k_z^2}\left(\omega\mu\nabla_T \times \hat{z}\,H_z + k_z\nabla_T E_z\right) \qquad (7.5.23)$$

Similarly, (7.5.21) and (7.5.22) also yield the transverse magnetic field as a function of E_z and H_z:

$$\overline{H}_T = \frac{-j}{\omega^2\mu\epsilon - k_z^2}\left(-\omega\epsilon\nabla_T \times \hat{z}\,E_z + k_z\nabla_T H_z\right) \qquad (7.5.24)$$

If we now substitute the $\hat{z}$ field components (7.5.16) (*TM* modes) or (7.5.18) (*TE* modes) into (7.5.23) and (7.5.24), we can completely determine the transverse electric and magnetic fields. The field solutions for rectangular waveguides are the same as those derived earlier in (7.4.13)–(7.4.18).

The rather general form of the waveguide wave equation (7.5.8) can be used to find field solutions for other (non-Cartesian) waveguide cross-sections. Consider the *circular waveguide* of radius a illustrated in Figure 7.16. We can solve the wave equation for either *TE* ($E_z = 0$) or *TM* modes ($H_z = 0$). Consider the *TM* modes first. Since the waveguide has a circular cross-section, we should represent the transverse Laplacian ∇_T^2 in terms of the cylindrical coordinates r and ϕ:

$$\nabla_T^2 E_z(r, \phi) = \frac{\partial^2 E_z}{\partial r^2} + \frac{1}{r}\frac{\partial E_z}{\partial r} + \frac{1}{r^2}\frac{\partial^2 E_z}{\partial \phi^2} \qquad (7.5.25)$$

We then write $E_z(r, \phi)$ as a product of functions of r and ϕ:

$$E_z(r, \phi) = R(r)\,F(\phi) \qquad (7.5.26)$$

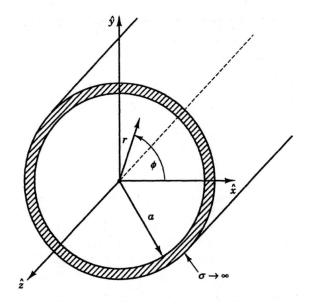

Figure 7.16 Circular waveguide of radius a.

Substituting (7.5.25) and (7.5.26) into (7.5.8), we find

$$r^2 \frac{R''}{R} + r \frac{R'}{R} + \frac{F''}{F} = r^2 \left(k_z^2 - \omega^2 \mu \epsilon\right) \tag{7.5.27}$$

where $R' = \partial R(r)/\partial r$, $R'' = \partial^2 R(r)/\partial r^2$, and $F'' = \partial^2 F(\phi)/\partial \phi^2$. We have also multiplied each term in (7.5.27) by r^2/RF.

Since F''/F is a function only of ϕ and $r^2 R''/R + r R'/R - r^2(k_z - \omega^2 \mu \epsilon)$ is a function only of r, each of these two functions must be equal to a constant so that (7.5.27) may remain valid for all values of r and ϕ. Therefore,

$$r^2 \left[\frac{1}{R} \frac{d^2 R}{dr^2} + \frac{1}{rR} \frac{dR}{dr} + \left(\omega^2 \mu \epsilon - k_z^2\right) \right] = v^2 \tag{7.5.28}$$

$$\frac{1}{F} \frac{d^2 F}{d\phi^2} = -v^2 \tag{7.5.29}$$

where v^2 is called the *separation constant*.

The solution to (7.5.29), the ϕ-dependent equation, is simply

$$F(\phi) = A \cos v\phi + B \sin v\phi \tag{7.5.30}$$

where $v = 0, 1, 2, \ldots$ to guarantee that $F(\phi)$ is single-valued; i.e., $F(\phi) = F(\phi + 2\pi n)$ for $n = 1, 2, 3, \ldots$ [6]

To solve (7.5.28), we first rearrange terms to get

$$\frac{d^2 R}{dr^2} + \frac{1}{r}\frac{dR}{dr} + \left(k_r^2 - \frac{v^2}{r^2}\right)R = 0 \qquad (7.5.31)$$

where we define

$$k_r^2 \equiv \omega^2 \mu \epsilon - k_z^2 \qquad (7.5.32)$$

The solution to the differential equation (7.5.31) is

$$R(r) = C J_v(k_r r) + D N_v(k_r r) \qquad (7.5.33)$$

where C and D are constants that must be determined by matching boundary conditions at $r = a$. The quantity $J_v(k_r r)$ is a Bessel function of the first kind of order v, and $N_v(k_r r)$ is a Bessel function of the second kind (Neumann function) of order v. Plots of $J_v(k_r r)$ and $N_v(k_r r)$ are given in Figure 7.17. Note that Bessel functions look vaguely sinusoidal with nonequally spaced zero crossings and nonconstant amplitudes. Although the constant D in (7.5.33) is generally nonzero in coaxial structures (where the region of interest does not include the origin), $D = 0$ for hollow circular waveguides because the Neumann function has a singularity at $r = 0$.

To match the *TM* mode boundary condition ($E_z = 0$ at $r = a$), we thus require the circular waveguide guidance condition $J_v(k_r a) = 0$, analogous to the guidance condition $\sin k_x d = 0$ for the parallel-plate guide. The arguments $k_r a$ for the first few zeros of the Bessel function $J_v(k_r r)$ ($v = 0, 1, 2$) are given in Table 7.2. The integer n in Table 7.2 refers to the nth null of the function $J_v(k_r r)$, excluding the null at $r = 0$ for $v \geq 1$. Because the *TM* modes are completely specified by the two integers v and n, they are designated as TM_{vn} modes, where the index v indicates the number of azimuthal variations and n the number of radial variations in the field pattern E_z. There are $2v$ nulls observed as we rotate through a full 2π radians (at fixed r), and $n - 1$ nulls observed as we move from $r = 0$ to $r = a$ (at fixed ϕ), not including the nulls at $r = 0$ or $r = a$. As an example, the TM_{23} mode has the guidance condition

$$k_r a = 11.6198$$

(see Table 7.2) or

$$k_z = \sqrt{\omega^2 \mu \epsilon - (11.6198/a)^2}$$

---

6. For waveguides that do not subtend every angle ϕ (e.g., a "pie-shaped" waveguide for which $0 \leq \phi \leq \phi_0 < 2\pi$), this condition is relaxed and v can take on noninteger values.

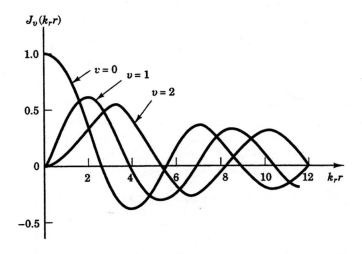

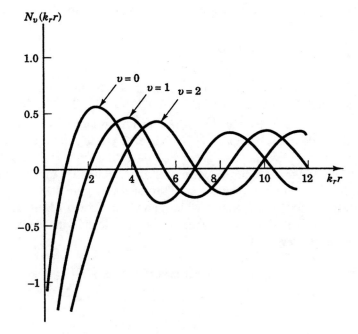

Figure 7.17   Bessel functions of the first and second kind, $J\,(k_r r)$ and $N\,(k_r r)$.

TABLE 7.2 Zeros of the Bessel function $J_\nu(k_r r)$.

$J_\nu(k_r a) = 0$	$n = 1$	$n = 2$	$n = 3$
$\nu = 0$	2.4048	5.5201	8.6537
$\nu = 1$	3.8317	7.0156	10.1735
$\nu = 2$	5.1356	8.4172	11.6198

and has four nulls for fixed r, $0 \leq \phi < 2\pi$. At any given angle, there are two nulls between $r = 0$ and $r = a$, not including either endpoint. Note that once E_z for the TM modes is found using (7.5.26), (7.5.30), and (7.5.33), the transverse fields can be found from (7.5.23) and (7.5.24) with $H_z = 0$.

The *TE* modes may be found by an entirely analogous procedure:

$$H_z(r, \phi) = \left(A \cos \nu\phi + B \sin \nu\phi\right)\left(C J_\nu(k_r r) + D N_\nu(k_r r)\right) \qquad (7.5.34)$$

Again, $D = 0$ for a hollow circular waveguide because $N_\nu(k_r r = 0) \to -\infty$ as $r \to 0$, and the fields should be everywhere finite. To ensure that H_z is single-valued, ν is an integer. The boundary condition that the tangential electric field is zero at the perfectly conducting surface $r = a$, or equivalently $\partial H_z / \partial n|_{r=a} = 0$, is given by

$$\left.\frac{\partial H_z(r, \phi)}{\partial r}\right|_{r=a} \propto J_\nu'(k_r a) = 0 \qquad (7.5.35)$$

where $J_\nu'(k_r r) = k_r (\partial J_\nu(k_r r)/\partial r)$. The arguments $k_r a$ for the first few zeros of the Bessel function derivatives are given in Table 7.3.

Again, we number the zeros of $J_\nu'(k_r r)$ with the index n and label the TE modes with the double subscript $TE_{\nu n}$. The integer ν refers to the azimuthal variations and n to radial variations. There are $n - 1$ nulls for $0 < r < a$ when ϕ is fixed and 2ν nulls for $0 < \phi \leq \phi$ when r is fixed. Representative field distributions for the $TE_{\nu n}$ and $TM_{\nu n}$ modes are illustrated in Figure 7.18. In general, a waveguide of arbitrary cross-section will support both TE and TM modes, each indexed by two parameters that specify the number of field nulls in the two transverse dimensions of the waveguide.

TABLE 7.3 Zeros of the Bessel function $J_\nu'(k_r r)$.

$J_\nu'(k_r a) = 0$	$n = 1$	$n = 2$	$n = 3$
$\nu = 0$	0.0000	3.8317	7.0156
$\nu = 1$	1.8412	5.3314	8.5363
$\nu = 2$	3.0542	6.7061	9.9695

Cross-section	Mid-plane section	Cut off wavelength (a = radius)

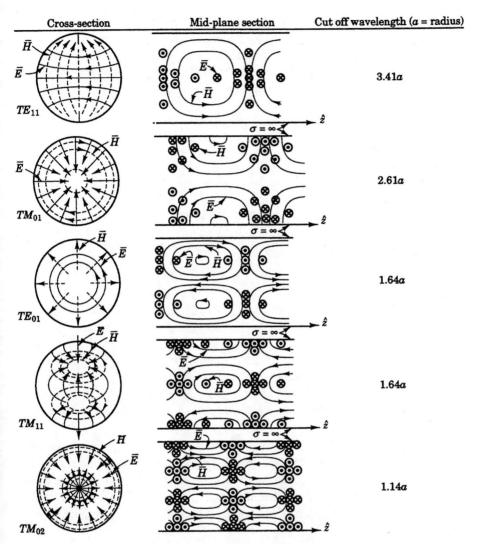

Figure 7.18 Modes of a circular waveguide.

7.6 SUMMARY

Waveguides are structures that confine and direct electromagnetic radiation. Solutions to the *parallel-plate waveguide* were obtained by studying the interference of incident and reflected waves from a single perfectly conducting plate. A second identical plate was then inserted parallel to the first plate (without disturbing the existing field) at any of the special locations where the boundary conditions of that field were still satisfied. This procedure determined the *guidance condition* for the waveguide. For the parallel-plate guide, this condition is $k_x d = m\pi$, $m = 1, 2, \ldots$ where each of the integers m corresponds to a TE_m ($E_z = 0$) or TM_m ($H_z = 0$) *mode* . These waveguide modes behave as *standing waves* in the plane transverse to the direction of propagation $\hat{z}$, and travel with phase and group velocities (in vacuum):

$$v_p = \frac{\omega}{k_z} = c \Big/ \sqrt{1 - \left(\frac{m\pi c}{\omega d}\right)^2}$$

$$v_g = \left(\frac{\partial k_z}{\partial \omega}\right)^{-1} = c \sqrt{1 - \left(\frac{m\pi c}{\omega d}\right)^2}$$

The TE_m and TM_m modes become *evanescent* below the *cut-off frequency* $\omega_m = m\pi c/d$, where cut-off occurs when $k_z = 0$. The *dispersion relation* for this waveguide is obtained from the wave equation, yielding

$$k_x^2 + k_z^2 = \omega^2 \mu \epsilon = \left(\frac{m\pi}{d}\right)^2 + k_z^2$$

The *dielectric-slab waveguide* also allows TE_m and TM_m mode propagation, where the guidance conditions are now frequency-dependent and involve solving a transcendental equation for $k_x d$. The cut-off frequency for the dielectric-slab waveguide corresponds to incident and reflected wave propagation at the *critical angle* θ_c within the guide. *Duality* is a useful way of instantly finding the *TM* modes once the *TE* modes are known for this waveguide because the boundary conditions are still valid when $\overline{E} \to \overline{H}$ and $\overline{H} \to -\overline{E}$ (unlike those of the parallel-plate guide).

In general, a possibly infinite superposition of modes propagates in a waveguide, but the coefficient of each mode may be easily found because the modes are *orthogonal* . Orthogonality allows us to mathematically filter out a single mode from the total *TE* or *TM* wave and determine the amplitude and phase of that mode. If we wish to know whether a particular mode is excited, we need only compute

$$P_m = -\frac{1}{2} \int_V \overline{E}_m \cdot \overline{J}^* \, dv$$

for that mode; if $P_m \neq 0$ then the mode is excited. An evanescent mode will satisfy $\text{Re}\{P_m\} = 0$.

Rectangular and *circular waveguides* are examples of tubular guides of uniform cross-section, whose $\overline{E}$ and $\overline{H}$ fields may be determined by the technique of *separation of variables* . Because the wave is constrained in the two-dimensional transverse plane, two indices are needed to specify a particular mode, e.g., TE_{mn}, TM_{mn}, TE_{vn}, or TM_{vn}. The first index determines the number of nulls in the field pattern along one coordinate axis, and the second index corresponds to the number of nulls along the other coordinate axis. For a rectangular waveguide of dimensions a and b, $k_x = m\pi/a$ and $k_y = n\pi/b$. For circular guides of radius a, $r = a$ is a zero of the Bessel function $J_\nu(k_r r)$ for the *TM* modes, and a zero of $J'_\nu(k_r r)$ for the *TE* modes, where ν is an integer corresponding to the number of azimuthal variations in the fields. Once E_z and H_z are determined for a tubular waveguide of uniform cross-section, the transverse $\overline{E}$ and $\overline{H}$ fields may easily be found. The rectangular and circular waveguide modes are cut-off at the frequency ω_{mn} or $\omega_{\nu n}$ where $k_z = 0$.

7.7 PROBLEMS

7.1.1 A certain 40-GHz parallel-plate waveguide can propagate energy in the $\pm z$ directions. The electric and magnetic field lines at $t = 0$ are illustrated in Figure 7.19 for a particular single mode.

(a) Which field lines are electric and which are magnetic (i.e., the $\hat{y}$-directed fields or the $\hat{x}$- and $\hat{z}$- fields)? Explain your reasoning. Is the power here flowing in the $+\hat{z}$ or $-\hat{z}$ direction, or both? What mode is this? (i.e., *TEM*, *TM₁*, *TE₂*, and so on.)

(b) If the waveguide width (in the $\hat{x}$ direction) is $d = 1$ cm, what is the resulting waveguide wavelength λ_g (cm)?

(c) What is the cut-off frequency (GHz) for this mode if $d = 1$ cm?

7.1.2 An air-filled (ϵ_0, μ_0) parallel-plate waveguide with 1-cm plate separation is being used to link a microwave source and antenna.

(a) What are the cut-off frequencies (Hz) for the TE_1, TE_2 and TM_1 modes?

(b) Assume the TM_1 mode is propagating at 20 GHz. What is the waveguide wavelength?

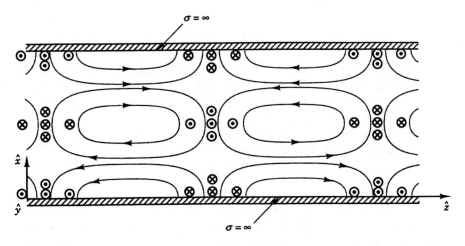

Figure 7.19   Electric and magnetic fields for a single mode of a parallel-plate waveguide.

(c) What are the phase and group velocities for the wave in part (b)? Does $v_g v_p = c^2$?

(d) Sketch $\overline{E}$ and $\overline{H}$ for the wave of part (b) at some time t.

(e) Sketch the corresponding time-dependent Poynting vector $\overline{S}(t) = \overline{E}(t) \times \overline{H}(t)$ for the same instant of time.

7.1.3 A certain air-filled parallel-plate waveguide is 10 cm wide in the $\hat{x}$ direction and has a plate separation of $d = 0.5$ cm.

(a) List the five lowest frequency modes (TE_m, TM_m) that can propagate on this line and compute their cut-off frequencies.

(b) What are the phase and group velocities (v_p and v_g) for the TM_2 mode at a frequency equal to 1.6 times the TM_2 cut-off frequency?

(c) What is the loss in this waveguide (dB/m) for the TM_2 mode at the frequency of part (b)? Express your answer in terms of σ (the conductivity of the waveguide), d, W (the plate width), and any other necessary parameters. Then convert your expression to a numerical value and see if it seems plausible.

7.1.4 Consider a parallel-plate waveguide propagating waves in the $\hat{z}$ direction and filled with a dielectric medium for $z > 0$. The dielectric medium has permittivity ϵ_1. The operating frequency is $30/2\pi$ GHz and the plate separation is d.

(a) Let $d = 2\sqrt{3}\pi$ cm and consider the empty waveguide with $\epsilon_1 = \epsilon_0$ (in the absence of the dielectric). Which TE_m and TM_m modes can propagate in this waveguide?

(b) Find expressions for the $\overline{E}$ and $\overline{H}$ fields for the TM_2 mode.

(c) Make field plots for $\overline{E}(x, z)$ and $\overline{H}(x, z)$ in the x–z plane at $t = 0$ for the TM_2 mode; $\hat{x}$ is perpendicular to the plates and $x = 0$ midway between them.

(d) What are the phase and group velocities for the TM_2 mode at this operating frequency?

(e) Let $\epsilon_1 = 3\epsilon_0$ and $d = 2\sqrt{3}\pi$ cm. For waves propagating in the $-\hat{z}$ direction, for which values of m will the TM_m modes be totally reflected at the dielectric boundary? Why?

(f) For which values of m can a propagating TM_m mode be totally transmitted at even one frequency (no reflection) and why?

7.1.5 The horizontal conducting surfaces formed by the ionosphere and the terrestrial surface can act as a parallel-plate waveguide propagating TE_m and TM_m modes. Assume the ionosphere can be modeled as a perfect conductor at $h = 100$ km altitude for the frequencies of interest here.

(a) What are the two lowest frequency modes that can propagate in this "waveguide," and what are their cut-off frequencies?

(b) What are the phase and group velocities (v_p and v_g) for the TE_1 mode at 1.6 kHz?

(c) A TM_1 wave of average power 10^{-3} W/m^2 is propagating in this waveguide at 1.6 kHz. What are the maximum values for $\overline{E}(\overline{r}, t)$ (V/m) and $\overline{H}(\overline{r}, t)$ (A/m)? (Hint: Express $\frac{1}{2} \int_0^h \hat{z} \cdot \text{Re}\{\overline{S}\} dx = 10^{-3}$ in terms of the maximum values of $\overline{E}$ and $\overline{H}$.)

(d) Sketch the field distribution $\overline{E}(\overline{r}, t)$ for the TE_1 mode at that instant of time when it is maximum. Assume $f = 1$ kHz, which is below cut-off.

7.1.6 An air-filled parallel-plate line has plate separation 1 mm and width 1 cm.

   (a) What modes (TE_m, TM_n) can propagate at 320 GHz?

   (b) What is the waveguide wavelength for the TE_1 mode at 200 GHz?

   (c) If the walls have conductivity $\sigma = 49 \times 10^6$ S/m, what is the approximate loss in the waveguide (dB/m) for the TE_1 mode at 200 GHz? (Hint: Recall Example 5.4.2, but evaluate P and P_d in terms of the fields instead.)

   (d) Sketch the electric and magnetic field distributions in the waveguide for the TE_2 mode at 200 GHz. Sketch each field at the instant of time when its magnitude is maximum and note that these two times are different.

7.1.7* In this problem, we examine the attenuation of guided TM waves due to wall loss caused by finite conductivity inside a parallel-plate waveguide $(\partial/\partial y = 0)$.

   (a) Write the field solutions for TM waves inside an air-filled parallel-plate waveguide with separation d in the $\hat{x}$ direction. Assume the plates are perfect conductors.

   (b) Suppose the plates are made of copper and have conductivity $\sigma \simeq 6 \times 10^7$ (S/m). Find the power dissipated in the conductors (W/m^2).

   (c) The amount of power lost to the conductors causes attenuation of the waves. Show that the attenuation coefficient is

$$\alpha = \frac{2}{d\sqrt{2\sigma/\omega\epsilon}\sqrt{1 - (k_x/k)^2}}$$

   where $k_x = m\pi/d$ and $k_0 = \omega\sqrt{\mu\epsilon}$.

7.2.1 A certain dielectric-slab waveguide of thickness d lies in the y–z plane centered at $x = 0$ with electric and magnetic fields:

$$\left.\begin{array}{l} \overline{H} = \hat{y}\, A\, \sin k_x x\, e^{-jk_z z} \\[4pt] \overline{E} = (\hat{x}\, B\, \sin k_x x - \hat{z}\, jC\, \cos k_x x)\, e^{-jk_z z} \end{array}\right\} \quad |x| \le d/2$$

$$\left.\begin{array}{l} \overline{H} = \hat{y}\, D\, e^{-\alpha x}\, e^{-jk_z z} \\[4pt] \overline{E} = (\hat{x}\, F\, e^{-\alpha x} - \hat{z}\, jG\, e^{-\alpha x})\, e^{-jk_z z} \end{array}\right\} \quad x > d/2$$

   (a) If $\lambda_0/2 < d < \lambda_0$, where λ_0 is the wavelength inside the dielectric $(\lambda_0 = 2\pi/\omega\sqrt{\mu\epsilon})$, what mode is propagating? (i.e., TE_m or TM_m with what mode index m?) (Hint: TE_m or TM_m modes have field distributions that resemble the corresponding TE_m or TM_m modes in metallic parallel-plate waveguides.)

   (b) Sketch the $\overline{E}$ fields at $t = 0$. Assume A, B, and so on, are real quantities.

   (c) The fields inside the slab can be thought of as trapped waves bouncing back and forth inside at an angle $\theta > \theta_c$, where θ_c is the critical angle. What is θ in terms of the given parameters?

7.2.2 Find the cut-off frequency for the TE_3 mode in a dielectric-slab waveguide with $\epsilon = 9\epsilon_0$ and thickness 1 cm, which is embedded in another dielectric with $\epsilon = 4\epsilon_0$.

7.2.3 The electric field inside a certain dielectric-slab waveguide (where $\epsilon = 9\epsilon_0$, $\mu = \mu_0$) is $\overline{E}(\bar{r}) = \hat{y}\, E_0\, \cos x\, e^{-jz}$ (V/m).

   (a) Is this a TE or TM mode?

   (b) What is the waveguide wavelength λ_z in meters?

   (c) What is ω in radians/s?

   (d) Which of the following straight wires carrying current at frequency ω could excite this mode? Each passes through the center of the waveguide: (i) a wire pointing only in the $\hat{x}$ direction, (ii) one pointing only in the $\hat{y}$ direction. Please explain briefly.

7.2.4 A certain dielectric-slab waveguide is $d = 9$ mm thick and is parallel to the y–z plane, centered at $x = 0$. The electric and magnetic fields are

$$\overline{E} = \hat{y} A \sin 1000x \, e^{-j1000z}$$

$$\overline{H} = \left[\hat{x} B \sin 1000x - \hat{z} jC \cos 1000x\right] e^{-j1000z} \quad\left.\right\} \quad |x| \le d/2$$

$$\overline{E} = \hat{y} D e^{-750x} e^{-j1000z}$$

$$\overline{H} = \left[\hat{x} F e^{-750x} - \hat{z} jG e^{-750x}\right] e^{-j1000z} \quad\left.\right\} \quad x > d/2$$

where A, B, C, D, F, and G are positive real constants.

(a) What are k_x, k_y, k_z and ω outside the slab in the free-space region?

(b) What are k_x, k_y, and k_z inside the slab (where $\epsilon \ne \epsilon_0$)? What is ϵ?

(c) What mode is propagating?

(d) The fields inside can be thought of as uniform plane waves bouncing back and forth between the surfaces at $x = \pm d/2$ at an angle $\theta > \theta_c$ with respect to the surface normal, where θ_c is the critical angle. What are θ and θ_c for the given waves?

(e) What is the cut-off frequency for this mode? What is θ when $\omega = \omega_c$?

7.2.5* Consider a grounded dielectric-slab waveguide consisting of an infinite ground plane at $x = 0$ surmounted by an infinite dielectric slab ϵ extending from $x = 0$ to $x = d$.

(a) *TM* waves are guided by the slab waveguide with the magnetic field:

$$\overline{H} = \hat{y} H_0 e^{-\alpha(x-d)} e^{-jk_z z} \qquad \text{for } x > d$$

$$\overline{H} = \hat{y} \left(A e^{jk_x x} + B e^{-jk_x x}\right) e^{-jk_z z} \text{cr} \qquad \text{for } 0 \le x \le d$$

Find the corresponding electric field in terms of H_0, α, k_z and d. Determine the amplitudes A and B.

(b) Obtain the transcendental equation for k_x as a function of frequency. Show how the equation may be solved graphically.

(c) Determine the cut-off frequency, i.e., the lowest frequency for which unattenuated propagation exists for the lowest order mode. Over what frequency range does only this mode propagate?

(d) Now consider *TE* waves with the electric field

$$\overline{E} = \hat{y} E_0 e^{-\alpha(x-d)} e^{-jk_z z} \qquad \text{for } x > d$$

$$\overline{E} = \hat{y} \left(A e^{jk_x x} + B e^{-jk_x x}\right) e^{-jk_z z} \qquad \text{for } 0 < x < d$$

Give the expressions for the corresponding magnetic field in terms of E_0, α, k_z, and d. Determine the amplitudes A and B.

(e) Repeat parts(b) and (c) for the *TE* waves.

7.3.1 A uniform current sheet $\overline{J}_s = \hat{x} J_{s0}$ (A/m) is radiating energy down the waveguide of Problem 7.1.3. What propagating modes will be excited at frequency of 1.6 times the TM_2 cut-off frequency?

7.3.2* An infinitely long air-filled parallel-plate waveguide has a plate separation of 1 cm and a width of 10 cm. A straight infinitesimal wire is located 1/3 cm above the bottom conductor and parallel to it. It is perpendicular to the $\hat{z}$ direction (the direction of wave propagation) and carries I A at 50 GHz.

(a) Which TE_m and TM_m modes can this wire excite?

(b) Which of the excited TE_m modes propagate?

(c) What is the electric field for the lowest order excited TE mode? Compute its amplitude in terms of I.

7.3.3* A parallel-plate air-filled waveguide 1 m wide with 2-cm plate separation in the $\hat{x}$ direction is excited with a 10 mA current in the $\hat{y}$ direction at $z = 0$. The current is uniformly distributed over a ribbon 1 cm high centered in the waveguide and extending the entire width of the waveguide.

(a) Will this ribbon excite TE or TM or both types of modes?

(b) Which propagating modes will be excited?

(c) Evaluate $\overline{E}(\overline{r}, t)$ for $z > 0$ for the lowest-frequency excited mode.

7.3.4* Consider an air-filled parallel-plate waveguide with plate separation b ($-b/2 < x < b/2$). Neglect fringing fields. A surface current sheet extending halfway across the guide ($0 < x < b/2$) radiates symmetrically in the $\pm\hat{z}$ directions at frequency $f = 1.2c/b$. The current sheet is described by $\overline{J}_s = \hat{x} J_{s0}$ and is located at $z = 0$.

(a) Find the propagating waveguide modes that are excited by this current source and determine equations for $\overline{E}$ and $\overline{H}$ for these modes.

(b) Find the total complex power $\int_A (\overline{E} \times \overline{H}^*) \cdot \hat{n} \, da$ generated by the current source and transferred into these modes.

(c) Find the time-averaged power supplied by the current source. (Assume the TM_p mode is the highest order mode for which k_z is real.)

7.4.1 A rectangular waveguide is filled with a lossless dielectric having $\epsilon = 9\epsilon_0$. The waveguide dimensions are 1×3 mm.

(a) What are the cut-off frequencies for the TE_{10}, TE_{20}, TE_{11} and TM_{11} modes?

(b) If the dielectric has $\epsilon = 9\epsilon_0 (1 - j10^{-3})$, what is α for the TE_{10} mode at cut-off (where $\overline{E}$ is proportional to $e^{-\alpha z}$)?

(c) If ω is $0.9\omega_{10}$, where ω_{10} is the cut-off frequency for the TE_{10} mode, what is α in this case?

7.4.2 Design a rectangular air-filled waveguide to transmit maximum power signals in the full range 3-6 GHz (i.e., find the waveguide dimensions a and b where $a \geq b$). (Hint: We want only one mode to propagate in this 3-6 GHz band.)

7.4.3* The Callahan Tunnel in Boston can be modeled as a rectangular waveguide 6×12 m in cross-section. Assume the tunnel is empty except for one small car. If that car's radio receives a signal 30 dB above the noise level at the mouth of the tunnel and the reception is just passable at 10 dB, then how far into the tunnel can the car travel and still receive a useable signal at (i) 100 MHz (the FM band)? (ii) 1 MHz (the AM band)? Assume a power cable runs the length of the tunnel and has picked up a 1 MHz signal outside the tunnel. Might this change your answer to part (ii)? Why or why not?

7.4.4 (a) A rectangular air-filled waveguide is 2×4 cm. List all the modes that can propagate at a frequency of 15 GHz and give their corresponding cut-off frequencies.

(b) Thin slots may be cut in the walls of waveguides without significantly perturbing the propagating waves provided that the slots do not intersect wall currents, i.e., are always parallel to the surface currents. For example, a thin slot in the $\hat{z}$ direction can be placed in the center of the top wall of a waveguide carrying waves in the $\hat{z}$ direction without perturbing the TE_{10}, TE_{30}, and many other modes. For each of the following waveguide modes, define all positions and orientations where thin slots may be cut without significantly perturbing the propagating mode. Use sketches as appropriate: (i) TE_{10}, (ii) TE_{11}, and (iii) TM_{21}.

7.4.5 Consider a coaxial cable with inner radius a and outer radius b where $b = a + \delta$ is approximately equal to a (δ is very small). We know that this cable can support a *TEM* wave that has no cut-off frequency. However, we wish to determine at what frequency higher-order modes will start to appear and what are their cut-off frequencies. Rigorously, this problem requires solving the wave equation in cylindrical coordinates, which involves using Bessel and Neumann functions. However, with the assumption that $a \simeq b$, we can get very accurate answers by approximating the space between the two conductors as a rectangular waveguide of width $R = 2\pi a \simeq 2\pi b$ and spacing δ. Note that the fields at each end of this "wrapped-around" waveguide must be identical.

(a) Give the first three modes in increasing order of cut-off frequencies by using the mode designation convention for rectangular waveguides.

(b) Determine these cut-off frequencies.

7.4.6 Guided waves in an air-filled rectangular waveguide of cross-section $a \times b$ in the x–y plane can be excited with a current wire at $z = 0$ described by $\overline{J}_s = \hat{x}\, I_0 \delta(y - d)$ (A/m) where $0 < x < a$, $0 < y < b$, and $0 < d < b$.

(a) Explain why only TE_{0n} modes can be excited by this current source.

(b) Let the excitation angular frequency be such that $4\pi/a > \omega\sqrt{\mu_0\epsilon_0} > 3\pi/a$. How many waveguide modes are propagating? Assume $a = 2b$.

(c) The magnetic fields for the TE_{0n} modes are

$$H_z = \sum_n A_{0n} \cos\frac{n\pi y}{b}\, e^{-jk_{zn}z} \qquad \text{for } z > 0$$

$$H_z = \sum_n A_{0n} \cos\frac{n\pi y}{b}\, e^{+jk_{zn}z} \qquad \text{for } z < 0$$

where

$$A_{0n} = j\, \frac{I_0 k_{yn}}{b k_{zn}} \sin\left(\frac{n\pi d}{b}\right)$$

$$k_{yn} = \frac{n\pi}{b}$$

$$k_{zn} = \sqrt{\omega^2 \mu_0 \epsilon_0 - \left(\frac{n\pi}{b}\right)^2}$$

Find the corresponding electric fields for these modes.

(d) Determine the time-averaged power supplied by the current to the TE_{01} mode.

7.4.7 A metallic rectangular air-filled waveguide has dimensions 1.5×0.75 cm in the $\hat{x}$ and $\hat{y}$ directions, respectively (see Figure 7.12 in Section 7.4).

(a) Determine the cut-off frequencies for the first five modes.

(b) Write the electric and magnetic fields for the TM_{11} mode.

(c) Separately sketch the $\hat{z}$ and the $\hat{x}$-$\hat{y}$ components of the electric field distribution of the TM_{11} mode in the x–y plane at cut-off.

7.4.8 One way to construct an attenuator is to insert a poorly conducting sheet $\sigma \ll \omega\epsilon_0$ (S/m) parallel to the electric field in a waveguide at distance d from the side wall ($0 < d < b$). The conducting sheet thickness is $\delta \ll b$ and the waveguide has dimensions a, b where $a \geq b$.

(a) Use a perturbation approach and calculate the power dissipated per meter for the propagating TE_{10} mode.

(b) What is the approximate attenuation coefficient (dB/m) in this waveguide as a function of frequency? What happens near the cut-off frequency?

7.4.9* An AM radio in an automobile cannot receive any signal when the car is inside a tunnel. Consider, for example, the Lincoln Tunnel under the Hudson River, which was built in 1939. Model the tunnel as a rectangular metallic waveguide of dimensions 6.55×4.19 m.

(a) Give the range of frequencies for which only the dominant mode, TE_{10}, may propagate.

(b) Explain why AM broadcast-band signals cannot be received (recall the government frequency allocations).

(c) Can FM broadcast signals be received? Above what frequencies?

7.5.1* Find the phase and group velocities for the (i) TE_{13}, (ii) TM_{22}, and (iii) TE_{21} modes of a metallic cylindrical waveguide. Sketch the field patterns for each of these modes in the x–y and x–z planes.

8

RESONATORS

8.1 INTRODUCTION

In each of the preceding chapters, we have discussed electromagnetic wave propagation in systems that are unbounded in at least one dimension, such as unbounded media, interfaces, transmission lines, and waveguides. In this chapter, we consider structures that are constrained in **all** dimensions such that electromagnetic energy cannot propagate but is instead trapped in one place, coupling only weakly to the external world. Such resonators exhibit strong frequency-dependent behavior within narrow frequency bands centered about discrete frequencies called *resonances*. Resonators can be used not only for energy storage, frequency measurement, and filtering, but also for impedance matching and the production of high-field strengths with limited input signals.

Resonators are easily made by closing the ends of any waveguide structure, such as a *TEM* line, confining electromagnetic radiation in all three dimensions. We recall that the parallel-plate waveguide, confined in one dimension and possessing one guidance condition, produced a one-dimensional standing wave pattern in the plane transverse to the direction of wave propagation. Likewise, rectangular and circular waveguides, confined in two dimensions, gave rise to two guidance conditions and exhibited two-dimensional standing-wave patterns in the plane perpendicular to wave motion. Therefore, we expect that a resonator will have three guidance conditions and will generate a standing-wave pattern in all three dimensions. Since each guidance condition constrains one component of the wave vector $\bar{k}$, the three resonator guidance conditions together limit the magnitude of $\bar{k}$. Since $\omega = |\bar{k}|/\sqrt{\mu\epsilon}$, it follows that ω is limited to discrete values called *resonant frequencies.* By contrast, waveguide modes may limit k_x and/or k_y but do not constrain k_z, where $\hat{z}$ is the direction of the waveguide axis. Therefore, k_z and ω may have any of a continuum of values in a waveguide.

Most common resonators are constructed by closing the ends of a rectangular or circular waveguide with conducting walls, forming a *cavity resonator*, or by reactively terminating the ends of a *TEM* transmission line, forming a *TEM resonator*. We shall discuss both of these structures in Sections 8.2 and 8.5, respectively. However, any structure that can trap oscillatory electromagnetic energy is a resonator that will possess an essentially infinite set of resonant frequencies.[1] Even a misshapen lump of dielectric in free space can be considered to be a resonator, although if its stored energy leaks away too rapidly, the sharp-frequency behavior of the system becomes "smeared" over a band of frequencies centered about each resonance.

Nonideal resonators lose energy gradually, either externally through coupling to the outside world, or internally by heat generation. After discussing ideal cavity resonators in Section 8.2, these slightly lossy resonators are considered in Section 8.3 together with methods for characterizing their nonideal resonance behavior using parameters such as the resonator quality factor Q. Each specific resonance can be closely modeled with a discrete element resonator circuit, treated in Section 8.4, while transmission line resonators, introduced in Section 8.5, provide a simple vehicle for understanding the transient behavior of resonators. Because exact analysis of nonideal resonators is difficult, Section 8.6 describes how realistic structures can often be approximated by perturbing their ideal counterparts, while Section 8.7 treats perturbations in a more general way. Sections 8.4 and 8.8 describe how resonators can be coupled to the external world in a wide variety of practical applications, many of which are described in Section 8.9. For example, resonators can be used as narrow-band filters that either pass or block energy within a bandwidth $\Delta\omega$ of a resonance, where $\Delta\omega$ is proportional to the rate at which the resonator loses energy at that frequency. Resonators can also function as transformers that couple energy into mismatched loads. In fact, some of the matching structures studied in Chapter 6 can be viewed as resonators with bandwidths varying inversely with the degree of mismatch.

8.2 CAVITY RESONATORS

We first discuss the *rectangular cavity resonator*, constructed from a waveguide of rectangular cross-section having a width a and height b ($a \geq b$) and shown in Figure 7.12. The waveguide is closed by two perfectly conducting plates located at

---

1. At very high frequencies, waves interact with media nonclassically; i.e., Maxwell's equations are not sufficient to describe wave behavior when quantum mechanical effects become important.

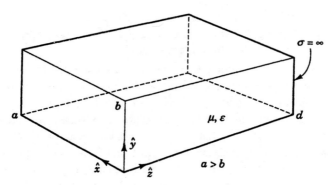

Figure 8.1 Rectangular waveguide resonator.

$z = 0$ and $z = d$ $(d \geq a)$, as illustrated in Figure 8.1, forming a rectangular parallelepiped. The convention that $d \geq a \geq b$ leads to unambiguous resonator mode designations, although care must be taken in reading the literature as not all texts use the same ordering of resonator side lengths. Since both *TE* and *TM* modes propagate in rectangular waveguides, *TE* and *TM* modes also exist in rectangular cavity resonators and are superpositions of positive and negative traveling *TE* or *TM* waveguide modes, respectively. Superimposed waves propagating in both directions are needed to satisfy the boundary conditions at the ends of the resonator:

$$E_x = E_y = 0 \qquad \text{at } z = 0, z = d \tag{8.2.1}$$

We may develop expressions for the *TE* electromagnetic fields within such a resonator by considering the electric field of a linear combination of $+\hat{z}$- and $-\hat{z}$-traveling TE_{mn} rectangular waveguide modes given by (7.4.13) and (7.4.14):

$$E_x = \frac{k_y}{k_0} E_0 \cos k_x x \sin k_y y \, (A \, e^{-jk_z z} + B \, e^{jk_z z})$$

$$E_y = -\frac{k_x}{k_0} E_0 \sin k_x x \cos k_y y \, (C \, e^{-jk_z z} + D \, e^{jk_z z}) \tag{8.2.2}$$

$$E_z = 0$$

where

$$k_x = m\pi/a \tag{8.2.3}$$

$$k_y = n\pi/b \tag{8.2.4}$$

$$k_z = \sqrt{\omega^2 \mu \epsilon - (m\pi/a)^2 - (n\pi/b)^2} \tag{8.2.5}$$

as was determined in (7.4.19).

By forcing the electric fields in (8.2.1) to equal zero at the $z = 0$ end of the waveguide, we find that $A = -B$ and $C = -D$, so both the E_x and E_y components of the TE_{mn} electric field are proportional to $\sin k_z z$. The additional constraint that the electric field in (8.2.1) is zero at the $z = d$ end of the waveguide forces $\sin k_z d = 0$, or

$$k_z d = p\pi \qquad p = 0, 1, 2\ldots \tag{8.2.6}$$

By requiring that $\nabla \cdot \overline{E} = 0$, we find that $A = C \equiv 1$ so that the arbitrary constant E_0 will have dimensions of V/m. Thus, the electric field of the TE_{mnp} mode is

$$E_x = \frac{k_y}{k_0} E_0 \cos k_x x \sin k_y y \sin k_z z$$

$$E_y = -\frac{k_x}{k_0} E_0 \sin k_x x \cos k_y y \sin k_z z \tag{8.2.7}$$

$$E_z = 0$$

The magnetic field of the TE_{mnp} mode is determined from Faraday's law (1.4.4)

$$H_x = \frac{j k_x k_z}{\eta k_0^2} E_0 \sin k_x x \cos k_y y \cos k_z z$$

$$H_y = \frac{j k_y k_z}{\eta k_0^2} E_0 \cos k_x x \sin k_y y \cos k_z z \tag{8.2.8}$$

$$H_z = \frac{j(k_x^2 + k_y^2)}{\eta k_0^2} E_0 \cos k_x x \cos k_y y \sin k_z z$$

where the three guidance conditions are

$$k_x = m\pi/a$$

$$k_y = n\pi/b \tag{8.2.9}$$

$$k_z = p\pi/d$$

from (8.2.3), (8.2.4) and (8.2.6). A quick check of these expressions for the magnetic field shows that they too satisfy the boundary condition that $H_z = 0$ at the ends of the rectangular cavity resonator.

The dispersion relation, obtained by substituting $\overline{E}$ or $\overline{H}$ into the wave equation (1.4.9) and using (8.2.9), is:

$$\omega^2 \mu \epsilon = k_x^2 + k_y^2 + k_z^2 = (m\pi/a)^2 + (n\pi/b)^2 + (p\pi/d)^2 \tag{8.2.10}$$

This key result means that a discrete, infinite set of TE resonances may be excited in the cavity. The resonant frequency of the TE_{mnp} mode is thus

$$\omega_{mnp} = \frac{1}{\sqrt{\mu\epsilon}} \sqrt{\left(\frac{m\pi}{a}\right)^2 + \left(\frac{n\pi}{b}\right)^2 + \left(\frac{p\pi}{d}\right)^2} \tag{8.2.11}$$

and is the **only** frequency for which that mode exists. If the dimensions of the resonator are such that two or more modes have exactly the same resonant frequency, then they are said to be *degenerate modes*. In our notation, the integers m and n are associated with the width and height of the waveguide (a and b), consistent with our previous conventions describing rectangular waveguide modes. The length d is associated with the integer p. Since it is not always clear which dimension of a cavity corresponds to its length, we choose $d \geq a \geq b$ as noted earlier.

A similar derivation yields the TM_{mnp} modes, where expressions for the magnetic field components can be determined from the TM_{mn} rectangular waveguide modes (7.4.16) and (7.4.17), with the z-dependence of H_x and H_y proportional to $\cos k_z z$ in order to match the boundary condition $E_y = E_x = 0$ at $z = 0$ and $z = d$, corresponding to

$$\frac{\partial H_x}{\partial z} = \frac{\partial H_y}{\partial z} = 0 \qquad \text{at } z = 0, z = d \qquad (8.2.12)$$

Both (8.2.1) and (8.2.12) are equivalent ways of stating that the tangential electric fields at the end walls are zero. The TM_{mnp} magnetic fields are thus

$$H_x = \frac{k_y}{k_0} H_0 \sin k_x x \cos k_y y \cos k_z z$$

$$H_y = -\frac{k_x}{k_0} H_0 \cos k_x x \sin k_y y \cos k_z z \qquad (8.2.13)$$

$$H_z = 0$$

and the TM_{mnp} electric field is then found from Ampere's law (1.4.5):

$$E_x = \frac{j\eta k_x k_z}{k_0^2} H_0 \cos k_x x \sin k_y y \sin k_z z$$

$$E_y = \frac{j\eta k_y k_z}{k_0^2} H_0 \sin k_x x \cos k_y y \sin k_z z \qquad (8.2.14)$$

$$E_x = \frac{j\eta (k_x^2 + k_y^2)}{k_0^2} H_0 \sin k_x x \sin k_y y \cos k_z z$$

Note that because the TM_{mnp} modes have the same guidance conditions (8.2.9) as the TE_{mnp} modes, they must also have the **same** resonant frequencies given by (8.2.11).

We find that the resonator field expressions (8.2.7), (8.2.8), (8.2.13) and (8.2.14) are zero for certain combinations of m, n, and p, meaning that these *trivial modes* can never be excited. For the rectangular cavity resonator, the trivial modes are TE_{00p}, TE_{mn0}, TM_{m0p}, and TM_{0np}. The nontrivial mode having the lowest useful resonant frequency is thus the TE_{101} mode, the fields for which are shown in Figure 8.2(a)–(b). The distribution of resonant frequencies for a typical

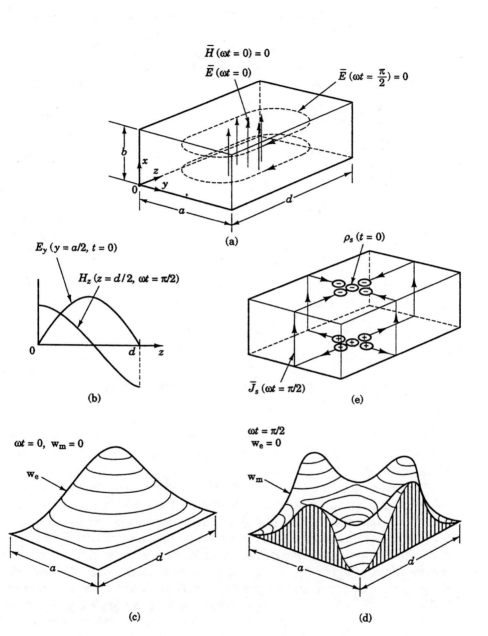

Figure 8.2   Fields, surface currents and charges, and energy densities in a rectangular waveguide resonator.

TABLE 8.1 Resonant Frequencies less than
30 GHz for a Rectangular Cavity
2 cm × 1 cm × 3 cm. Superscripts 1–8
represent Degenerate Modes

mnp	TE Modes (GHz)	TM Modes (GHz)
101	9.01	—
102	12.50	—
011	15.81[1]	—
201	15.81[1]	—
110	—	16.77[2]
103	16.77[2]	—
111	17.50[3]	17.50[3]
012	18.03[4]	—
202	18.03[4]	—
112	19.53[5]	19.53[5]
210	—	21.21[6]
013	21.21[6]	—
203	21.21[6]	—
211	21.79[7]	21.79[7]
301	23.05	—
212	23.45[8]	23.45[8]
302	24.62	—
310	—	27.04

cavity resonator 1 × 2 × 3 cm is given in Table 8.1. Note that many modes are degenerate (e.g., TE_{011} and TE_{201} have exactly the same frequency, as do TM_{110} and
TM_{103}, and so on.)

The expression for the TE electric fields in (8.2.7) is purely real (if E_0 is real),
implying that they reach maximum strength at $t = 0$ and vanish one-quarter cycle
later. The TE magnetic fields in (8.2.8) are purely imaginary and are therefore zero
at $t = 0$; one-quarter of a cycle later they reach maximum amplitude. Thus at $t =
0$, all of the energy in the cavity is stored in the form of electric energy concentrated
near the center of the cavity, as suggested by Figure 8.2(c). When the magnetic
energy peaks one-quarter cycle later, it is concentrated at the walls, illustrated
in Figure 8.2(d). Thus, during every cycle there are two complete interchanges
between purely electric and purely magnetic energy storage. When the energy is
purely electric, the boundary conditions are satisfied by surface charges distributed
appropriately at the top and bottom surfaces of the cavity. When the energy is
purely magnetic, the boundary conditions are matched by surface currents flowing
in the walls, reversing the surface charge distribution in preparation for the instant
one-quarter cycle later when the energy will again be stored electrically but the

fields will point in the opposite direction. These charge and current distributions are pictured in Figure 8.2(e).

The electromagnetic fields and surface currents for four of the rectangular cavity resonator modes are sketched in Figure 8.3. The fields for all other resonator modes are simply combinations of an appropriate number of "unit cells" of one of these modes (or TE_{011}, which is not shown because it is merely a rotated version of TE_{101}). For example, the TE_{232} mode consists of twelve TE_{111} unit cells stacked in a $2 \times 3 \times 2$ array. The appropriate directions of the fields in adjacent cells may be established by noting that the magnetic field lines must circulate without terminating ($\nabla \cdot \overline{B} = 0$), and the electric field lines may terminate only on surface charges since $\nabla \cdot \overline{D} = \rho$ and ρ is zero inside the cavity. The boundary condition $\hat{n} \times \overline{H} = \overline{J}_s$ applies everywhere on the resonator surface, relating the direction of the surface current $\overline{J}_s$ and the adjacent magnetic field. With these observations, the field patterns are easily determined for any rectangular resonator mode.

The electromagnetic field distributions pictured in Figure 8.3 are not all truly distinct: TE_{101}, TE_{011}, and TM_{110} are identical except for a rotation of axes. The TE_{111} and TM_{111} are related by duality. In particular, the TE_{111} fields of (8.2.7) and (8.2.8) are transformed into the TM_{111} fields of (8.2.13) and (8.2.14) and vice versa, simply by making the substitutions $\overline{E} \to \overline{H}$, $\overline{H} \to -\overline{E}$, $\mu \leftrightarrow \epsilon$, **and** $x \to x - a/2$, $y \to y - b/2$, $z \to z - d/2$. It is necessary to shift the coordinate systems because the resonator boundary conditions are not truly dual. The tangential electric fields and normal magnetic fields must be zero at the perfectly conducting surfaces of the resonator regardless of whether the fields are TE or TM. Thus only two truly distinct topologies are represented in Figure 8.3, those corresponding to the $\{101\}$ and $\{111\}$ sets of modes.

The electric and magnetic fields are $90°$ out of phase not only for the TE modes, as noted earlier, but also for the TM modes: $\overline{S} = \overline{E} \times \overline{H}^*$ is purely imaginary. Therefore, the time-averaged Poynting power $\langle \overline{S} \rangle = \frac{1}{2}\text{Re}\{\overline{S}\}$ is zero and there is no net power flow in the resonator; the waves are trapped in the cavity and cannot propagate.

If there is no loss in the cavity, the Poynting theorem can be used to prove the obvious fact that the total amount of energy in the cavity does not change with time. We recall the time-dependent Poynting theorem from (1.6.8):

$$-\int_V \overline{E}(t) \cdot \overline{J}(t)\, dv = \frac{d}{dt} \int_V \left(\frac{1}{2}\mu |\overline{H}(t)|^2 + \frac{1}{2}\epsilon |\overline{E}(t)|^2 \right) dv$$
$$+ \oint_A \left(\overline{E}(t) \times \overline{H}(t) \right) \cdot \hat{n}\, da$$

$$(8.2.15)$$

The electric field and current are never nonzero at the same location within the cavity, since $\overline{E}$ is zero in the perfectly conducting walls where $\overline{J}$ flows, and $\overline{J}$ is zero in the lossless dielectric where $\overline{E}$ is nonzero. Therefore, $\overline{E}(t) \cdot \overline{J}(t)$ is always

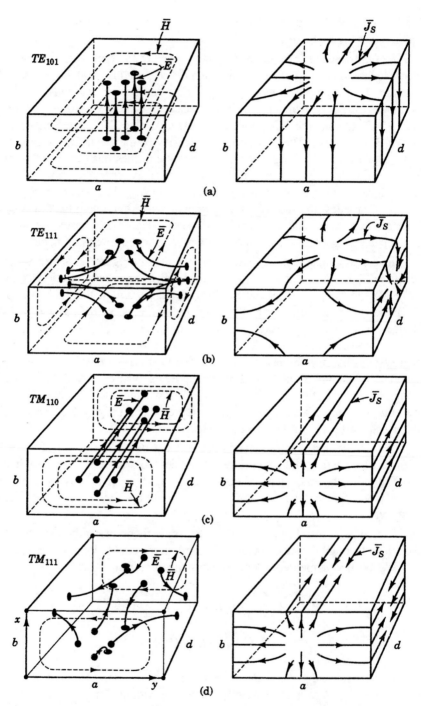

Figure 8.3   Four basic modes of a rectangular resonator.

zero for the lossless resonator cavity. The integral of the Poynting vector $\overline{E}(t) \times \overline{H}(t)$ over the resonator surface is also zero because $\overline{E}$ is everywhere perpendicular to the cavity walls so $\overline{E} \times \overline{H}$ is parallel to them. Therefore, (8.2.15) reduces to

$$0 = \frac{d}{dt} \int_V \left(\frac{1}{2} \mu |\overline{H}(t)|^2 + \frac{1}{2} \epsilon |\overline{E}(t)|^2 \right) dv = \frac{d}{dt} \left(\mathrm{w}_m(t) + \mathrm{w}_e(t) \right) \qquad (8.2.16)$$

and we see, as expected, that the total stored energy is constant even though it alternates between storage in electric and magnetic forms at intervals of one-quarter of a cycle, as suggested earlier in Figure 8.2 for the TE_{101} mode.

An alternative way to prove the equality of the time-averaged electric and magnetic stored energies uses the complex Poynting theorem, which we recall from (1.6.30) is:

$$-\frac{1}{2} \int_V (\overline{E} \cdot \overline{J}^*) \, dv = 2j\omega \left[\langle \mathrm{w}_m \rangle - \langle \mathrm{w}_e \rangle \right] + \frac{1}{2} \oint_A (\overline{E} \times \overline{H}^*) \cdot \hat{n} \, da \qquad (8.2.17)$$

The desired result follows directly because $\overline{E} \cdot \overline{J}^*$ and the surface integral of $\overline{E} \times \overline{H}^*$ are both zero for the same reasons noted earlier, leaving

$$\langle \mathrm{w}_e \rangle = \langle \mathrm{w}_m \rangle \qquad (8.2.18)$$

Finally, we can prove the equality in (8.2.18) by direct evaluation of these time-averaged stored energies, using

$$\langle \mathrm{w}_e \rangle = \frac{1}{4} \int_V \epsilon |\overline{E}|^2 \, dv \qquad (8.2.19)$$

$$\langle \mathrm{w}_m \rangle = \frac{1}{4} \int_V \mu |\overline{H}|^2 \, dv \qquad (8.2.20)$$

For a general TE_{mnp} mode where m, n, and p are all nonzero integers,

$$\begin{aligned}
\langle \mathrm{w}_e \rangle &= \frac{1}{4} \int_V \epsilon \left(|E_x|^2 + |E_y|^2 + |E_z|^2 \right) dv \\
&= \frac{\epsilon}{4} \int_0^d \int_0^b \int_0^a \left(\frac{k_y^2}{k_0^2} |E_0|^2 \cos^2 k_x x \sin^2 k_y y \sin^2 k_z z \right. \\
&\quad \left. + \frac{k_x^2}{k_0^2} |E_0|^2 \sin^2 k_x x \cos^2 k_y y \sin^2 k_z z \right) dx \, dy \, dz
\end{aligned} \qquad (8.2.21)$$

Because each squared sinusoid in (8.2.21) is integrated over an integral number of half periods and has average value equal to 0.5, the integral along any resonator dimension is just one-half its length.[2] Therefore, (8.2.21) becomes

$$\langle w_e \rangle = \frac{\epsilon}{4} \frac{(k_x^2 + k_y^2)}{k_0^2} |E_0|^2 \frac{abd}{8} \tag{8.2.22}$$

for the TE_{mnp} modes (m, n, $p > 0$). The time-averaged stored magnetic energy for the same mode is similarly computed to be

$$\langle w_m \rangle = \frac{1}{4} \int_V \mu (|H_x|^2 + |H_y|^2 + |H_z|^2) \, dv$$

$$= \frac{\mu}{4} \left| \frac{E_0}{\omega \mu k_0} \right|^2 (k_x^2 k_z^2 + k_y^2 k_z^2 + (k_x^2 + k_y^2)^2) \frac{abd}{8}$$

$$= \frac{\mu}{4} \left| \frac{E_0}{\omega \mu k_0} \right|^2 (k_x^2 + k_y^2)(k_x^2 + k_y^2 + k_z^2) \frac{abd}{8}$$

$$= \frac{\epsilon}{4} |E_0|^2 \frac{(k_x^2 + k_y^2)}{k_0^2} \frac{abd}{8} \tag{8.2.23}$$

where the dispersion relation $\omega^2 \mu \epsilon = k_x^2 + k_y^2 + k_z^2$ has been used to simplify (8.2.23). As expected, the time-averaged electric and magnetic energies are equal for such TE modes, as may be observed by comparing (8.2.22) and (8.2.23). Expressions for the time-averaged electric and magnetic energy stored in the TE_{m0p} and TE_{0np} modes are found in a similar fashion

$$\langle w_e \rangle = \langle w_m \rangle = \frac{\epsilon}{4} |E_0|^2 \frac{k_x^2}{k_0^2} \frac{abd}{4} \qquad (TE_{m0p} \text{ mode})$$

$$\langle w_e \rangle = \langle w_m \rangle = \frac{\epsilon}{4} |E_0|^2 \frac{k_y^2}{k_0^2} \frac{abd}{4} \qquad (TE_{0np} \text{ mode}) \tag{8.2.24}$$

as are the stored energies of the TM modes.

The stored energy distribution for the TE_{101} mode can also be related to a simple discrete element LC resonator, from which it evolves morphologically as suggested in Figure 8.4. The magnetic energy associated with the inductor L is equivalent to the excess magnetic energy near the current-carrying walls, while the electric energy associated with the capacitor C is equivalent to the excess electric energy in the high electric field zone in the center of the resonator (see

---

2. For example, the identity $\cos^2 \theta = \frac{1}{2} + \frac{1}{2} \cos 2\theta$ may be used to show that

$$\int_0^a \cos^2 \left(\frac{m\pi x}{a} \right) dx = \int_0^a \left[\frac{1}{2} + \frac{1}{2} \cos \left(\frac{2m\pi x}{a} \right) \right] dx = \begin{cases} a/2, & m \neq 0 \\ a, & m = 0 \end{cases}$$

and

$$\int_0^a \sin^2 \left(\frac{m\pi x}{a} \right) dx = \int_0^a \left[1 - \cos^2 \left(\frac{m\pi x}{a} \right) \right] dx = \begin{cases} a/2, & m \neq 0 \\ 0, & m = 0 \end{cases}$$

with similar results for the y and z integrations.

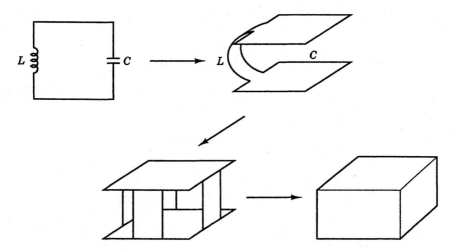

Figure 8.4   Evolution of a cavity resonator from an LC resonator.

also Figure 8.2). By spreading the current over a large wall area, the resistive losses associated with typical inductor coils are greatly reduced, and by closing the energy cavity, radiation losses are prevented. Thus cavity resonators are desirable in very high frequency applications (above $\simeq 1$ GHz), where skin-depth effects can introduce substantial series resistance in typical inductors. Unlike a discrete LC resonator with a fixed resonance at $\omega = 1/\sqrt{LC}$, a cavity has a large number of allowed resonant frequencies given by (8.2.11).

Sometimes it is important to know how densely the resonant modes of a cavity resonator are spaced, or how many modes exist within some frequency band or below some maximum frequency. From (8.2.11) we find that the resonant frequency f_{mnp} is

$$f_{mnp} = \frac{\omega_{mnp}}{2\pi} = \sqrt{\left(\frac{mc}{2a}\right)^2 + \left(\frac{nc}{2b}\right)^2 + \left(\frac{pc}{2d}\right)^2}$$

$$= \sqrt{f_x^2 + f_y^2 + f_z^2} \qquad (8.2.25)$$

for cavities enclosing vacuum, where $f_x = mc/2a$, $f_y = nc/2b$, and $f_z = pc/2d$. Equation (8.2.25) suggests a simple geometric relationship between the resonator frequency and the mode numbers m, n, and p, as illustrated in Figure 8.5. The resonator frequency is simply the square root of the sum of the squares of its three components, each associated with its own modal number. For each triplet of mode numbers $\{mnp\}$, there exists both a TE and a TM resonator mode with identical frequencies given by (8.2.25). To determine how many resonant frequencies exist below some maximum frequency f_0, we need only count the number of grid points within a spherical segment of radius f_0. Only one-eighth of the sphere is

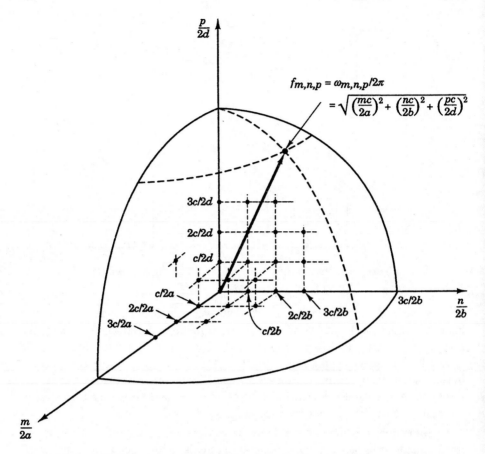

Figure 8.5 Frequency dependence of mode density for a rectangular resonator.

relevant since f_x, f_y, and f_z must all be positive. If the volume of the unit cell $\Delta f_x \Delta f_y \Delta f_z = c^3/8abd$ associated with each grid point is much smaller than the volume of the eighth sphere $(1/8)(4\pi f_0^3/3)$, then the number of resonator modes with frequencies less than f_0 is approximately

$$n_0 \simeq 2 \times \frac{\pi f_0^3/6}{c^3/8abd} = \frac{8\pi V}{3\lambda_0^3} \qquad (8.2.26)$$

where the factor of two accounts for the two modes (*TE* and *TM*) that exist for each grid point. The volume of the resonator is $V = abd$ and $\lambda_0 = c/f_0$. Equation (8.2.26) is a good approximation when λ_0^3 is much less than V. The number of resonances δn between f_0 and $f_0 + \delta f$ can be found by differentiating (8.2.26), yielding

$$\delta n \simeq \left(\frac{8\pi V}{c\lambda_0^2}\right) \delta f \qquad (8.2.27)$$

Thus, the number of resonances within a bandwidth δf about f_0 is proportional to $1/\lambda_0^2$ and to f_0^2.

8.3 LOSS IN RESONATORS

As the energy in a resonator oscillates back and forth between its electric and magnetic forms, some of this energy typically heats the resonator or escapes to the outside environment. If the total electromagnetic energy of the nth mode w_{Tn} decays over time, then the resonator is said to be lossy, and w_{Tn} has the approximate form[3]

$$w_{Tn}(t) = w_{Tn}(t = 0)\, e^{-\omega_n t/Q_n} \qquad (8.3.1)$$

where ω_n is the resonant frequency of the nth mode, and the *quality factor* Q_n is a dimensionless constant measuring how well the resonator stores energy. This quality factor is defined by (8.3.1) as the average number of radians over which the stored energy w_{Tn} decays by a factor of $1/e$. This relationship is also suggested in Figure 8.6. Note that we are considering only the total energy loss with **time** here, not attenuation in space, irrelevant in resonators in which no time-averaged power flows.

The quality factor Q_n typically differs for each mode, with larger Q-values associated with higher-quality resonances and smaller losses; Q becomes infinite in the absence of loss. (Ideal, lossless, rectangular resonators were described in Section 8.2, where it was shown that the total stored energy did not change in time.)

The quality factor Q_n may also be obtained by computing P_n, the power lost by the nth mode, averaged over one cycle. From (8.3.1), we see that

$$P_n \simeq -\frac{dw_{Tn}}{dt} = \frac{\omega_n w_{Tn}}{Q_n} \qquad (8.3.2)$$

Rearranging (8.3.2) gives

$$Q_n = \frac{w_{Tn}}{P_n/\omega_n} = \frac{\text{total energy of the } n^{th} \text{ mode}}{\text{average energy lost per radian of the } n^{th} \text{ mode}} \qquad (8.3.3)$$

Although the average dissipated power decreases exponentially with time, $P_n(t)$ may have a sinusoidal component peaking twice per cycle. If the loss is dominated by current flowing in the resonator walls, the dissipated power will peak when the magnetic energy is maximum and will be zero when the wall currents become zero. Alternatively, if the loss is associated exclusively with dielectric conduction within the cavity, it will peak twice per cycle when the electric energy storage is a maximum, exactly out of phase with any loss associated with wall

---

3. In this and subsequent sections, the single index n denotes a specific resonator mode. In the case of a rectangular cavity resonator, the $n = 1$ mode might actually represent the TE_{101} mode while the TE_{102} mode might be designated by $n = 2$.

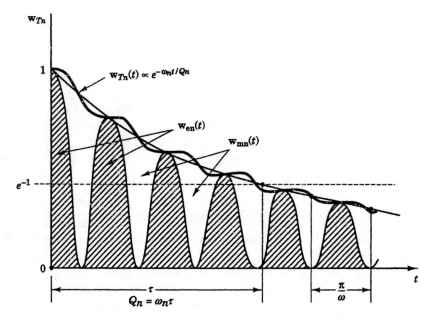

Figure 8.6   Energy of a lossy resonator as a function of time.

currents. This power ripple effect is illustrated in Figure 8.6 for the case where wall losses dominate.[4]

Up until now, we have tacitly assumed that ω_n is a real frequency. In the presence of loss, however, we will have to represent resonances by complex frequencies, commonly denoted by s_n, where

$$s_n = j\omega_n - \alpha_n \qquad (8.3.4)$$

and both ω_n and α_n are now explicitly real quantities. In the absence of loss, $\alpha_n = 0$ so that $e^{s_n t} = e^{j\omega_n t}$ as usual. The time dependence of the electric and

---

4. The power ripple effect is caused by a slight additional phase shift between electric and magnetic energy storage. To first order, we can represent the energies as

$$w_e(t) \simeq w_{Tn}(t = 0)\, e^{-\omega_n t/Q_n}\, \cos^2(\omega_n t + \phi_n)$$

$$w_m(t) \simeq w_{Tn}(t = 0)\, e^{-\omega_n t/Q_n}\, \sin^2(\omega_n t + \phi_n + \delta_n)$$

where $\delta_n \ll 1$. The total energy of the lossy resonator is thus

$$w_{Tn}(t) \simeq w_{Tn}(t = 0)\, e^{-\omega_n t/Q_n}\, [1 + \delta_n \sin 2(\omega_n t + \phi_n)]$$

where terms of order δ_n^2 have been neglected. The ripples can now be seen to vary about the exponentially decaying envelope at twice the resonant frequency of the mode.

magnetic field is then

$$\overline{E}_n(\overline{r}, t) = \mathrm{Re}\left\{\overline{E}_n(\overline{r})\, e^{s_n t}\right\} = e^{-\alpha_n t}\, \mathrm{Re}\left\{\overline{E}_n(\overline{r})\, e^{j\omega_n t}\right\}$$

$$\overline{H}_n(\overline{r}, t) = \mathrm{Re}\left\{\overline{H}_n(\overline{r})\, e^{s_n t}\right\} = e^{-\alpha_n t}\, \mathrm{Re}\left\{\overline{H}_n(\overline{r})\, e^{j\omega_n t}\right\} \tag{8.3.5}$$

Because the average electric and magnetic field energies are proportional to the field intensities $|\overline{E}_n(\overline{r}, t)|^2$ and $|\overline{H}_n(\overline{r}, t)|^2$, respectively, we see from (8.3.5) that both are proportional to $e^{-2\alpha_n t}$ and that the total resonator energy is just

$$w_{Tn}(t) = w_{Tn}(t = 0)\, e^{-2\alpha_n t} \tag{8.3.6}$$

By comparing (8.3.1) and (8.3.6), we observe an important relationship between α_n and Q_n:

$$Q_n = \frac{\omega_n}{2\alpha_n} \tag{8.3.7}$$

When more than one mode of a resonator is excited, the average decay rate of each mode is usually independent of the amplitudes of the other modes. This independence requires that the interacting modes be orthogonal in space or time; usually the frequencies are simply different. In unusual cases, two degenerate modes with the same frequency can superimpose coherently at a resistive element and thus interact. If two resonant frequencies are close but not exactly the same, they may interact at the beat frequency (the difference between the two resonances). The likelihood that two modes might interact is increased as the loss is concentrated at one point in the resonator instead of being distributed throughout.

To determine the Q of a particular resonance, we need simply to evaluate (8.3.3). Unfortunately, exact computation is difficult because power dissipation depends explicitly on the electric and magnetic fields, but the field distributions also depend on how much power is dissipated. Although exact solutions can be found in very special cases, a more general approach to determining electromagnetic losses involves using the perturbation methods discussed in Section 8.7. We have already used perturbation techniques informally in Sections 4.2, 4.4 and 7.4 where we calculated power dissipation at a lossy planar interface and in lossy waveguides, and we shall continue with this informal approach for the remainder of this section.

The essence of all *perturbation methods* is to find a quantity that differs from its lossless value by only a small percentage, and then to use the lossless (unperturbed) value of that quantity in all subsequent calculations, with often quite acceptable accuracy. Because the **percentage** change of the quantity must be small, its unperturbed value cannot be zero.

To illustrate these perturbation methods, we first calculate the quality factor for a rectangular resonator having lossless (perfectly conducting) walls but filled with a slightly lossy dielectric with conductivity σ_i, where the loss tangent $\sigma_i/\omega\epsilon$ is much less than unity. The exact fields in this resonator will be represented by the phasors $\overline{E}_n$, $\overline{H}_n$ and so on, while the fields that would arise in an identically shaped, perfectly insulating resonator are $\overline{E}_n^{(0)}$, $\overline{H}_n^{(0)}$, given by (8.2.7) and (8.2.8) for the TE_{mnp} modes and (8.2.13) and (8.2.14) for the TM_{mnp} modes. If σ_i is sufficiently small,[5] we expect that the actual and lossless fields can be used somewhat interchangeably. The total energy stored in the nth mode is then

$$w_{Tn} = 2\langle w_e \rangle = \frac{1}{2} \int_V \epsilon |\overline{E}_n|^2 \, dv \simeq \frac{1}{2} \int_V \epsilon |\overline{E}_n^{(0)}|^2 \, dv \tag{8.3.8}$$

where V is the resonator volume and $\overline{E}_n$ has been replaced by $\overline{E}_n^{(0)}$. The average power dissipated in one cycle in the nth mode is

$$P_n = \frac{1}{2} \int_V \overline{E}_n \cdot \overline{J}_n^* \, dv \tag{8.3.9}$$

where we may choose to express $\overline{J}_n = \sigma_i \overline{E}_n \simeq \sigma_i \overline{E}_n^{(0)}$ or $\overline{E}_n = \overline{J}_n/\sigma_i \simeq \overline{J}_n^{(0)}/\sigma_i$. Because the current inside the cavity is zero if the cavity is perfectly insulating ($\sigma_i = 0$), $\overline{J}_n^{(0)}$ is zero and is thus an unacceptable quantity on which to use perturbation techniques; the change from $\overline{J}_n^{(0)} = 0$ to $\overline{J}_n \neq 0$ is an infinite percentage change! In contrast, $\overline{E}_n^{(0)}$ and $\overline{E}_n$ differ only by a small percentage. We therefore represent P_n by

$$P_n \simeq \frac{1}{2} \int_V \sigma_i |\overline{E}_n^{(0)}|^2 \, dv \tag{8.3.10}$$

which necessarily vanishes as $\sigma_i \to 0$; no power is dissipated in a lossless resonator. Equation (8.3.10) is combined with (8.3.8) and (8.3.3) to find Q_n:

$$Q_n \simeq \frac{\omega_n w_{Tn}}{P_n} \simeq \frac{\omega_n \frac{1}{2} \int_V \epsilon |\overline{E}_n^{(0)}|^2 \, dv}{\frac{1}{2} \int_V \sigma_i |\overline{E}_n^{(0)}|^2 \, dv} = \frac{\omega_n \epsilon}{\sigma_i} \tag{8.3.11}$$

This remarkably simple result does not require any knowledge of the resonator fields (except insofar as ω_n must be calculated) because both energy storage and dissipation are proportional to $|E_n|^2$. Therefore, the time it takes for the energy of the nth mode to decay by $1/e$ is the same for all modes, since $\tau_n = Q_n/\omega_n = \epsilon/\sigma_i$ is just the charge relaxation time of the lossy dielectric! Note that Q_n becomes infinite as σ_i vanishes, and that our earlier assumption that $\sigma_i/\omega\epsilon \ll 1$ automatically guarantees that Q_n will be a large number, as it must be if perturbation techniques are to succeed. Often, we calculate Q **assuming** perturbation methods to be valid. If Q is sufficiently large at the end of the calculation, we can be fairly sure that our original assumption was justified.

---

5. Note that if σ_i is large instead, the current $\overline{J}_n \simeq \sigma_i \overline{E}_n^{(0)}$ will be nonnegligible. This current will give rise to a magnetic field correction by Ampere's law, which will in turn lead to an electric field correction by Faraday's law and to an additional current by Ohms' law. This secondary current will in turn produce new $\overline{E}$ and $\overline{H}$ fields, and this process will continue indefinitely. Only when loss is small can these additional higher-order self-consistent field correction terms be neglected.

A more usual loss mechanism in a cavity resonator is power dissipation in the resonator walls, caused by an imperfectly conducting cavity closure. In order for perturbation techniques to be applicable to this case, we consider resonator walls, each with large but finite conductivity σ_w, where $\sigma_w/\omega\epsilon \gg 1$. Once again, the total unperturbed energy in the resonator is given by (8.3.8) and the average power dissipated in the nth mode is given by (8.3.9), where this time $\frac{1}{2}\int_V \overline{E}_n \cdot \overline{J}_n^* \, dv$ must be computed within the resonator walls.

We again must decide whether to approximate $\overline{E}_n \cdot \overline{J}_n^*$ as $\sigma_w|\overline{E}_n^{(0)}|^2$ or $|\overline{J}_n^{(0)}|^2/\sigma_w$. Because the electric field inside perfectly conducting resonator walls is zero, the electric field is not a "slightly" perturbed quantity even if $\overline{E}_n$ is small. On the other hand, $\overline{J}_n^{(0)}$ is not zero in the walls of a perfect conductor because the discontinuity in the tangential magnetic field at the cavity surface gives rise to a surface current, and $\overline{J}_n$ differs only slightly from $\overline{J}_n^{(0)}$ if the wall conductivity is large. We thus approximate the dissipated power as

$$P_n \simeq \frac{1}{2}\int_{V'} \frac{|\overline{J}_n^{(0)}|^2}{\sigma_w} \, dv \tag{8.3.12}$$

which also has the required property that $P_n \to 0$ as $\sigma_w \to \infty$. The volume V' is the volume of the conducting walls of the resonator in which the surface current flows, **not** the entire resonator volume V.

Because the resonator surface is an excellent conductor, most of the current flows within a skin depth $\delta = \sqrt{2/\omega_n\mu\sigma_w}$ of the surface. We can therefore make the usual approximation

$$\overline{J}_n^{(0)} \simeq \overline{J}_{sn}^{(0)}/\delta \tag{8.3.13}$$

which was discussed in Sections 4.2 and 4.4. For computational purposes, all of the current is assumed to flow uniformly within the thickness δ. The mathematical justification for this approximation formed the basis of Example 4.2.2.

Combining (8.3.12) and (8.3.13) yields

$$P_n \simeq \frac{1}{2\sigma_w}\int_0^\delta \int_{A'} \left|\frac{\overline{J}_{sn}^{(0)}}{\delta}\right|^2 dv$$
$$= \frac{1}{2\sigma_w\delta}\int_{A'} |\overline{J}_{sn}^{(0)}|^2 \, da \tag{8.3.14}$$

where the volume integral V' is separated into an area integral over the resonator surface A' and an integral over the skin depth δ. We can rewrite $\overline{J}_{sn}^{(0)}$ in terms of the unperturbed magnetic field $\overline{H}_n^{(0)}$ by using Ampere's boundary condition $\overline{J}_{sn}^{(0)} = \hat{n} \times \overline{H}_n^{(0)}$. Equation (8.3.14) finally becomes

$$P_n \simeq \frac{1}{2\sigma_w\delta}\int_{A'} |\hat{n} \times \overline{H}_n^{(0)}(\text{cavity surface})|^2 \, da \tag{8.3.15}$$

where $\hat{n}$ is the normal to the cavity surface A' pointing into the cavity.

We shall use (8.3.15) to find Q_n for a rectangular resonator with lossy walls ($\sigma_w \neq \infty$) and a lossless dielectric ($\sigma_i = 0$). For simplicity, only the TE_{101} mode will be considered. We start by finding $P_{TE_{101}}$ in each of the six cavity walls; by symmetry the power dissipated in each pair of opposite walls is the same. Therefore,

$$
\begin{aligned}
P_{TE_{101}} \simeq 2 \times \Bigg[& \frac{1}{2\sigma_w\delta} \int_0^d \int_0^b \left| \hat{x} \times \overline{H}_{TE_{101}}^{(0)}(x=0) \right|^2 dy\, dz \\
& + \frac{1}{2\sigma_w\delta} \int_0^d \int_0^a \left| \hat{y} \times \overline{H}_{TE_{101}}^{(0)}(y=0) \right|^2 dx\, dz \qquad (8.3.16) \\
& + \frac{1}{2\sigma_w\delta} \int_0^b \int_0^a \left| \hat{z} \times \overline{H}_{TE_{101}}^{(0)}(z=0) \right|^2 dx\, dy \Bigg]
\end{aligned}
$$

where

$$
\overline{H}_{TE_{101}} = \hat{x} \frac{j\pi^2 E_0}{ad\eta k_0^2} \sin\left(\frac{\pi x}{a}\right)\cos\left(\frac{\pi z}{d}\right) + \hat{z} \frac{j\pi^2 E_0}{a^2\eta k_0^2} \cos\left(\frac{\pi x}{a}\right)\sin\left(\frac{\pi z}{d}\right) \quad (8.3.17)
$$

from (8.2.8) and (8.2.9). Combining (8.3.16) and (8.3.17) yields

$$
\begin{aligned}
P_{TE_{101}} \simeq \frac{1}{\sigma_w\delta} \Bigg[& \int_0^d \int_0^b \left| \frac{\pi^2 E_0}{a^2\eta k_0^2} \right|^2 \sin^2\left(\frac{\pi z}{d}\right) dy\, dz \\
& + \int_0^d \int_0^a \left| \frac{\pi^2 E_0}{ad\eta k_0^2} \right|^2 \sin^2\left(\frac{\pi x}{a}\right)\cos^2\left(\frac{\pi z}{d}\right) dx\, dz \\
& + \int_0^d \int_0^a \left| \frac{\pi^2 E_0}{a^2\eta k_0^2} \right|^2 \cos^2\left(\frac{\pi x}{a}\right)\sin^2\left(\frac{\pi z}{d}\right) dx\, dz \\
& + \int_0^b \int_0^a \left| \frac{\pi^2 E_0}{ad\eta k_0^2} \right|^2 \sin^2\left(\frac{\pi x}{a}\right) dx\, dy \Bigg] \\
= \frac{1}{\sigma_w\delta} & \left| \frac{\pi^2 E_0}{2a^2 d\eta k_0^2} \right|^2 \left[ad(a^2 + d^2) + 2b(a^3 + d^3) \right] \qquad (8.3.18)
\end{aligned}
$$

From (8.2.24), we find that the total stored energy of the TE_{101} mode is

$$
w_{TE_{101}} \simeq \frac{\epsilon\pi^2 |E_0|^2 abd}{8k_0^2 a^2} \qquad (8.3.19)
$$

which means that (8.3.18) and (8.3.19) may be substituted into (8.3.3) to yield

$$
Q_{TE_{101}} \simeq \frac{\omega_{TE_{101}}\langle w_{TE_{101}}\rangle}{P_{TE_{101}}} \simeq \frac{abd}{\delta} \left[\frac{a^2 + d^2}{ad(a^2 + d^2) + 2b(a^3 + d^3)} \right] \qquad (8.3.20)
$$

where we have simplified (8.3.20) by using the dispersion relation, resonator $(\pi/a)^2 + (\pi/d)^2 = \omega_{TE_{101}}^2 \mu\epsilon$ for the TE_{101} mode.

Note that Q depends upon cavity geometry and is inversely proportional to the skin depth $\delta = \sqrt{2/\omega_n \mu \sigma_w}$. Therefore, $Q \propto \sqrt{\omega_n \sigma_w}$ becomes infinite as the walls of the resonator become lossless or as the frequency of the nth mode becomes infinite.

If the resonator has **both** lossy walls and a lossy dielectric, we can easily recompute the new Q for this system if we know the Q-factors for each type of loss separately. The resonant frequency ω_n and the total energy stored in the resonator w_{Tn} are the same for any type of small loss, and the total power dissipated is just the sum of the powers dissipated in the dielectric and walls. Therefore, if $Q_{ni} = \omega_n w_{Tn}/P_{ni}$ is the quality factor for a lossy dielectric resonator, and $Q_{nw} = \omega_n w_{Tn}/P_{nw}$ is the quality factor for a lossy wall resonator, then the total Q_n for a resonator with both lossy walls and dielectric is

$$Q_n \simeq \frac{\omega_n w_{Tn}}{P_{ni} + P_{nw}} = \left(\frac{1}{Q_{ni}} + \frac{1}{Q_{nw}}\right)^{-1} \qquad (8.3.21)$$

Note that the total Q_n is less than either Q_{ni} or Q_{nw} because two types of resonator loss mechanisms are present. In practice, losses in most cavity resonators result primarily from imperfectly conducting cavity walls.

Example 8.3.1

A certain microwave oven manufacturer wants to sell the smallest oven possible that will be able to cook a 10-cm hamburger oriented in any direction. The operating frequency of the oven is 1200 MHz.[6] What oven cavity dimensions should he use such that at least one resonance falls at or below 1200 MHz?

Solution: The resonant frequencies of the rectangular oven are

$$f_{mnp} = \frac{c}{2}\sqrt{\left(\frac{m}{a}\right)^2 + \left(\frac{n}{b}\right)^2 + \left(\frac{p}{d}\right)^2}$$

where $d \geq a \geq b$. It is clear that the smaller the values of a, b, and d we choose, the larger the minimum resonant frequency will be, so

$$f_{min} = f_{TE_{101}} = \frac{c}{2}\sqrt{\left(\frac{1}{a}\right)^2 + \left(\frac{1}{d}\right)^2} = 1200 \text{ MHz} \qquad (1)$$

We also have the constraint that the area $A = ad$ is to be as small as possible, so we can rewrite (1) as a function of the variables A and a and solve for A. Minimizing A by setting $\partial A/\partial a = 0$ yields the intuitive result that the area (and hence the volume) of the resonator is smallest when $a = d$. Substituting $a = d$ into (1) and solving for a yields

$$a = d = 17.7 \text{ cm}$$

and therefore $b = 10.0$ cm to comply with the requirement that the minimum dimension be

---

6. Actual microwave ovens operate at the government-allowed 2450 MHz, but this is a nonstandard model.

at least 10 cm. The total volume of the oven is then

$$V = (17.7 \text{ cm})^2 \times 10.0 \text{ cm} = 3125 \text{ cm}^3$$

Note that only the TE_{101} mode can resonate in this oven; all of the higher-order modes are cut off. This is actually not a very good design for a microwave oven, because the fields are strong only at the center of the oven. If the oven can support a large number of modes having a variety of field distributions and the source frequency and/or oven geometry varies (e.g., from moving fan blades or a rotating carousel), then food will tend to be more evenly cooked.

Example 8.3.2

The microwave oven described in Example 8.3.1 has dimensions $17.7 \times 17.7 \times 10.0$ cm and is resonant in the TE_{101} mode at 1200 MHz. A hamburger of dimensions $8.0 \times 8.0 \times 1.0$ cm having complex permittivity $\epsilon = \epsilon_0(4 - j0.1)$ is placed in the center of the oven. The wall conductivity is 4×10^6 S/m. What is the resonator Q resulting from the hamburger alone? If the oven is empty and the 400 W transmitter power is dissipated in the oven walls, what is the peak electric field strength in the cavity? Will the oven arc?

Solution: In order to determine whether perturbation techniques may be used for this problem, we first compute the hamburger loss tangent $\sigma_b/\omega\epsilon$. Because $\epsilon - j\sigma_b/\omega = 4\epsilon_0 - j0.1\epsilon_0$, $\epsilon = 4\epsilon_0$ and $\sigma_b/\omega = 0.1\epsilon_0$ so $\sigma_b/\omega\epsilon = 0.025$, sufficiently less than unity so that the burger just slightly perturbs the cavity fields that exist when it is not present. Therefore, we can assume that the electric field in the hamburger is approximately uniform and nearly equal to the unperturbed electric field at the center of the oven. From (8.2.7), we find that this electric field in the burger is

$$\overline{E}_b^{(0)} \simeq \hat{y}\, \frac{\pi E_0}{k_0 a} \quad \text{(V/m)} \tag{1}$$

where (8.2.7) has been evaluated at $x = a/2$, $y = b/2$, and $z = d/2$. The power dissipated in the burger is then given by (8.3.10)

$$P_b \simeq \frac{1}{2} \int_V \sigma_b \left|\overline{E}_b^{(0)}\right|^2 dv \simeq \frac{\sigma_b}{2} \left(\frac{\pi}{k_0 a}\right)^2 |E_0|^2 V_b \quad \text{(W)} \tag{2}$$

where $V_b = 64$ cm^3 is the volume of the meat. The total stored electromagnetic energy in the oven is given by (8.2.24) for the TE_{101} mode:

$$w_T \simeq 2\langle w_e \rangle = \frac{\epsilon}{8} \left(\frac{\pi}{k_0 a}\right)^2 |E_0|^2 abd \quad \text{(J)} \tag{3}$$

Combining (2) and (3) yields a value for Q:

$$Q_b \simeq \frac{\omega w_T}{P_b} = \frac{\omega\epsilon}{4\sigma_b} \left(\frac{abd}{V_b}\right) = \frac{1}{4 \times 0.025} \left(\frac{3125}{64}\right) \cong 488$$

If the microwave power is instead completely dissipated in the oven walls, we can use (8.3.18) directly to solve for the field at the center of the oven, $|\overline{E}_b| \simeq \pi |E_0|/k_0 a$ from (1). Because only the TE_{101} mode is excited, the electric field at the center of the oven is also the maximum field. We rewrite (8.3.18) as

$$P = \frac{1}{\sigma_w \delta} \left| \frac{\pi^2 E_b}{\eta a k_0^2} \right|^2 \left(1 + \frac{2b}{a} \right) \quad \text{(W)}$$

where we have substituted $d = a$, and thus

$$|\overline{E}_b| = \left(\frac{\omega \mu a}{\pi} \right)^2 \sqrt{\frac{2P\sigma_w \delta}{1 + 2b/a}} \quad \text{(V/m)} \tag{4}$$

Since $\omega = 2\pi \times 1200 \times 10^6 = 7.54 \times 10^9$ rad/s and $\delta = \sqrt{2/\omega\mu\sigma_w} = 7.26 \times 10^{-6}$ m, we can substitute these parameters into (4) to yield

$$|E_b| = 2.98 \times 10^5 \text{ V/m}$$

which is an insufficiently large field to cause arcing except near sharp metallic objects (which should never be placed in a microwave oven). Arcing occurs at field strengths of approximately 3×10^6 V/m, depending on humidity and other factors. If the walls were perfectly conducting and 400 W could be continuously supplied, the electric field intensity at the center of the empty oven would in time become infinitely large. In reality, the oven would arc first, dissipating power by other means.

8.4 DISCRETE-ELEMENT RESONATOR CIRCUITS

Cavity resonators exhibit an infinite number of resonances, making it difficult to model their global behavior in a simple way. The same problem exists for **all** distributed electromagnetic structures where the electric and magnetic energy is not stored exclusively in discrete elements such as inductors or capacitors, nor is power necessarily dissipated in discrete resistive elements. Fortunately, we are generally interested in the frequency response of such devices near specific, isolated resonances, for which equivalent discrete-element circuits can be found. These equivalent circuit models are attractive for several reasons—they not only enable us to view resonators with all the intuition developed from circuit theory, but they are also often easier to use when discussing the coupling of resonators to the external world. However, we should be aware that the equivalence between discrete RLC circuit resonators and cavity resonators is approximate and applies **only near the specific resonant frequency of the discrete-element resonator**. Each of the infinite number of cavity resonances must therefore be modeled by a separate equivalent resonator circuit. Because modeling resonators near resonance as simple RLC circuits is such a powerful technique, we devote this section to reviewing the behavior of these discrete circuits.

Consider the series RLC circuit connected to a voltage source, illustrated in Figure 8.7. Because we care only about the behavior of this circuit near resonance, we focus on the frequency domain, assuming that all of the signals in the circuit have time dependence e^{st} where s is the complex frequency

$$s = j\omega - \alpha, \qquad \alpha \geq 0 \tag{8.4.1}$$

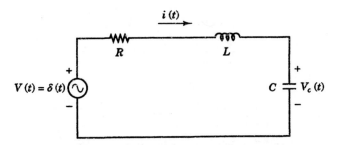

Figure 8.7   Series *RLC* discrete-element resonator.

The current $I(s)$ flowing through each element in the resonator is just the Laplace transform of the time-dependent current $i(t) = \text{Re}\{I(s) e^{st}\}$ and the voltage drops across the resistor, inductor, and capacitor are $V_R(s)$, $V_L(s)$, and $V_C(s)$, respectively. Each frequency-dependent voltage $V(s)$ in the circuit is also related to a corresponding time-dependent voltage $v(t)$ by the relation $v(t) = \text{Re}\{V(s) e^{st}\}$. The complex quantities $V(s)$ and $I(s)$ are exactly analogous to the phasors $\overline{E}$ and $\overline{H}$ defined by (1.4.2) and (1.4.3). The only difference is that here we are not limited to real frequencies.

The voltage/current relations for these three elements are

$$
\begin{array}{llll}
v_R(t) & = & R\,i(t) & V_R(s) & = & R\,I(s) \\
v_L(t) & = & L\,di(t)/dt & V_L(s) & = & sL\,I(s) \\
v_C(t) & = & \left(\int i(t)\,dt\right)/C & V_C(s) & = & I(s)/sC
\end{array}
\qquad (8.4.2)
$$

where the first expression in each pair is the time-domain representation and the second is the frequency-domain relation between voltage and current. Kirchhoff's voltage law then states that the source voltage $V_S(s)$ is the sum of the voltage drops across each of the series elements, so

$$
\begin{aligned}
V_S(s) &= \left(R + sL + \frac{1}{sC}\right) I(s) \\
&= \left(\frac{s^2 + sR/L + 1/LC}{s/L}\right) I(s)
\end{aligned}
\qquad (8.4.3)
$$

Equation (8.4.3) may be factored using the quadratic formula, resulting in

$$
V_S(s) = \frac{(s - s_1)(s - s_2)}{s/L} I(s)
\qquad (8.4.4)
$$

where

$$
\begin{aligned}
s_{1,2} &= -\frac{R}{2L} \pm j\sqrt{\frac{1}{LC} - \frac{R^2}{2L}} \\
&= -\alpha_0 \pm j\sqrt{\omega_0^2 - \alpha_0^2}
\end{aligned}
\qquad (8.4.5)
$$

and we have suggestively defined the new variables

$$\omega_0 \equiv 1/\sqrt{LC} \tag{8.4.6}$$

$$\alpha_0 \equiv R/2L \tag{8.4.7}$$

For this RLC circuit to exhibit strong resonant behavior, the amount of power dissipated in the circuit must be small. Since α_0 is proportional to R, α_0 must also be small and therefore $\alpha_0 \ll \omega_0$, which means that the resonant frequency $\sqrt{\omega_0^2 - \alpha_0^2}$ can be approximated by ω_0. We immediately see that the natural, complex frequency of this circuit is approximately $s \cong -\alpha_0 + j\omega_0$, which means that the resonant frequency in this circuit is near ω_0 and the rate at which power is dissipated is proportional to α_0. All voltages and currents in this circuit have time-dependencies given approximately by $e^{-\alpha_0 t} e^{j\omega_0 t}$ so the stored energies are proportional to $|e^{-\alpha_0 t} e^{j\omega_0 t}|^2 = e^{-2\alpha_0 t}$. If we use the definition of Q in (8.3.3) and the definition of ω_0 and α_0 in (8.4.6) and (8.4.7), the quality factor for this resonance is

$$Q = \frac{\omega_0}{2\alpha_0} = \frac{\omega_0 L}{R} = \frac{1}{R}\sqrt{\frac{L}{C}} \tag{8.4.8}$$

where Q becomes infinite as the lossy element R vanishes; (8.4.8) is exactly the same as (8.3.7). We may also find Q directly from its definition in (8.3.3)

$$Q = \frac{\omega_0 w_T}{P} = \frac{\omega_0 \, 2\langle w_m(s)\rangle}{P(s)} = \frac{\omega_0 \, 2 \cdot \frac{1}{4}L|I(s)|^2}{\frac{1}{2}R|I(s)|^2} = \frac{\omega_0 L}{R} \tag{8.4.9}$$

where the power dissipated in the resistor is just $P(s) = \frac{1}{2}R|I(s)|^2$ and the total energy stored in this circuit is twice the magnetic energy $\langle w_m(s)\rangle = \frac{1}{4}L|I(s)|^2$.

We should now like to find the power dissipation in this resonator as a function of frequency. Using the definition of $P(s)$ and the relations between current and source voltage given by (8.4.3)–(8.4.4), we find

$$P(s) = \frac{1}{2}R|I(s)|^2 = \frac{1}{2}R|V_S(s)|^2 \left|\frac{s/L}{(s-s_1)(s-s_2)}\right|^2$$

$$= \frac{1}{2}R|V_S(s)|^2 \left|\frac{s/L}{(s+\alpha_0 - j\omega_0)(s - \alpha_0 + j\omega_0)}\right|^2 \tag{8.4.10}$$

which is written in terms of the source voltage $V_S(s)$ with the approximation that $\sqrt{\omega_0^2 - \alpha_0^2} \cong \omega_0$. Both the dissipated power and the complex admittance $I(s)/V_S(s)$ approach infinity whenever the complex frequency s approaches either of the *frequency-plane poles* at s_1 or s_2. The location of these admittance poles

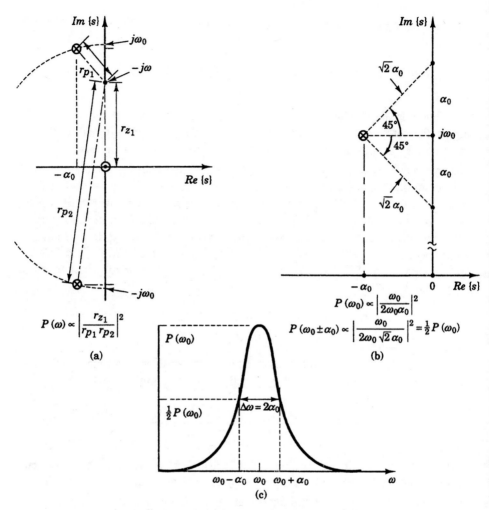

Figure 8.8   *RLC* resonator in the complex plane.

in the *s-plane* or *complex frequency plane* is illustrated in Figure 8.8(a). In addition to the poles at s_1 and s_2, there is also a *frequency-plane zero* at $s = 0$ in both admittance and the power dissipation function (8.4.10).[7]

Because we must drive the series circuit with a real frequency ω (equivalent to a wholly imaginary complex frequency $s = j\omega$), we are confined to s values that lie along the imaginary axis in the complex frequency plane and therefore cannot get closer than the distance α_0 to either pole. The complex admittance of

---

7. Note that the admittance poles become impedance zeros and vice versa because impedance and admittance are reciprocal functions.

this circuit is never infinite, therefore, unless $\alpha_0 = 0$, corresponding to a perfect resonator with $R = 0$ and $Q \to \infty$.

The driving frequency which minimizes the distance to the pole at $s_1 = -\alpha_0 + j\sqrt{\omega_0^2 - \alpha_0^2}$ is $\omega = \sqrt{\omega_0^2 - \alpha_0^2} \simeq \omega_0$, which means that maximum power is dissipated in the resistor R at resonance. This power can be approximated by substituting $s = j\omega_0$ into (8.4.10), yielding

$$P(\omega_0) = \frac{R}{2}|V_S(\omega_0)|^2 \left| \frac{j\omega_0/L}{\alpha_0(2j\omega_0 + \alpha_0)} \right|^2$$

$$\approx \frac{R}{2}|V_S(\omega_0)|^2 \left| \frac{j\omega_0/L}{\alpha_0(2j\omega_0)} \right|^2$$

$$\approx \frac{|V_S(\omega_0)|^2}{2R} \tag{8.4.11}$$

where $\alpha_0 = R/2L$ has been used to simplify (8.4.11) and the factor $2j\omega_0 + \alpha_0$ has been approximated by $2j\omega_0$. The value of maximum power dissipation given by (8.4.11) is the power that would be dissipated if the reactances of the inductor and capacitor canceled each other exactly, leaving an input impedance $Z = R$. This occurs when $j\omega L + 1/j\omega C = 0$, or when $\omega = 1/\sqrt{LC} = \omega_0$, i.e., approximately at resonance.

If we move away from resonance to $\omega_0 \pm \alpha_0$, the distance to the $-\alpha_0 + j\omega_0$ pole goes from α_0 to $\sqrt{2}\alpha_0$ as seen graphically in Figure 8.9(b). Because this pole dominates (8.4.10), the dissipated power drops to one-half its peak power, shown by substituting $s = j(\omega_0 \pm \alpha_0)$ into (8.4.10):

$$P(\omega_0 \pm \alpha_0) = \frac{R}{2}|V_S(\omega_0 \pm \alpha_0)|^2 \times$$

$$\left| \frac{j(\omega_0 \pm \alpha_0)/L}{[j(\omega_0 \pm \alpha_0) + \alpha_0 - j\omega_0][j(\omega_0 \pm \alpha_0) + \alpha_0 + j\omega_0]} \right|^2$$

$$\approx \frac{R}{2}|V_S(\omega_0)|^2 \left| \frac{j\omega_0/L}{(\alpha_0 \pm j\alpha_0)(2j\omega_0)} \right|^2 \approx \frac{R}{4}|V_S(\omega_0)|^2 \approx \frac{1}{2}P(\omega_0) \tag{8.4.12}$$

The frequencies $\omega = \omega_0 \pm \alpha_0$ are therefore called the *half-power frequencies* of the resonance, and we define $\Delta\omega \cong 2\alpha_0$ as the *half-power bandwidth* or full width at half maximum (FWHM).

Because $\alpha_0 \ll \omega_0$, the circuit is a high-quality resonator and $P(\omega)$ is sharply peaked about ω_0, as suggested in Figure 8.8(c). As the poles approach the imaginary axis in the complex s-plane, $2\alpha_0$ becomes smaller and the resonance peak is narrowed. As $Q \to \infty$, the resonance becomes a delta function, and **only** the resonant frequency itself can be excited, with zero driving point impedance. The width of this resonance is closely related to its quality factor Q. From (8.4.8),

$$Q = \frac{\omega_0}{2\alpha_0} \cong \frac{\omega_0}{\Delta\omega} = \frac{f_0}{\Delta f} \tag{8.4.13}$$

The dual circuit to a series RLC resonator driven by a voltage source is a parallel GLC resonator driven by a current source, as illustrated in Figure 8.9. The current and voltage in this resonator are related by

$$\begin{aligned} I_S(s) &= \left(G + sC + \frac{1}{sL}\right) V(s) \\ &= \left(\frac{s^2 + sG/C + 1/LC}{s/C}\right) V(s) \end{aligned} \tag{8.4.14}$$

which is completely analogous to (8.4.3) if V and I are interchanged along with $G \leftrightarrow R$ and $L \leftrightarrow C$. The resonant frequency $\omega_0 = 1/\sqrt{LC}$ is not affected when L and C are interchanged, but the value of α_0 becomes

$$\alpha_0 = \frac{G}{2C} \tag{8.4.15}$$

for the parallel circuit; the quality factor is then

$$Q = \frac{\omega_0}{2\alpha_0} = \frac{\omega_0 C}{G} = \frac{1}{G}\sqrt{\frac{C}{L}} \tag{8.4.16}$$

This value of Q can also be found from its definition

$$Q = \frac{\omega_0 w_T}{P} = \frac{\omega_0 2\langle w_e(s)\rangle}{P(s)} = \frac{\omega_0 2 \cdot \frac{1}{4} C |V(s)|^2}{\frac{1}{2} G |V(s)|^2} = \frac{\omega_0 C}{G} \tag{8.4.17}$$

which is just the dual of (8.4.9). If we recall that the Q of a cavity resonator with lossy dielectric is $\omega_0 \epsilon / \sigma_i$ from (8.3.11), we find a plausible relation between C, G, ϵ, and σ_i:

$$\frac{\epsilon}{\sigma_i} = \frac{C}{G} \tag{8.4.18}$$

A similar relationship between cavity resonators and RLC circuits is not so readily

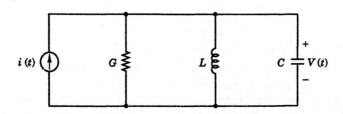

Figure 8.9 Parallel GLC resonator circuit.

found, since the specific mode and actual structure of the cavity are important in determining resistive losses.

We now describe the coupling of a series RLC or parallel GLC resonator to an external circuit, in preparation for Section 8.8 in which we shall discuss coupling by modeling the resonator by an equivalent circuit. Figure 8.10(a) depicts a discrete-element RLC resonator coupled to a Thevenin equivalent circuit modeled as a voltage source V_{Th} in series with an impedance Z_{Th}, representing **any** passive, external circuit.

Because there are now two elements in the combined circuit that dissipate power, we specify power lost in the internal circuit as P_I and power lost in the external circuit as P_E. The quality factors for the internal and external circuits are also specified explicitly as the *internal Q*, Q_I and *external Q*, Q_E

$$Q_I \equiv \frac{\omega w_T}{P_I} \tag{8.4.19}$$

$$Q_E \equiv \frac{\omega w_T}{P_E} \tag{8.4.20}$$

where ω is the resonant frequency and w_T is the total energy stored in the resonator. The total, or *loaded Q*, Q_L is the quality factor that results from both loss

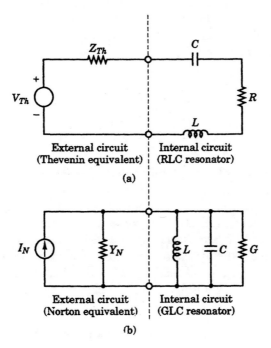

Figure 8.10 Coupling between circuit resonators and external sources.

mechanisms:

$$Q_L = \frac{\omega w_T}{P_I + P_E} = \left[\frac{1}{Q_I} + \frac{1}{Q_E} \right]^{-1} \qquad (8.4.21)$$

In the resonator circuits discussed to this point, Q_E was assumed to be infinite so that Q_L and Q_I could be used interchangeably.

If the maximum possible amount of power is dissipated in the resonator from the external circuit, then the resonator is said to be *critically coupled* or *matched* to the external circuit. It is easy to show that critical coupling occurs when the resistive parts of the internal and external circuits are equal and the reactive parts of the two circuits cancel. In this case, the same amount of power is dissipated in both circuits so $P_I = P_E$, $Q_I = Q_E$, and $Q_L = Q_I/2$ (see (6.3.12)).

We can find the value of Z_{Th} necessary to critically couple both halves of the circuit in Figure 8.10(a) at resonance by noting that the impedance $Z(\omega)$ looking into the RLC circuit is $Z(\omega) = R + j\omega L + 1/j\omega C$. At the resonant frequency $\omega_0 = 1/\sqrt{LC}$, the resonator impedance is $Z(\omega_0) = R$. The Thevenin impedance that matches this resonator at resonance is therefore the resistor $R_{Th} = R$.[8]

Overcoupling occurs when more power is dissipated in the external circuit than in the resonator so that $P_E > P_I$ and $Q_E < Q_I$. *Undercoupling* is the reverse situation where $P_E < P_I$ and $Q_E > Q_I$. For the circuit in Figure 8.10(a), an overcoupled circuit is one in which $R_{Th} > R$ and an undercoupled circuit implies $R_{Th} < R$.

We can therefore define Q_I, Q_E, and Q_L for this circuit by the following relations

$$Q_I = \omega_0 L/R$$
$$Q_E = \omega_0 L/R_{Th} \qquad (8.4.22)$$
$$Q_I = \omega_0 L/(R + R_{Th})$$

which follow from (8.4.8) and the definition of critical coupling.

A similar coupling between an external circuit and a parallel GLC resonator is illustrated in Figure 8.10(b), where the external Thevenin equivalent circuit is replaced by a Norton equivalent circuit, consisting of a current source I_N in parallel with an admittance Y_N. Because the two circuits in Figure 8.10 are complete duals of one another, the entire discussion above can be repeated with the substitutions $V_{Th} \to I_N$, $Z_{Th} \to Y_N$, $R \to G$, and $L \to C$. Critical-coupling again occurs when the parallel GLC resonator is matched to the external circuit, or $Y_N = G_N = G$.[9]

---

8. Strictly speaking, it is also possible to have a reactive part of the Thevenin impedance as long as $X_{Th}(\omega_0) = 0$.

9. Or, more generally, when $Y_N = G + jB$ where $B(\omega_0) = 0$.

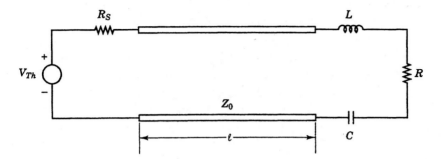

Figure 8.11   Power coupling to a RLC resonator in Example 8.4.1.

This circuit has quality factors Q_I, Q_E, and Q_L given by

$$Q_I = \omega_0 C / G$$
$$Q_E = \omega_0 C / G_{\text{Th}} \tag{8.4.23}$$
$$Q_I = \omega_0 C / (G + G_{\text{Th}})$$

which result from (8.4.16) and the definition of critical coupling.

Example 8.4.1

We wish to couple the maximum amount of power (at resonance) to a series RLC resonator via a transmission line connected to a series resistor R_s, shown in Figure 8.11. Find a relationship between R_s, Z_0, ℓ, and R. Determine R_s in terms of Z_0 and R when $\ell = \lambda/2$ and $\ell = \lambda/4$.

Solution: At resonance, the RLC circuit has a purely real impedance R. For maximum power transfer to occur, the resonator must be critically coupled and the Thevenin equivalent circuit should be simply a voltage source in series with a resistor having the value $Z_{\text{Th}} = R_{\text{Th}} = R$. We can find the Thevenin impedance of the external circuit by looking back into the TEM line and shorting the voltage source. From (6.1.14), we find

$$Z_{\text{Th}} = Z_0 \left(\frac{R_s + j Z_0 \tan k\ell}{Z_0 + j R_s \tan k\ell} \right) \tag{1}$$

which is purely real when $\tan k\ell = n\pi$ ($\ell = 0, \lambda/2, \lambda, \ldots$) or when $\tan k\ell = (n + 1/2)\ell$ ($\ell = \lambda/4, 3\lambda/4, \ldots$). If ℓ is a multiple of $\lambda/2$, then $Z_{\text{Th}} = R_s$ and so $R_s = R$ is necessary for critical coupling. If, instead, ℓ is an odd multiple of $\lambda/4$, $Z_{\text{Th}} = Z_0^2/R_s$ so $R_s = Z_0^2/R$ is the appropriate choice for the series resistance.

8.5 TRANSMISSION LINE RESONATORS

The one-dimensional TEM-line resonator is another important resonating structure, formed by terminating a section of TEM transmission line at both ends. Such resonators exhibit all the subtleties of resonator behavior without the additional complexities arising from a three-dimensional structure. Slightly less simple than the series and parallel circuit resonators described in the previous section, the *TEM*-line resonators possess an infinite number of resonant frequencies instead of just a single resonance at $\omega = 1/\sqrt{LC}$.

We first discuss a *TEM*-line resonator formed from a piece of uniform *TEM* transmission line of length ℓ shorted at both ends, as illustrated in Figure 8.12.

The voltage and current for such a resonator are given by the usual forward- and backward-propagating wave solutions (5.3.45) and (5.3.49):

$$V(z) = V_+ e^{-jkz} + V_- e^{jkz}$$
$$I(z) = Y_0(V_+ e^{-jkz} - V_- e^{jkz}) \tag{8.5.1}$$

The short-circuit boundary condition $V(z = 0) = 0$ may be applied to (8.5.1) to show that $V_+ = -V_-$. Therefore,

$$V(z) = -2jV_+ \sin kz$$
$$I(z) = 2Y_0 V_+ \cos kz \tag{8.5.2}$$

Since the voltage in (8.5.2) is also forced to vanish at $z = \ell$, then $\sin k\ell = 0$ or

$$k\ell = n\pi, \qquad n = 0, 1, 2, \ldots \tag{8.5.3}$$

Therefore, the wavenumber k may take on a series of discrete values indexed by n such that $k_n = n\pi/\ell$.[10] Because $k = \omega\sqrt{LC} = \omega/v$, the resonator can only sustain voltages and currents at the resonant frequencies

$$\omega_n = \frac{k_n}{\sqrt{LC}} = \frac{n\pi}{\ell\sqrt{LC}} = \frac{n\pi v}{\ell}, \qquad n = 0, 1, 2, \ldots \tag{8.5.4}$$

where $v = 1/\sqrt{LC}$ is the speed of wave propagation on the transmission line.

---

10. In this case, where both ends are shorted, we see that $\ell = n\lambda_n/2$. The same result follows for *TEM* resonators open-circuited at both ends, whereas the length ℓ of a resonator shorted at one end and open-circuited at the other is an odd multiple of $\lambda_n/4$.

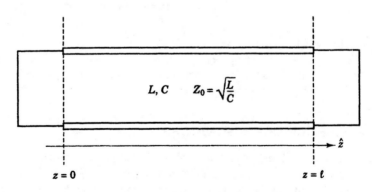

Figure 8.12   Short-circuit *TEM*-line resonator.

Note that one of these modes ($n = 0$) has zero frequency, corresponding to a steady (DC) current $I(z) = 2Y_0 V_+$ flowing in a loop and purely magnetic energy storage since $V(z)$ is everywhere zero for this mode.

We can also find the resonant frequency for *TEM*-line resonators terminated with reactive (capacitive and/or inductive) loads. Consider the case where the *TEM*-line resonator of Figure 8.7 is terminated at $z = \ell$ with an inductor L instead of a short-circuit, while the $z = 0$ end remains a short-circuit. The impedance at $z = \ell$ is then

$$Z(z = \ell) = \frac{V(z = \ell)}{I(z = \ell)} = \frac{-2jV_+ \sin k\ell}{2Y_0 V_+ \cos k\ell} = -jZ_0 \tan k\ell \qquad (8.5.5)$$

using the voltage and current in (8.5.2), which satisfy $V(z = 0) = 0$. But the impedance at $z = \ell$ must also be equal to the inductor impedance $j\omega L$, resulting in the transcendental equation

$$\omega L = -Z_0 \tan\left(\frac{\omega \ell}{v}\right) \qquad (8.5.6)$$

which has an infinite number of solutions ω_n. In this case, the resonant frequencies must be obtained either graphically or numerically. If $L \to 0$, corresponding to a short-circuit impedance, then $\tan(\omega \ell / v)$ must also vanish, recovering the resonant frequencies $\omega_n = n\pi v / \ell$ of (8.5.4). In the opposite limit, where $L \to \infty$ and the resonator is open-circuited at $z = \ell$, $\tan(\omega_n \ell / v) \to \infty$ and $\omega_n = (n + 1/2)\pi v / \ell$.

Once the currents and voltages are found for a *TEM*-line resonator, it is straightforward to calculate the Q of any resonance. Suppose we insert a small series resistor R into the short-circuited *TEM*-line resonator at $z = a$ as shown in Figure 8.13(a). We assume that R is small enough that the current flowing through the transmission line is not seriously perturbed from its original value. Since we have already determined in (5.3.50) that a lossy *TEM* line has the approximate impedance

$$Z \approx \sqrt{\frac{j\omega L + R}{j\omega C + G}} \qquad (8.5.7)$$

we find that $R \ll \omega L$ must be true near resonance for perturbation methods to be successfully applied. Since $\omega L \approx Z_0$ near resonance, this condition is equivalent to $R \ll Z_0$. The power dissipated in the resistor is thus

$$P_n = \frac{1}{2} R \, |I_n(z = a)|^2 \approx \frac{1}{2} R \, |I_n^{(0)}(z = a)|^2$$

where $I_n^{(0)}(z = a) = -2jY_0 V_{+n} \cos k_n a$ is the unperturbed current that would exist if the resistor were not present. Therefore,

$$P_n \simeq \frac{1}{2} R |I_n^{(0)}(z = a)|^2 = \frac{1}{2} R \, |2Y_0 V_{+n}|^2 \cos^2\left(\frac{n\pi a}{\ell}\right) \qquad (8.5.8)$$

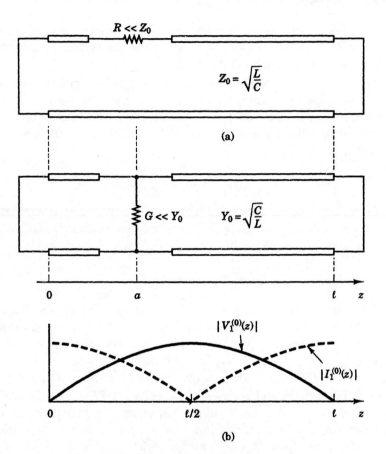

Figure 8.13   Loss in a transmission line resonator.

The total energy stored in this resonator is twice its time-averaged magnetic energy

$$w_{Tn} = 2 \cdot \frac{1}{4} \int_0^\ell L |I_n^{(0)}(z)|^2 \, dz = \frac{L}{2} \int_0^\ell |2Y_0 V_{+n}|^2 \cos^2 k_n z \, dz$$

$$= \frac{\ell}{4} L |2Y_0 V_{+n}|^2 \tag{8.5.9}$$

where we have written w_{Tn} in terms of $I_n^{(0)}(z)$ to simplify the calculation of Q. An expression for the quality factor of this resistive-wall resonator results from its definition in (8.3.3):

$$Q_n \simeq \frac{\omega_n w_{Tn}}{P_n} = \frac{n\pi Z_0}{2R} \frac{1}{\cos^2(n\pi a/\ell)} \tag{8.5.10}$$

It is clear that Q is large when R/Z_0 is much less than unity. However, Q is also

large if we put the resistor at a current null on the ideal *TEM* line $[\cos(n\pi a/\ell) = 0]$
so that no power can be dissipated. In this case Q would be very large even if
$R = Z_0$!

We can also perturb the *TEM*-line resonator by connecting a small conductance G in parallel across the resonator at $z = a$, as shown in Figure 8.13(b). This
resonator is the dual of the one just described. That is, we can repeat the entire
discussion of the small R resonator above but interchange the roles of $I \leftrightarrow V$,
$Z_0 \leftrightarrow Y_0$, and $R \leftrightarrow G$ to describe the small G resonator.

If $G \ll Y_0$, very little current will flow through the conducting shunt and the
voltage will not differ much from its unperturbed value $V_n^{(0)}(z = a)$. This suggests
that we write

$$P_n \simeq \frac{1}{2} G |V_n^{(0)}|^2 = \frac{1}{2} G |2V_{+n}|^2 \sin^2 \left(\frac{n\pi a}{\ell} \right) \tag{8.5.11}$$

using (8.5.2) and (8.5.3). The total stored energy w_{Tn} is given by (8.5.9) so the
quality factor for the nth mode of the conducting-shunt resonator is

$$Q_n \simeq \frac{\omega_n w_{Tn}}{p_n} = \frac{n\pi Y_0}{2G} \frac{1}{\sin^2(n\pi a/\ell)} \tag{8.5.12}$$

If the shunt is placed close to either end of the resonator (at $a \simeq 0$ or $a \simeq \ell$), then
$Q \to \infty$ even for large G because the voltage is small at either end of the resonator
so very little power can be dissipated. Note that the ratios Z_0/R and Y_0/G are
both important measures of resonance quality, but do not uniquely determine Q.

8.6 TRANSIENTS IN RESONATORS

Transients are significant only for limited intervals of time, having signal components at a wide range of frequencies. The more limited the time interval associated with the transient, the broader the range of frequencies that are present, as a
Fourier or Laplace transform reveals. In order to understand the transient behavior
of resonators, we express the voltage and current as the superposition of an infinite
number of terms, each at a distinct resonant frequency of the system. As an example, we revisit the short-circuited *TEM*-line resonator of length ℓ with voltage and
current given by (8.5.2). An arbitrary waveform is constructed by superposing the
voltages and currents of each mode using (8.5.1)–(8.5.4)

$$V(z,t) = \sum_{n=0}^{\infty} V_n(z,t) = \text{Re} \left\{ \sum_{n=0}^{\infty} A_n \sin k_n z \, e^{j\omega_n t} \right\} \tag{8.6.1}$$

$$I(z,t) = \sum_{n=0}^{\infty} I_n(z,t) = \text{Re} \left\{ \sum_{n=0}^{\infty} j Y_0 A_n \cos k_n z \, e^{j\omega_n t} \right\} \tag{8.6.2}$$

where $A_n = -2jV_{+n}$ is the complex coefficient of the nth mode of this resonator
and has yet to be determined explicitly.

As an example of how A_n might be found, consider an air-filled *TEM* transmission line of length ℓ initially open-circuited at both ends and charged to V_0 volts with zero current everywhere on the line. At $t = 0^+$, both ends of the resonator are suddenly short-circuited and the structure becomes the resonator of Figure 8.12 with voltage and current governed by (8.6.1) and (8.6.2). To evaluate A_n, we set $V(z, t = 0^+) = V_0$ and $I(z, t = 0^+) = 0$. This latter boundary condition can be written as

$$I(z, t = 0^+) = \text{Re} \left\{ \sum_{n=0}^{\infty} j Y_0 A_n \cos k_n z \right\} = 0 \tag{8.6.3}$$

where (8.6.3) implies that A_n is a real coefficient for all n; that is, $\text{Im}\{A_n\} = 0$. Applying the voltage boundary condition at $t = 0^+$ to (8.6.1) yields

$$V(z, t = 0^+) = \text{Re} \left\{ \sum_{n=0}^{\infty} A_n \sin k_n z \right\} = V_0 \tag{8.6.4}$$

but since $A_n \sin k_n z$ is purely real here, the $\text{Re}\{\}$ operator in (8.6.4) may be dropped. Therefore,

$$V_0 = \sum_{n=0}^{\infty} A_n \sin k_n z, \qquad k_n = n\pi/\ell \tag{8.6.5}$$

The orthogonality relation for sinusoids discussed in Section 7.3 may now be used to determine A_n. Recalling from (7.3.10) that

$$\int_0^\ell \frac{2}{\ell} \sin\left(\frac{n\pi z}{\ell}\right) \sin\left(\frac{m\pi z}{\ell}\right) dz = \delta_{nm} = \begin{cases} 1 & n = m \\ 0 & n \neq m \end{cases} \tag{8.6.6}$$

we multiply both sides of (8.6.5) by $(2/\ell) \sin k_m z$ and integrate from 0 to ℓ, yielding

$$\int_0^\ell \frac{2V_0}{\ell} \sin k_m z \, dz = \sum_{n=0}^{\infty} A_n \int_0^\ell \frac{2}{\ell} \sin\left(\frac{n\pi z}{\ell}\right) \sin\left(\frac{m\pi z}{\ell}\right) dz$$

$$= \sum_{n=0}^{\infty} A_n \delta_{nm}$$

$$= A_m \tag{8.6.7}$$

Therefore,

$$A_m = \int_0^\ell \frac{2V_0}{\ell} \sin k_m z \, dz = \frac{2V_0}{m\pi} \left[1 - (-1)^m \right] \tag{8.6.8}$$

because $\cos k_m \ell = (-1)^m$. We can thus re-express A_n as

$$A_n = \begin{cases} 4V_0/n\pi, & n \text{ odd} \\ 0 & n \text{ even} \end{cases} \tag{8.6.9}$$

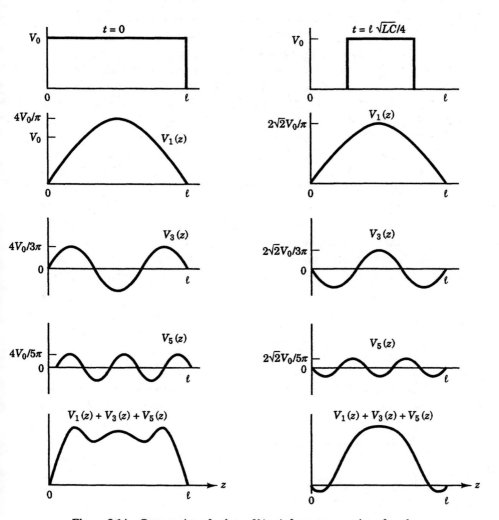

Figure 8.14   Construction of voltage $V(z, t)$ from a summation of modes.

so the voltage on this line finally becomes

$$V(z, t) = \text{Re} \left\{ \sum_{n \text{ odd}}^{\infty} \frac{4V_0}{n\pi} \sin\left(\frac{n\pi z}{\ell}\right) e^{j\omega_n t} \right\}$$

$$= \sum_{n \text{ odd}}^{\infty} \frac{4V_0}{n\pi} \sin\left(\frac{n\pi z}{\ell}\right) \cos\left(\frac{n\pi ct}{\ell}\right) \qquad (8.6.10)$$

The expression for $V(z, t)$ given by (8.6.10) is illustrated in Figure 8.14 at $t = 0^+$ and at $t = \ell/4c$, together with the contributions from $V_n(z, t) = (4V_0/n\pi) \sin(n\pi z/\ell) \cos(n\pi ct/\ell)$ for $n = 1, 3, 5$.

In the limit where the summation includes all of the odd modes, the resulting voltage $V(z, t)$ becomes a simple traveling step function that can be determined in an alternative manner using the methods of time-domain analysis presented in Sections 5.8 and 6.8. We recall that the boundary conditions at $t = 0^+$ determine the initial forward- and backward-propagating voltage pulses, both of magnitude $V_0/2$, as shown in Figure 8.15. Each pulse propagates independently, and the short-circuit boundary conditions at $z = 0$ and $z = \ell$ imply that the reflection coefficients Γ_L and Γ_S at each end of the resonator are both equal to -1. Figure 8.15 plots the voltage $V(z, t)$ at several instants of time. (A further explanation of time-domain methods can be found in Section 6.8.) The energy oscillates between its electric and magnetic forms. At $t = 0, \ell/c, 2\ell/c, \ldots$ the current I is zero everywhere and the energy is purely electric, whereas at $t = \ell/2c, 3\ell/2c, \ldots$ the voltage is zero everywhere and the energy is purely magnetic.

Either method (modal decomposition or time-domain analysis) may be used to analyze a *TEM* resonator. While the time-domain method is more straightforward and intuitive for simple pulse shapes on a dispersionless *TEM* line, the

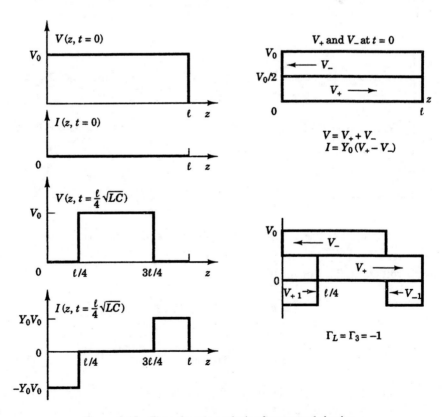

Figure 8.15   Time-domain analysis of resonator behavior.

frequency-domain method of decomposing voltage and current into a set of orthogonal modes is more general and gives additional insight into resonator behavior.

This direct approach to solving problems where the initial conditions are specified is outlined below:

(1) Expand the field expressions in terms of modes, where the resonant frequencies ω_n are determined by the spatial boundary conditions of the resonator.

(2) Set these sums of modes with unknown coefficients equal to the initial conditions for $V(z, t)$ and $I(z, t)$.

(3) Solve for the complex amplitude of every mode by using an appropriate orthogonality relation that singles out each mode individually.

(4) Solve for $V(z, t)$ and $I(z, t)$ by summing over all modes, each at its own characteristic frequency $s_n = j\omega_n - \alpha_n$.

8.7* PERTURBATION OF RESONATORS

We have shown in Sections 8.2 and 8.5 that a lossless cavity or *TEM*-line resonator possesses a (possibly infinite) number of undamped resonant frequencies ω_n. In Sections 8.3–8.5, we saw how small dissipative perturbations could broaden these resonances and even shifted them slightly from ω_0 to $\sqrt{\omega_0^2 - \alpha_0^2}$ [see (8.4.5)]. In Section 8.5 we also saw how a reactive element could be added to a *TEM* resonator to produce a new set of resonant frequencies shifted somewhat from the set produced without the reactive element. (The example used was an inductor at the end of a shorted *TEM* line.) In each case, it is convenient to think of the unperturbed resonant frequency $s_n = j\omega_n$ altered to a new resonant frequency $s'_n = j\omega'_n - \alpha_n$ where $\omega'_n \approx \omega_n$ and $\alpha_n \approx \omega_n/2Q_n \approx P_n/2w_{Tn}$. In general, dissipative losses have their greatest effect on α_n, while reactive (lossless) perturbations principally affect ω_n. In this section, we develop a more rigorous approach to calculating changes in the resonant frequency s_n as small losses and reactances are introduced. These generalizations will also permit us to calculate the effects of denting the walls of cavity resonators or introducing small lumps of dielectric, magnetic, or conducting materials into a cavity.

We shall begin our derivation of an expression for frequency perturbation by applying Maxwell's equations simultaneously to a perfectly conducting resonant cavity and its almost-identical (but lossy) twin. Consider a perfectly conducting closed cavity filled with a medium having permittivity $\epsilon_A(\overline{r})$ and permeability $\mu_A(\overline{r})$. The cavity is resonant at frequency $s = j\omega_0$ and satisfies Maxwell's equations

$$\nabla \times \overline{E}_A = -j\omega_0 \mu_A \overline{H}_A \qquad (8.7.1)$$

$$\nabla \times \overline{H}_A = j\omega_0 \epsilon_A \overline{E}_A \qquad (8.7.2)$$

where $\overline{J}_A$ is zero inside the cavity since $\sigma_A = 0$ and no currents have been applied. If we now alter this cavity slightly, either by adding loss to the walls or medium and/or changing the cavity shape, the new cavity satisfies the equations

$$\nabla \times \overline{E}_B = -s\mu_B\overline{H}_B \tag{8.7.3}$$

$$\nabla \times \overline{H}_B = s\epsilon_B\overline{E}_B + \overline{J}_B \tag{8.7.4}$$

where $\overline{E}_B$ and $\overline{H}_B$ are the perturbed fields with values very close to $\overline{E}_A$ and $\overline{H}_A$, respectively, and $\overline{J}_B$ is a perturbed current introduced by adding loss to the cavity, inserting a modulated electron beam, and so on. As before, the perturbed time-harmonic fields are proportional to e^{st} where s is generally a complex frequency. The real part of s is proportional to the rate at which energy is lost from the resonator and the imaginary part of s is the frequency at which the fields resonate.

We now use the vector identity $\nabla \cdot (\overline{A} \times \overline{B}) = (\nabla \times \overline{A}) \cdot \overline{B} - (\nabla \times \overline{B}) \cdot \overline{A}$ to compute

$$\begin{aligned}
\nabla \cdot (\overline{E}_A^* \times \overline{H}_B + \overline{E}_B \times \overline{H}_A^*) &= (\nabla \times \overline{E}_A^*) \cdot \overline{H}_B - (\nabla \times \overline{H}_B) \cdot \overline{E}_A^* \\
&\quad + (\nabla \times \overline{E}_B) \cdot \overline{H}_A^* - (\nabla \times \overline{H}_A^*) \cdot \overline{E}_B \\
&= j\omega_0\mu_A\overline{H}_A^* \cdot \overline{H}_B - s\epsilon_B\overline{E}_B \cdot \overline{E}_A^* - \overline{J}_B \cdot \overline{E}_A^* \\
&\quad - s\mu_B\overline{H}_B \cdot \overline{H}_A^* + j\omega_0\epsilon_A\overline{E}_A^* \cdot \overline{E}_B
\end{aligned} \tag{8.7.5}$$

where (8.7.1)–(8.7.4) have been used. If we integrate (8.7.5) over the volume of the lossless resonator V_A, we can use Gauss's divergence theorem (4.1.1) to convert the left side of (8.7.5) to a closed-surface integral over the area of the unperturbed resonator A_A:

$$\oint_{A_A} (\overline{E}_A^* \times \overline{H}_B + \overline{E}_B \times \overline{H}_A^*) \cdot \hat{n}_A \, da = (j\omega\mu_A - s\mu_B) \int_{V_A} (\overline{H}_A^* \cdot \overline{H}_B) dv$$
$$+ (j\omega\epsilon_A - s\epsilon_B) \int_{V_A} (\overline{E}_A^* \cdot \overline{E}_B) dv - \int_{V_A} (\overline{J}_B \cdot \overline{E}_A^*) \, dv \tag{8.7.6}$$

Because the electric field $\overline{E}_A$ is perpendicular to the surface of the lossless resonator, $\hat{n}_A \times \overline{E}_A = 0$ and the first term on the left side of (8.7.6) vanishes. We can then solve (8.7.6) for $s - j\omega_0$, yielding

$$s - j\omega_0 = -\left[\oint_{A_A} (\overline{E}_B \times \overline{H}_A^*) \cdot \hat{n}_A \, da + \int_{V_A} (\overline{J}_B \cdot \overline{E}_A^*) \, dv + \right.$$
$$\left. j\omega_0(\mu_A - \mu_B) \int_{V_A} (\overline{H}_A^* \cdot \overline{H}_B) \, dv + j\omega_0(\epsilon_A - \epsilon_B) \int_{V_A} (\overline{E}_A^* \cdot \overline{E}_B) \, dv \right] \Big/ \tag{8.7.7}$$
$$\left[\int_{V_A} (\mu_B\overline{H}_A^* \cdot \overline{H}_B + \epsilon_B\overline{E}_A^* \cdot \overline{E}_B) \, dv \right]$$

where (8.7.7) is still an exact equation. We now make the observation that since $\mu_B \simeq \mu_A$, $\epsilon_B \simeq \epsilon_A$, $\overline{E}_B \simeq \overline{E}_A$, and $\overline{H}_B \simeq \overline{H}_A$, the denominator of the right side of

(8.7.7) may be written:

$$\int_{V_A} (\mu_B \overline{H}_A^* \cdot \overline{H}_B + \epsilon_B \overline{E}_A^* \cdot E_B) \, dv \simeq 4 \int_{V_A} \left(\frac{1}{4} \mu_A |\overline{H}_A|^2 + \frac{1}{4} \epsilon_A |\overline{E}_A|^2 \right) dv$$
$$\simeq 4 \left[\langle w_m \rangle + \langle w_e \rangle \right]$$
$$\simeq 4 w_T \tag{8.7.8}$$

where (8.2.19) and (8.2.20) have been substituted. Note that we cannot approximate $\epsilon_B \simeq \epsilon_A$ and $\mu_B \simeq \mu_A$ in the numerator of (8.7.7) or two terms will completely vanish. Combining (8.7.7) and (8.7.8) yields the approximate difference between the perturbed and unperturbed frequencies:

$$s - j\omega_0 \simeq -\frac{1}{4 w_T} \left[\oint_{A_A} (\overline{E}_B \times \overline{H}_A^*) \cdot \hat{n}_A \, da + \int_{V_A} (\overline{J}_B \cdot \overline{E}_A^*) \, dv \right.$$
$$\left. + j\omega_0(\mu_A - \mu_B) \int_{V_A} (\overline{H}_A^* \cdot \overline{H}_B) dv + j\omega_0(\epsilon_A - \epsilon_B) \int_{V_A} (\overline{E}_A^* \cdot \overline{E}_B) dv \right] \tag{8.7.9}$$

The first two terms on the right side of (8.7.9) are proportional to the radiated and dissipated (real) power, respectively, so we see that

$$\text{Re}\{s - j\omega_0\} = \text{Re}\{j\omega' - \alpha - j\omega_0\} = -\alpha = -\frac{1}{2 w_T} \text{Re}\{P\} \tag{8.7.10}$$

As discussed in the beginning of this section, $\alpha_n \simeq P_n/2w_{Tn}$, which is certainly consistent with (8.7.10). The second two terms on the right side of (8.7.9) correspond to changes in reactive or stored (imaginary) power and thus reflect a shift in the actual resonant frequency $\text{Im}\{s - j\omega_0\} = \omega' - \omega_0$. Because $\epsilon_B \simeq \epsilon_A$ and $\mu_B \simeq \mu_A$, the shift in frequency is generally small, as it must be if the perturbation approximation is to be valid. Equation (8.7.9) is called the *resonator perturbation equation*.

We now apply (8.7.9) to the resonator of Figure 8.16(a), which has a thin, flat dielectric ϵ_B introduced at the conducting wall of a lossless resonator with fields $\overline{E}_A$, $\overline{H}_A$ such that the surface of the dielectric is approximately perpendicular to $\overline{E}_A$. At the dielectric boundary, the normal displacement field $\overline{D}$ must be continuous and thus $\epsilon_B \overline{E}_B \simeq \epsilon_A \overline{E}_A$ inside the dielectric. We assume that the field inside the dielectric is almost all normally directed because the dielectric layer is so thin and flat. The surface integral $\oint_{A_A} (\overline{E}_B \times \overline{H}_A^*) \cdot \hat{n}_A \, da$ therefore vanishes because $\overline{E}_B$ is parallel to $\overline{E}_A$ and $\hat{n}_A \times \overline{E}_A = 0$ at the surface of the perfectly conducting resonator. Likewise, $\int_{V_A} (\overline{J}_B \cdot \overline{E}_A) \, dv = 0$ because no current $\overline{J}_B$ is produced by the dielectric. Because $\mu_B = \mu_A$ and $\epsilon_B = \epsilon_A$ everywhere except where the dielectric is located, the only nonvanishing term in (8.7.9) is

$$s - j\omega_0 \simeq \frac{j\omega_0}{4 w_T} (\epsilon_A - \epsilon_B) \int_{V_A} (\overline{E}_A^* \cdot \overline{E}_B) \, dv$$

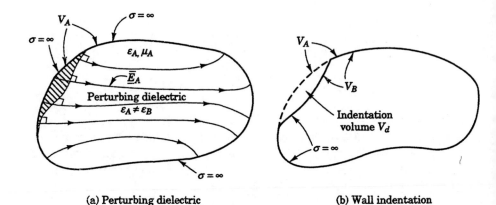

(a) Perturbing dielectric   (b) Wall indentation

Figure 8.16   Gentle perturbations of cavity resonators.

$$\simeq \frac{j\omega_0}{4w_T}(\epsilon_A - \epsilon_B)\frac{\epsilon_A}{\epsilon_B}\int_{V_D}|\overline{E}_A|^2\,dv \qquad (8.7.11)$$

where V_D is the volume of the dielectric. Because (8.7.11) is purely imaginary, only the resonant frequency of the cavity shifts. If $\epsilon_B > \epsilon_A$, then the resonant frequency of the cavity is reduced slightly below ω_0 without affecting the rate of decay of energy in the resonator.

If the cavity wall were instead indented, removing the volume V_D as shown in Figure 8.16(b), this perturbation could be modeled by allowing the dielectric constant ϵ_B in Figure 8.16(a) to approach infinity so that $\overline{E}_B \to 0$, and (8.7.11) would become

$$s - j\omega_0 \simeq -\frac{j\omega_0\epsilon_A}{4w_T}\int_{V_D}|\overline{E}_A|^2\,dv \qquad (8.7.12)$$

But this frequency shift is not exactly correct, because μ_B must accurately model the "perfect conductor" in which $\overline{B}_B = 0$. Since $\overline{H}_B \simeq \overline{H}_A$ as a result of tangential magnetic field continuity at the dielectric surface, μ_B must equal zero so that $\overline{B}_B = \mu_B\overline{H}_B = 0$. The resonator perturbation equation (8.7.9) thus includes an additional term that vanished when $\mu_B \simeq \mu_A$ for the dielectric perturbation, and the frequency shift is more accurately represented by

$$s - j\omega_0 \simeq \frac{j\omega_0}{w_T}\int_{V_D}\left(\frac{1}{4}\mu_A|\overline{H}_A|^2 - \frac{1}{4}\epsilon_A|\overline{E}_A|^2\right)dv \qquad (8.7.13)$$

or

$$\frac{\Delta\omega}{\omega_0} = \frac{\langle\Delta w_m\rangle - \langle\Delta w_e\rangle}{w_T} \qquad (8.7.14)$$

where the only contributions to (8.7.13) and (8.7.14) are from the last two terms of (8.7.9). If the cavity indentation occurs at a place where the unperturbed stored electric energy is larger than the unperturbed stored magnetic energy, then the

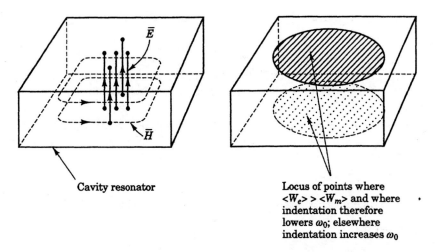

Cavity resonator

Locus of points where
$<W_e> > <W_m>$ and where
indentation therefore
lowers ω_0; elsewhere
indentation increases ω_0

Figure 8.17   Cavity perturbation by indentation.

resonant frequency is reduced, and vice versa. Negative indentations (outward protrusions of the cavity) of course have the opposite effect. Figure 8.17 shows the locus of points where indentation would lower ω_0 for the TE_{101} mode.

The reduction of the cavity resonant frequency caused by indenting the top and bottom walls can also be understood by referring to Figure 8.3, where it is clear that such indentation would increase the capacitance C by decreasing the space between the top and bottom capacitor plates in the equivalent LC resonator, thereby reducing $\omega_0 = 1/\sqrt{LC}$.

Yet another way to understand cavity indentation is to assume that the cavity contains a fixed number of photons n each having energy $w = h\omega_0/2\pi$, where h is Planck's constant ($h = 6.625 \times 10^{-34}$ J · s) so that the total energy w_{Tn} in the cavity is $nw = nh\omega_0/2\pi$. The electric field lines of the TE_{101} mode pull at the top and bottom walls of the cavity with an average force density $\frac{1}{4}\epsilon \left|\overline{E}\right|^2$ (N/m^2), and if the cavity yields slightly, the photons do work and collectively lose energy Δw_T. Since the number of photons is conserved for processes slow compared to ω_0^{-1}, ω_0 is necessarily reduced. The energy shift Δw_T and the frequency shift $\Delta\omega$ are therefore directly proportional:

$$\frac{nh\Delta\omega}{2\pi} = \Delta w_T$$

or

$$\Delta\omega = \frac{2\pi \Delta w_T}{nh} = \frac{\omega_0 \Delta w_T}{w_T} \qquad (8.7.15)$$

The energy shift Δw_T also depends on the average outwardly directed magnetic forces, $\frac{1}{4}\mu \left|\overline{H}\right|^2$ N/m^2. At places where this force density exceeds the inward force density of electric origin, indentation would increase ω_0, and at places where these two forces balance, such indentation would have little effect.

Example 8.7.1

A cavity resonator has slightly lossy walls of conductivity σ. Find a simple expression for the complex resonant frequency s for this resonator, where ω_0 would be the resonant frequency if $\sigma = \infty$.

Solution: We consider V_A in (8.7.9) to exclude the walls where $\overline{J}_B$ is nonzero, and thus only the $(\overline{E}_B \times \overline{H}_A^*) \cdot \hat{n}_A$ term requires evaluation. The electric field $\overline{E}_B$ parallel to the wall is given by

$$\overline{E}_B \cong \eta_w \hat{n}_A \times \overline{H}_A$$

where $\hat{n}_A$ is the surface normal vector and the wall impedance follows from (3.4.15)

$$\eta_w \cong \sqrt{\frac{\omega_0 \mu}{2\sigma}}(1+j)$$

with $\sigma \gg \omega_0 \mu/2$ required for the perturbation method to be useful.

Therefore, (8.7.9) becomes

$$s - j\omega_0 \cong -\frac{1}{4w_T} \oint_{A_A} (\overline{E}_B \times \overline{H}_A^*) \cdot \hat{n}_A \, da$$

$$\cong -\frac{1}{4w_T} \sqrt{\frac{\omega_0 \mu}{2\sigma}}(1+j) \oint_{A_A} |\overline{H}_A|^2 da$$

where A_A is the area of the inside surface of the resonator. The real and imaginary parts are both perturbed equally by the highly conducting but lossy walls, and the resonant frequency is reduced.

Example 8.7.2

A small dielectric sphere having radius R much smaller than a wavelength and permittivity ϵ_B is introduced into a cavity resonant at frequency ω_0. What is the percentage change in the resonant frequency as a result of the perturbation? How does the frequency change if the sphere is perfectly conducting as opposed to perfectly insulating?

Solution: Without the sphere, the cavity has an electric field $\overline{E}_A$ assumed to be roughly uniform over the small region where the sphere will be introduced. Because the sphere is much smaller than a wavelength in radius, we can find the nearfields of the sphere quasistatically, i.e., by solving Laplace's equation rather than the more complicated wave equation. The electric potential is thus

$$\begin{aligned}
\Phi_B &= -E_A r \cos\theta + \frac{C_1 R^3}{r^2}\cos\theta & r > R \\
\Phi_B &= C_2 r \cos\theta & r < R
\end{aligned} \tag{1}$$

where Φ_B satisfies $\nabla^2 \Phi_B = 0$ in spherical coordinates. Far away from the sphere, $\Phi_B \rightarrow -E_A r \cos\theta = -E_A z$, which is the potential of the uniform field $\overline{E}_A = \hat{z} E_A$ without the sphere. (We assume that the sphere is sufficiently small that the electric field is approximately uniform in its vicinity and orient the $\hat{z}$-axis in the direction of the field.) The constant coefficients C_1 and C_2 are found by enforcing continuity of tangential $\overline{E}$ and normal

$\overline{D}$ at $r = R$

$$-\frac{1}{r}\frac{\partial \Phi_B(r > R)}{\partial \theta}\bigg|_{r=R} = -\frac{1}{r}\frac{\partial \Phi_B(r < R)}{\partial \theta}\bigg|_{r=R} \rightarrow -E_A + C_1 = C_2$$

$$-\epsilon_A \frac{\partial \Phi_B(r > R)}{\partial r}\bigg|_{r=R} = -\epsilon_B \frac{\partial \Phi_B(r < R)}{\partial r}\bigg|_{r=R} \rightarrow \epsilon_A(E_A + 2C_1) = -\epsilon_B C_2$$

which leads to $C_1 = -E_A(\epsilon_A - \epsilon_B)/(2\epsilon_A + \epsilon_B)$ and $C_2 = -3\epsilon_A E_A/(2\epsilon_A + \epsilon_B)$. If we neglect the $1/r^2$ field outside the sphere, then

$$\overline{E}_B \approx \frac{3\epsilon_A}{2\epsilon_A + \epsilon_B}\overline{E}_A, \quad r < R$$

$$\overline{E}_B \approx \overline{E}_A, \qquad\qquad r > R$$

We also neglect contributions to $\Delta\omega$ from changes in magnetic energy. The resonator perturbation equation (8.7.9) thus becomes

$$s - j\omega \simeq \frac{j\omega_0}{4w_T}(\epsilon_A - \epsilon_B)\int_{V_{\text{sphere}}}(\overline{E}_A^* \cdot \overline{E}_B)\,dv$$

$$= \frac{j\omega}{4w_T}\frac{3\epsilon_A(\epsilon_A - \epsilon_B)}{2\epsilon_A + \epsilon_B}|\overline{E}_A(\overline{r})|^2 \frac{4\pi R^3}{3}$$

where $\overline{r}$ is the coordinate where the sphere is located. Since $\text{Im}\{s - j\omega\} = \omega' - \omega_0 = \Delta\omega$, we can also write the frequency shift as

$$\frac{\Delta\omega}{\omega_0} \simeq -\frac{\epsilon_A(\epsilon_B - \epsilon_A)}{2\epsilon_A + \epsilon_B}\frac{\pi R^3|\overline{E}_A(\overline{r})|^2}{w_T}$$

By monitoring the frequency change $\Delta\omega$ caused by a spherical dielectric probe as a function of $\overline{r}$, it is possible to experimentally determine the approximate intensity distribution of the electric field associated with any given resonant frequency ω_0.

If the sphere is perfectly conducting, $\epsilon_B \rightarrow \infty$, and the frequency shift becomes

$$\frac{\Delta\omega}{\omega_0}\bigg|_{\epsilon_B \rightarrow \infty} \simeq -\frac{\epsilon_A \pi R^3|\overline{E}_A(\overline{r})|^2}{w_T} = -3\frac{\langle\Delta w_e\rangle}{w_T}$$

where $\langle\Delta w_e\rangle = \frac{1}{4}\epsilon_A|\overline{E}_A|^2 \cdot \frac{4}{3}\pi R^3$. If the sphere is insulating, $\epsilon_B = 0$ and

$$\frac{\Delta\omega}{\omega_0}\bigg|_{\epsilon_B = \infty} \simeq +\frac{3}{2}\frac{\langle\Delta w_e\rangle}{w_T}$$

When the sphere has a dielectric constant greater than that of the empty cavity, it increases the overall cavity capacitance and thereby lowers the resonant frequency ($\Delta\omega < 0$). Conversely, the insulating probe lowers the capacitance of the cavity so the resonant frequency increases and $\Delta\omega$ is positive. Note that a dielectric sphere would be pulled into the high-field regions of a cavity because the system energy is then less and because of forces on the dielectric surface charges. System energy reductions also correspond to lowered resonant frequencies.

Example 8.7.3

A rectangular cavity resonant in the TE_{101} mode has a metal tuning screw of volume V_s located at the position of maximum field strength $\overline{E}_M$, as illustrated in Figure 8.18.

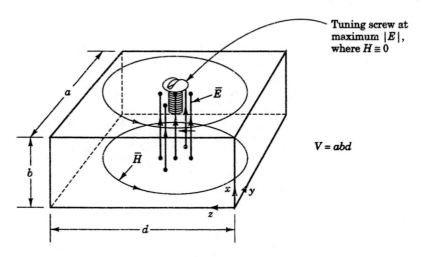

Figure 8.18   Spherical probe of a cavity resonator in Example 8.7.2.

This screw lowers the unperturbed resonant frequency f_0 by Δf. Approximately what is $\Delta f / f_0$?

Solution: If this screw is perfectly conducting $(\epsilon_B \to \infty)$, then we may approximate it by a hemispherical protrusion of radius R and volume $V_s = 2\pi R^3 / 3$. The quasistatic approach to finding the fields near the screw is developed in Example 8.7.2, where we showed that

$$\frac{\Delta \omega}{\omega_0} = \frac{\Delta f}{f_0} \cong -\frac{3\langle \Delta w_e \rangle}{w_T}$$

and $\langle \Delta w_e \rangle \cong \frac{1}{4}\epsilon_A \left| \overline{E}_M \right|^2 V_s$. From (8.2.24), we find the average electric and magnetic fields in the cavity to be

$$w_T = \langle w_e \rangle + \langle w_m \rangle = \frac{1}{8}\epsilon_A |\overline{E}_M|^2 V$$

where $\overline{E}_M = \hat{y}\,(k_x / k_0) E_0$ is the maximum field strength of the TE_{101} mode and V is the resonator volume. Therefore,

$$\frac{\Delta f}{f_0} \cong -\frac{3\left(\frac{1}{4}\epsilon_A |\overline{E}_M|^2 V_s\right)}{\frac{1}{8}\epsilon_A |\overline{E}_M|^2 V} \cong -\frac{6V_s}{V}$$

For a perturbation approach to be successful, $V_s \ll V$, which in turn implies that the frequency shift $\Delta f / f_0$ is small.

8.8* RESONATOR COUPLING TO EXTERNAL CIRCUITS

The energy storage and frequency-sensitive properties of resonators make them very useful in a wide variety of applications, requiring that they be coupled to the external world. In this section, we shall derive the input characteristics of an arbitrary resonator connected to a lossless *TEM* line. At special places on this line, the resonator will look just like a series or parallel RLC resonator circuit. Of course, the resonator must be operating near one of its resonances for a simple equivalent circuit to be a useful model, and each resonance will be characterized by a different equivalent circuit. From a general description of resonator impedance that will rely heavily on the energy theorem, we shall specialize to a *TEM*-line resonator and will make use of the Smith chart to describe its behavior near resonance. Section 8.9 will discuss specific examples and applications of resonator circuits.

We consider an arbitrary high-Q resonator connected via a lossless *TEM* line to the outside world, as shown in Figure 8.19. The energy theorem, presented in Section 1.7, is a useful and general way to understand the frequency behavior of this resonator near a specific resonance.[11] We recall that

$$\frac{1}{4} \oint_{A'} \left(\frac{\partial \overline{E}}{\partial \omega} \times \overline{H}^* + \overline{E}^* \times \frac{\partial \overline{H}}{\partial \omega} \right) \cdot \hat{n}\, da = -j \mathrm{w}_T$$

$$-\frac{1}{4} \int_V \left(\frac{\partial \overline{E}}{\partial \omega} \cdot \overline{J}^* + \overline{E}^* \cdot \frac{\partial \overline{J}}{\partial \omega} \right) dv$$

(8.8.1)

where w_T is the total energy stored in the resonator. The surface integral A' surrounds the resonator of volume V and cuts across its *TEM* feed line as shown in

---

11. The following algebra may be avoided, if desired, by moving directly to the final equations for the impedance of an arbitrary resonator in (8.8.10) and (8.8.11).

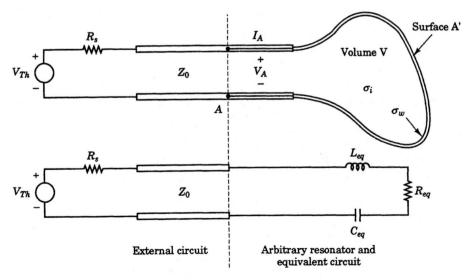

Figure 8.19   Coupling from a *TEM* line to an arbitrary resonator.

Figure 8.19. The point on the *TEM* line where this happens is arbitrary and indicated by the subscript A. The resonator may contain a slightly lossy dielectric with conductivity σ_i ($\sigma_i/\omega\epsilon \ll 1$) and/or have slightly lossy walls with conductivity σ_w ($\sigma_w/\omega\epsilon \gg 1$), but the *TEM* line feed is assumed to be lossless.

Because the walls of the resonator are almost perfectly conducting, the electric field is approximately normal to these walls so $\partial\overline{E}/\partial\omega \times \overline{H}^*$ and $\overline{E}^* \times \partial\overline{H}/\partial\omega$ are both parallel to the resonator surface. Therefore, the surface integral in (8.8.1) vanishes **except** where the surface A' crosses the feed line at A. Here, $\partial\overline{E}/\partial\omega \times \overline{H}^*$ and $\overline{E}^* \times \partial\overline{H}/\partial\omega$ point **along** the transmission line axis since $\overline{E}$ and $\overline{H}$ lie in planes perpendicular to this axis. (See the discussion in Section 5.2 and 5.3.) We have suggested in Section 5.3 that $\int_{A'}(\overline{E} \times \overline{H}^*) \cdot \hat{n}\, da$ is the total power VI^* (W) on the *TEM* line. A similar argument gives the following expression for the left side of (8.8.1) in terms of the voltage and current at A [12]

$$\frac{1}{4}\oint_{A'}\left(\frac{\partial\overline{E}}{\partial\omega} \times \overline{H}^* + \overline{E}^* \times \frac{\partial\overline{H}}{\partial\omega}\right) \cdot \hat{n}\, da = -\frac{1}{4}\left(\frac{\partial V_A}{\partial\omega}I_A^* + V_A^*\frac{\partial I_A}{\partial\omega}\right) \qquad (8.8.2)$$

The negative sign arises because the surface normal points in the opposite direction to the power flowing into the resonator.

The second term on the right side of (8.8.1) is nonzero only in the lossy regions of the resonator because $\overline{E}$ and $\overline{J}$ are never simultaneously nonzero in a lossless resonator or *TEM* line. As usual, we try a perturbation approach to find this term. Inside the resonator, the conduction current $\overline{J}$ is approximately $\sigma_i\overline{E}^{(0)}$, while within the resonator walls of volume V_w, the electric field $\overline{E}$ is nearly $\overline{J}^{(0)}/\sigma_i$. Dropping the superscript (0) hereafter, we find

---

12. The surface integral in (8.8.2) consists of two orthogonal line integrals proportional to current and voltage:

$$\frac{1}{4}\oint_{A'}\left(\frac{\partial\overline{E}}{\partial\omega} \times \overline{H}^* + \overline{E}^* \times \frac{\partial\overline{H}}{\partial\omega}\right) \cdot \hat{n}\, da$$

$$= \frac{1}{4}\int_C\left(\frac{\partial\overline{E}}{\partial\omega} \cdot d\overline{\ell}\right)\oint_{C'}\left(\overline{H}^* \cdot d\overline{s}\right) - \frac{1}{4}\int_C\left(\overline{E}^* \cdot d\overline{\ell}\right)\oint_{C'}\left(\frac{\partial\overline{H}^*}{\partial\omega} \cdot d\overline{s}\right)$$

$$= -\frac{1}{4}\frac{\partial V_A}{\partial\omega}I_A^* + \frac{1}{4}V_A^*\frac{\partial I_A}{\partial\omega}$$

where $da = d\ell\, ds$.

$$\frac{1}{4} \int_V \left(\frac{\partial \overline{E}}{\partial \omega} \cdot \overline{J}^* + \overline{E}^* \cdot \frac{\partial \overline{J}}{\partial \omega} \right) dv = \frac{1}{4} \int_V \left(\sigma_i^* \frac{\partial \overline{E}}{\partial \omega} \cdot \overline{E}^* + \sigma_i \overline{E}^* \cdot \frac{\partial \overline{E}}{\partial \omega} \right) dv$$

$$+ \frac{1}{4} \int_{V_w} \left(\frac{1}{\sigma_w} \frac{\partial \overline{J}}{\partial \omega} \cdot \overline{J}^* + \frac{1}{\sigma_w^*} \overline{J}^* \cdot \frac{\partial \overline{J}}{\partial \omega} \right) dv$$

$$= \frac{\sigma_i}{2} \int_V \left(\overline{E}^* \cdot \frac{\partial \overline{E}}{\partial \omega} \right) dv + \frac{1}{2\sigma_w} \int_{V_w} \left(\overline{J}^* \cdot \frac{\partial \overline{J}}{\partial \omega} \right) dv$$

$$(8.8.3)$$

The conductivities σ_i and σ_w are assumed to be real.

Because Maxwell's equations are linear, V_A and I_A are both proportional to $\overline{E}$, $\overline{H}$, and $\overline{J}$ inside the resonator where the various proportionality constants are frequency-dependent.[13] Therefore, we can relate the volume integrals of $\overline{E}^* \cdot \partial\overline{E}/\partial\omega$ and $\overline{J}^* \cdot \partial\overline{J}/\partial\omega$ in (8.8.3) to the voltage and current at A:

$$\frac{1}{2}\sigma_i \int_V \left(\overline{E}^* \cdot \frac{\partial \overline{E}}{\partial \omega} \right) dv \equiv \frac{1}{2} G(\omega) V_A^* \frac{\partial V_A}{\partial \omega} \qquad (8.8.4)$$

$$\frac{1}{2\sigma_w} \int_{V_w} \left(\overline{J}^* \cdot \frac{\partial \overline{J}}{\partial \omega} \right) dv \equiv \frac{1}{2} R(\omega) I_A^* \frac{\partial I_A}{\partial \omega} \qquad (8.8.5)$$

where $G(\omega) \propto \sigma_i$ and $R(\omega) \propto 1/\sigma_w$. These relations (8.8.4) and (8.8.5) **define** the values of $R(\omega)$ and $G(\omega)$, which may be complex. If we now substitute (8.8.2)–(8.8.5) into the energy theorem (8.8.1), we are left with

$$\frac{1}{4} \left(\frac{\partial V_A}{\partial \omega} I_A^* + V_A^* \frac{\partial I_A}{\partial \omega} \right) = j \mathrm{w}_T + \frac{1}{2} R I_A^* \frac{\partial I_A}{\partial \omega} + \frac{1}{2} G V_A^* \frac{\partial V_A}{\partial \omega} \qquad (8.8.6)$$

We define the impedance looking into the resonator at A to be $Z_A \equiv V_A/I_A$, which means that we can rewrite the left side of (8.8.6) as

$$\frac{1}{4} \left(\frac{\partial Z_A}{\partial \omega} |I_A|^2 + Z_A I_A^* \frac{\partial I_A}{\partial \omega} + Z_A^* I_A^* \frac{\partial I_A}{\partial \omega} \right)$$

$$= \frac{1}{4} \frac{\partial Z_A}{\partial \omega} |I_A|^2 + \frac{1}{2} R_A I_A^* \frac{\partial I_A}{\partial \omega} \qquad (8.8.7)$$

where Z_A has a resistive part R_A and a reactive part X_A $(Z_A = R_A + jX_A)$ and so $Z_A + Z_A^* = 2R_A$. We now substitute (8.8.7) into (8.8.6) and separate the resulting

---

13. We exclude the singular case of an ideal lossless resonator where V_A or I_A is zero.

equation into its real and imaginary parts:

$$
\begin{aligned}
\text{Re}: \quad \tfrac{1}{4}(\partial R_A/\partial \omega)\,|I_A|^2 &= \tfrac{1}{2}\,\text{Re}\,\{(R - R_A)I_A^*\,\partial I_A/\partial \omega\} + \\
&\quad \tfrac{1}{2}\,\text{Re}\,\{GV_A^*\,\partial V_A/\partial \omega\} \\
\text{Im}: \quad \tfrac{1}{4}\partial X_A/\partial \omega\,|I_A|^2 &= w_T + \tfrac{1}{2}\,\text{Im}\,\{(R - R_A)I_A^*\,\partial I_A/\partial \omega\} + \\
&\quad \tfrac{1}{2}\,\text{Im}\,\{GV_A^*\,\partial V_A/\partial \omega\}
\end{aligned}
\tag{8.8.8}
$$

Up until now, the location A on the *TEM* feed line has been somewhat arbitrary, but we now choose A such that the real part of (8.8.8) vanishes near resonance. This implies that $\partial R_A/\partial \omega$ is zero and

$$
R_A(\omega_0) = \left. \frac{\text{Re}\,\{GV_A^*(\partial V_A/\partial \omega)\} + \text{Re}\,\{RI_A^*(\partial I_A/\partial \omega)\}}{\text{Re}\,\{I_A^*(\partial I_A/\partial \omega)\}} \right|_{\omega=\omega_0}
\tag{8.8.9}
$$

This choice of terminal location substantially simplifies the expressions for R_A and X_A near resonance, as we shall see. Note that if $\sigma_i = G = 0$, then $R_A(\omega_0) = R$.[14]

Near resonance, the second and third terms on the right side of (8.8.8) can be neglected because the total energy stored in the resonator must be much larger than terms proportional to dissipated power if the resonator is to have a large Q.[15] Equation (8.8.8) thus simplifies considerably, yielding equations for both the real and imaginary parts of the resonator impedance $Z_A = R_A + jX_A$:

$$
\left. \frac{\partial R_A}{\partial \omega} \right|_{\omega \approx \omega_0} = 0
\tag{8.8.10}
$$

$$
\left. \frac{\partial X_A}{\partial \omega} \right|_{\omega \approx \omega_0} = \frac{4w_T}{|I_A|^2}
\tag{8.8.11}
$$

Equations (8.8.10) and (8.8.11) are **only** true near resonance and at the location A where (8.8.9) is valid.

At a resonance ω_0, the impedance $R_A + jX_A$ is purely resistive, since maximum power is dissipated in the resonator at resonance, and we find

$$
\begin{aligned}
R_A(\omega_0) &= R_{eq} \\
X_A(\omega_0) &= 0
\end{aligned}
\tag{8.8.12}
$$

---

14. It is almost always possible to find a location on the *TEM* line where (8.8.9) is satisfied, as we see by appealing to a Smith chart. A useful resonator cannot dissipate much power if it is to have a sharp frequency response near resonance, so its reflection coefficient off-resonance must be close to unity; most of the power fed to it becomes trapped. The off-resonance input impedance directly at the entrance to the resonator must therefore lie very near the $|\Gamma| = 1$ circle on the Smith chart. This circle with $|\Gamma| \approx 1$ intersects almost every circle of constant real part near the point $\Gamma = 1$. Only those small circles near $\Gamma = 1$ that have resistances greater than $\text{Re}\,\{(1 + \Gamma)/(1 - \Gamma)\}$ cannot be reached at some point on the *TEM* feed line (by rotating about the origin of the Smith chart with constant radius $|\Gamma| \approx 1$). If (8.8.9) can not be satisfied at some position A, then we generally do not have a useful resonator.

15. Note that if $G = 0$ so that $R_A = R$, these terms vanish automatically.

The equivalent resistance R_{eq} of the resonator is just $R_A(\omega_0)$, given by (8.8.9). Combining (8.8.10)–(8.8.12) thus yields a very general expression for the impedance Z_A near resonance

$$Z_A(\omega \approx \omega_0) = R_{eq} + j\frac{4\langle w_T\rangle}{|I_{A0}|^2}(\omega - \omega_0) \qquad (8.8.13)$$

where I_{A0} is the current at A where $\omega \approx \omega_0$ and $\langle w_T\rangle$ is the total stored energy at resonance. Equation (8.8.13) is the input impedance of a series resonator circuit![16]

Because the real and imaginary parts of $Z_A(\omega \approx \omega_0)$ are equal at the half-power frequencies $\omega = \omega_0 \pm \alpha_0$,[17] we substitute $\omega - \omega_0 = \alpha_0$ into (8.8.13), yielding

$$R_{eq} = \frac{4\langle w_T\rangle}{|I_{A0}|^2}\alpha_0 \qquad (8.8.14)$$

which means that the internal quality factor of this resonator is

$$Q_I = \frac{\omega_0 w_T}{P_I} = \frac{\omega_0\langle w_T\rangle}{\frac{1}{2}R_{eq}|I_{A0}|^2} = \frac{\omega_0}{2\alpha_0} = \frac{\omega_0}{\Delta\omega} \qquad (8.8.15)$$

We have thus demonstrated the connection between Q_I, ω_0, and $\alpha_0 = \Delta\omega/2$ for an **arbitrary** resonator.

It is customary to express the reactance $X(\omega)$ in (8.8.13) as the sum of the reactances of an equivalent capacitor C_{eq} and inductor L_{eq}, so

$$jX(\omega) = j\frac{4\langle w_T\rangle}{|I_{A0}|^2}(\omega - \omega_0) \equiv j\omega L_{eq} + \frac{1}{j\omega C_{eq}} \qquad (8.8.16)$$

---

16. The impedance of an RLC resonator is

$$Z(\omega) = R + j\omega L + \frac{1}{j\omega C}$$

which can be represented near resonance $(\omega = \omega_0 + \delta\omega)$ as

$$Z(\omega_0 + \delta\omega) = R + j(\omega_0 + \delta\omega)L + \frac{1}{j(\omega_0 + \delta\omega)C}$$

$$\approx R + j\omega_0 L - \frac{1}{j\omega_0 C} + j\delta\omega L + \frac{j}{\omega_0^2 C}\delta\omega$$

$$= R + 2jL\delta\omega$$

where $\omega_0 = 1/\sqrt{LC}$. Since $4w_T/|I_{A0}|^2 = 2L$, we see that (8.8.13) also characterizes this discrete-element series resonator.

17. If the real and imaginary parts of Z_A are equal, we see from (8.8.13) that $Z_A \approx R_{eq}(1 + j)$, which means that the power dissipated in the resonator is $V_S^2/4R_{eq}$ in contrast to the power dissipation at resonance ($V_S^2/2R_{eq}$) when $Z_A = R_{eq}$. Therefore, we are justified in calling the frequency for which $Z_A = R_{eq}(1 + j)$ the half-power frequency.

Because the reactance must be zero at $\omega = \omega_0$ from (8.8.13), we see at once that

$$\omega_0 = \frac{1}{\sqrt{L_{eq}C_{eq}}} \tag{8.8.17}$$

If we take the derivative of both sides of (8.8.16) and evaluate the result at resonance, we find that

$$\frac{\partial X(\omega)}{\partial \omega}\bigg|_{\omega=\omega_0} = \frac{4\langle w_T \rangle}{|I_{A0}|^2} = L_{eq} + \frac{1}{\omega_0^2 C_{eq}} = 2L_{eq} \tag{8.8.18}$$

so we finally arrive at

$$L_{eq} = \frac{2\langle w_T \rangle}{|I_{A0}|^2} = \frac{R_{eq}Q_I}{\omega_0}$$
$$C_{eq} = \frac{1}{\omega_0^2 L_{eq}} = \frac{|I_{A0}|^2}{2\omega_0^2 \langle w_T \rangle} = \frac{1}{\omega_0 R_{eq} Q_I} \tag{8.8.19}$$

for the values of the three equivalent series circuit elements.

We have therefore shown that **any resonator structure can be replaced by an equivalent series RLC circuit as seen from a special location on the external TEM line where (8.8.9) is valid.** This location is called the *series resonance plane.* If we move a distance $\lambda/4$ away from this plane, we rotate halfway around the Smith chart so impedances become admittances and vice versa. The equivalent series RLC circuit thus becomes an equivalent parallel GLC circuit with $G_{eq} = 1/R_{eq}$ and the same values for C_{eq} and L_{eq}

$$Y_A = G_{eq} + j\omega C_{eq} + \frac{1}{j\omega L_{eq}} \tag{8.8.20}$$

where (8.8.20) is valid only at the *parallel resonance plane* for ω_0 near resonance. We can check that (8.8.20) is correct by multiplying $Z_A Y_A$; the result should equal unity. The series and parallel resonance planes therefore alternate and are spaced $\lambda/4$ apart. It is of course possible to define a frequency-dependent impedance from **any** location on the TEM feed line, but only the series and parallel resonance planes have simple interpretations.

When analyzing or designing cavity, TEM, or other resonators, it is often more convenient not to evaluate L_{eq} or C_{eq} but to work directly in terms of Q_I and Q_E. Consider the impedance $Z_A(\omega)$ near resonance for the arbitrary resonator characterized by (8.8.13), where we now assume it is coupled to a matched (Thevenin or Norton) source of impedance Z_0. The normalized impedance of the resonator $Z_{An}(\omega)$ is then

$$Z_{An}(\omega) = \frac{R_{eq}}{Z_0} + j\frac{4w_T}{Z_0|I_{A0}|^2}(\omega - \omega_0) \tag{8.8.21}$$

But $R_{eq}/Z_0 = Q_E/E_I$ from (8.4.22), and we have shown in (8.8.14) that $4w_T/$

$|I_{A0}|^2 = R_{eq}/\alpha_0 = 2R_{eq}Q_I/\omega_0$. Therefore, we can rewrite (8.8.21) as

$$Z_{An} = \frac{Q_E}{Q_I} + 2jQ_E\frac{(\omega - \omega_0)}{\omega_0} \qquad (8.8.22)$$

which is simply a function of Q_I, Q_E, ω, and ω_0. Note that $R_{An} = 1$ if $Q_I = Q_E$ and the resonator is matched or critically coupled.

A similar expression for the normalized admittance of a parallel resonance is found by duality, where $R_{eq} \rightarrow G_{eq}$, $Z \rightarrow Y$ and $I \rightarrow V$:

$$Y_{An}(\omega) = \frac{G_{eq}}{Y_0} + j\frac{4w_T}{Y_0|V_{A0}|^2}(\omega - \omega_0) \qquad (8.8.23)$$

We then rewrite this admittance in terms of the internal and external Qs in (8.4.23), giving

$$Y_{An} = \frac{Q_E}{Q_I} + 2jQ_E\frac{(\omega - \omega_0)}{\omega_0} \qquad (8.8.24)$$

which is identical in form to (8.8.22). Again $G_{An} = 1$ if $Q_I = Q_E$.

Example 8.8.1

A resonator consists of an air-filled *TEM* line of length ℓ and characteristic impedance Z_0 terminated by a very small resistance $R_L \ll Z_0$. This *TEM* line is connected in series to a reactance jX, which is in turn connected in series to an external *TEM* line having Thevenin impedance Z_0 as illustrated in Figure 8.20. If this resonator is to be critically

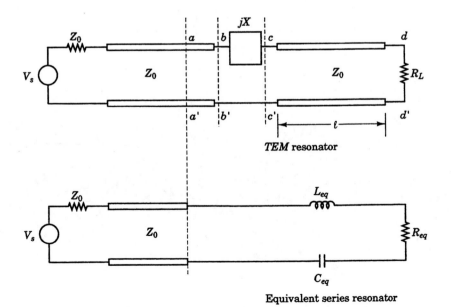

Figure 8.20   *TEM* resonator coupled to *TEM* line.

coupled to the external circuit, what values of ℓ and X are necessary? What values of R_{eq}, L_{eq} and C_{eq} are needed to model this resonator as a discrete-element series resonator with frequency ω_0, and where is the series resonance plane?

Solution: If the resonator is to be critically coupled, the power dissipated in the external circuit (with impedance Z_0) must be the same as that dissipated in the resonator; i.e., $R_{eq} = Z_0$ and the normalized impedances $Z_{n,aa'}(\omega_0) = Z_{n,bb'}(\omega_0) = 1$ where the planes aa' and bb' are given in Figure 8.20. Therefore, $Z_{n,cc'}(\omega_0) = 1 - jX_n$ ($X_n = X/Z_0$) and $Z_{n,cc'}$ lies on the unit resistance circle on the Smith chart at resonance. We find ℓ by rotating the normalized load impedance $Z_{n,dd'} = R_L/Z_0$ about the Smith chart center to the unit circle as shown in Figure 8.21(a). Two possible values of ℓ are possible for any frequency ω_0 depending on whether X is capacitive or inductive. Because $R_L/Z_0 \ll 1$, $\ell \simeq \lambda/4$ for the lowest resonant frequency mode. We now use the analytic expression for the normalized

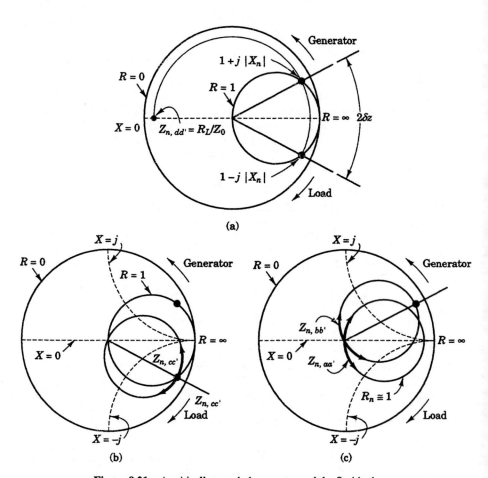

(a)

(b)   (c)

Figure 8.21 A critically coupled resonator and the Smith chart.

impedance $Z_{n,cc'}(\omega_0)$ in (6.1.5), given that the normalized load is R_L/Z_0,

$$Z_{n,cc'}(\omega_0) = \frac{(R_L/Z_0) + j\tan k_0\ell}{1 + j(R_L/Z_0)\tan k_0\ell} = 1 - jX_n \tag{1}$$

from which we find that $\tan k_0\ell = \pm\sqrt{Z_0/R_L}$ and $X_n \approx \mp\sqrt{Z_0/R_L}$ after equating real and imaginary parts of (1). Because $\tan k_0\ell$ is a very large number, ℓ is close to $\lambda/4$ as expected. If we arbitrarily choose ℓ to be slightly larger than a quarter wavelength ($\ell = \lambda/4 + \delta z$), we find that $\tan k_0(\lambda/4 + \delta z) = -\cot k_0\delta z \approx (k_0\delta z)^{-1} = -\sqrt{Z_0/R_L}$. In this case, the lumped reactance connecting the two lines is positive: $X = Z_0X_n = Z_0\sqrt{Z_0/R_L}$. The possible design choices for ℓ and X are thus

$$\ell = \frac{\lambda}{4} \pm \frac{\lambda}{2\pi}\sqrt{\frac{R_L}{Z_0}}$$

and

$$X = \pm Z_0\sqrt{\frac{Z_0}{R_L}} \tag{2}$$

The values of ℓ and X above perfectly match the resonator to the external *TEM* line at resonance; off-resonance, the resonator is either undercoupled or overcoupled. Because $Z_{n,cc'}(\omega) = |\Gamma_L|e^{2j\omega\ell/c}$, it is clear that $Z_{n,cc'}(\omega)$ lies along a circle centered at $\Gamma = 0$ with radius Γ_L. As ω increases (decreases) past resonance, the impedance is rotated in the clockwise (counterclockwise) direction an additional amount past the $\text{Re}\{Z_n\} = 1$ circle. For small changes in ω about ω_0, this path lies approximately along another circle as shown in Figure 8.21(b) for the case where X and δz are both positive. With some algebra, it is possible to show the locus of points representing $Z_{n,bb'}(\omega_0 \pm \Delta\omega)$ in Figure 8.21(c). These points can be made to coincide with the $\text{Re}\{Z_n\} = 1$ circle by rotating it δz meters toward the load when X is inductive. From this last rotation, however, we obtain the impedance of a matched series RLC circuit having $Z_n = 1 + 2jQ\Delta\omega/\omega_0$ [see (8.8.22)]. The series resonance plane at aa' is therefore $\delta z = (\lambda/2\pi)\sqrt{R_L/Z_0}$ meters from bb' when X is positive, and $\lambda/2 - \delta z$ meters from bb' when X is negative.

The resonator Q can be calculated by finding the power dissipated in R_L ($R_L|I_0|^2/2$) and the energy stored in the *TEM* line ($L|I_0|^2\ell/4$)

$$Q_I = \frac{\omega_0\langle w_T\rangle}{P} = \frac{\omega_0 L\ell}{2R_L} = \frac{\pi Z_0}{4R_L} \tag{3}$$

because $\omega_0 = ck_0 = 2\pi c/\lambda$, $\ell \approx \lambda/4$, $c = 1/\sqrt{LC}$, and $Z_0 = \sqrt{L/C}$. Since the resonator is critically coupled, we have already noted that $R_{eq} = Z_0$ and $Q_E = Q_I$. We find L_{eq} and C_{eq} from (8.8.19):

$$L_{eq} = \frac{R_{eq}Q_I}{\omega_0} = \frac{\pi Z_0^2}{4R_L\omega_0}$$

$$C_{eq} = \frac{1}{\omega_0 R_{eq}Q_I} = \frac{4R_L}{\pi\omega_0 Z_0^2} \tag{4}$$

8.9* DESIGN AND APPLICATIONS OF RESONATORS

Resonators are widely used for signal filtering, impedance matching, energy storage, and frequency measurement. They can be implemented in discrete-element circuits, *TEM* transmission lines, waveguides of all types, and cavities, as well as being incorporated on partially reflecting surfaces or distributed throughout a medium. Resonance can also be an unwanted side effect of some electrical systems. For example, any of the impedance matching circuits discussed in Chapter 6 behave as resonators if forced to couple badly mismatched impedances—the resulting narrow-band behavior can pose problems for some applications.

Designing resonators for specific applications is a straightforward exercise, provided that inaccuracies introduced by the perturbation techniques used to compute losses are acceptable. If the resonator Q is large, perturbation methods are usually successful. Most resonators can be characterized by a few simple parameters: resonant frequency, half-power bandwidth, impedance at resonance, and possibly the ability to accommodate a load impedance mismatch at resonance. Related parameters include the resonator energy decay time, maximum energy storage for a given input power, and power dissipation.

Resonators match loads because they permit field strengths to build internally until large enough to dissipate the amount of power equal to the power available from the source. When matching low impedance loads, the resonator fields increase until the **current** flowing through the load is adequate to ensure that all the incident power is dissipated rather than reflected. If no power is reflected, then the load is matched. Resonators match high impedance loads similarly, except that the load is placed so that the **voltage** or electric fields across it are sufficiently large. The same argument also makes it clear that the circuits needed to match badly mismatched loads will behave as resonators within limited bandwidths because the field strengths and therefore the stored energy must be large. But since energy storage is directly proportional to the resonator Q for fixed input power [see (8.8.15)], badly mismatched circuits require increasingly narrow-band matching circuits.

Resonators can also be used for temporary energy storage. For example, energy can be built up slowly and then released in a burst by throwing a switch, or it can be introduced suddenly into a resonator and then released slowly. A typical application is a device that powers an arc discharge. Here, the resonator energy builds slowly until the spark gap breaks down and the resonator discharges its power through the low-impedance plasma at the desired resonator frequency. Another example is a very large enclosed superconducting cavity that stores energy in its dominant mode for later use. Large cavities have low resonant frequencies, and superconducting cavities have very high Qs, so energy can be stored for hours or days, depending on the cavity size. If the superconducting walls of a cavity could tolerate magnetic fields up to one tesla (10,000 gauss), then the peak energy storage density would be $\frac{1}{4}\mu_0|\overline{B}|^2$ or $10^7/16\pi$ J/m^3. Fifteen cubic meters of

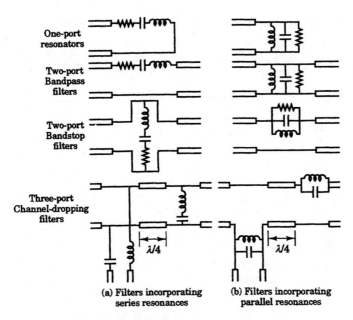

Figure 8.22   Topologies of resonant filters.

vacuum in such a cavity would rival the storage capacity of a car battery, roughly 1 kWh. Unfortunately, the corresponding electric fields of $\sim 3 \times 10^8$ V/m would cause dielectric breakdown and arcing in air, limiting use of this technique.

The major use of resonators is for filtering signals. Filters can pass or block narrow bands of frequencies, or can shunt narrow bands to side circuits while permitting the rest of the band to pass unaltered from one circuit to another. These three circuit types are called *band-pass filters*, *band-stop filters*, and *channel-dropping filters*, respectively, and examples are discussed later in this section. Bandpass and band-stop filters can be implemented in one- or two-port configurations, while channel-dropping filters need two or three ports. A series or parallel *RLC* circuit can be a one-port resonator, as it can be driven by a single *TEM* line. If two *TEM* lines are connected to the resonator, then the filter has two ports. Some of these configurations are illustrated in Figure 8.22.

Channel-dropping filters are usually implemented in three-port configurations, with the desired narrow band extracted from one port while the remainder of the power passes out the other output power. The input port is typically approximately matched over the entire band. Channel-dropping filters are often cascaded to provide many output ports, one for each frequency band of interest. They are employed when the simpler alternative of connecting the output ports in series or parallel transfers power too inefficiently. They can also be implemented by combining lossless band-stop or band-pass filters with three-port ferrite isolators, which pass energy from the input port to the output port but pass energy coming from the output port into a side port instead. This combination of an isolator and a band-pass filter forms a three-port circuit, which also can be cascaded to form a larger circuit with one input and multiple outputs covering each frequency band of interest.

All of these filters can be designed using discrete-element circuits, *TEM* lines, or waveguides and cavities. Alternatively, they may be implemented by depositing specially shaped resonant structures on planar interfaces to manipulate uniform plane waves. Resonant beam-splitters, for example, can be designed to selectively reflect a desired frequency band in a specific direction.

Cavity resonators can be coupled to waveguides by cutting holes in their top, side, or end walls, where the degree of coupling is determined by the size, shape, and position of the holes. If the holes are small, then coupling occurs predominantly via the magnetic fields that leak between the two structures, and the associated wall currents that also pass through the holes. For larger holes, the role of the electric field increases, and additional directivity may be imparted to the emerging waves in the second waveguide.

A cavity resonator attached to the wall of a waveguide by a small connecting hole can have a profound effect on the radiation passing through the waveguide. If the cavity Q_I is sufficiently large compared to Q_E, the fields in the cavity can build in strength so that despite the small size of the hole, they can radiate in both directions down the waveguide, largely canceling the forward-propagating wave by reflecting it losslessly toward the source. This can occur inadvertently when a section of oversized waveguide connects two smaller guides. The middle section may support modes that cannot escape through the small guides, which thus become resonant. If dimensional imperfections couple these trapped resonant modes to the propagating waveguide mode, sharp partial nulls may result in the pass-band of the waveguide.

Cavities can also be coupled to coaxial cables and other *TEM* structures by soldering one conductor to the outside of the cavity and bringing the other wire into the cavity as an antenna. Several antenna configurations are possible, including a short-dipole antenna (a short wire protruding into the cavity plus its reflection in the conducting wall through which it enters), or a small loop antenna, usually constructed by soldering the wire to the cavity wall. The loop then can intercept the magnetic fields associated with the cavity modes to which coupling is desired. If the joint between the open-ended *TEM* line and cavity is capable of rotation, then the loop can be oriented to intercept only the desired amount of magnetic flux. Figure 8.23 illustrates some of these coupling techniques.

To design a resonator circuit to perform any of these functions, it is necessary only to select the desired configuration, resonant frequency, bandwidth, and impedance mismatch at resonance before and after the resonator is in place. Other

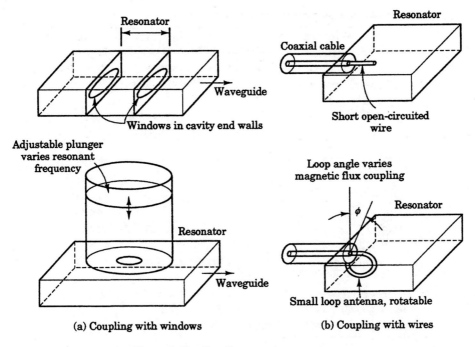

Figure 8.23   Coupling to cavity resonators.

considerations such as tunability, power-handling capability, broad-band behavior, and size limitations may also affect design decisions. A few key equations, summarized in Table 8.2, are the most useful for resonator design. Subsequent examples should help clarify how these formulas are used.

Example 8.9.1

Design a one-port RLC resonator that absorbs all incident power at 1 MHz from a 100-Ω TEM line, but resembles a short circuit otherwise. The total resonant circuit also is to have a half-power bandwidth of 10 kHz.

Solution: We choose a parallel RLC resonator (also called a GLC resonator) since its admittance $Y_n(\omega)$ given by (8.8.24) is equal to Q_E/Q_I at resonance and is large for $\omega \neq \omega_0$. Therefore, this resonator resembles a short circuit off resonance. (A series resonator resembles an open circuit.) For the parallel RLC circuit to match the TEM line, $Y_n(\omega_0) = 1$, which means that $Q_I = Q_E$ and $R = 1/G = Z_0 = 100 \ \Omega$. From (8.4.13), we find the loaded Q

$$Q_L = \frac{f_0}{\Delta f_0} = \frac{1 \text{ MHz}}{10 \text{ kHz}} = 100$$

TABLE 8.2 Resonator Design Equations.

Equation	Application	Eq. No.
$\omega_n = (n + 1/2)\,\pi v/\ell$	*TEM* line open at one end; closed at the other	(8.5.4)
$\omega_n = n\pi v/\ell$	*TEM* line open- or short-circuited at both ends	(8.5.4)
$Q_n = \omega_n \langle w_{Tn} \rangle / P_n$	All resonators	(8.3.3), (8.3.7)
$Q_L^{-1} = Q_I^{-1} + Q_E^{-1}$ $\quad = \omega_n/2\alpha_n = \omega_n/\Delta\omega_n$	All resonators	(8.4.21) (8.4.13)
$\omega_0 = 1/\sqrt{L_{eq}C_{eq}}$	Series, parallel *RLC* circuits	(8.4.6), (8.8.17)
$Q_I = \omega_0 L_{eq}/R_{eq}$	Series circuits	(8.4.8), (8.8.19)
$Q_I = \omega_0 C_{eq}/G_{eq}$	Parallel circuits	(8.4.16)
$Z_n = Q_E/Q_I + 2jQ_E(\omega - \omega_0)/\omega_0$	Series circuits	(8.8.22)
$Y_n = Q_E/Q_I + 2jQ_E(\omega - \omega_0)/\omega_0$	Parallel circuits	(8.8.24)

so $Q_I = Q_E = 2Q_L = 200$ from (8.4.21). We find C and L from (8.8.19) with the duality-inspired substitutions $L \leftrightarrow C$ and $R \to G$:

$$C = \frac{GQ_I}{\omega_0} = \frac{0.01 \times 200}{2\pi \times 10^6} = 0.32 \ \mu\text{F}$$

$$L = \frac{1}{\omega_0 GQ_I} = \frac{1}{2\pi \times 10^6 \times 0.01 \times 200} = 80 \ \text{nH}$$

Example 8.9.2

Design a two-port band-stop series RLC filter that passes only 90 percent of its input power at 1 MHz and that has a half-power bandwidth equal to 10 kHz. This filter is to be connected between two 100-Ω *TEM* lines so that the out-of-band power passes down the lines unperturbed. Such a circuit can be used to determine frequency by varying C and L and noting at what point the output power briefly diminishes by about 10 percent.

Solution: The series band-stop configuration shown in Figure 8.22(a) is the basis for this circuit design. Off-resonance, the RLC resonator is essentially open-circuited so the two-port appears to be a 100-Ω Thevenin circuit driving a 100-Ω load. All power is transferred to the load and the system is matched:

$$P(\omega \neq \omega_0) = \frac{|V_{\text{Th}}|^2}{8Z_0} \tag{1}$$

On resonance, the RLC circuit can be replaced by the resistance R, and the power dissipated in the second *TEM* line is

$$P(\omega_0) = \frac{Z_0 |V_{\text{Th}}|^2}{2(2Z_0 + Z_0^2/R)^2} \qquad (2)$$

where (2) results from calculating the power dissipated in Z_0 in the parallel Z_0, R combination, which is connected in series to another Z_0 impedance. Since $P(\omega_0) = 0.9P(\omega \neq \omega_0)$, we find that $R = 924 \ \Omega$. The loaded Q is found from (8.4.13)

$$Q_L = \frac{f_0}{\Delta f} = 100$$

which is related to the inductance of the series circuit by

$$L = \frac{RQ_I}{\omega_0} = \frac{R_L Q_L}{\omega_0} = (R + Z_0/2)\frac{Q_L}{\omega_0} = 0.016H$$

so $C = 1.63$ pF. Because loaded Q is specified, the total resistive load $R_L = R + Z_0/2$ must be used in the calculation of the resonator parameters.

Example 8.9.3

Repeat Example 8.9.1 using an air-filled *TEM* resonator instead of a discrete-element *RLC* configuration.

Solution: One approach is to use the solution of Example 8.9.1 to find R_{eq}, L_{eq}, and C_{eq} for the equivalent parallel *TEM* resonator circuit, and use (8.4.23) to determine Q_E and Q_I. The *TEM* resonator may then be designed using the perturbation techniques described in Sections 8.5 and 8.7. A more direct approach is taken here, bypassing the unnecessary intermediate step of finding the equivalent circuit elements. If the *TEM* line is air-filled and either shorted or opened at both ends, then its length ℓ must be a multiple of a half wavelength. Because the wavelength is long, we choose the lowest nonzero mode ($\ell = 150$ m) and short-circuit the *TEM* line at both ends to minimize radiation losses. (If this line length were still unacceptably long, we could use a quarter-wave resonator 75 m long, which is shorted at one end and opened at the other.) For the resonator to be critically coupled at resonance, we need $Q_I = Q_E = 2Q_L = 200$ since $Q_L = 100$ to produce a 10-kHz half-power bandwidth at 1 MHz.

To relate these Qs to the resonator requires some decisions about its configuration. Assume that the *TEM* line is lossless, so the only loss in the resonator is the resistor R_L, which shunts the line at a distance d_2 from one end, as illustrated in Figure 8.24. Assume the resonator is fed by a parallel *TEM* line of impedance Z_0 at a distance d_1 from its other end. If we choose R_L equal to Z_0, then we need only compute Q and d once, since $Q = Q_I = Q_E$ and $d = d_1 = d_2$. From (8.5.12), we see that

$$Q_I = \frac{\pi R_L}{2Z_0 \sin^2(\pi d_2/\ell)}$$

which means that $d_2 = 4.24$ m.

It only remains to find the length of the connecting line to the first parallel resonance plane, where the resonator appears close to a short-circuit off resonance. If we consider the resonator to consist of a 4.24-m length of shorted line in parallel with a much longer piece,

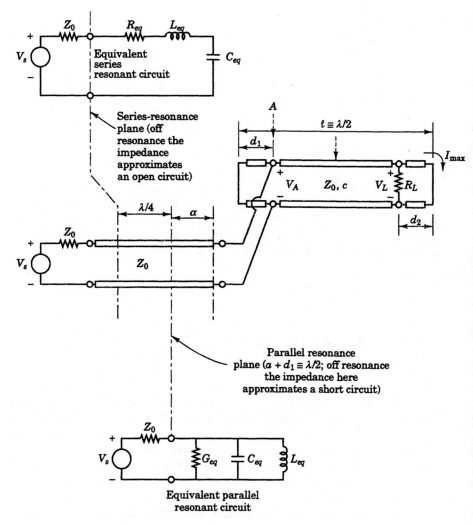

Figure 8.24   One-port *TEM* resonator in Example 8.9.3.

then at most frequencies well-removed from resonance, the total admittance of this parallel combination is approximately that of the short segment. Because this large admittance transforms to a short circuit $\lambda/2 = 150$ m from the end of the resonator, the parallel resonance plane is at $150 - 4.24 = 145.76$ m from the end of the external circuit. (See also Example 8.8.1 for another discussion of the locations of the series and parallel resonance planes.) Note that d and ℓ are unchanged if we wish to find the series resonance plane at which the resonator appears to be an open circuit. We merely move $\lambda/4 = 75$ m to the right or left of the parallel resonance plane.

Example 8.9.4

Use a critically coupled air-filled *TEM* resonator of the same general form described in Example 8.9.3 to show that if $R_L = 400Z_0$, the bandwidth of the matching circuit is necessarily narrow.

Solution: The power dissipated in R_L is $\sim |V_L|^2/2R_L$, where V_L is the unperturbed line voltage across the load and the load is d_2 m from the short-circuited end of the line, as illustrated above. At resonance, this power must equal the input power $|V_S|^2/2Z_0$ where V_S is the source voltage: $|V_L| = 20|V_S|$ although the peak voltage on the resonator may be even larger. To maximize the bandwidth of the matching circuit, we wish to minimize $Q_L = Q_I/2 = Q_E/2$. But the energy theorem tells us that $Q_L = \omega_0 \langle w_T \rangle / P$, and since ω_0 and P are fixed, we must minimize $\langle w_T \rangle = \frac{1}{4}C\ell|V_{max}|^2$ to minimize Q_L which means minimizing $|V_{max}|$. This is best done by placing the R_L at the center of the resonator so that $d_2 = 75$ m and $|V_L| = |V_{max}|$. Substituting these values in (8.4.16) yields $Q_I = \omega_0 C\ell R_L/2 = \pi R_L/2Z_0 = 628$, which means that the half-power bandwidth of this matching circuit can be no greater than $f_0/Q_L \simeq 1.6$ kHz.

Example 8.9.5

Design a critically coupled cavity band-pass filter for a resonance frequency of 10 GHz and a Q_L of 1000. Assume it is made of coin silver. What is the insertion loss (power lost in the resonator walls) for your design?

Solution: Although we know that Q_I is greatest when we use large resonators with maximum volume-to-surface ratios, here we shall assume that the resonator is constructed simply by placing two windows across a waveguide to form a cavity resonant at 10 GHz, as illustrated in Figure 8.23. If the dimensions of this waveguide are 1×2 cm, then from (8.2.11), the unperturbed cavity length d for a 10-GHz TE_{101} resonance is 2.27 cm. Since $Q_E = Q_I = 2Q_L = 2000$ for critical coupling, the windows in the end walls that connect the lossless resonator to the input and output waveguides must have the same electrical characteristics and are normally identical. It only remains to find the size, shape, and positions of the holes in the end walls that yield the desired Q_L; this is often done empirically by drilling them progressively larger until the desired performance is obtained. (To account for cavity wall losses, the output hole would be slightly smaller to preserve $Q_E = Q_I$.) Calculating the window shapes from first principles is difficult, although powerful electromagnetic computer software packages are making these calculations easier.

Even though we may not know the sizes of the windows, we can estimate an insertion loss by computing Q_w associated with the wall losses P_w. For silver, $\sigma \simeq 5 \times 10^7 \; \Omega^{-1}\text{m}^{-1}$ so the skin depth is approximately 0.712 μm, and (8.3.20) can be used to find the unperturbed $Q_w \simeq 7210$. The fraction of power lost in the walls is simply $P_I/(P_w + P_I) = Q_w^{-1}/(Q_w^{-1} + Q_I^{-1})$ or about 22 percent. This is sufficiently high to suggest using a cavity with a cross-section larger than the waveguide.

Example 8.9.6

A certain infrared *TEM* resonator can be modeled as an infinite dielectric slab ($\epsilon = 9\epsilon_0$) $\ell = 10$ cm thick bounded on one side by a mirror with conductivity $\sigma = 10^7 \; \Omega^{-1}\text{m}^{-1}$ and on the other side by a lossless multilayer dielectric mirror designed to reflect 99 percent of the $\lambda = 100$ μm radiation incident upon it. Approximately what resonator mode is being excited, and what are Q_E, Q_I, and Q_L for the mode? What is the width $\Delta\omega$ of this

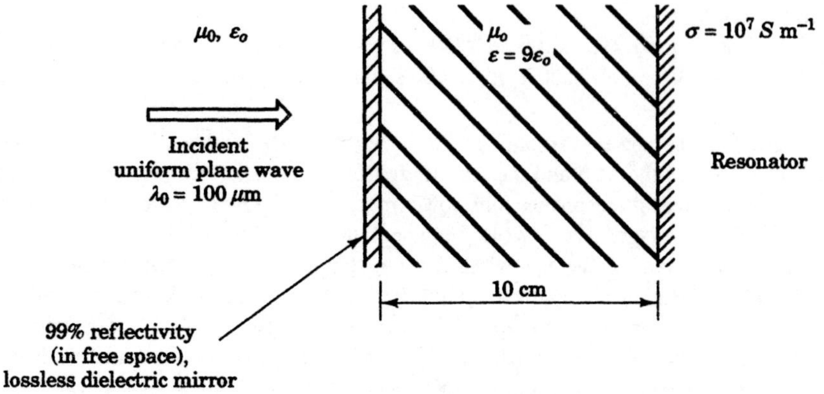

Figure 8.25 Infrared *TEM* resonator in Example 8.9.6.

resonance, and how does it compare to the spacing between resonances $\omega_{n+1} - \omega_n$? At resonance, what fraction of the power is absorbed by the resonator? What reflectivity is needed in the dielectric mirror to couple all incident power into the metallic mirror? This resonator is pictured in Figure 8.25.

Solution: In the dielectric, $\lambda = \lambda_0/\sqrt{\epsilon/\epsilon_0} = \lambda_0/3$. The mode number n follows from $n\lambda/2 = \ell$, so $n \cong 2\ell/\lambda = 0.2/(\lambda_0/3) = 0.2 \times 3/10^{-4} = 6000$. From (8.4.20)

$$Q_E = \omega_0 \mathrm{w}_T/P_E$$

where P_E is the power lost externally through the lossless mirror. The total energy W_T in this resonator can be considered to be composed of one stream moving to the right and another approximately equal one moving to the left, each at velocity $v = 1/\sqrt{9\epsilon_0\mu_0} \cong 10^8$ (m/s). Each stream conveys power $P_+ = P_-$, where $P_E = 10^{-2}P_-$. The energy is these two streams is $\langle \mathrm{w}_T \rangle$:

$$\langle \mathrm{w}_T \rangle = (P_+ + P_-)\ell/v \cong 2P_- 10^{-1}/10^8 = 2 \times 10^{-9} P_-$$

and

$$Q_E = \frac{\omega_0 \langle \mathrm{w}_T \rangle}{P_E} = \frac{2\pi \times 10^{12} \times 2 \times 10^9\, P_-}{10^{-2}\, P_-} = 1.26 \times 10^6$$

The power per unit area P_I lost internally in the metal mirror can be found from (4.4.29)

$$P_I = \frac{|E_0|^2}{\eta^2}\sqrt{\frac{2\omega_0\mu_0}{\sigma}}$$

and the energy stored in the slab is $\langle \mathrm{w}_T \rangle = \epsilon |E_0|^2 \ell/4$, so

$$Q_I = \frac{\omega_0 \langle \mathrm{w}_T \rangle}{P_I} = \ell\sqrt{\frac{\sigma\omega_0\mu_0}{8}} = 3.14 \times 10^5$$

Therefore, $Q_L = (Q_I^{-1} + Q_E^{-1})^{-1} = 2.5 \times 10^5$ and the bandwidth is $\Delta\omega = \omega_0/Q_L = 2.5 \times 10^7$ rad/s. The frequency difference between modes is $\delta\omega \simeq \omega_0/6000 = 10^9$ rad/s, which is much greater than $\Delta\omega$.

At resonance, the fraction of power absorbed can be found using a reference plane for a parallel resonance

$$\frac{Q_E}{Q_I} = \frac{G_{eq}}{Y_0} = \frac{1}{Z_n}$$

$$|\Gamma|^2 = \left|\frac{Z_n - 1}{Z_n + 1}\right|^2 = \left|\frac{Q_I - Q_E}{Q_I + Q_E}\right|^2 = 0.36$$

so only 74 percent of the incident power is absorbed at resonance. To achieve 100 percent absorption, we set $Q_I = Q_E$ by decreasing the reflectivity γ of the dielectric mirror from $\gamma = 0.99$ to

$$\gamma = 1 - \frac{4000\pi}{Q_I} = 0.96$$

8.10 SUMMARY

A *resonator* is any structure that can trap electromagnetic energy, although useful (high-quality) resonators also prevent this energy from leaking away too quickly. In accordance with the *energy theorem* (1.7.6), resonators exhibit strongly frequency-dependent behavior near one or more discrete frequencies called *resonances*. Resonators are used for energy storage, frequency measurement, filtering, impedance matching, and the production of high field strengths from modest input sources.

A *rectangular cavity resonator*, formed by closing the ends of a rectangular waveguide, can support TE_{mnp} and TM_{mnp} modes, where m, n, and p are integers indicating how many half wavelengths can fit between the resonator walls of lengths a, b, and d ($b \le a \le d$), respectively. Both the TE_{mnp} and TM_{mnp} modes resonate at the frequency f_{mnp}, where

$$f_{mnp} = \frac{c}{2}\sqrt{\left(\frac{m}{a}\right)^2 + \left(\frac{n}{b}\right)^2 + \left(\frac{p}{d}\right)^2}$$

Resonators can also be fabricated by terminating *TEM* lines with open or short circuits, or with any highly reflective (mostly reactive) impedance. For a *TEM* line of length ℓ short-circuited at both ends, resonances occur for $f_n = nv/2\ell$. Circuits containing both inductive and capacitive reactances typically possess one or more resonances, as do dielectric waveguides of finite length and almost any arbitrarily shaped conducting or nonconducting structure.

The *quality factor* Q of a particular resonance mode is dependent on the rate at which that mode loses energy, where for the nth mode,

$$Q_n = \frac{w_{Tn}}{P_n/\omega_n} = \frac{\text{total energy of the } n^{th} \text{ mode}}{\text{average energy lost per radian of the } n^{th} \text{ mode}}$$

The *complex frequency* of the nth resonance is commonly denoted s_n, where

$$s_n = j\omega_n - \alpha_n$$

and $Q_n = \omega_n / 2\alpha_n$.

We distinguish three types of quality factors: *internal*, *external*, and *loaded*, represented by Q_I, Q_E, and Q_L, respectively. The internal Q is related to power dissipation within the resonant structure, while the external Q depends on the composition and configuration of the external circuit that couples to the resonator. The loaded Q is a measure of the total power dissipated by all loss mechanisms in the resonator circuit, so that

$$\frac{1}{Q_L} = \frac{1}{Q_I} + \frac{1}{Q_E}$$

When a resonator is connected to an external circuit, the *half-power bandwidth* of the resonance is $\Delta\omega = \omega/Q_L$, so high-quality resonators have narrow bandwidths.

The loss in any particular resonator near any particular resonance can be easily approximated using *perturbation techniques*. The voltages, currents, and/or electromagnetic fields are first computed for a lossless structure at the resonant frequency of interest. The power lost per frequency cycle is then estimated by using these unperturbed fields to compute the power that would be dissipated if loss were present. Dissipation in high conductivity elements is estimated using the unperturbed currents, which do not change appreciably as loss is added, whereas unperturbed voltages must be calculated to estimate power dissipation in low-conductivity structures. Only the ratio of energy stored to power dissipated is needed to compute Q.

At a specific resonance, an arbitrary resonator may be modeled by a (non-unique) *equivalent discrete-element RLC circuit*, where a different circuit may be necessary to describe each of the distinct resonant frequencies. A specific resonance can be modeled by either a series or parallel RLC circuit, depending on whether the complex impedance at resonance is small or large, respectively. In both cases, the impedance is purely real at resonance. If we imagine a Thevenin equivalent source driving a series RLC resonator for which the Thevenin resistance R_{Th} is equal to R, then all available power is transferred to the resonator at the resonant frequency $\omega_0 = 1/\sqrt{LC}$ (equivalently, $|\Gamma(\omega_0)| = 0$). Such a resonator is said to be *critically coupled* since its internal $Q_I = \omega_0 L/R$ is equal to Q_E. If the cavity or *TEM* line resonator is instead modeled by a parallel RLC circuit driven by an external Norton equivalent circuit having the Norton resistance R_N equal to R, the resonator is again critically coupled; in this case $Q_I = Q_E = \omega_0 C/G$. Resonators for which Q_E is less than Q_I are said to be *undercoupled*, while those for which $Q_E > Q_I$ are *overcoupled*. In either case there is imperfect power transfer at resonance.

It is often useful to tune resonators about their natural resonant frequencies by slightly modifying their structures. As loss is added to a lossless resonator, the resonant frequency $s = j\omega_0$ typically becomes complex ($s = j\omega + \alpha$), where both the frequency shift $\omega - \omega_0$ and the quality factor $Q = \omega/2\alpha$ can be computed using the *resonator perturbation equation* (8.7.9). Resonators can also be used in nonsinusoidal steady-state configurations. For example, any closed electromagnetic structure that is energized and then allowed to behave naturally after a specific time will have energy resident in each of its many resonant modes simultaneously. Each mode oscillates with a decaying exponential behavior corresponding to its own complex resonant frequency s_n. The total voltage, current, and field distributions are then linear superpositions of those associated with the separate orthogonal modes. Since each of these modes can be analyzed individually, the behavior of the total system can also be studied.

Design of resonators is straightforward. We start with the desired resonant frequencies and bandwidths and choose between *band-pass*, *band-stop*, and *channel-dropping* resonator configurations. Once the desired mismatch at resonance is specified, perturbation techniques are used to compute the corresponding internal, external, and loaded quality factors. In some cases, the dimensions of the resonator must be altered or lower loss materials chosen in order to achieve the desired values for the internal Q. These quality factors can then be related by perturbation theory to the various resonator parameters and to its external coupling structure.

8.11 PROBLEMS

8.2.1

(a) What are the two lowest resonant frequencies for a perfectly conducting cubical resonator 1 cm on a side filled with dielectric having $\epsilon = 16\epsilon_0$?

(b) Note that several modes share each of these frequencies. List all modes corresponding to each of these two frequencies (TE_{mnp} and TM_{mnp}), where for the purposes of this question, we let m, n, p correspond to the $\hat{x}$-, $\hat{y}$-, and $\hat{z}$-axes, respectively.

8.2.2 A rectangular evacuated superconducting cavity is $2 \times 3 \times 4$ m.

(a) What are the four lowest resonant frequencies of this cavity, and what are their mode designations (e.g., TM_{112})?

(b) The largest magnetic field this superconductor can stand before going "normal" is 20 Wb/m^2 (200,000 gauss, or 20 tesla). What is the maximum total energy that can be stored in this cavity at its lowest resonant frequency? Neglect electrical breakdown phenomena.

(c) A good car battery can store a nominal 1 kWh. Compare the kWh volume storage efficiencies (J/m^3) of these two media.

(d) Show all places and orientations where thin straight slots might be placed in the cavity wall without interfering significantly with the TE_{101} mode.

(e) Repeat part (d) for the TM_{111} mode.

8.3.1 (a) If the conductivity of the dielectric having $\epsilon = 4\epsilon_0$ in a 1-cm cubical resonator is $\sigma = 10^{-2}$ S/m, what is the Q of the TE_{101} mode?

   (b) What is the half-power bandwidth (Hz) of this resonance?

8.3.2 A rectangular perfectly conducting cavity resonator is $1 \times 2 \times 3$ cm in size.

   (a) What are the three TE_{mnp} modes with the lowest resonant frequencies, and what are these frequencies ω_{mnp}?

   (b) Sketch the magnetic field distribution for the TE_{102} mode at some instant of time when it is nonzero. Also sketch the corresponding surface currents on the resonator walls.

   (c) If the cavity is filled with a dielectric $\epsilon = (4 - j10^{-3})\epsilon_0$, then what is (i) the decay constant $\alpha_{TE_{102}}$ at the resonant frequency $\omega_{TE_{102}}$? ($s_{TE_{102}} = j\omega_{TE_{102}} - \alpha_{TE_{102}}$), (ii) $Q_{TE_{102}}$?

8.4.1 A series RLC resonator has an impedance $Z(s)$. Let $R = 2\ \Omega$, $C = 3$ farads, $L = 4$ henries.

   (a) Sketch the locations (quantitatively) of the poles and zeros of $Z(s)$ in the s plane.

   (b) This resonator is driven by a unit-step voltage source. What is the resulting current $I(t)$? Give an equation for $I(t)$ and then sketch it.

8.5.1 Find all the resonant frequencies for each of these TEM resonators:

   (a) Air-filled, open circuited at both ends, 2 m long.

   (b) Same as part (a), but short-circuited at one end.

   (c) Same as part (a), but connected in a closed loop with a single twist (i.e., the top wire is connected to the bottom wire of the other end, and vice versa).

8.5.2 Consider an air-filled 100-Ω TEM resonator 1 m long, which is short-circuited at both ends.

   (a) What are the resonant frequencies f_n (Hz)?

   (b) Assume $f = f_1$, the lowest nonzero resonant frequency. What is the resonator Q if one of the short circuits is replaced by a 1-Ω resistor? Here we use the unperturbed current to find I_R, and $\frac{1}{2}R\,|I_R|^2$ to find the approximate power dissipated, where I_R is the current through the resistor R. Why don't we use the unperturbed resistor voltage instead?

   (c) What are the Q and lowest nonzero complex resonant frequency $f_1 = j\omega_0 - \alpha$ if $R = 10,000\ \Omega$? Is the dissipated power here approximately $\frac{1}{2}|V_R|^2/R$ or $\frac{1}{2}R\,|I_R|^2$, where V_R and I_R are the unperturbed voltage and current at the load?

8.5.3 A nearly lossless resonator is constructed from a 3-m TEM line characterized by ϵ_0, μ_0, $Z_0 = 100\ \Omega$. One end is shorted and the other is an open circuit ($R_L \to \infty$).

   (a) List the two lowest nonzero resonant frequencies (Hz) for this resonator.

   (b) If the load resistor $R_L = 10^4\ \Omega$ is placed across the open circuit, what is the approximate internal Q_I for this resonator when the free-space wavelength λ is 12 m? Give a numerical answer.

8.6.1 Consider a TEM resonator short-circuited at both ends, ℓ meters long, and characterized by L (H/m) and C (F/m).

   (a) What are all the resonant frequencies ω_n?

(b) Show that the total time-averaged electric and magnetic energies $\langle W_e \rangle$ and $\langle W_m \rangle$ are equal at any nonzero resonant frequency.

(c) At $t = 0$ the voltage is everywhere zero and the current is one ampere on the right half of the line ($\ell/2 < z < \ell$) and zero elsewhere ($0 < z < \ell/2$). Decompose the current step into a sum of modes and evaluate the first two modes in this series at $t = 0$ and $t = \ell\sqrt{LC}$ (the lowest mode is DC).

(d) These two lowest modes have resonant frequencies $\omega_0 = 0$ and ω_1. They are shifted approximately to $s_0 = j\omega_0 - \alpha_0$ and $s_1 = j\omega_1 - \alpha_1$ by the insertion in series of a small perturbing loss $R = 10^{-2}\sqrt{L/C}$ at one end of the line. Evaluate α_0 and α_1.

8.7.1 A rectangular cavity $1 \times 2 \times 3$ cm is filled with a dielectric having permittivity $4\epsilon_0$ and permeability μ_0.

(a) List the four lowest distinct resonant frequencies of the cavity and their corresponding mode designations (TE_{mnp}, TM_{mnp}).

(b) What is the resonant frequency ω_{mnp} for this cavity corresponding to the fields sketched in Figure 8.3(d)?

(c*) We wish to tune this cavity by appropriately hitting it from the outside with a ballpeen hammer. Indicate the approximate regions where hitting the cavity increases the fundamental resonant frequency, and the regions where the resonant frequency decreases. Compute the location of the boundary between the two regions on the 1×2 cm surface; i.e., find the places where hitting the resonator has no effect on the resonant frequency of this mode.

8.7.2 (a) If a rectangular metal waveguide 1.5×0.75 cm in cross-section is closed with perfectly conducting plates at $z = 0$ and $z = 3$ cm, the TM_{11} mode at cutoff becomes the fundamental resonator field TM_{110}. What is the resonant frequency?

(b*) If the resonator is pushed in at the center of the front wall at $z = 0$ or the back wall at $z = 3$ cm, will the resonant frequency for TM_{110} mode become lower or higher?

8.7.3 Consider a resonator made of a section of coaxial line with length d and short-circuited with conducting plates at both ends. It is filled with dielectric having $\epsilon = 9\epsilon_0$.

(a) What are the electric and magnetic fields for the TEM modes?

(b) Let $d = 10$ cm. What is the lowest nonzero resonant frequency (which is associated with the TEM_1 mode)?

(c) Plot the magnitude of the electric field and the magnitude of the magnetic field as functions of z for the TEM_2 mode.

(d*) Suppose the TEM_2 mode is excited inside the resonator, and an inward perturbation of the outside wall, achieved by using a perfectly conducting screw, is made near the center ($z = d/2$) of the resonator. Will the resonant frequency of that mode be lowered or raised?

8.8.1* An air-filled 1-m long TEM resonator open-circuited at both ends is made of 100-Ω

line and is connected to a 100-Ω resistor located in series with the line 1 cm from one end.

(a) What is the Q of the lowest nonzero resonance (ω_1) of this structure (where the line is $\sim \lambda/2$ long)? Use perturbation techniques.

(b) A 100-Ω *TEM* line is then connected in series with this line one centimeter from the other end, connecting the resonator to an external matched load. What are the external and total Qs (Q_I and Q_L) for this resonator at ω_1?

(c) What is the 3-dB bandwidth (Hz) of this coupled resonator, as perceived externally?

(d) The feed line connecting to the resonator has a length D such that the resonance is a series resonance. What value of D has this property?

(e) What is the impedance Z_R of this series resonator at ω_1?

(f) What is $Z_R(\omega)$?

(g) Find an equivalent RLC resonator circuit valid for $\omega \cong \omega_1$, giving numerical values for R_{eq}, L_{eq}, and C_{eq}.

8.8.2* Design a *TEM* resonator of the general form employed in Problem 8.8.1, but using different dimensions. Assume the 100-Ω air-filled *TEM* line is lossless ($R = G = 0$). We want a band-pass filter centered at 100 MHz and having a half-power (3-dB) bandwidth of 100 kHz. All the power should be coupled to the 200-Ω load at resonance.

(a) What is one possible value for ℓ, the length of the *TEM* resonator?

(b) If the 200-Ω load R_L is connected in series with the *TEM* line d_2 meters from one end of the resonator, and the feed line ($Z_0 = 100$ Ω) is connected in series d_1 meters from the other end, what are d_1 and d_2? Give numerical values using the perturbation approximation.

(c) What is the voltage-amplification factor for this resonator, i.e., the ratio of $|V_{max}|$ inside the resonator to $|V|$ on the feed line at resonance?

(d) Approximately what is D, the length of the feed line for the resonator to appear to be a series RLC circuit?

(e) Find the values of L_{eq}, C_{eq}, and R_{eq} at this series resonance plane.

8.8.3* A *TEM* resonator consists of a 1-m long air-filled coaxial cable with $Z_0 = 50$ Ω that is short-circuited at both ends.

(a) What are the inductance L (H/m) and capacitance C (F/m) for this line?

(b) A 1-Ω resistor is placed across one end of the lossless resonator.

    (i) What are the two lowest resonant frequencies (ω_1 and ω_2) of this lossless resonator?

    (ii) What is Q_I of the lossy resonance ω_1?

    (iii) This resonator is now connected to the external world by a 50-Ω *TEM* transmission line, connected across the 1 Ω resistor. What is Q_E for ω_1?

    (iv) What is Q_L for ω_1 now?

(c) For the resonator of part (b-iii), approximately where is the nearest parallel-resonance reference plane on the feed line? Where is the nearest series-resonance reference plane?

(d) Find the equivalent circuit for the series resonance ω_1 (as viewed from the series resonance reference plane); i.e., find R_{eq}, L_{eq}, and C_{eq}.

(e) Find the equivalent circuit for the parallel resonance ω_1 (as viewed from the parallel resonance reference plane); i.e., find G_{eq}, L_{eq}, and C_{eq}.

(f) What is $\Delta\omega$, the half-power bandwidth for the ω_1 resonance (when coupled to the given feed line)?

8.9.1* A home TV set is being bothered by a strong interfering carrier signal that is near a channel of interest. The question is how well a filter constructed from commercially available 300-Ω twin-lead *TEM* cable can filter out this carrier at 200 MHz. Assume the losses in twin-lead cable are dominated by resistive losses in the two parallel wires, each of which has diameter of 1 mm and a bulk conductivity of 5×10^7 (S/m); the wires are separated by 1 cm. Assume that the line is effectively air-filled and characterized by ϵ_0, μ_0; its impedance is 300 Ω.

(a) If we characterize this line by L (H/m), C (F/m), and R (Ω/m), approximately what are L, C, and R (at 200 MHz)?

(b) Power transmitted along this line decays as $P = P_0 e^{-\alpha_0 z}$. At 200 MHz what is the decay length α_0^{-1} (meters)?

(c) For a 200-MHz resonator formed by short-circuiting the ends of this cable, what is the internal Q_I? For fixed frequency, does it depend on mode number n, where n is the length of this resonator in units of $\lambda/2$? Explain briefly.

(d) Assume such a *TEM* resonator is used as a two-port band-stop filter (see Figure 8.22) to reject the unwanted carrier at 200 MHz, and suppose it is used to provide a parallel resonance as illustrated in Figure 8.24. To provide 10-dB of carrier rejection at 200 MHz, what must G_{eq} be?

(e) For the value of G_{eq} found in part (d) plus the maximum value of Q_I found in part (c), (i) what is the resulting value of Q_E for this resonator? What then is the (ii) maximum value of Q_L and (iii) minimum value of $\Delta\omega$ we can achieve here for 10-dB rejection?

(f) Repeat parts (d) and (e) for 20-dB rejection. Note the trade-off between rejection and $\Delta\omega$.

(g) Complete the resonator design for 20-dB rejection at 200 MHz. That is, find the lengths d_1 and a for $n = 1$. Which of these answers (d_1 and a) depend significantly on mode number n? Explain briefly.

9

ANTENNAS

9.1 INTRODUCTION

Until now we have been discussing the propagation of electromagnetic waves from one location to another principally by means of a guiding structure, usually a transmission line or waveguide. But from Chapter 2, we also recall that waves can radiate directly through space. *Antennas* act as transducers, coupling waves traveling in space to waves traveling in guides. Often the antenna will be designed to receive or transmit a single mode of the guiding structure.

Because most physical antennas can couple to an infinite number of plane waves coexisting in space, our traditional methods for solving boundary value problems are much too cumbersome to use for an arbitrarily shaped antenna. The basic problem is that we do not know the current distribution $\overline{J}$ on the antenna that satisfies all boundary conditions. Therefore, except for a few particularly simple examples, antennas are usually analyzed in an approximate fashion. We simply guess a physically reasonable current distribution for the antenna and then use (2.2.15) to find the vector potential field $\overline{A}$. From $\overline{A}$ we can use (2.1.22) and (1.2.2) to find the radiated fields $\overline{E}$ and $\overline{H}$. It is then necessary to check that these electric and magnetic fields at least approximately satisfy boundary conditions (4.1.13) and (4.1.16) at the surface of the antenna. In a more sophisticated calculation, these lowest-order electric and magnetic fields can be used to correct the assumed current distribution $\overline{J}$, leading to further accuracy in $\overline{E}$ and $\overline{H}$.

The simplest antenna is an isolated Hertzian dipole, discussed extensively in Section 2.3. Properties of the Hertzian dipole calculated in Section 2.3 include the antenna gain $G(\theta, \phi)$, characterizing the angular distribution of radiated power, and the radiation resistance R_{rad}, which relates the total power radiated by the antenna to the current flowing at its terminals. Section 9.2 discusses the radiation properties of Hertzian dipoles and other simple antennas, including short (but not infinitesimal) dipoles and half-wave dipoles. A more general treatment of long wire antennas is found in Section 9.7. Circuit models of antennas are explored in Section 9.3, followed by a discussion of antennas arrays and images in Section 9.4. The receiving and transmitting properties of antennas are discussed in Section 9.5, followed by a derivation of the reciprocity theorem in Section 9.6. Section 9.8 concludes with simple diffraction theory and a discussion of common aperture antennas.

9.2 SIMPLE DIPOLE ANTENNAS

In this section we shall assume that the current distribution $\overline{J}(\overline{r}')$ is specified on an isolated antenna in free space, enabling us to calculate the vector potential $\overline{A}(\overline{r})$ from the superposition integral (2.2.15):

$$\overline{A}(\overline{r}) = \int_{V'} \frac{\mu_0 \overline{J}(\overline{r}')}{4\pi |\overline{r} - \overline{r}'|} e^{-jk|\overline{r} - \overline{r}'|} dv' \qquad (9.2.1)$$

The antenna coordinate $\overline{r}'$ connects the origin and the infinitesimal volume element dv' through which current $\overline{J}(\overline{r}')$ flows and ranges over the entire volume of the antenna V'. The observer measuring the vector potential is located at position $\overline{r}$; these antenna and observer coordinates are illustrated in Figure 9.1(a). The electric and magnetic fields in the free space around the antenna are then related to

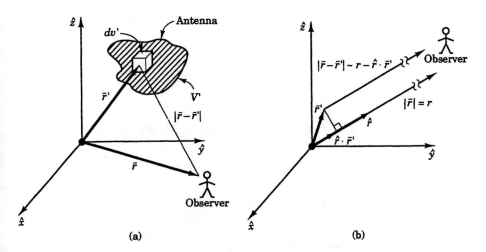

Figure 9.1 Observer and antenna coordinates.

the vector potential (9.2.1) by Faraday's and Ampere's laws (2.1.22) and (1.4.5):

$$\overline{H}(\overline{r}) = \frac{1}{\mu_0} \left(\nabla \times \overline{A}(\overline{r}) \right) \tag{9.2.2}$$

$$\overline{E}(\overline{r}) = \frac{1}{j\omega\epsilon_0} \left(\nabla \times \left(\nabla \times \overline{A}(\overline{r}) \right) \right) \tag{9.2.3}$$

Equations (9.2.1)–(9.2.3) are **exact** expressions for the fields surrounding an antenna provided that $\overline{J}(\overline{r}')$ is known. However, we are often interested only in fields far from the antenna because farfield radiation conveys power through space. (By contrast, fields near the antenna tend to store electric and magnetic energy.)

In the *far field*, defined as the region where $r = |\overline{r}| \gg |\overline{r}'|$ and $r \gg \lambda/2\pi$, (9.2.1) may be considerably simplified by making the *Fraunhofer farfield approximation*, also discussed in Section 2.4 for a two-element dipole array. If we assume that the observer is sufficiently far from the antenna, then

$$|\overline{r} - \overline{r}'| \simeq r - \hat{r} \cdot \overline{r}' \tag{9.2.4}$$

because $\overline{r} - \overline{r}'$ is approximately parallel to $\overline{r}$. The geometrical justification for (9.2.4) is given in Figure 9.1(b).[1] This approximation for $|\overline{r} - \overline{r}'|$ is substituted directly into the phase term $e^{-jk|\overline{r}-\overline{r}'|}$ in (9.2.1), which is sensitive to changes in $\overline{r}'$ as the antenna coordinate ranges over the entire antenna volume. Because the function $1/|\overline{r} - \overline{r}'|$ in (9.2.1) is insensitive to changes in $\overline{r}'$ when $|\overline{r}| \gg |\overline{r}'|$, we further approximate $|\overline{r} - \overline{r}'|^{-1}$ as r^{-1}. The farfield vector potential is therefore

$$\overline{A}_{\text{ff}}(\overline{r}) = \frac{\mu_0}{4\pi r} e^{-jkr} \int_{V'} \overline{J}(\overline{r}') e^{jk\hat{r}\cdot\overline{r}'} \, dv' \tag{9.2.5}$$

We make the important observation that (9.2.5) looks like a radially propagating plane wave so that the substitution $\nabla \rightarrow -jk\hat{r}$ can be made in (9.2.2) and (9.2.3).[2]

---

1. We can also derive (9.2.4) mathematically by writing

$$|\overline{r} - \overline{r}'| = \sqrt{(\overline{r} - \overline{r}') \cdot (\overline{r} - \overline{r}')}$$
$$= \sqrt{r^2 - 2\overline{r} \cdot \overline{r}' + |\overline{r}'|^2}$$

After factoring an r from this expression and recalling that the Taylor expansion for $\sqrt{1-x}$ is $1 - \frac{1}{2}x$ if x is small ($r \gg |\overline{r}'|$), we find that

$$|\overline{r} - \overline{r}'| \cong r - \frac{\overline{r} \cdot \overline{r}'}{r} + \frac{|\overline{r}'|^2}{2r}$$

This last term does not contribute to the phase of $e^{jk|\overline{r}-\overline{r}'|}$ if $k|\overline{r}'|^2/2r \ll \pi/8$ or $r \gg 8|\overline{r}'|^2/\lambda$ where the choice of maximum phase variation $\pi/8$ is somewhat arbitrary.

2. We have tacitly assumed that $\overline{A}_{\text{ff}} = \overline{A}_0 e^{-jkr}$ where $\overline{A}_0$ is a constant when we replace ∇ by $-jk\hat{r}$. Because $\overline{A}_0$ is proportional to $1/r$, ∇ should actually be replaced by $(-jk - r^{-1})\hat{r}$. However, in the far field ($kr \gg 1$ or $r \gg \lambda/2\pi$), the additional r^{-1} term can be neglected.

The farfield electric and magnetic fields are thus

$$\overline{H}_{ff}(\overline{r}) = \frac{\nabla \times \overline{A}_{ff}(\overline{r})}{\mu_0} = \frac{-jk\hat{r} \times \overline{A}_{ff}(\overline{r})}{\mu_0} \tag{9.2.6}$$

and

$$\overline{E}_{ff}(\overline{r}) = \frac{\nabla \times \overline{H}_{ff}(\overline{r})}{j\omega\epsilon_0} = -\eta_0 \hat{r} \times \overline{H}_{ff}(\overline{r})$$

$$= j\omega\hat{r} \times \left(\hat{r} \times \overline{A}_{ff}(\overline{r})\right) \tag{9.2.7}$$

In the farfield region ($r \gg |\overline{r}'|$ and $r \gg \lambda/2\pi$), we conclude that $\overline{E}_{ff}(\overline{r})$ and $\overline{H}_{ff}(\overline{r})$ are both perpendicular to the direction of wave propagation $\hat{r}$ and also to each other, which is the hallmark of a uniform plane wave. The ratio of $|\overline{E}_{ff}|$ to $|\overline{H}_{ff}|$ is just $\eta_0 = 377\ \Omega$ in free space, and the Poynting vector $\overline{S}_{ff} = \overline{E}_{ff} \times \overline{H}_{ff}^*$ is $\hat{r}$-directed.

We now find the farfield vector potential (9.2.5) for several dipole antennas, the simplest being the *Hertzian dipole* discussed in Section 2.3. A $\hat{z}$-directed Hertzian dipole radiator consists of two reservoirs of equal and opposite charge q located on the $\hat{z}$-axis and separated by an infinitesimal distance d such that the electric dipole moment qd is finite. If the charge is oscillating between the reservoirs at frequency ω, a uniform current $I = j\omega q$ exists in the dipole. The corresponding current density is then $\overline{J}(\overline{r}') = \hat{z}\, I d\, \delta(\overline{r}')$ as illustrated in Figure 2.2.[3] Substituting this expression for $\overline{J}(\overline{r}')$ into (9.2.1) or (9.2.5) yields

$$\overline{A}(\overline{r}) = \overline{A}_{ff}(\overline{r}) = \hat{z}\, \frac{\mu_0 I d}{4\pi r} e^{-jkr}$$

$$= (\hat{r}\cos\theta - \hat{\theta}\sin\theta)\frac{\mu_0 I d}{4\pi r} e^{-jkr} \tag{9.2.8}$$

The farfield vector potential of the Hertzian dipole is therefore also the exact vector potential. Equations (9.2.6) and (9.2.7) give

$$\overline{E}_{ff}(\overline{r}) = \hat{\theta}\, \frac{j\eta_0 k I d}{4\pi r} e^{-jkr} \sin\theta \tag{9.2.9}$$

$$\overline{H}_{ff}(\overline{r}) = \hat{\phi}\, \frac{jk I d}{4\pi r} e^{-jkr} \sin\theta \tag{9.2.10}$$

where we see explicitly that $\overline{E}_{ff}(\overline{r})$ and $\overline{H}_{ff}(\overline{r})$ are both perpendicular to $\hat{r}$ and $|\overline{E}_{ff}|/|\overline{H}_{ff}| = \eta_0$. Section 2.3 contains an alternate derivation of these results and includes an illustration of the way in which spherical coordinates are defined in Figure 2.3.

---

3. We shall henceforth refer to the product Id as the *dipole moment*, which is just $j\omega$ times the electric dipole moment qd.

The time-averaged Poynting power for the far fields radiated by a Hertzian dipole can then easily be computed as

$$\langle \overline{S}_{\text{ff}}(\overline{r}) \rangle = \frac{1}{2} \text{Re} \left\{ \overline{E}_{\text{ff}}(\overline{r}) \times \overline{H}_{\text{ff}}^*(\overline{r}) \right\} = \hat{r} \frac{\eta_0}{2} \left| \frac{kId}{4\pi r} \right|^2 \sin^2 \theta \qquad (9.2.11)$$

which for a $\hat{r}$-propagating uniform plane wave is also given by $\langle \overline{S}_{\text{ff}} \rangle = \hat{r} |\overline{E}_{\text{ff}}|^2 / 2\eta_0$. The total power radiated by the Hertzian dipole is thus

$$P_{\text{rad}} = \int_0^{2\pi} d\phi \int_0^{\pi} r^2 \sin\theta d\theta \ \langle \overline{S}_{\text{ff}}(\overline{r}) \cdot \hat{r} \rangle = \frac{\eta_0}{12\pi} |kId|^2 \qquad (9.2.12)$$

and the radiation resistance R_{rad} of the Hertzian dipole is

$$R_{\text{rad}} = \frac{P_{\text{rad}}}{|I|^2/2} = \frac{\eta_0}{6\pi} (kd)^2 \cong 20 (kd)^2 \qquad (9.2.13)$$

The *radiation resistance* is an important intrinsic property of any antenna, describing the behavior of that antenna as a component in an equivalent electrical circuit. We shall develop this viewpoint further in Section 9.3; for now we merely recognize that for an antenna to transmit or receive efficiently, its radiation resistance must be nonzero. The radiation resistance is **not** a measure of loss in the antenna; the Hertzian dipole is lossless but has a nonzero R_{rad} given by (9.2.13). Rather, the radiation resistance is analogous to the real characteristic impedance Z_0 of a transmission line.

The *antenna gain* $G(\theta, \phi)$ is defined as the ratio of the farfield power density $\langle \overline{S}_{\text{ff}} \cdot \hat{r} \rangle$ radiated in the (θ, ϕ) direction to the power density radiated by an isotropic antenna that would emit a uniform flux $\langle \overline{S}_{\text{isotropic}} \cdot \hat{r} \rangle = P/4\pi r^2$. Therefore,

$$G(\theta, \phi) = \frac{\langle \overline{S}_{\text{ff}}(\overline{r}) \cdot \hat{r} \rangle}{P/4\pi r^2} \qquad (9.2.14)$$

where P is the total power available to the antenna at its input. Because some of this power is lost in a mismatched, dissipative antenna, the full transmitter power P is not generally radiated and $P_{\text{rad}} \leq P$.[4] The ratio P_{rad}/P is called the *antenna efficiency* η_{rad}, where $0 \leq \eta_{\text{rad}} \leq 1$ and $\eta_{\text{rad}} = 1$ only for a lossless, matched antenna. The efficiency η_{rad} is typically slightly less than unity. Another useful antenna function is therefore the *directivity* $D(\theta, \phi)$, where

$$D(\theta, \phi) = \frac{\langle \overline{S}_{\text{ff}} \cdot \hat{r} \rangle}{P_{\text{rad}}/4\pi r^2} = \frac{G(\theta, \phi)}{\eta_{\text{rad}}} \qquad (9.2.15)$$

---

4. The self-impedance of a mismatched antenna differs from the impedance of the guiding structure. For maximum power deposition, the impedance of the guiding structure should be equal to the complex conjugate of the antenna's self-impedance; such a configuration is *matched*.

The directivity is simply the gain of a lossless, matched antenna and it has the useful property that

$$\int_0^{2\pi} d\phi \int_0^\pi \sin\theta d\theta \, D(\theta,\phi) = 4\pi \qquad (9.2.16)$$

for all antennas, as can easily be shown by combining (9.2.12) and (9.2.15). The *radiation pattern* $p(\theta,\phi)$ is another function proportional to the gain that is sometimes useful:

$$p(\theta,\phi) = \frac{G(\theta,\phi)}{\max\{G(\theta,\phi)\}} = \frac{D(\theta,\phi)}{\max\{D(\theta,\phi)\}} \qquad (9.2.17)$$

The radiation pattern is therefore just the gain pattern normalized to unity. Because $G(\theta,\phi)$ is measured experimentally, it is the most widely utilized of the antenna functions. As we shall be discussing primarily lossless, matched antennas in this chapter, directivity and gain are used somewhat interchangeably here.

Substituting (9.2.11) and (9.2.12) into (9.2.14) yields the gain of a lossless, matched Hertzian dipole

$$G(\theta,\phi) = \frac{3}{2}\sin^2\theta \qquad (9.2.18)$$

which is independent of wavelength and dipole moment Id. The radiation pattern for the Hertzian dipole is illustrated in Figure 2.4(c). The power is radiated primarily in the plane transverse to the dipole with nulls along the wire axis.

A *short-dipole antenna* has a length d, which, while not infinitesimal, is still much shorter than a wavelength ($d \ll \lambda/2\pi$), and a radius $R \ll d$. The current distribution is no longer necessarily uniform in such an antenna, but may vary along its length from $-d/2$ to $d/2$, where we cannot predict the exact distribution a priori. Hence, the current distribution for an ideal ($R \to 0$) $\hat{z}$-directed short-dipole antenna is

$$\overline{J}(\overline{r}') = \hat{z}\, I(z')\delta(x')\delta(y') \qquad (9.2.19)$$

where $\delta(x')$ is a one-dimensional Dirac delta function such that

$$\int_{-\infty}^\infty f(x-x')\delta(x')dx' = f(x)$$

for any analytic function $f(x)$. Note that the two terminals that connect the short dipole to its external circuit are located at $z = 0^+$ and $z = 0^-$ as illustrated in Figure 9.2, which means that the dipole is *center-fed*. Substituting (9.2.19) into (9.2.5) yields the farfield vector potential

$$\overline{A}_{\mathrm{ff}}(\overline{r}) = \hat{z}\,\frac{\mu_0}{4\pi r}\,e^{-jkr} \int_{-d/2}^{d/2} I(z')\,e^{jkz'\cos\theta}\,dz' \qquad (9.2.20)$$

where $\hat{r}\cdot\overline{r}' = \hat{r}\cdot z'\hat{z} = z'\cos\theta$. But because this dipole is short compared to a wavelength ($kd \ll 1$), the phase factor $e^{jkz'\cos\theta}$ is essentially constant and approximately unity for the entire range of the integral. Therefore,

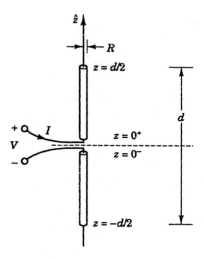

Figure 9.2   Short-dipole antenna.

$$\overline{A}_{\text{ff}}(\overline{r}) \cong \hat{z}\,\frac{\mu_0}{4\pi r}\,e^{-jkr}\int_{-d/2}^{d/2} I(z')dz' \tag{9.2.21}$$

is the farfield vector potential for a short dipole.

If we define an *effective dipole length* d_{eff} by the relation

$$I\,d_{\text{eff}} \equiv \int_{-d/2}^{d/2} I(z')dz' \tag{9.2.22}$$

we can rewrite (9.2.21) as

$$\overline{A}_{\text{ff}}(\overline{r}) \cong \hat{z}\,\frac{\mu_0 I\,d_{\text{eff}}}{4\pi r}\,e^{-jkr} \tag{9.2.23}$$

which is the same expression as that of the Hertzian dipole (9.2.8) with d replaced by d_{eff}. If the current distribution for the short dipole is uniform, then $d_{\text{eff}} = d$ (Hertzian dipole case).

It is not easy to calculate the exact current distribution of a short dipole because it has finite length. A physically reasonable guess for the current is a distribution that goes to zero at the free (open-circuited) ends of the dipole wires, and the current distribution is often assumed to be triangular when $d \ll \lambda/2\pi$, as shown in Figure 9.3.[5] If the magnitude of the current at the center of the dipole ($z' = 0$) is I and the current at $z' = \pm d/2$ is zero, then the effective dipole length for a triangular current distribution is

----

5. A more detailed discussion of the current distribution on long wires is deferred to Section 9.7.

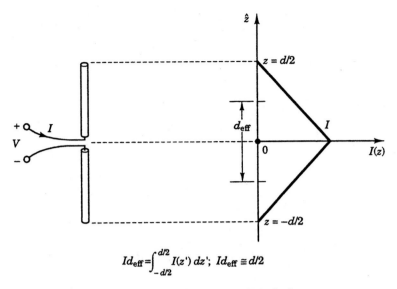

$$Id_{\text{eff}} = \int_{-d/2}^{d/2} I(z')\, dz'; \quad Id_{\text{eff}} \cong d/2$$

Figure 9.3   Current distribution on a short-dipole antenna.

$$d_{\text{eff}} = \frac{1}{I} \int_{-d/2}^{d/2} I\left(\frac{d/2 - |z'|}{d/2}\right) dz' = \frac{d}{2} \tag{9.2.24}$$

Note that $d_{\text{eff}} \le d \ll \lambda/2\pi$ must be true for a short-dipole antenna.

Because the expression for the farfield vector potential of a short dipole (9.2.23) is essentially identical to that of the Hertzian dipole (9.2.8), the entire discussion of Hertzian dipoles also applies to short dipoles with the substitution $d \to d_{\text{eff}}$ in (9.2.8)–(9.2.13). The gain of a short dipole is equal to that of a Hertzian dipole because its antenna pattern is independent of dipole length, but the radiation resistance of the short dipole is $\sim 20(kd_{\text{eff}})^2$.

The half-wave dipole illustrated in Figure 9.4 has a length $d = \lambda/2$, which is no longer small compared to a wavelength. This is the first wire antenna for which the propagation delays between signals originating from different parts of the wire cannot be neglected as they were for short and Hertzian dipoles. The farfield vector potential of the half-wave dipole is thus given by (9.2.20) where $kz' \cos\theta$ is now not constrained to be much smaller than unity. We approximate the current on this antenna as

$$I(z') = \hat{z}\, I \cos kz' \tag{9.2.25}$$

for $-\lambda/4 \le z' \le \lambda/4$. This current has a maximum value at $z' = 0$ and goes to zero at the (open-circuited) ends of the wire. A more careful analysis of the current distributions on long wire antennas is given in Section 9.7 by treating the wires as approximate *TEM* lines.

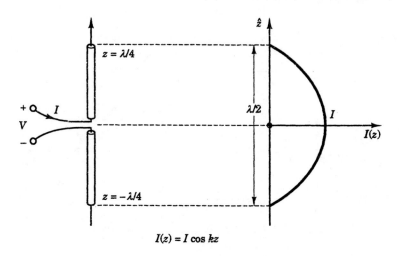

$$I(z) = I \cos kz$$

Figure 9.4 Current distribution on a half-wave dipole.

If we substitute (9.2.25) into (9.2.20), we find that

$$\overline{A}_{ff}(\overline{r}) = \hat{z}\, \frac{\mu_0}{4\pi r}\, e^{-jkr} \int_{-\lambda/4}^{\lambda/4} I \cos kz'\, e^{jkz'\cos\theta}\, dz'$$

$$= \hat{z}\, \frac{\mu_0}{4\pi r}\, e^{-jkr} \int_{-\lambda/4}^{\lambda/4} \frac{I}{2} \left(e^{jkz'} + e^{-jkz'} \right) e^{jkz'\cos\theta}\, dz'$$

$$= \hat{z}\, \frac{\mu_0 I}{4\pi r}\, e^{-jkr} \left[\frac{e^{jkz'(1+\cos\theta)}}{2jk(1+\cos\theta)} + \frac{e^{jkz'(-1+\cos\theta)}}{2jk(-1+\cos\theta)} \right]_{-\lambda/4}^{\lambda/4}$$

$$= \hat{z}\, \frac{\mu_0 I}{2k\pi r}\, e^{-jkr} \frac{\cos(\frac{\pi}{2}\cos\theta)}{\sin^2\theta} \qquad (9.2.26)$$

where some of the intervening algebra has been omitted. The farfield electric field of the half-wave dipole is easily found from (9.2.7)

$$\overline{E}_{ff}(\overline{r}) = j\omega\hat{r} \times \left[\hat{r} \times \overline{A}_{ff}(\overline{r}) \right]$$

$$= \hat{\theta}\, \frac{j\eta_0 I}{2\pi r}\, e^{-jkr} \frac{\cos(\frac{\pi}{2}\cos\theta)}{\sin\theta} \qquad (9.2.27)$$

where $\hat{r} \times \hat{z} = -\hat{\phi}\sin\theta$ and $\hat{r} \times (-\hat{\phi}) = \hat{\theta}$. The gain of the half-wave dipole is proportional to the square of (9.2.27) and is sketched in Figure 9.5 along with that of a Hertzian dipole for comparison. In both cases, the farfield electric field is a $\hat{\theta}$-polarized, radially propagating plane wave. Note that by (9.2.27) and L'Hôpital's rule, $\overline{E}_{ff}(\theta = 0) = 0$.

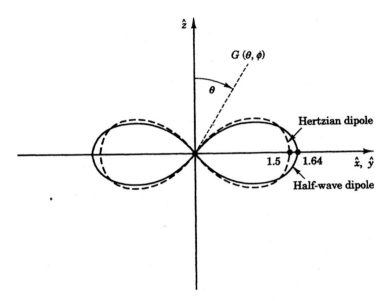

Figure 9.5 Gain of half-wave and Hertzian dipoles.

The total power radiated by this half-wave dipole is therefore

$$P_{rad} = \int_0^{2\pi} d\phi \int_0^{\pi} r^2 \sin\theta d\theta \frac{1}{2\eta_0} \left| \frac{\eta_0 I}{2\pi r} \right|^2 \frac{\cos^2(\frac{\pi}{2}\cos\theta)}{\sin^2\theta}$$

$$= \frac{\eta_0}{4\pi} |I|^2 \int_0^{\pi} \frac{\cos^2(\frac{\pi}{2}\cos\theta)}{\sin\theta} d\theta$$

$$= 36.5|I|^2 \tag{9.2.28}$$

where the θ-integral in (9.2.28) is computed numerically to be 1.22 and the radiation resistance is therefore $2P_{rad}/|I|^2 = 73\ \Omega.^6$ The gain is then found from (9.2.14). For a lossless, matched half-wave dipole antenna, it is

$$G(\theta, \phi) = 1.64 \frac{\cos^2(\frac{\pi}{2}\cos\theta)}{\sin^2\theta} \tag{9.2.29}$$

The half-wave dipoles are slightly more directive ($D_{max} = 1.64$) than Hertzian or short dipoles ($D_{max} = 1.5$), which are in turn more directive than isotropic radiators ($D = 1$).

This section has demonstrated that the farfield vector potential $\overline{A}_{ff}(\overline{r})$ [and

---

6. The 73 Ω radiation resistance of a half-wave dipole can easily be matched to broad-band active circuits, making these dipole antennas much more attractive than short dipoles with hard-to-match impedances ($R_{rad} \ll 1$).

therefore $\overline{E}_{ff}(\overline{r})$ and $\overline{H}_{ff}(\overline{r})$] may be readily calculated if the current distribution on an antenna is known. From $\overline{E}_{ff}(\overline{r})$ or $\overline{H}_{ff}(\overline{r})$ the useful antenna quantities $G(\theta, \phi)$, P_{rad}, and R_{rad} may then be computed. The next section discusses the transmitting properties of arbitrary antennas from a more general perspective, showing how an antenna can be modeled by a circuit.

Example 9.2.1

A 1-m dipole antenna is to transmit emergency signals at a wavelength of 2 m to a search-and-rescue satellite in a geosynchronous orbit approximately 36,000 km away. If the receiver on the satellite can detect electric field strengths of 1 $\mu V/m$ and we neglect the effects of ground reflections, what is the minimum possible antenna terminal current sufficient to permit detection by the satellite? What is the total power radiated by the antenna in this case?

Solution: A satellite in a geosynchronous orbit always remains in the same position with respect to Earth; i.e., the period of its orbit is exactly one day. This special orbit occurs at 36,000 km from the Earth's surface.[7] Therefore, if the search-and-rescue satellite is directly overhead from the emergency transmitter, it will remain overhead even as the Earth rotates.

We recognize that the 1-m dipole is one-half wavelength long, so the half-wave dipole antenna has a farfield electric field pattern given by (9.2.27), maximum when $\theta = \pm\pi/2$ relative to the dipole axis. In order to minimize transmitter power, we orient the emergency dipole so that it transmits maximally toward the satellite. For a satellite directly overhead, this means orienting the antenna parallel to the ground. The magnitude of the farfield electric field in this case is then $|\overline{E}_{ff}| = |\eta_0 I/2\pi r|$ so

$$I_{min} = \frac{2\pi r E_{min}}{\eta_0} = \frac{2\pi \times 3.6 \times 10^7 \text{ m} \times 10^{-6} \text{ V/m}}{377 \text{ }\Omega} = 0.60 \text{ A}$$

The total power radiated by this dipole is then just

$$P_{rad} = \frac{1}{2} R_{rad}|I_{min}|^2 = \frac{1}{2} \times 73 \text{ }\Omega \times (0.60 \text{ A})^2 = 13.1 \text{ W}$$

where we recall that the half-wave dipole has a radiation resistance of 73 Ω.

9.3 CIRCUIT PROPERTIES OF TRANSMITTING ANTENNAS

The analysis and design of systems incorporating antennas becomes much simpler if the antennas can be modeled by simple equivalent circuits that describe how voltages and currents are related at the terminals of the antennas. Unless an antenna is

---

7. The radius of the geosynchronous orbit can be calculated from simple classical mechanics. The magnitude of the force on the satellite exerted by the Earth is $|\overline{f}| = GmM/R^2 = m\omega^2 R$, where $\omega = 2\pi/T$ is the angular frequency of the satellite and T is the time necessary for the satellite to orbit, i.e, one day. The masses m and $M = 5.98 \times 10^{24}$ kg refer to the satellite and Earth, respectively, and $G = 6.67 \times 10^{-11}$ N $\cdot$ m^2/kg^2 is the universal gravitation constant. Therefore, $R = (GMT^2/4\pi^2)^{1/3} = 42,260$ km. But R is the distance from the satellite to the center of the Earth, so the geosynchronous orbit lies at 42,260 - 6370 = 35,890 km, where 6370 km is the radius of the Earth.

embedded in a nonlinear medium, which is beyond the scope of this text, the antenna as seen from its terminals can **always** be modeled by a complex impedance $Z = R + jX$ where R includes dissipative (lossy) as well as radiative effects and the reactance jX corresponds to energy stored in the near fields of the antenna. We have already defined the radiation resistance

$$R_{\text{rad}} = \frac{P_{\text{rad}}}{|I|^2/2} \tag{9.3.1}$$

where P_{rad} is the total power radiated into the far field of the antenna. The radiation resistance R_{rad} is an important contributor to the real part of the antenna resistance R.

The complex impedance of an arbitrary antenna is found using the same techniques employed earlier for transmission line circuits and resonators. The complex Poynting theorem is applied to the general antenna in Figure 9.6. We recall from (1.6.29) that the integral form of the complex Poynting theorem is

$$-\frac{1}{2} \int_V \overline{E} \cdot \overline{J}^* \, dv = \frac{1}{2} \oint_A (\overline{E} \times \overline{H}^*) \cdot \hat{n} \, da$$

$$+ \frac{j\omega}{2} \int_V (\overline{H}^* \cdot \overline{B} - \overline{E} \cdot \overline{D}^*) \, dv \tag{9.3.2}$$

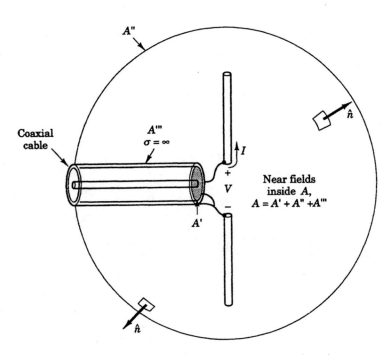

Figure 9.6 Closed surface used to calculate antenna impedance.

Even before we attempt to apply (9.3.2) to a real antenna, we can predict which circuit quantities will result from each of the terms. The $\overline{E} \cdot \overline{J}^*$ term in (9.3.2) describes how power is dissipated in the antenna (R_d), while the $\overline{E} \times \overline{H}^*$ term describes both how power is supplied to the antenna (Z) and how it is carried into the far field (R_{rad}). The terms $\overline{H}^* \cdot \overline{B} - \overline{E} \cdot \overline{D}^*$ describe how energy is stored in the near fields of the antenna (jX).

To apply (9.3.2) to a general antenna, we define a closed surface $A = A' + A'' + A'''$ made up of three subsurfaces enclosing a volume V, as shown in Figure 9.6. Surface A' intercepts all of the power flowing into the antenna by cutting across the feed line near the antenna terminals. Surface A'' encloses the antenna at a sufficiently large radius r that all of the reactive near fields are contained within A''. The surface A''' connects A' to A'' and is coincident with the perfectly conducting surface that forms the external guiding structure.

By integrating the Poynting power $\overline{E} \times \overline{H}^*$ over the surface A' incorporating the antenna terminals, we find

$$\int_{A'} \frac{1}{2} (\overline{E} \times \overline{H}^*) \cdot \hat{n}\, da = -\frac{1}{2} V I^* = -\frac{1}{2} Z |I|^2 \qquad (9.3.3)$$

which will enable us to find the antenna impedance $Z = R + jX$. Equation (9.3.3) simply equates the total power flowing into the antenna, found by integrating the Poynting power over the surface through which the antenna is fed, with the traditional circuit expression for input power. The voltage and current measured at the antenna terminals are V and I, respectively, and their ratio is defined as the complex impedance $Z = V/I$. The negative sign in (9.3.3) occurs because the surface normal to A' points **away** from the dipole but the power flows into the antenna for positive $\frac{1}{2} \text{Re}\{V I^*\}$.

We now integrate the Poynting power over surface A'' and A''':

$$\int_{A''} \frac{1}{2} (\overline{E} \times \overline{H}^*) \cdot \hat{n}\, da = \frac{1}{2} R_{rad} |I|^2 \qquad (9.3.4)$$

$$\int_{A'''} \frac{1}{2} (\overline{E} \times \overline{H}^*) \cdot \hat{n}\, da = 0 \qquad (9.3.5)$$

Equation (9.3.4) follows directly from the definition of radiation resistance in (9.3.1) since the integral of $\frac{1}{2} \text{Re}\{\overline{E} \times \overline{H}^*\}$ represents the total power radiated into the far field of the antenna. Because A'' is sufficiently far from the antenna, $\overline{S} = \overline{E} \times \overline{H}^*$ has only an $\hat{r}$-directed real part. In the farfield the fields are just those of a uniform, outwardly radiating plane wave as described in Section 9.2. The integral of $\overline{E} \times \overline{H}^*$ over the guiding structure A''' is zero because $\overline{E}$ is normal to both the conductor and A''', and therefore $\overline{E} \times \overline{H}^*$ is perpendicular to $\hat{n}'''$.

We now combine (9.3.2)–(9.3.5) to obtain a simple expression for the complex impedance of the antenna as seen from its terminals:

$$Z = R + jX = R_{\text{rad}} + \frac{1}{|I|^2} \left[\int_V \overline{E} \cdot \overline{J}^* \, dv + j\omega \int_V (\overline{H}^* \cdot \overline{B} - \overline{E} \cdot \overline{D}^*) \, dv \right]$$
$$(9.3.6)$$

The physical significance of each of these terms is easy to interpret. The resistive part of the antenna terminal impedance R_{rad} has already been associated with the total power radiated away by the antenna. The second term in (9.3.6) arises from conductive losses in the antenna structure as well as in the medium immediately surrounding it. Although these losses are normally quite small, it can happen that the radiation resistance of a short dipole is sufficiently small that the dissipative losses are comparable to the radiated power and significant mismatch problems occur.[8] Thus, although the effect of dissipative losses can be reduced by using low-loss materials, it is generally easier to use antennas for which R_{rad} is large. The total resistive impedance of the antenna is thus $R = R_{\text{rad}} + R_d$, where

$$R_d = \frac{1}{|I|^2} \text{Re} \left\{ \int_V \overline{E} \cdot \overline{J}^* \, dv \right\} \qquad (9.3.7)$$

The antenna reactance jX comes from the energy stored in the near fields of the antenna as well as from reactive currents $\overline{J}$ (currents that are 90° out of phase with the electric field). We find jX by taking the imaginary part of (9.3.6):

$$X = \frac{\omega}{|I|^2} \int_V (\overline{H}^* \cdot \overline{B} - \overline{E} \cdot \overline{D}^*) \, dv + \frac{1}{|I|^2} \text{Im} \left\{ \int_V \overline{E} \cdot \overline{J}^* \, dv \right\} \qquad (9.3.8)$$

When the medium is nondispersive so that $\overline{B} = \mu \overline{H}$ and $\overline{D} = \epsilon \overline{E}$, (9.3.8) can be expressed in terms of the average stored electric and magnetic energies using (1.6.31) and (1.6.32)

$$X = \frac{4\omega}{|I|^2} (\langle w_m \rangle - \langle w_e \rangle) + \frac{1}{|I|^2} \text{Im} \left\{ \int_V \overline{E} \cdot \overline{J}^* \, dv \right\} \qquad (9.3.9)$$

Note that stored magnetic energy contributes a positive (inductive) value of X, while stored electric energy contributes a negative (capacitive) value.[9] Hertzian and short-dipole antennas store energy predominantly in their electric near field so their reactance is capacitive [as may be shown by calculating (9.3.9) explicitly using the full electric and magnetic fields (2.3.8) and (2.3.7)]. Conversely, a small loop antenna is inductive, as is shown below in Example 9.3.1. This approach to finding the complex impedance Z applies to **any** antenna for which the antenna current distribution $\overline{J}(\overline{r}')$ and therefore $\overline{E}(\overline{r})$ and $\overline{H}(\overline{r})$ are known.

---

8. A 1-MHz short-dipole antenna with $d_{\text{eff}} = 0.5$ m has a radiation resistance $R_{\text{rad}} \cong 20(kd_{\text{eff}})^2 = 0.0088 \ \Omega$.

9. The reactive elements that typify electric and magnetic energy storage devices are capacitors ($X = -1/\omega C$) and inductors ($X = \omega L$), respectively.

Example 9.3.1

Calculate the gain G and radiation resistance R_{rad} of the small loop antenna illustrated in Figure 9.7(a), where $a \ll \lambda/2\pi$. Repeat the problem for the case where the antenna has N turns in the loop.

Solution: In order to find the farfield $\overline{E}_{ff}(\overline{r})$, we first compute $\overline{A}_{ff}(\overline{r})$ using (9.2.5) for a $\hat{\phi}$-directed current I flowing in a small loop of radius a:

$$\overline{A}_{ff}(\overline{r}) = \frac{\mu_0}{4\pi r} e^{-jkr} \int_{V'} \overline{J}(\overline{r}') e^{jk\hat{r}\cdot\overline{r}'} dv'$$

$$= \frac{\mu_0}{4\pi r} e^{-jkr} \int_0^{2\pi} \hat{\phi}' I e^{jk\hat{r}\cdot\overline{r}'} a \, d\phi' \tag{1}$$

Because the current is azimuthally symmetric, it does not matter at which value of ϕ

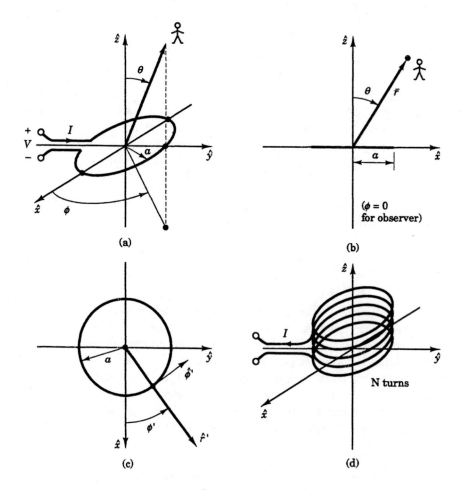

Figure 9.7   Small loop antenna in Example 9.3.1.

we observe the far field, so we choose $\phi = 0$. In this case, $\hat{r} = \hat{x} \sin\theta + \hat{z} \cos\theta$ and $\bar{r}' = \hat{x} a \cos\phi' + \hat{y} a \sin\phi'$, as illustrated in Figure 9.7(b)-(c). We represent $\hat{\phi}'$ in Cartesian coordinates as $\hat{\phi}' = -\hat{x} \sin\phi' + \hat{y} \cos\phi'$, and $\overline{A}_{ff}(r, \theta, \phi = 0)$ becomes

$$\overline{A}_{ff}(r, \theta, \phi = 0) = \frac{\mu_0 I a}{4\pi r} e^{-jkr} \int_0^{2\pi} (-\hat{x} \sin\phi' + \hat{y} \cos\phi') e^{jka \sin\theta \cos\phi'} d\phi' \quad (2)$$

The integral in (2) is trivial for $a \ll \lambda/2\pi$ $(ka \ll 1)$ because we can approximate $e^{jka \sin\theta \cos\phi'}$ by $1 + jka \sin\theta \cos\phi'$. We then integrate ϕ' from 0 to 2π, noticing that

$$\int_0^{2\pi} \sin\phi' \, d\phi' = \int_0^{2\pi} \cos\phi' \, d\phi' = \int_0^{2\pi} \sin\phi' \cos\phi' \, d\phi' = 0$$

and

$$\int_0^{2\pi} \cos^2\phi' \, d\phi' = \pi$$

Equation (2) thus becomes

$$\overline{A}_{ff}(r, \theta, \phi = 0) \cong \hat{y} \frac{j\mu_0 k I \pi a^2}{4\pi r} e^{-jkr} \sin\theta$$

where $\hat{y}$ is actually $\hat{\phi}$ when $\phi = 0$. The farfield vector potential is thus

$$\overline{A}_{ff}(r, \theta, \phi) \cong \hat{\phi} \frac{j\mu_0 k I \pi a^2}{4\pi r} e^{-jkr} \sin\theta \quad (3)$$

and the farfield $\overline{E}_{ff}$ is computed from (9.2.7)

$$\overline{E}_{ff}(\bar{r}) = j\omega\hat{r} \times \left(\hat{r} \times \overline{A}_{ff}(\bar{r})\right) = \hat{\phi} \frac{\eta_0 k^2 I \pi a^2}{4\pi r} e^{-jkr} \sin\theta \quad (4)$$

which looks very similar to the farfield magnetic field of the Hertzian dipole in (9.2.10). In fact, a small current loop is the dual of the short electric dipole and is often called a magnetic dipole. Because the farfield power for both electric and magnetic dipoles is proportional to $\sin^2\theta$, the lossless, matched gain of both dipoles is simply $G(\theta, \phi) = (3/2) \sin^2\theta$ where the 3/2 coefficient is found from (9.2.16). The radiated power is found from (9.2.12)

$$P_{\text{rad}} = \int_0^{2\pi} d\phi \int_0^{\pi} r^2 \sin\theta d\theta \frac{1}{2\eta_0} \left| \frac{\eta_0 k^2 I \pi a^2}{4\pi r} \right|^2 \sin^2\theta$$

$$= \frac{\pi}{12} \eta_0 (ka)^4 |I|^2 \quad \text{W} \quad (5)$$

and so the radiation resistance of this loop is

$$R_{\text{rad}} = \frac{P_{\text{rad}}}{|I|^2/2} = \frac{\pi}{6} \eta_0 (ka)^4 \cong 3.1 \times 10^5 \, (a/\lambda)^4 \, \Omega \quad (6)$$

Note that the radiation resistance of the electric dipole is $R_{\text{rad}} = 20 \, (kd)^2 \cong 790 \, (d/\lambda)^2$ so the radiation resistances of the two dipoles are different functions of wavelength and antenna size.

If the loop has N turns as shown in Figure 9.7(d), then the current flowing is $\hat{\phi} NI$, and every I in (1)–(5) can be replaced by NI. The electric and magnetic fields are thus N times larger than before, and P_{rad} is N^2 times its original value. The radiation resistance of the N-turn loop is just $R_{rad} = 2P_{rad}/|I|^2 \cong 3.1 \times 10^5 N^2 (a/\lambda)^4$, but the gain remains unchanged. It is easier to obtain large values of R_{rad} with a current loop than with a comparable-size short dipole, and loop antennas are often found in small, portable AM/FM radios. Loading the loop with a ferrite ($\mu \gg \mu_0$) also increases R_{rad}. Thomas Edison once built such an antenna to detect radio emission from the sun at audio frequencies using a very long coil wound around the surface of a large iron ore deposit. Failure to observe any radio emissions resulted in part from reflection by the ionospheric plasma (then unknown) of all solar radiation emitted below the 10-MHz ionospheric plasma frequency.

9.4 DIPOLE ARRAYS AND IMAGES

In Section 2.4 we discussed radiation from two-element arrays of Hertzian dipoles and noted that even with just two elements, a wide variety of radiation patterns could be generated by varying the separation distance and the current magnitudes and phases of the dipoles. This section will discuss arrays of more than two identical elements, which can be useful for synthesizing arbitrary radiation patterns.

Because Maxwell's equations are linear, the total electric and magnetic fields radiated by an array of dipoles can easily be found by superposing the fields generated by individual dipole elements in the array, provided we know the current distribution on each dipole. We shall neglect the mutual coupling of individual dipole elements, discussed in Section 9.5, because interactions between dipoles that can perturb the assumed currents are not important for well-spaced dipole arrays with fixed excitations.

We start by calculating the total electric field that would be measured by an observer located at (r, θ, ϕ) far from an array of N $\hat{z}$-directed short dipoles located in a finite region around the origin. The total farfield electric field can be found by summing (9.2.9) over each of the N dipoles

$$\overline{E}_{ff}(r, \theta, \phi) = \sum_{i=0}^{N-1} \hat{\theta} \frac{j\eta_0 k I_i d_i}{4\pi r_i} e^{-jkr_i} \sin \theta_i \qquad (9.4.1)$$

where the dipole moment (maximum current times the effective dipole length) of the ith dipole is $I_i d_i$, the distance between the observer and the ith dipole is $r_i \equiv |\overline{r} - \overline{a}_i|$, and the angle between the $\hat{z}$-axis and the vector $\overline{r}_i$ is θ_i.

If the observer is located sufficiently far away from each dipole, we can make the Fraunhofer farfield approximation (9.2.4)

$$r_i \cong r - \hat{r} \cdot \overline{a}_i \qquad (9.4.2)$$

and

$$\theta_i \cong \theta \qquad (9.4.3)$$

which can be seen geometrically from Figure 9.1(b) with $\overline{a}_i$ replacing $\overline{r}'$. Since $k|\overline{a}_i|$ is not necessarily small compared to unity, each dipole contributes a different phase to the far field $\overline{E}_{ff}(\overline{r})$ when the array diameter exceeds $\cong \lambda/2\pi$, and (9.4.2) must be used to approximate the phase of each dipole. The factor $1/r_i$ in (9.4.1) is simply replaced by $1/r$ in the farfield region as described in Section 9.2.

In the Fraunhofer region, (9.4.1) may therefore be approximated as

$$\overline{E}_{\text{ff}}(r, \theta, \phi) \cong \left[\hat{\theta} \, \frac{j\eta_0 k I d}{4\pi r} \, e^{-jkr} \sin\theta \right] \left[\sum_{i=0}^{N-1} w_i \, e^{jk\hat{r} \cdot \bar{a}_i} \right] \qquad (9.4.4)$$

where the complex weighting factor $w_i \equiv I_i d_i / I d$ characterizes the excitation of the ith element, and all of the terms common to each dipole have been factored outside the summation sign. The suggestive way in which (9.4.4) has been written makes a general property of identical-element arrays quite evident; the farfield of such an array can always be written as a product of two functions called element and array factors.

The *element factor* $\overline{\mathcal{E}}(r, \theta, \phi)$ is just the electric field of a single radiating element, given by

$$\overline{\mathcal{E}}(r, \theta, \phi) = \hat{\theta} \, \frac{j\eta_0 k I d}{4\pi r} \, e^{-jkr} \sin\theta \qquad (9.4.5)$$

for dipole elements where the dipole moment $I d$ is located at the origin. The element factor contains no information about the spatial distribution or relative phasing of the array elements.

The *array factor* $F(\theta, \phi)$ is a scalar quantity characterizing the relative magnitudes and phases of dipole moments w_i as well as the location of elements in the array, but does not include any information about the radiation pattern of a single element. The array factor is also independent of the distance r from the observer to the origin. For the previously discussed N-dipole array with elements located at $\bar{a}_i$,

$$F(\theta, \phi) = \sum_{i=0}^{N-1} w_i \, e^{jk\hat{r} \cdot \bar{a}_i} \qquad (9.4.6)$$

and we can rewrite (9.4.4) very generally as

$$\overline{E}_{\text{ff}}(r, \theta, \phi) = \overline{\mathcal{E}}(r, \theta, \phi) F(\theta, \phi) \qquad (9.4.7)$$

where $\overline{\mathcal{E}}(r, \theta, \phi)$ may represent the far field of **any** type of radiating element. However, an important warning about the use of (9.4.7) must be emphasized—in order to decompose a total farfield radiation pattern into element and array factors, **all** of the elements in the array must be identical types of radiators (e.g., all short dipoles) that are identically oriented (e.g., all $\hat{z}$-directed). The only differences allowed between elements in this type of array are in the magnitudes and phases of the current excitations.

Once the total farfield electric field is known, it is a simple matter to compute the time-averaged Poynting vector and thus the gain and radiation pattern. The time-averaged Poynting vector is simply

$$\langle|\overline{S}(r, \theta, \phi)|\rangle = \frac{1}{2\eta_0}|\overline{E}_{\text{ff}}(\overline{r})|^2$$

$$= \frac{1}{2\eta_0}|\overline{\mathcal{E}}(r, \theta, \phi)|^2|F(\theta, \phi)|^2 \qquad (9.4.8)$$

and therefore the gain $G(\theta, \phi)$ is given by

$$G(\theta, \phi) = K|\overline{\mathcal{E}}(\theta, \phi)|^2|F(\theta, \phi)|^2 \qquad (9.4.9)$$

where K is a constant to be evaluated. For a lossless matched antenna, K is found by integrating (9.4.9) over all values of the solid angle $d\Omega = \sin\theta d\theta d\phi$ and setting the result equal to $4\pi\eta_{\text{rad}}$ where $\eta_{\text{rad}} = 1$ if the antenna structure is lossless and matched. The factor $\overline{\mathcal{E}}(\theta, \phi)$ is just the angular part of $\overline{\mathcal{E}}(r, \theta, \phi)$. For $\hat{z}$-directed Hertzian and short-dipole antennas, $|\overline{\mathcal{E}}(\theta, \phi)|^2 = \sin^2\theta$, while for half-wave dipoles $|\overline{\mathcal{E}}(\theta, \phi)|^2 = \cos^2[(\pi/2)\cos\theta]/\sin^2\theta$.

We wish to emphasize that although an antenna radiation pattern is three-dimensional and exists for all angles (θ, ϕ), we usually plot a two-dimensional projection of the pattern. Often the observer is in either the x–y plane ($\theta = \pi/2$) or the x–z plane ($\phi = 0$), making it critical to correctly identify the geometry of any given antenna.

We have already described radiation patterns for two-dipole arrays in Section 2.4, and we may therefore use (9.4.5) and (9.4.6) to confirm that the radiation pattern observed in the x–y plane ($\theta = \pi/2$) for two $\hat{z}$-directed Hertzian dipoles located on the $\hat{x}$-axis at $x = \pm D/2$ is given by (2.4.8). The geometry of this two-element array is shown in Figure 2.6. From (9.4.6), we compute the array factor to be

$$F(\theta = \pi/2, \phi) = \sum_{i=0}^{1} \text{w}_i\, e^{jk\hat{r}\cdot\overline{a}_i}$$

$$= 1\, e^{jk\hat{r}\cdot(-\hat{x}D/2)} + A\, e^{j\alpha}\, e^{jk\hat{r}\cdot(\hat{x}D/2)}$$

$$= e^{-jk\frac{D}{2}\cos\phi} + A\, e^{j\alpha}\, e^{jk\frac{D}{2}\cos\phi}$$

$$= e^{-jk\frac{D}{2}\cos\phi}\left(1 + A\, e^{j(\alpha+kD\cos\phi)}\right) \qquad (9.4.10)$$

where $\overline{a}_0 = -\hat{x}\, D/2$, $\overline{a}_1 = \hat{x}\, D/2$, $\text{w}_0 = 1$, and $\text{w}_1 = A\, e^{j\alpha}$. The angle between the $\hat{x}$-axis and the vector $\hat{r}$ in the x–y plane is just the usual polar coordinate ϕ. We also recall the angular part of the element factor here is $|\overline{\mathcal{E}}(\theta, \phi)|^2 = \sin^2\theta$ for a single dipole. But as θ is constant for an observer in the x–y plane, the element factor does not contribute any angular dependence to the total radiation pattern and therefore (9.2.17) and (9.4.10) may be combined for this two-element array, yielding

$$p(\theta = \pi/2, \phi) = \frac{\left|1 + A\,e^{j(\alpha + kD\cos\phi)}\right|^2}{\max\left|1 + A\,e^{j(\alpha + kD\cos\phi)}\right|^2}$$

$$= \frac{1 + A^2 + 2A\cos(kD\cos\phi + \alpha)}{(1 + A)^2}$$

which is identical to (2.4.8) as expected.

Another interesting type of array is a linear, equally spaced collection of N $\hat{z}$-directed dipoles, each with the same dipole moment magnitude but with a constant difference in the phase excitation between adjacent dipoles. If the dipoles are equally spaced along the $\hat{x}$-axis so that $\bar{a}_i = \hat{x}\,ia$ ($i = 0, 1, \ldots, N-1$), the weighting factors w_i are given by

$$w_i = e^{ji\psi} \tag{9.4.11}$$

where a and ψ are constants. This linear array is depicted in Figure 9.8. If the observer lies in the positive x–z plane ($x > 0$), then $\phi = 0$ and $\hat{r} \cdot \bar{a}_i = ia\sin\theta$. The array factor may be written as

$$F(\theta, \phi = 0) = \sum_{i=0}^{N-1} w_i\,e^{jk\hat{r}\cdot\bar{a}_i} = \sum_{i=0}^{N-1} e^{ji\psi}\,e^{jkia\sin\theta}$$

$$= \sum_{i=0}^{N-1}\left[e^{j(\psi + ka\sin\theta)}\right]^i \tag{9.4.12}$$

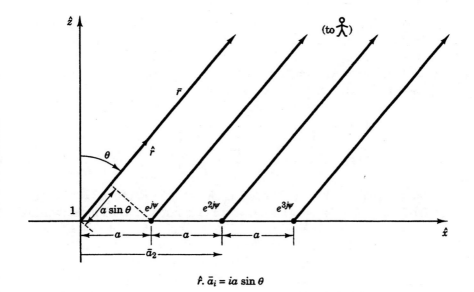

$\hat{r} \cdot \bar{a}_i = ia\sin\theta$

Figure 9.8 Linear dipole array.

Recalling that the geometric series $\sum_{i=0}^{N-1} r^i$ is equal to $(1 - r^N)/(1 - r)$, we evaluate (9.4.12):

$$F(\theta, \phi = 0) = \frac{1 - e^{jN(\psi + ka \sin\theta)}}{1 - e^{j(\psi + ka \sin\theta)}}$$

$$= \frac{e^{j\frac{N}{2}(\psi + ka \sin\theta)} \sin\frac{N}{2}(\psi + ka \sin\theta)}{e^{j\frac{1}{2}(\psi + ka \sin\theta)} \sin\frac{1}{2}(\psi + ka \sin\theta)} \tag{9.4.13}$$

If we include the angular part of the element factor for a Hertzian dipole $(\sin^2\theta)$, we can use (9.4.9) to find the gain of this lossless linear phased array

$$G(\theta, \phi = 0) \propto \sin^2\theta \, \frac{\sin^2\frac{N}{2}(\psi + ka \sin\theta)}{\sin^2\frac{1}{2}(\psi + ka \sin\theta)} \tag{9.4.14}$$

which is sketched in Figure 9.9 for $N = 7$, $ka = \pi$, and $\psi = -\pi/2$. Note that the peak gain is in the direction θ_{max} where $\psi + ka \sin\theta_{max} = 0$ or $\theta_{max} = -\sin^{-1}(\psi/ka)$. For the array in Figure 9.9, $\theta_{max} = 30°$ or $150°$. The direction of the antenna beam can thus be scanned by varying ψ, where the relationship between ψ and θ_{max} is approximately linear for small values of ψ. If we wanted to find the radiation pattern for this array in the x–y plane, the $\sin^2\theta$ term in (9.4.14) would be replaced by $\sin^2(\pi/2) = 1$ and $\hat{r} \cdot \bar{a}_i$ would be equal to $ia \cos\phi$ because $\hat{r}$ would now lie in the x–y plane. The gain for this observer would then be

$$G(\theta = \pi/2, \phi) \propto \frac{\sin^2\frac{N}{2}(\psi + ka \cos\phi)}{\sin^2\frac{1}{2}(\psi + ka \cos\phi)} \tag{9.4.15}$$

Another way to view the construction of an array pattern is by using phasors, suggested in Figure 9.10 for the same linear array of $N = 7$ dipoles where $\psi = -\pi/2$ and $a = \lambda/2$ $(ka = \pi)$. For an observer located in the x–y plane $(\theta = \pi/2)$, each phasor contribution to the array factor is plotted in the complex plane by representing it as a vector with both magnitude and phase. When $\phi = \pi/3$ for the seven-element linear array in Figure 9.10(a), $\psi + ka \cos\phi = 0$ and each of the $N = 7$ phasors equals unity

$$F(\theta = \pi/2, \phi = \pi/3) = \sum_{i=0}^{N-1} e^{j(\psi + ka \cos\phi)} \Bigg|_{\phi = \pi/3} = N$$

as shown in Figure 9.10(b). On the other hand, when $\psi + ka \cos\phi = 2\pi/N$ $(\phi = 38.2°$ for $N = 7)$, the phasors geometrically add to zero and a null in the radiation pattern results, seen in Figure 9.10(c). A further null results at $\psi + ka \cos\phi = -4\pi/7$ $(94.1°)$, as shown in Figure 9.10(d). Any intermediate value of the pattern may be found by summing the seven unit-length vectors, each successively rotated by $\psi + ka \cos\phi$ radians, and then squaring the magnitude of the resulting total vector as indicated in Figure 9.10(e). If the total pattern has an angular dependence resulting from the element factor, this must be included in the total radiation pattern

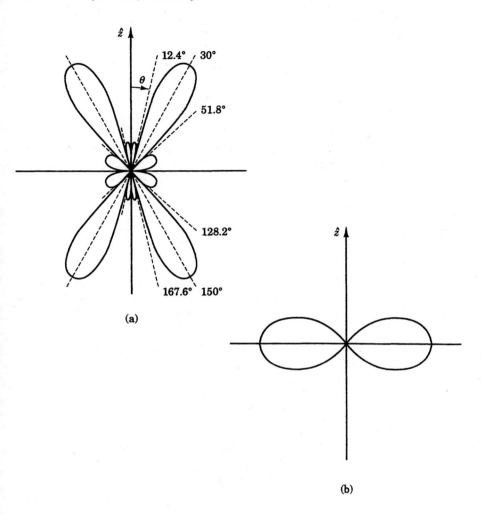

(a)

(b)

Figure 9.9   Gain of linear phased array; $N = 7$, $ka = \pi$, $\psi = -\pi/2$.

too. This phasor addition method is more general than the visible window method that follows because the phasor terms in the array factor need not be constrained to have equal magnitudes and linear phase relationships. **Any** set of N phasors may be geometrically added to give a total vector whose magnitude squared is proportional to the gain.

An alternative way to visualize the construction of these linear array patterns is the *visible window method* suggested in Figure 9.11. If we define $u = (\psi + ka \cos \phi)/2$, then we may rewrite the gain of a uniformly excited linear array in the x–y plane (9.4.15) as

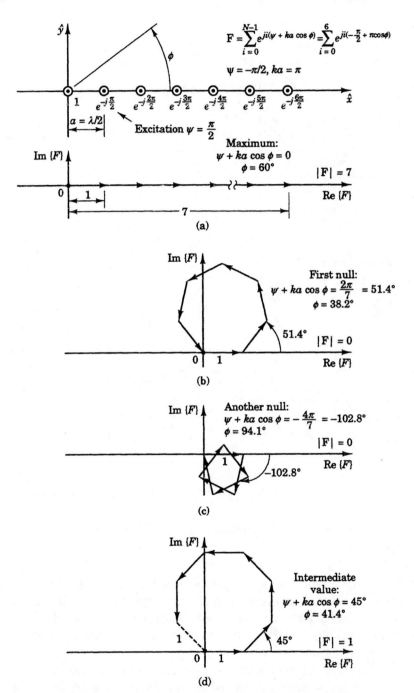

Figure 9.10 Construction of antenna pattern by phasor addition.

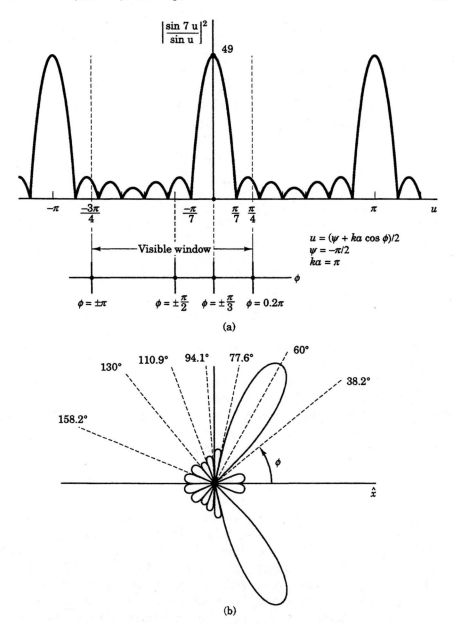

Figure 9.11   Visible window method of linear array construction.

$$G(u) \propto \frac{\sin^2 Nu}{\sin^2 u} \qquad (9.4.16)$$

where the periodic factor $\sin^2 Nu / \sin^2 u$ is plotted as a function of u in Figure 9.11(a) for the $N = 7$ array. Note that $\sin^2 Nu / \sin^2 u = N^2$ for $u = 0, \pm\pi$, $\pm 2\pi, \ldots$ and that the nulls in this expression occur at $u = m\pi/N$ where m is an integer not equal to $0, \pm N, \pm 2N, \ldots$. Because $\cos\phi = (2u - \psi)/ka$ is bounded between ± 1, only those values of u for which $(\psi - ka)/2 \le u \le (\psi + ka)/2$ can be part of the radiation pattern. These values of u are contained in the visible window. Values of u outside this visible window yield imaginary values of ϕ. Therefore, the radiation pattern is easily sketched in Figure 9.11(b) by "wrapping" the visible region into the angular plot between $\phi = 0$ and $\phi = 2\pi$. If the gain has an extra angular term arising from the element factor, as in (9.4.14), the radiation pattern for the array factor must be multiplied by that of the element factor, illustrated in Figure 9.9(b). A useful way of describing the radiation pattern is to plot the gain in decibels as a function of angle, where

$$G_{dB}(\theta, \phi) = 10 \log_{10} G(\theta, \phi) \qquad (9.4.17)$$

Figure 9.12 gives dB plots of the radiation patterns (9.4.14) and (9.4.15) for $N = 7$, $ka = \pi$ and $\psi = -\pi/2$.

By varying ψ, it is possible to shift the visible window for u to the right or left, changing the angle at which the array radiates most of its power. By increasing the element spacing and/or decreasing the array wavelength, the size of the visible window increases, resulting in a narrowed main beam with more sidelobes. The visible window can therefore be a useful design tool for uniformly excited linear arrays by helping visualize how a change in array specifications will affect the resulting radiation pattern.

We conclude this section with a discussion of antenna images, because yet another way to form antenna arrays is with perfectly conducting mirrors. Consider the simplest case of a point charge $+q$ located a distance h over a perfectly conducting ground plane, as shown in Figure 9.13(a). We seek to replace the ground plane at $z = 0$ with an equivalent charge distribution so that the fields in the upper half plane ($z > 0$) remain unchanged. Superposition of fields from the original point charge and the charge distribution equivalent to the conducting plane then enable us to find the fields everywhere for $z > 0$. In order that the point charge plus equivalent charge distribution give rise to the same fields for $z > 0$ as the point charge plus ground plane, both sources must:

(1) contain identical charge distributions for $z > 0$ so that Poisson's equation (2.1.6) has identical sources in the region of interest.

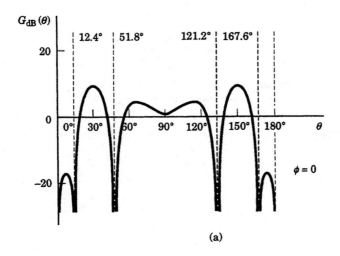

(a)

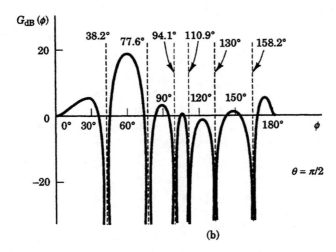

(b)

Figure 9.12   Gain in decibels versus angle for $N = 7$ linear array.

(2) satisfy the same boundary conditions in the region $z \geq 0$.

We can then invoke uniqueness, discussed in Section 1.8 to show that only **one** field can satisfy both (1) and (2). Both sets of sources thus give rise to the same field for $z > 0$.

The first condition implies that the equivalent charge distribution is located beneath the ground plane $(z \leq 0)$. The second implies that $\hat{z} \times \overline{E} = 0$ at $z = 0$ since no tangential $\overline{E}$ field exists at the surface of a perfect conductor. Both of these constraints are satisfied by an image charge distribution consisting of a point charge of equal magnitude and opposite sign $(-q)$ located a distance h below the

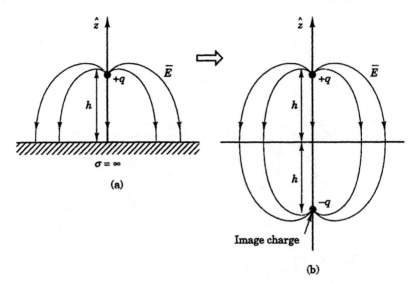

Figure 9.13 Image of a point charge over a ground plane.

conducting plane (directly under the $+q$ charge), as illustrated in Figure 9.13(b). [10] Note that the image charge does **not** produce the correct field for $z \leq 0$ because we have put a source in this region and Poisson's equation is no longer identical for both problems in the region $z < 0$. But we already know the true electric field below the conducting plane: $\overline{E}(z < 0) = 0$.

We can now generalize this argument to an arbitrary charge distribution $\rho(\overline{r}) = \rho(x, y, z)$. By superposition, the image distribution $\rho_{\text{im}}(\overline{r})$ that is used to replace a perfectly conducting plane at $z = 0$ is just $\rho_{\text{im}}(\overline{r}) = -\rho(x, y, -z)$. Since $\rho(\overline{r})$ also gives rise to a current distribution $\overline{J}(\overline{r})$ by the conservation of charge equation (1.2.5), it follows that the image current $\overline{J}_{\text{im}}(\overline{r})$ is just

$$\overline{J}_{\text{im}}(\overline{r}) = -\overline{J}(x, y, -z) = -\hat{x}\, J_x(x, y, -z) - \hat{y}\, J_y(x, y, -z) + \hat{z}\, J_z(x, y, -z)$$
$$(9.4.18)$$

Note that $\hat{z}$ is replaced by $-\hat{z}$ as well as z by $-z$ in (9.4.18), as suggested by Figure 9.14. This figure also depicts how the image currents enforce the boundary condition that only tangential magnetic fields exist at the conductor surface.

Equation (9.4.18) simply means that the ground plane at $z = 0$ under an $\hat{x}$-directed dipole at $(0, 0, h)$ may be replaced by an equal-magnitude, $-\hat{x}$-directed

---

10. We have not discussed the boundary condition at infinity for the region $z > 0$, but would like the electric field to vanish as $\overline{r} \to \infty$. Since this boundary condition is satisfied by the image charge, we see that the boundary conditions on the **entire** surface enclosing the upper half plane are obeyed. An alternative set of equivalent charges would be those actually present as surface charge on the conductor at $z = 0$. This choice does not simplify the problem, however.

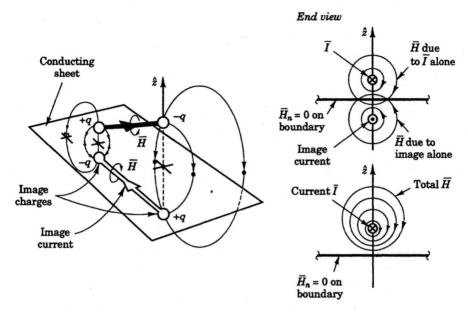

Figure 9.14   Image currents and charges equivalent to a conducting plane.

(180° out of phase) image dipole at $(0, 0, -h)$. Likewise, the image of a $\hat{y}$-directed dipole is $-\hat{y}$-directed, whereas the image of a $\hat{z}$-oriented dipole is $+\hat{z}$-oriented (in phase). These special cases are illustrated in Figure 9.15.

If there is more than one ground plane in the problem, then multiple images are necessary, their locations determined by symmetry and boundary conditions. For example, the 90° corner reflector in Figure 9.16 can be represented by three images in addition to the true dipole, while the 120° corner reflector needs five additional images. In each case, the original source in addition to all the images must yield fields that satisfy the boundary conditions imposed by the original mirrors. Angles that do not divide evenly into 360° require an infinite number of images, making the image technique much less useful.

Example 9.4.1

An antenna array is composed of two vertical dipoles driven by currents 1 and $A\,e^{j\psi}$. The resulting antenna gains are plotted in polar coordinates in the $x-y$ plane in Figure 9.17. For each pattern, suggest a possible orientation of the two dipoles, the dipole separation distance d, and values for A and ψ.

Solution: The circular pattern in Figure 9.17(a) is produced by a single dipole, so $A = 0$. The figure-eight pattern in Figure 9.17(b) has two nulls, so the two dipoles must have equal magnitudes to be able to interfere completely destructively; $A = 1$. However, there are two possible orientations that yield pattern (b). The first is if the two dipoles are

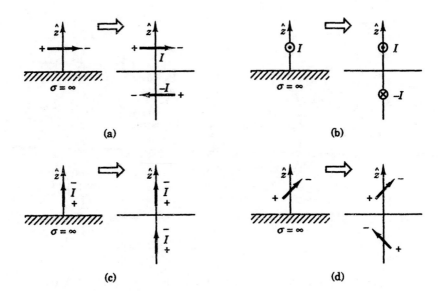

Figure 9.15   Images of $\hat{x}$-, $\hat{y}$- and $\hat{z}$-directed dipoles over a ground plane.

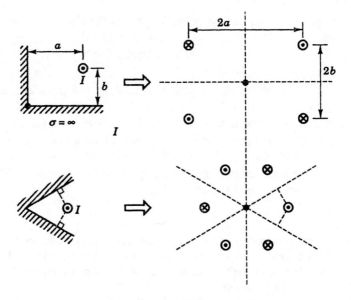

Figure 9.16   90° and 120° corner reflectors.

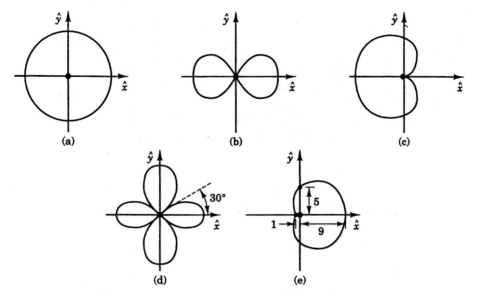

Figure 9.17   Two-dipole antenna array in Example 9.4.1.

located on the $\hat{x}$-axis, in which case they must be in phase ($\psi = 0$) to interfere construc-tively along the $\hat{y}$-axis. In this case, they must also be spaced $d = \lambda/2$ apart to produce destructive interference along the $\hat{x}$-axis. Note that $d \neq 3\lambda/2, 5\lambda/2$, and so on because these larger separations would produce patterns with additional nulls. The other orienta-tion that produces this antenna pattern is when the dipoles are oriented along the $\hat{y}$-axis. In this case they must be completely out of phase ($\psi = \pi$) to interfere destructively along the $\hat{x}$-axis and are again separated by $d = \lambda/2$ to produce a single maximum along the $\pm\hat{y}$-axis.

The cardioid pattern in Figure 9.17(c) has a null: $A = 1$. It also has only one null, as opposed to the two nulls in Figure 9.17(b), so $d = \lambda/4$. The dipoles must be oriented on the $\hat{x}$-axis and be 90° out of phase ($\psi = \pi/2$) so that the $+j$ dipole adds constructively in the $-\hat{x}$-direction and destructively in the $+\hat{x}$-direction.

The cloverleaf pattern in Figure 9.17(d) has four nulls ($A = 1$) and four maxima, twice the number in Figure 9.17(b), so $d = \lambda$. Whether the dipoles are oriented along the $\hat{x}$-axis or the $\hat{y}$-axis, they must be in phase ($\psi = 0$) to interfere constructively along both axes. If we assume that the dipoles lie along the $\hat{x}$-axis, then at 30° the two dipole differ in phase by $\lambda/2$ and a null results as desired. The $\hat{y}$ orientation requires $\phi = 60°$ to yield a null and thus does not produce the correct pattern.

The final pattern [Figure 9.17(e)] has no nulls so $A \neq 1$ but has one maximum and one minimum, so we suspect that $d = \lambda/4$. The dipoles must lie along the axis of symmetry of the pattern ($\hat{x}$-axis). If the maximum gain is 9 and the minimum gain is 1, then $A = 2$ ($|1 + 2|^2 = 9$; $|1 - 2|^2 = 1$). If the two phasors are to add constructively along the $+\hat{x}$-axis, then $\psi = -\pi/2$, which is the opposite phase angle to that in Figure 9.17(b) where the maximum gain is along the $-\hat{x}$-axis. In the $\pm\hat{y}$ directions, the gain is $|1 \pm 2j|^2 = 5$ as specified.

Example 9.4.2

One thousand identically excited Hertzian dipoles parallel to the $\hat{z}$-axis are equally spaced along the $\hat{x}$-axis over a distance D with separation $a = \lambda/10$ between dipoles and $Id = 1$ A·m, where $\lambda = 1$ m. In what direction ϕ_{max} is the maximum flux density $\text{Re}\{\overline{S}(r, \theta, \phi)\}$ radiated and what is its value (W/m^2)? The flux is zero at $\phi_{null} = \phi_{max} \pm \Delta\phi$, where ϕ is the angle in the x–y plane measured from the $\hat{x}$-axis. What is the value of $\Delta\phi$ for the first null?

Solution: Since all of the dipoles are excited in phase, they interfere constructively only in the y–z plane; ϕ_{max} is thus equal to $\pm\pi/2$. By superposition, the farfield $\overline{E}(r, \theta, \phi_{max})$ is just the sum of $N = 1000$ in-phase dipoles. From (9.2.9),

$$|\overline{E}(r, \theta, \phi_{max} = \pi/2)| = N \left| \frac{\eta_0 k I d}{4\pi r} \right| \sin\theta \tag{1}$$

which is a maximum when $\theta = \pi/2$. The Poynting power density in free space is given by (9.4.8)

$$\text{Re}\{\overline{S}(r, \theta, \phi)\} = \frac{1}{2\eta_0} \left| \overline{E}(r, \theta, \phi) \right|^2 \quad \text{(W/m}^2\text{)} \tag{2}$$

so that the maximum flux density is just

$$\text{Re}\left\{ \overline{S}\left(r, \frac{\pi}{2}, \frac{\pi}{2}\right) \right\} = \frac{1}{2\eta_0} \left| \overline{E}\left(r, \frac{\pi}{2}, \frac{\pi}{2}\right) \right|^2$$

$$= \frac{1}{2\eta_0} \left| \frac{N\eta_0 k I d}{4\pi r} \right|^2 = \frac{\eta_0}{8} \left(\frac{N I d}{r \lambda} \right)^2$$

$$= \frac{4.71 \times 10^7}{r^2} \quad \text{(W/m}^2\text{)}$$

We can find $\Delta\phi$ by considering the array in Figure 9.18 to be 500 pairs of dipoles such that each pair interferes destructively. If waves from dipoles 1 and 501 arrive one-half wavelength out of phase, then they will cancel, as will the waves from pairs 2 and 502, 3 and 503, and so on. From Figure 9.18, we see that $(Na/2) \sin \Delta\phi = \lambda/2$ or $\Delta\phi \cong \lambda/Na = 0.01$ radians at the first null. We can also solve this problem by using (9.4.15) directly with $\psi = 0$ since all the dipoles are excited in phase

$$\text{Re}\{\overline{S}(r, \theta, \phi)\} = \frac{|\overline{E}_0(r, \theta, \phi)|^2}{2\eta_0} \left| \frac{\sin \frac{N}{2}(ka \cos\phi)}{\sin \frac{1}{2}(ka \cos\phi)} \right|^2 \tag{3}$$

and $\overline{E}_0(r, \theta, \phi)$ is the field of a single dipole. Equation (3) is valid when r is sufficiently large compared to the array "diameter" $D \simeq Na$. Later we show that this requires $r \gg 2D^2/\lambda$. It is clear that (3) is a maximum when $ka \cos\phi_{max} = 0$ or $\phi_{max} = \pm\pi/2$, and is zero when $(Nka/2) \cos\phi_{null} = (Nka/2) \sin \Delta\phi = \pi$ or $\Delta\phi \cong 2\pi/Nka = \lambda/Na \simeq \lambda/D$ as before.

Example 9.4.3

If we are given N equally spaced $\hat{z}$-directed dipoles radiating in phase with binomial excitation coefficients $w_i = N!/i!(N - i)!$, what is the radiation pattern for this array in the x–y plane if the separation distance a between dipoles is $a = \lambda/2$?

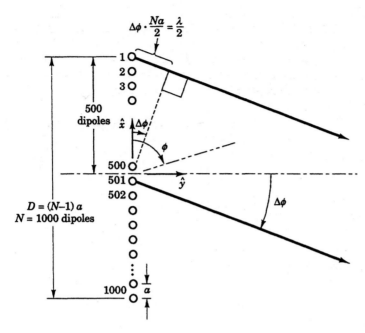

Figure 9.18   One thousand-element array in Example 9.4.2.

Solution: The array factor for the N dipoles is given by (9.4.6):

$$F_N(\theta = \pi/2, \phi) = \sum_{i=0}^{N-1} w_i\, e^{jk\bar{r}\cdot\bar{a}_i}$$

$$= \sum_{i=0}^{N-1} \frac{N!}{i!(N-1)!}\, e^{ji\pi\cos\phi} \tag{1}$$

where $ka = \pi$ since $a = \lambda/2$. But we recall that the *binomial theorem* is

$$(a+b)^N = \sum_{i=0}^{N-1} \frac{N!}{i!(N-i)}\, a^i b^{N-i} \tag{2}$$

Letting $a = e^{j\pi\cos\phi}$ and $b = 1$, we use (2) to evaluate (1):

$$F_N(\theta = \pi/2, \phi) = \left(1 + e^{j\pi\cos\phi}\right)^N$$

$$= e^{j\frac{N}{2}\pi\cos\phi}\left(e^{j\frac{\pi}{2}\cos\phi} + e^{-j\frac{\pi}{2}\cos\phi}\right)^N$$

and therefore

$$G_N(\theta = \pi/2, \phi) \propto |F(\theta = \pi/2, \phi)|^2 \propto \left[\cos\left(\frac{\pi}{2}\cos\phi\right)\right]^{2N} \tag{3}$$

This antenna pattern is useful because it possesses a single main lobe at $\phi = \pm\pi/2$ and has no sidelobes, which becomes increasingly directive as N becomes large (compare $p_1(\phi)$, $p_2(\phi)$, and $p_3(\phi)$ in Figure 9.19).

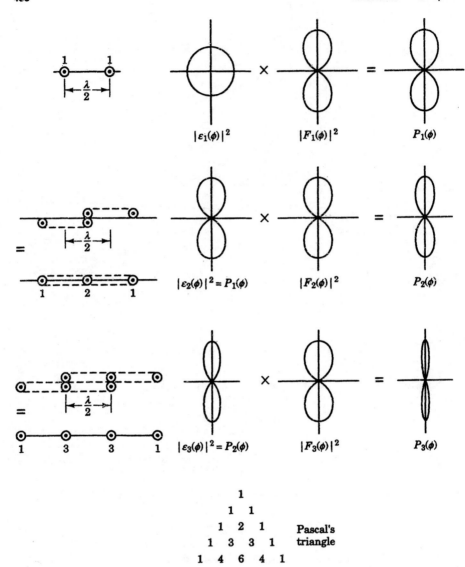

$$
\begin{array}{ccccc}
&&1&&\\
&1&&1&\\
1&&2&&1\\
\end{array}
$$

Figure 9.19 Pattern multiplication technique.

An alternative way of viewing this array invokes the ability to separate a radiation pattern into element and array factors using the *pattern multiplication* technique. Consider two $\hat{z}$-directed dipoles of equal magnitude and phase located at $x = \pm\lambda/4$. The radiation pattern in the x–y plane for this pair of dipoles may be written as

$$p_1(\theta = \pi/2, \phi) \propto |\overline{\mathcal{E}}(\theta, \phi)|^2 |F(\theta, \phi)|^2 \propto \left|1 + e^{j\frac{k\lambda}{2}\cos\phi}\right|^2 \propto \cos^2\left(\frac{\pi}{2}\cos\phi\right) \qquad (4)$$

since $\overline{\mathcal{E}}(\theta = \pi/2, \phi) = 1$. Let us now consider this pair of dipoles as a single new type of element with an element factor given by (4) and create an array with two of these new elements separated again by a half-wavelength. This is equivalent to three of the single dipoles, two of unit magnitude at $x = \pm\lambda/2$ and one of twice unit magnitude at $x = 0$, as illustrated in Figure 9.19 (a 1:2:1 excitation). Because we can express the total radiation pattern $p_2(\theta, \phi)$ as the product of element and array factors, we find that

$$p_2(\theta = \pi/2, \phi) = \cos^2\left(\frac{\pi}{2}\cos\phi\right)\cos^2\left(\frac{\pi}{2}\cos\phi\right) = \cos^4\left(\frac{\pi}{2}\cos\phi\right) \qquad (5)$$

where element factor and array factors for the two-dipole element are both now given by (4).[11] It is not difficult to see that the 1:2:1 antenna could itself be considered as an individual radiating element, which when separated by $\lambda/2$ from another such element would give a radiation pattern proportional to $\cos^6\left(\frac{\pi}{2}\cos\phi\right)$ since the element and array factors for this new antenna would be given by (5) and (4), respectively. This new antenna could also be thought of as two unit magnitude dipoles at $x = \pm3\lambda/4$ and two 3-unit magnitude dipoles at $x = \pm\lambda/4$ (a 1:3:3:1 excitation). If this process were continued, the next array would have excitations in the ratio 1:4:6:4:1 where these excitation coefficients are just the binomial coefficients suggested by Pascal's triangle in Figure 9.19 and given by $w_i = N!/i!(N-i)!$. Each successive array is thus the convolution of the preceding array with a pair of unit impulses separated by $\lambda/2$. This process of pattern multiplication thus generates the same binomial weighting coefficients and radiation pattern $p_N = \cos^{2N}[(\pi/2)\cos\phi]$ found in (3).

Example 9.4.4

A short dipole of effective length d_{eff} has a complex impedance $Z = R_{\text{rad}} + 1/j\omega C$ at frequency ω. One-half of this dipole is mounted near and perpendicular to an infinite ground plane at $z = 0$, as illustrated in Figure 9.20. What is the complex impedance $R'_{\text{rad}} + 1/j\omega C'$ of this new (monopole) antenna near ω?

Solution: Notice that the fields and currents are identical for both the short-dipole and the monopole-over-a-ground-plane antennas when $z > 0$. For the short dipole, the power radiated above and below the $z = 0$ plane is the same, but the monopole antenna does not radiate for $z < 0$. Therefore, $P'_{\text{rad}} = P_{\text{rad}}/2$, where P_{rad} is radiated power. Since P_{rad} is proportional to R_{rad}, we see that

$$R'_{\text{rad}} = R_{\text{rad}}/2$$

From (9.3.8), we also see that $X' = X/2$ because the volume integral of the electric and magnetic energies is nonzero only for $z > 0$ in the monopole case, while for the original dipole it extends over all space. The capacitance of the monopole is therefore twice that of the dipole: $C' = 2C$.

---

11. This combination process can be viewed as *convolution* of the original dipole pair excitation with two impulses spaced $\lambda/2$ apart. The Fourier relationship between spatial excitation functions and antenna patterns suggests that convolution of the excitation functions transforms into multiplication of the corresponding antenna patterns, as is seen here.

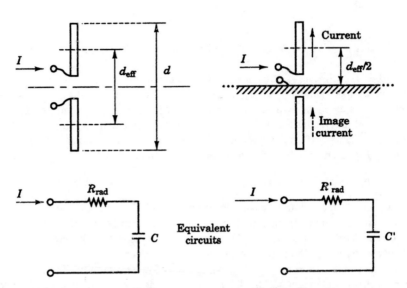

Figure 9.20   Short dipole over a ground plane in Example 9.4.4.

The advantage of mounting a short-dipole wire antenna over a ground plane is that the effective length of the wire antenna becomes twice what it would be for the same wire isolated in space. Therefore, if the two antennas are driven with identical current sources, the wire over the ground plane will radiate four times as much power (into half the space) as the same isolated wire. Such half-wave dipole antennas are frequently used on automobiles where the ground plane is formed by the body of the car and is not exactly planar.

Example 9.4.5

A Hertzian dipole of length d pointing in the direction $\hat{x} + \hat{y}$ is located at the point $(\ell/\sqrt{2}, \ell/\sqrt{2}, 0)$ and is surrounded by a 90° corner reflector comprised of two perfectly conducting half-planes at $x = 0$, $y \geq 0$ and $y = 0$, $x \geq 0$ as shown in Figure 9.21(a). Find the appropriate image dipoles with which to replace the 90° corner reflector and sketch the resulting radiation patterns in the x–y plane for $\ell = \lambda/2$ and $\ell = \lambda$.

Solution: The location and orientation of the three additional image dipoles are shown in Figure 9.21(b) and are given below

$$\text{real dipole}: \quad (\ell/\sqrt{2}, \ell/\sqrt{2}, 0) \text{ at } \hat{x} + \hat{y}$$
$$\text{image 1}: \quad (\ell/\sqrt{2}, -\ell/\sqrt{2}, 0) \text{ at } -\hat{x} + \hat{y}$$
$$\text{image 2}: \quad (-\ell/\sqrt{2}, \ell/\sqrt{2}, 0) \text{ at } \hat{x} - \hat{y}$$
$$\text{image 3}: \quad (-\ell/\sqrt{2}, -\ell/\sqrt{2}, 0) \text{ at } -\hat{x} - \hat{y}$$

where (9.4.18) was used.

It is convenient to consider a rotated coordinate system x'-y' as shown in Figure 9.21(b), where $\hat{x}'$ lies along the axis of image 1 and $\hat{y}'$ lies along the real dipole axis; $\hat{x}' \cdot \hat{r} \equiv \cos\alpha$.

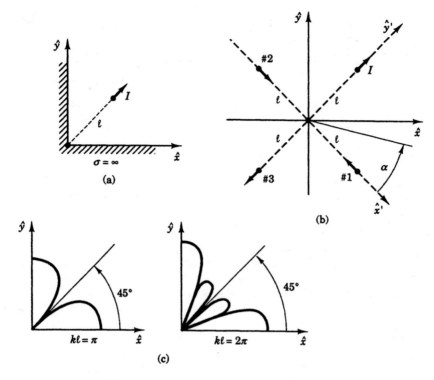

Figure 9.21 Dipole and its three image dipoles in Example 9.4.5.

Recalling that the electric field for a single Hertzian dipole located at the origin and oriented in the $+\hat{z}$ direction is

$$\overline{E}_{\text{ff}}(r, \theta) = \hat{\theta}\,\frac{j\eta k I d}{4\pi r}\,e^{-jkr}\sin\theta$$

we see that the far field for the $\hat{y}'$-directed real dipole is just

$$\overline{E}_{\text{ff}}(r, \alpha) = -\hat{\alpha}\,\frac{j\eta k I d}{4\pi r}\,e^{-jk(r-\ell\sin\alpha)}\cos\alpha$$

where $(\pi/2 - \alpha)$ is the angle between the dipole axis and the observer, and the distance between the dipole and the origin is just ℓ. The total electric field for this corner reflector antenna is then the sum of the farfield $\overline{E}_{\text{ff}}$ associated with the original dipole and each image:

$$\begin{aligned}
\overline{E}_{\text{ff}}(r, \alpha) = -\hat{\alpha}\,\frac{j\eta k I d}{4\pi r}\,e^{-jkr}\,\Big[& e^{jk\ell\sin\alpha}\cos\alpha + e^{jk\ell\cos\alpha}\sin\alpha \\
& -e^{-jk\ell\cos\alpha}\sin\alpha - e^{-jk\ell\sin\alpha}\cos\alpha \Big]
\end{aligned} \tag{1}$$

The four terms in (1) correspond to the real dipole and images 1–3, respectively. By combining terms, (1) may be simplified to yield

$$\overline{E}_{\text{ff}}(r, \alpha) = -\hat{\alpha} \frac{\eta k I d}{2\pi r} e^{-jkr} [\cos\alpha \sin(k\ell \sin\alpha) + \sin\alpha \sin(k\ell \cos\alpha)]$$

which is only valid in the quarter-space $x \geq 0$, $y \geq 0$ (i.e., $\pi/4 \leq \alpha \leq 3\pi/4$); $\overline{E}_{\text{ff}} = 0$ elsewhere.

As expected, there is a null along the axis of the dipole ($\alpha = \pi/2$) and that the electric field has only a normal component at the surface of the planar conductors. Plots of $G(\alpha) \propto |\overline{E}_{\text{ff}}|^2$ are given in Figure 9.21(c) for $\ell = \lambda/2$ and $\ell = \lambda$. Notice that as ℓ increases, additional lobes are created in the radiation pattern.

Example 9.4.6

A linear array of N evenly spaced equal magnitude $\hat{z}$-directed dipoles is located along the $\hat{x}$-axis as shown in Figure 9.8. We wish to synthesize a radiation pattern obeying the following specifications:

(a) The single main lobe for $0 \leq \phi \leq 180°$ is a maximum at $\phi = 60°$.

(b) One of the two nulls separating the main lobe and the first sidelobe is at $\theta = 70°$. What is the minimum number of dipoles N, and the difference in phase ψ and distance a between adjacent dipoles if the array is driven at 10 GHz? What is the location of the other first null? What is the maximum sidelobe level in dB for this array?

Solution: The gain for an N-dipole array in the x–y plane ($\theta = \pi/2$) is given by (9.4.15), where $w_i = e^{ji\psi}$ ($i = 0, 1, \ldots, N-1$) for the ith dipole and $\overline{a}_i = \hat{z} i a$:

$$G(\theta = \pi/2, \phi) \propto \frac{\sin^2 \frac{N}{2}(\psi + ka\cos\phi)}{\sin^2 \frac{1}{2}(\psi + ka\cos\phi)} \tag{1}$$

The gain is plotted in Figure 9.22(a) and is constrained to be a maximum at $\theta = 60°$, so $\frac{1}{2}(\psi + ka\cos 60°) = 0$ or

$$\psi = -ka/2 \tag{2}$$

where a remains to be found.

We also know that $\phi = 70°$ is a first null so that $\frac{N}{2}(\psi + ka\cos 70°) = \pm\pi$ which is combined with (2) to yield a number for the product Nka:

$$Nka = \frac{2\pi}{\cos 60° - \cos 70°} = 39.77 \tag{3}$$

However, we also wish to ensure that no part of the main lobes centered at $\frac{N}{2}(\psi + ka\cos\phi) = \pm N\pi$ are part of the visible window for this array, which means that

$$\frac{N}{2}|\psi \pm ka| \leq (N-1)\pi \tag{4}$$

as can be seen from Figure 9.22(a). The negative sign in (4) gives the more stringent inequality since ψ is negative. Combining (2)–(4) yields

$$N \geq 1 + \frac{3Nka}{4\pi} = 10.49 \tag{5}$$

Therefore, N must be at least 11 to ensure that the pattern has only one main lobe. If $N = 11$, then from (3), we find $ka = 39.77/11 = 3.62$ or $a = 0.58\lambda$. If $f = 10$ GHz, then

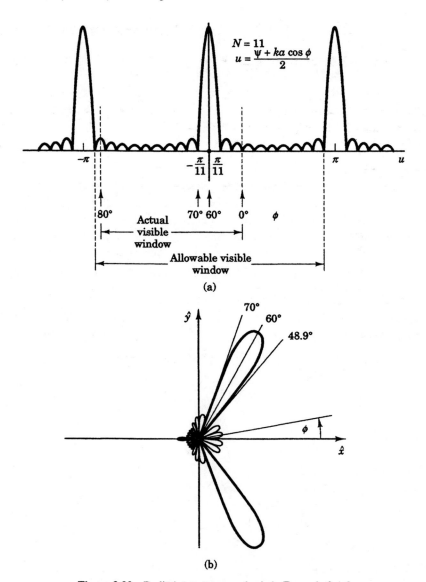

Figure 9.22   Radiation pattern synthesis in Example 9.4.6.

$\lambda = c/f = 3$ cm so $a = 1.73$ cm. We then use (2) to find that $\psi = -103.6°$. The other null adjacent to the main lobe is given by

$$\frac{N}{2}(\psi + ka \cos \phi_{\text{null}}) = +\pi$$

which means that $\phi_{\text{null}} = 48.9°$. The main lobe thus spans $21.1°$ and is not quite symmetric about its maximum. The radiation pattern is plotted in Figure 9.22(b), with other nulls given by $\frac{N}{2}\psi + ka \cos \phi_{\text{null}} = \pm n\pi$.

A programmable calculator can be used to find the locations and heights of the local maxima of the first sidelobes from (1). These local maxima are found at $\frac{1}{2}(\psi + ka\cos\phi) = \pm 0.41$, which corresponds to $\phi = 43.4°$ and $74.1°$. The height of the local maxima is 6.04 compared to the $11^2 = 121$ height of the main lobe at $\phi = 60°$. Therefore, the first sidelobes are $10\log_{10}(6.04/121) = -13.0$ dB down from the main lobe.

9.5 RECEIVING PROPERTIES OF ANTENNAS

In the previous sections of this chapter we have discussed the transmitting properties of Hertzian and short-dipole antennas, which couple the wave energy in an external circuit to radiated power in free space. But the same antennas can also be used to couple free space radiation to the external circuit. A transducer used for this purpose is called a *receiving antenna*. In this section, we will show that the receiving properties of an antenna are closely linked to its transmitting properties, but to see this we shall have to invoke the concept of reciprocity.

In order to understand reciprocity for coupled antennas, we consider the following two experiments, schematically illustrated in Figure 9.23. Any two arbitrarily oriented antennas are located within a finite region of space. In the first experiment, we drive antenna 1 with a current I_1 and open-circuit the terminals of antenna 2 so that $I_2 = 0$. If we then measure the voltage V_2 induced across the terminals of antenna 2, we can calculate a transfer impedance Z_{21}, defined by

$$Z_{21} \equiv \left.\frac{V_2}{I_1}\right|_{I_2=0} \tag{9.5.1}$$

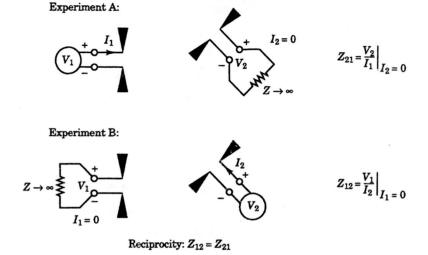

Figure 9.23   Calculation of mutual impedances Z_{12} and Z_{21}.

In the second experiment, we interchange the roles of antennas 1 and 2. We drive antenna 2 with a current I_2 and measure the voltage V_1 across the open-circuited terminals of antenna 1. This defines a second transfer impedance Z_{12}, where

$$Z_{12} \equiv \left. \frac{V_1}{I_2} \right|_{I_1=0} \tag{9.5.2}$$

The *reciprocity theorem* simply states that for a reciprocal medium (such as free space) [12]

$$Z_{12} = Z_{21} \tag{9.5.3}$$

A proof of (9.5.3) follows directly from Maxwell's equations and is presented in Section 9.6.

Similarly, self-impedances Z_{11} and Z_{22} may be calculated using

$$Z_{11} = \left. \frac{V_1}{I_1} \right|_{I_2=0} \tag{9.5.4}$$

and

$$Z_{22} = \left. \frac{V_2}{I_2} \right|_{I_1=0} \tag{9.5.5}$$

where these are **not** the impedances of antennas 1 and 2 in isolation but depend implicitly on the configuration of the two antennas.

Because Maxwell's equations are linear, (9.5.1)–(9.5.2) and (9.5.4)–(9.5.5) may be superposed to yield the circuit relations:

$$\begin{aligned} V_1 &= Z_{11}I_1 + Z_{12}I_2 \\ V_2 &= Z_{21}I_1 + Z_{22}I_2 \end{aligned} \tag{9.5.6}$$

We can also represent (9.5.6) in matrix form as

$$\begin{bmatrix} V_1 \\ V_2 \end{bmatrix} = \begin{bmatrix} Z_{11} & Z_{12} \\ Z_{21} & Z_{22} \end{bmatrix} \begin{bmatrix} I_1 \\ I_2 \end{bmatrix} \tag{9.5.7}$$

where $Z_{12} = Z_{21}$ for reciprocal media (or equivalently, $\overline{\overline{Z}} = \overline{\overline{Z}}^T$). The circuit model of the two antennas is illustrated in Figure 9.24, with induced voltages represented by dependent Thevenin voltage sources. The equivalent circuit for antenna 1 is thus the Thevenin voltage $V_{\text{Th}}^{(1)} = Z_{12}I_2$ in series with self-impedance Z_{11}, with a similar equivalent circuit for antenna 2 found by interchanging labels 1 and 2.

Once the equivalent circuits for the two antennas are known, it is a simple matter to determine the power received by either antenna, which we shall find to be directly proportional to the antenna gain. We begin by considering antenna 1 to be

---

12. The exact meaning of a reciprocal medium will be clarified in Section 9.6.

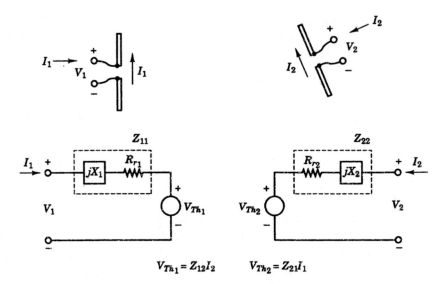

$$V_{Th_1} = Z_{12}I_2 \qquad V_{Th_2} = Z_{21}I_1$$

Figure 9.24   Equivalent circuits for two coupled antennas.

any arbitrary transmitting antenna, while for simplicity the receiving antenna 2 is a short dipole. The Thevenin voltage for the receiving antenna is thus

$$V_{Th}^{(2)} = Z_{21}I_1 = -\int_2 \overline{E}_A \cdot d\overline{\ell}_2 \tag{9.5.8}$$

where $\overline{E}_A$ is the electric field generated by experiment A [depicted in Figure 9.23(a)] in which antenna 1 is transmitting and antenna 2 is open-circuited; $d\overline{\ell}_2$ is any path connecting the terminals of antenna 2.

We wish to relate $\overline{E}_A$ to $\overline{E}_1$, where $\overline{E}_1$ is the electric field that would be observed if the receiving antenna did not exist. Since the short dipole is electrically small, we expect that $\overline{E}_1$ is approximately a uniform field in the vicinity of the dipole and that $\overline{E}_A = \overline{E}_1 + \overline{E}_{\text{scatt}}$ differs from $\overline{E}_1$ only near the dipole. Since the dipole is small compared to a wavelength, we analyze its scattered nearfield $\overline{E}_{\text{scatt}}$ quasistatically as illustrated in Figure 9.25 (i.e., we neglect wave propagation in the immediate vicinity of the dipole and solve Laplace's equation instead of the wave equation for the near fields). We observe from Figure 9.25 that the voltage $V_{Th}^{(2)}$ measured between the open-circuited terminals of antenna 2 is just the difference in voltage between the equipotential surfaces passing through the centers of each dipole arm, and that **these particular equipotential contours are not appreciably changed by the presence of the dipole.** Therefore, the voltage measured is

$$V_{Th}^{(2)} = -\int_2 \overline{E}_A \cdot d\overline{\ell}_2 = -\overline{E}_1 \cdot \frac{\overline{d}_2}{2} = -|\overline{E}_1| d_2^{\text{eff}} \cos \psi \tag{9.5.9}$$

where $\overline{d}_2^{\text{eff}} = \overline{d}_2/2$ is the effective electrical length of a short dipole having physical

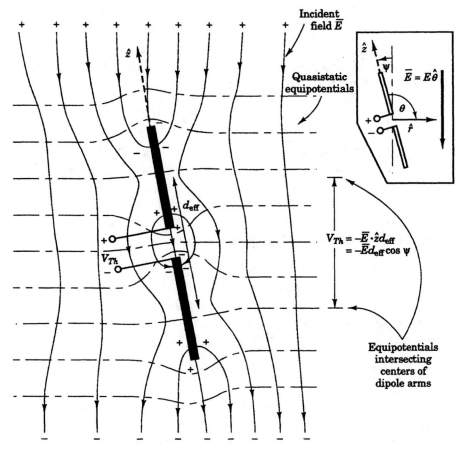

Figure 9.25   Thevenin voltage V_{Th} of a Hertzian dipole.

length d_2 and a triangular current distribution; ψ is the angle between $\overline{E}_1$ and the dipole axis. [13]

It has already been shown in Section 6.3 [see (6.3.12)] that maximum power is received by antenna 2 when the impedance Z_{L2} of transmission line 2 is matched to Z_{22}, or $Z_{L2} = Z_{22}^* = R_{22} - jX_{22}$. The maximum received power P_2 is thus

$$P_2 = \frac{|V_{Th}^{(2)}|^2}{8R_{22}} = \frac{|\overline{E}_1|^2(d_2^{\text{eff}})^2 \cos^2 \psi}{8 \cdot 20k^2d_2^2} \qquad (9.5.10)$$

---

13. Triangular current distributions are expected for thin short wires because of their *TEM*-like behavior, discussed further in Section 9.7.

where $R_{22} = 20(kd_2^{\text{eff}})^2$ for a short dipole and no dissipative effects are included since the dipole is located in free space and is perfectly conducting.

We now define an *effective antenna area* $A_{\text{eff}}(\theta, \phi)$ for the receiving antenna such that the maximum power it receives from direction (θ, ϕ) is equal to $A_{\text{eff}}(\theta, \phi)$ times the transmitted power flux

$$P_2 = A_{\text{eff}}(\theta, \phi)|\overline{S}_1(\overline{r})|^2 = A_{\text{eff}}(\theta, \phi)\frac{|\overline{E}_1(\overline{r})|^2}{2\eta_0} \qquad (9.5.11)$$

where A_{eff} has units of m^2. The effective area can be thought of as the "capture cross-section" of the antenna, or the equivalent area from which all power transmitted from direction (θ, ϕ) could be extracted. In general, there are two independent effective antenna areas for a given direction of received radiation—one for each of two specified orthogonal polarizations.

Combining (9.5.10) and (9.5.11), we find that

$$A_{\text{eff}}(\theta, \phi) = \frac{\eta_0 \cos^2 \psi}{80k^2} = \frac{3\lambda^2}{8\pi} \cos^2 \psi \qquad (9.5.12)$$

for the receiving short dipole. But $\theta = \psi - \pi/2$ as shown in Figure 9.25, since $\hat{\theta}$ is referenced from the axis of the receiving dipole so that $\hat{\theta}$ is parallel to the transmitted field $\overline{E}_1$. Therefore, (9.5.12) can also be expressed as

$$A_{\text{eff}}(\theta, \phi) = \frac{\lambda^2}{4\pi}\left(\frac{3}{2}\sin^2\theta\right) = \frac{\lambda^2}{4\pi}G(\theta, \phi) \qquad (9.5.13)$$

where the second equality has only been demonstrated for a short dipole. We would now like to show that the simple and very useful relation $A_{\text{eff}}(\theta, \phi) = (\lambda^2/4\pi)G(\theta, \phi)$ applies to **all** antennas located in reciprocal media; it is one of the most fundamental equations governing antennas.

Consider the arbitrary antenna shown in Figure 9.26, characterized by an effective area $A_1(\theta, \phi)$ and gain $G_1(\theta, \phi)$ coupled to a short-dipole antenna with $A_2 = (\lambda^2/4\pi)G_2$ from (9.5.13). The powers received by antennas 1 and 2 are $P_{\text{rec}}^{(1)}$ and $P_{\text{rec}}^{(2)}$, while the transmitted powers are $P_{\text{tr}}^{(1)}$ and $P_{\text{tr}}^{(2)}$. For algebraic simplicity, the loads receiving power from the antennas are properly matched so that the reactive elements in the antenna self-impedances and loads cancel: $Z_{L1} = Z_{11}^*$ and $Z_{L2} = Z_{22}^*$. In this case, the power received by antenna 1 is

$$P_{\text{rec}}^{(1)} = \frac{|Z_{12}I_2|^2}{8R_{11}} = \left(\frac{P_{\text{tr}}^{(2)}}{4\pi r^2}\right)G_2 A_1 \qquad (9.5.14)$$

where the first equality is a restatement of (9.5.10) with the subscripts 1 and 2 interchanged. The factor $P_{\text{tr}}^{(2)}/4\pi r^2$ is the isotropic power flux radiated by antenna 2, which must be multiplied by the gain of that antenna $G_2(\theta, \phi)$ to yield the actual power flux (W/m^2) transmitted by antenna 2. Using (9.5.11), this transmitted flux

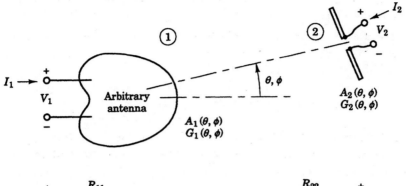

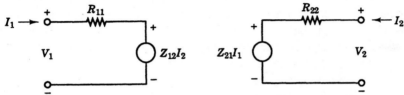

Figure 9.26 Coupling between two conjugate-matched antennas obeying reciprocity, $Z_{12} = Z_{21}$.

is then multiplied by the effective area of antenna 1 (A_1) to give the total power received by that antenna. By a similar argument,

$$P_{rec}^{(2)} = \frac{|Z_{21}I_1|^2}{8R_{22}} = \left(\frac{P_{tr}^{(1)}}{4\pi r^2} \right) G_1 A_2 \qquad (9.5.15)$$

We now take the ratio of (9.5.14) to (9.5.15), recalling that $P_{tr}^{(1)} = \frac{1}{2}R_{11}|I_1|^2$ and $P_{tr}^{(2)} = \frac{1}{2}R_{22}|I_2|^2$:

$$\frac{A_2}{A_1} = \frac{G_2}{G_1} \left| \frac{Z_{21}}{Z_{12}} \right|^2 \qquad (9.5.16)$$

For a reciprocal medium, $Z_{12} = Z_{21}$, and (9.5.16) becomes

$$\frac{A_1(\theta, \phi)}{G_1(\theta, \phi)} = \frac{A_2(\theta, \phi)}{G_2(\theta, \phi)} = \frac{\lambda^2}{4\pi} \qquad (9.5.17)$$

where (9.5.13) has been used to evaluate A_2/G_2 for the short-dipole antenna. Equation (9.5.17) thus shows that the ratio of effective antenna area to antenna gain is a constant independent of the type of antenna! That is, the receiving area of an antenna is directly proportional to its transmitting gain, with a proportionality constant depending only on wavelength. The link between transmitting and receiving properties of antennas is thus expressed most succinctly by (9.5.17). Although this derivation assumed that the receiving antennas had matched loads to maximize power transfer, (9.5.17) is valid for **any** loads. Both the gain and effective area are comparably diminished by any mismatch that exists. [14]

The effective receiving area of an arbitrary antenna is thus simply calculated from its gain $G(\theta, \phi)$. Fortunately, computation of $G(\theta, \phi)$ is generally straight-forward if the current distribution on the antenna can be accurately estimated. Note that the effective receiving area of an antenna is **not** the same as its physical area (in general). The most dramatic example is that of a Hertzian dipole, which has an infinitesimally small physical area but a maximum effective area of $3\lambda^2/8\pi$, equivalent to a circle of radius $\sim 0.2\lambda$. Antennas that have physical dimensions of many wavelengths may have effective areas closer to their physical sizes.

Inert objects may also intercept power, reradiating it in various directions depending on the shape, composition, and orientation of the object as well as on the wavelength and polarization of the incident radiation. One way to characterize this scattering ability is in terms of a *scattering cross-section* σ, which is the area necessary for an ideal object to scatter the same amount of power isotropically toward the same observer as the real scatterer. Therefore, σ may depend on the relative positions of the object, transmitter, and receiver.

For example, if a transmitter/receiver transmits P_{tr} W toward an object r meters away having scattering cross-section σ m^2, the power "captured" by this object is $(P_{\text{tr}}G/4\pi r^2)\sigma$, where $P_{\text{tr}}G/4\pi r^2$ is just the power flux that arrives at the object if the transmitter has gain G. If this received power is now reradiated isotropically by the scatterer, the transmitter/receiver receives $A/4\pi r^2$ of the scattered power where A is the effective receiving area of the transmitter/receiver. But since G and A are related by $A/G = \lambda^2/4\pi$, the total power received back at the transmitter is

$$P_{\text{rec}} = \left(\frac{P_{\text{tr}}G\sigma}{4\pi r^2}\right)\left(\frac{A}{4\pi r^2}\right) = \frac{\sigma(\lambda G)^2 P_{\text{tr}}}{(4\pi)^3 r^4}$$

which is the classic *radar range equation* for a lossless medium. Note that the received power is inversely proportional to the fourth power of the range r.

Example 9.5.1

A portable radio operating at 1 MHz contains a short-dipole antenna 10 cm long and can detect a maximum open-circuit voltage of 1 mV. A station 10 km away consists of a single vertical "half-dipole" tower $\lambda/4 = 75$ m high. The ground effectively acts as a conducting plane, creating an image tower so the total effective length of this antenna is $\lambda/2 = 150$ m. The maximum gain of this antenna is thus $2 \times 1.64 = 3.28$ but the gain in the direction of the radio is $G_{\text{tr}} = 3$. (See Example 9.4.4 for a discussion of how a monopole-over-a-ground-plane behaves compared with a dipole in free space.) What is the maximum possible receiving area A_{eff} for the radio antenna? What is the minimum transmitter power that can be detected by the radio? What is the maximum A'_{eff} if the antenna is connected

---

14. Provided that all media are reciprocal, even within the antenna structure.

to a receiver with a tuning coil that resonates with the antenna capacitance, but adds a series resistance $R_s = 0.01$ Ω to the load?

Solution: A 1-MHz signal has wavelength $\lambda = c/f = 3 \times 10^8$ m s$^{-1}/10^6$ m = 300 m. From (9.5.13), the maximum possible effective receiving area for the matched dipole is

$$A_{rec}(\theta = 0) = \frac{\lambda^2}{4\pi} G_{rec}(\theta = 0) = \frac{3\lambda^2}{8\pi} = \frac{3(300\,\text{m})^2}{8\pi} = 10,743\,\text{m}^2 \tag{1}$$

equivalent to a square 104 m on a side. This vividly demonstrates that the capture cross-section of an antenna can be much larger than its physical dimensions.

The power received by the radio P_{rec} is related to the transmitted power by

$$P_{rec} = \left(\frac{P_{tr}}{4\pi r^2}\right) G_{tr} A_{rec} \tag{2}$$

from (9.5.15), where maximum power is received when the load impedance of the receiving antenna is matched to its self-impedance: $Z_L = Z_{rec}^*$. This maximum received power is just

$$P_{rec} = \frac{|V_{Th}^{rec}|^2}{8 R_{rad}} = \frac{|V_{Th}^{rec}|^2}{160(kd_{eff})^2} \tag{3}$$

from (9.5.10). Combining (1)–(3) yields an expression for the transmitter power

$$\begin{aligned}
P_{tr} &\geq \frac{4\pi r^2 P_{rec}}{G_{tr} A_{rec}} = \frac{4\pi r^2}{G_{tr}} \left(\frac{|V_{Th}^{rec}|^2}{160(kd_{eff})^2}\right)\left(\frac{8\pi}{3\lambda^2}\right) \\
&= \frac{1}{60 G_{tr}}\left(\frac{r|V_{Th}^{rec}|}{d_{eff}}\right)^2 = \frac{1}{60\,\Omega \times 3}\left(\frac{10^4\,\text{m} \times 10^{-3}\,\text{V}}{0.05\,\text{m}}\right)^2 \\
&= 222\,\text{W}
\end{aligned}$$

where $d_{eff} \cong d/2 = 0.05$ m for the short-dipole receiver, $|V_{Th}^{rec}| = 10^{-3}$ V, and $r = 10^4$ m. Because A_{rec} may be smaller than $3\lambda^2/8\pi$, the transmitter power may have to be larger than 222 W.

If the dipole impedance is not matched to Z_L, then power will not be as efficiently received by the dipole and the new effective area A_{rec}' will be much less than A_{rec}. If the tuning coil has impedance $Z_L = R_s - jX$ and the self-impedance of the antenna is $Z_{rec} = R_{rad} + jX$, then the power dissipated in the load R_L is

$$P_{rec}' = \frac{R_L |V_{Th}^{rec}|^2}{2(R_L + R_{rad} + R_s)^2} \tag{4}$$

where $R_{rad} = 20(kd_{eff})^2 = 20(2\pi \times 0.05/300)^2 = 2.2 \times 10^{-5}$ Ω and $R_L = 0.01$ Ω. We find the value of the receiver resistance R_L that maximizes P_{rec}' by differentiating (4) with respect to R_L and setting the result equal to zero. The value of R_L which maximized P_{rec}' is $R_L = R_{rad} + R_s \simeq 0.01$ Ω. We can thus find an expression for A_{rec}' from (2) and (3):

$$A'_{rec} = \frac{4\pi r^2 P'_{rec}}{P_{tr} G_{tr}} = \frac{P'_{rec}}{P_{rec}} A_{rec} = \frac{4 R_L R_{rad}}{(R_L + R_{rad} + R_s)^2} A_{rec}$$

$$= \frac{R_{rad}}{R_{rad} + R_s} A_{rec} = \frac{2.2 \times 10^{-5} \; \Omega}{2.2 \times 10^{-5} \; \Omega + 0.01 \; \Omega} \times 10,743 \; m^2$$

$$= 0.24 \; m^2 \ll A_{rec}$$

The gain of the receiving antenna is then $G'_{rec} = 4\pi A'_{rec}/\lambda^2 = 3.4 \times 10^{-5}$, and the receiving circuits are pictured in Figure 9.27 for both the matched and mismatched loads. This performance is so poor that longer dipoles or ferrite-loaded loop antennas are generally employed instead.

Example 9.5.2

What is the mutual impedance $Z_{12} = Z_{21}$ between two short dipoles with effective lengths $d_1, d_2 \ll \lambda/2\pi$ located a distance R apart and oriented in the $\bar{d}_1$ and $\bar{d}_2$ directions, respectively?

Solution: We assume that R is sufficiently large that the dipoles are in each others' far fields. If we drive the first dipole with current I_1, then it transmits a farfield $\overline{E}_1$ at $\bar{r} = \overline{R}$

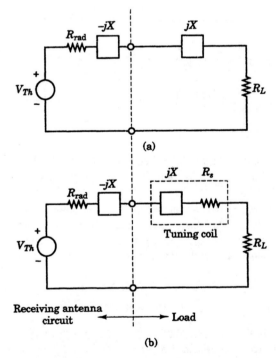

(a)

(b)

Figure 9.27   Antenna receiving circuits in Example 9.5.1.

$$\overline{E}_1(\overline{r} = \overline{R}) = \hat{\theta}_1 \frac{j\eta_0 k I_1 d_1}{4\pi R} e^{-jkR} \sin\theta_1 \tag{1}$$

where θ_1 is the angle between $\overline{d}_1$ and $\overline{R}$, as illustrated in Figure 9.28. If the second dipole is open-circuited ($I_2 = 0$), the voltage measured across the dipole terminals is $V_{\text{Th}}^{(2)}$. Because the receiving dipole 2 is much smaller than a wavelength in size, the electric field $\overline{E}_1$ is not appreciably perturbed except near the dipole itself. Equation (9.5.9) may then be used to compute $V_{\text{Th}}^{(2)}$:

$$V_{\text{Th}}^{(2)} = -\int_2 \overline{E}_1 \cdot d\overline{\ell}_2 = -(\hat{\theta}_1 \cdot \overline{d}_2) \frac{j\eta_0 k I_1 d_1}{4\pi R} e^{-jkR} \sin\theta_1 \tag{2}$$

If we define θ_2 as the angle between $\overline{d}_2$ and $\overline{R}$, then $\hat{\theta}_1 = \hat{\theta}_2$ and $-\hat{\theta}_1 \cdot \overline{d}_2 = d_2 \sin\theta_2$, as suggested in Figure 9.28. We thus compute the mutual impedance Z_{21} from (9.5.1) to be

$$Z_{21} = \frac{V_2}{I_1}\bigg|_{I_2=0} = \frac{j\eta_0 k d_1 d_2}{4\pi R} e^{-jkR} \sin\theta_1 \sin\theta_2 \tag{3}$$

Note that interchanging the labels 1 and 2 in the right side of (3) does not change the mutual impedance, which suggests that $Z_{21} = Z_{12}$. Also note that there is no coupling between the dipoles if either of them is oriented along $\overline{R}$, which is not surprising because the radiation pattern has a null along the dipole axis. Likewise, maximum coupling occurs when $\theta_1 = \theta_2 = \pi/2$ where the radiation pattern for each dipole is a maximum.

If dipole 2 is short-circuited instead, then the transfer impedance measured is

$$Z_t = \frac{V_1}{I_2}\bigg|_{V_2=0} = \frac{Z_{12}Z_{21} - Z_{11}Z_{22}}{Z_{21}}$$

as can be shown by combining the circuit equations (9.5.6) when $V_2 = 0$. This transfer impedance is clearly **not** equal to the mutual impedance $Z_{12} = Z_{21}$.

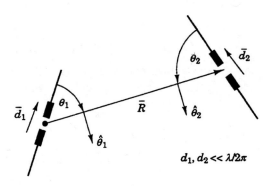

Figure 9.28   Mutual impedance of two dipoles in Example 9.5.2.

Example 9.5.3

A 1.5-cm dipole antenna is to be used in a hand-held radar gun at 10 GHz (X-band) to discourage speeders. The antenna transmits a 10-kW peak power signal that induces current on the metal body of an automobile a distance r away and causes it to reradiate a new signal that is received by the original antenna. If the car has a scattering cross-section of 1 m^2 and the radar can measure the Doppler shift of return signals greater than 0.01 nW, how far away must the car be to escape detection?

Solution: The 10-GHz radar has a wavelength $\lambda = c/f = 3 \times 10^8/10^{10} = 3$ cm, and because the dipole antenna is 1.5 cm long, the radar contains a half-wave dipole with maximum gain $G_{tr} = 1.64$. If the radar transmits P_{tr} W, the car will receive $P_{tr}G_{tr}\sigma/4\pi r^2$ W where $P_{tr}G_{tr}/4\pi r^2$ is the power flux at the car and σ is the scattering cross-section or area of power capture by the car. The car reradiates this power isotropically, so the radar receives $A_{tr}/4\pi r^2$ of this scattered power, or $P_{rec} = P_{tr}G_{tr}\sigma A_{tr}/16\pi^2 r^4 = P_{tr}G_{tr}^2\lambda^2\sigma/64\pi^3 r^4$ and the maximum radar range is

$$r = \left(\frac{G_{tr}^2\lambda^2\sigma P_{tr}}{64\pi^3 P_{rec}} \right)^{1/4}$$

$$= \left(\frac{1.64^2 \times (0.03 \text{ m})^2 \times 1 \text{ m}^2 \times 10^5 \text{ W}}{64\pi^3 \times 10^{-11} \text{ W}} \right)^{1/4} = 332 \text{ m}$$

This radar can therefore detect speeding cars up to ~0.4 km. An actual radar gun is much more directive, with a gain of about 20, so that less transmitter power is necessary. A pulsed radar gun might transmit 100 W peak power, yielding a maximum range of about 200 m.

9.6* RECIPROCITY THEOREM

The reciprocity theorem is widely invoked in analyses of almost any linear electromagnetic system or circuit and was used in the previous section to relate the transmitting and receiving behavior of a single antenna. Consider a current source $\overline{J}_A$ giving rise to electric and magnetic fields $\overline{E}_A$ and $\overline{H}_A$ in a linear environment. The time-harmonic forms of Faraday's and Ampere's laws are then

$$\nabla \times \overline{E}_A = -j\omega\overline{B}_A \tag{9.6.1}$$

$$\nabla \times \overline{H}_A = j\omega\overline{D}_A + \overline{J}_A \tag{9.6.2}$$

from (1.4.4) and (1.4.5). We can rewrite (9.6.1) and (9.6.2) for a different current source $\overline{J}_B$ in the same environment, generating fields $\overline{E}_B$ and $\overline{H}_B$:

$$\nabla \times \overline{E}_B = -j\omega\overline{B}_B \tag{9.6.3}$$

$$\nabla \times \overline{H}_B = j\omega\overline{D}_B + \overline{J}_B \tag{9.6.4}$$

If we then dot $\overline{H}_B$ into both sides of (9.6.1) and $\overline{E}_A$ into both sides of (9.6.4), we find that

$$(\nabla \times \overline{E}_A) \cdot \overline{H}_B - (\nabla \times \overline{H}_B) \cdot \overline{E}_A = \nabla \cdot (\overline{E}_A \times \overline{H}_B)$$

$$= -j\omega\overline{B}_A \cdot \overline{H}_B - j\omega\overline{D}_B \cdot \overline{E}_A - \overline{E}_A \cdot \overline{J}_B \tag{9.6.5}$$

where the first equality results from the vector identity $\nabla \cdot (\overline{A} \times \overline{B}) = (\nabla \times \overline{A}) \cdot \overline{B} - \overline{A} \cdot (\nabla \times \overline{B})$. Likewise, we can dot $\overline{E}_B$ into both sides of (9.6.2) and $\overline{H}_A$ into both sides of (9.6.3):

$$(\nabla \times \overline{E}_B) \cdot \overline{H}_A - (\nabla \times \overline{H}_A) \cdot \overline{E}_B = \nabla \cdot (\overline{E}_B \times \overline{H}_A)$$
$$= -j\omega \overline{B}_B \cdot \overline{H}_A - j\omega \overline{D}_A \cdot \overline{E}_B - \overline{E}_B \cdot \overline{J}_A \qquad (9.6.6)$$

Subtracting (9.6.6) from (9.6.5) then gives the equation

$$\nabla \cdot (\overline{E}_A \times \overline{H}_B - \overline{E}_B \times \overline{H}_A) = (\overline{E}_B \cdot \overline{J}_A - \overline{E}_A \cdot \overline{J}_B) +$$
$$j\omega (\overline{B}_B \cdot \overline{H}_A - \overline{B}_A \cdot \overline{H}_B) + j\omega (\overline{D}_A \cdot \overline{E}_B - \overline{D}_B \cdot \overline{E}_A) \qquad (9.6.7)$$

which is valid for **any** linear medium. If the last two terms of (9.6.7) vanish, so that

$$\overline{B}_B \cdot \overline{H}_A = \overline{B}_A \cdot \overline{H}_B \qquad (9.6.8)$$

and

$$\overline{D}_B \cdot \overline{E}_A = \overline{D}_A \cdot \overline{E}_B \qquad (9.6.9)$$

then the medium is reciprocal. Equations (9.6.8) and (9.6.9) are clearly valid in free space, for which $\overline{B}_i = \mu_0 \overline{H}_i$ and $\overline{D}_i = \epsilon_0 \overline{E}_i$ $(i = A, B)$. If the medium is characterized by the more general tensor constitutive relations $\overline{B}_i = \overline{\overline{\mu}} \cdot \overline{H}_i$ and $\overline{D}_i = \overline{\overline{\epsilon}} \cdot \overline{E}_i$ $(i = A, B)$, then (9.6.8) and (9.6.9) can be written as $\overline{H}_A \cdot \overline{\overline{\mu}} \cdot \overline{H}_B = \overline{H}_B \cdot \overline{\overline{\mu}} \cdot \overline{H}_A$ and $\overline{E}_A \cdot \overline{\overline{\epsilon}} \cdot \overline{E}_B = \overline{E}_B \cdot \overline{\overline{\epsilon}} \cdot \overline{E}_A$. Since $\overline{H}_B \cdot \overline{\overline{\mu}} \cdot \overline{H}_A = \overline{H}_A \cdot \overline{\overline{\mu}}^T \cdot \overline{H}_B$ and $\overline{E}_B \cdot \overline{\overline{\epsilon}} \cdot \overline{E}_A = \overline{E}_A \cdot \overline{\overline{\epsilon}}^T \cdot \overline{E}_B$, a *reciprocal medium* must have the property that $\overline{\overline{\mu}} = \overline{\overline{\mu}}^T$ and $\overline{\overline{\epsilon}} = \overline{\overline{\epsilon}}^T$.[15]

For a reciprocal medium such as free space, (9.6.7) may be simplified:

$$\nabla \cdot (\overline{E}_A \times \overline{H}_B - \overline{E}_B \times \overline{H}_A) = \overline{E}_B \cdot \overline{J}_A - \overline{E}_A \cdot \overline{J}_B \qquad (9.6.10)$$

We then integrate (9.6.10) over the entire medium, where Gauss's divergence theorem (4.1.1) is used to convert the volume integral over the divergence of $\overline{E}_A \times \overline{H}_B - \overline{E}_B \times \overline{H}_A$ to a surface integral at infinity:

$$\oint_{A_\infty} (\overline{E}_A \times \overline{H}_B - \overline{E}_B \times \overline{H}_A) \cdot \hat{n}\, da = \int_{V_\infty} (\overline{E}_B \cdot \overline{J}_A - \overline{E}_A \cdot \overline{J}_B)\, dv \qquad (9.6.11)$$

We showed in (9.2.6) and (9.2.7) that the farfield electric and magnetic fields are both perpendicular to the radial vector $\hat{r}$ and also to each other. Therefore,

---

15. Most media are reciprocal, the principal examples of nonreciprocal media being ferrites and magnetized plasmas (gyrotropic media) discussed in Sections 3.6 and 3.7. As we recall from (3.6.19) and (3.7.10), the permittivity or permeability tensor in these cases is Hermitian, where a Hermitian tensor has the property that $\overline{\overline{H}} = (\overline{\overline{H}}^T)^*$. Because gyrotropic media have complex off-diagonal permeability and/or permittivity elements, $(\overline{\overline{H}}^T)^* \neq \overline{\overline{H}}^T$ and these media are nonreciprocal.

$$\left(\overline{E}_A \times \overline{H}_B - \overline{E}_B \times \overline{H}_A\right)_{\text{ff}} = -\eta(\hat{r} \times \overline{H}_A) \times \overline{H}_B + \eta(\hat{r} \times \overline{H}_B) \times \overline{H}_A$$

$$= -\eta\overline{H}_A(\hat{r} \cdot \overline{H}_B) + \eta\overline{H}_B(\hat{r} \cdot \overline{H}_A)$$

$$= 0 \tag{9.6.12}$$

in the farfield of $\overline{J}_A$ and $\overline{J}_B$ where the vector identity $\overline{A} \times (\overline{B} \times \overline{C}) = \overline{B}(\overline{A} \cdot \overline{C}) - \overline{C}(\overline{A} \cdot \overline{B})$ has been used. The farfield electric fields $\overline{E}_A$ and $\overline{E}_B$ have been represented in terms of the farfield magnetic fields using (9.2.7), and $\hat{r} \cdot \overline{H}_A = \hat{r} \cdot \overline{H}_B = 0$ from (9.2.6).

Equation (9.6.11) thus becomes

$$\int_{V_\infty} (\overline{E}_B \cdot \overline{J}_A - \overline{E}_A \cdot \overline{J}_B)\, dv = 0 \tag{9.6.13}$$

for a reciprocal medium, where V_∞ spans the entire medium. However, we wish to apply (9.6.13) to a system of two interacting antennas and we may therefore replace V_∞ by the actual volume through which current flows.[16] We also replace the integrand of (9.6.13) using (9.6.10), so that

$$\int_{V_1+V_2} \nabla \cdot (\overline{E}_A \times \overline{H}_B - \overline{E}_B \times \overline{H}_A)\, dv = 0 \tag{9.6.14}$$

which may again be converted to a surface integral by Gauss's divergence theorem:

$$\oint_{A_1+A_2} (\overline{E}_A \times \overline{H}_B - \overline{E}_B \times \overline{H}_A) \cdot \hat{n}\, da = 0 \tag{9.6.15}$$

The areas A_1 and A_2 are any closed surfaces that enclose all of the current on antennas 1 and 2, respectively. Equations (9.6.10), (9.6.13) and (9.6.15) are statements of the *reciprocity theorem* in its differential and integral forms.

It is crucial not to confuse the field subscripts A and B with the antenna subscripts 1 and 2; $\overline{J}_A$ does **not** refer to the current on antenna 1 nor $\overline{J}_B$ to antenna 2. The A and B subscripts instead refer to the two distinct experiments described at the beginning of Section 9.5, **both** of which include antennas 1 and 2 and the same reciprocal medium. In experiment A, in which antenna 1 is driven by a current I_1 and the terminals of antenna 2 are open-circuited, the electric and magnetic fields everywhere are $\overline{E}_A$ and $\overline{H}_A$. The fields $\overline{E}_B$ and $\overline{H}_B$ are created by driving antenna 2 with current I_2 and open-circuiting antenna 1 in experiment B. Equation (9.6.15), which relates the two experiments, is thus the basis from which

---

16. If the medium is lossy, $\overline{J}_A = \overline{J}_{A0} + \sigma\overline{E}_A$ and $\overline{J}_B = \overline{J}_{B0} + \sigma\overline{E}_B$ where the total current $\overline{J}_A$ is now explicitly written as the sum of an applied current $\overline{J}_{A0}$ and an ohmic conduction current $\sigma\overline{E}_A$, and similarly for $\overline{J}_B$. But the ohmic terms cancel in (9.6.13), leaving only the applied currents to be integrated over the two antenna volumes $V_1 + V_2$. If the conductivity were represented by a tensor $\overline{\overline{\sigma}}$ instead of the scalar σ, $\overline{E}_B \cdot \overline{\overline{\sigma}} \cdot \overline{E}_A - \overline{E}_A \cdot \overline{\overline{\sigma}} \cdot \overline{E}_B$ would vanish if $\overline{\overline{\sigma}} = \overline{\overline{\sigma}}^T$. Therefore, $\overline{\overline{\sigma}} = \overline{\overline{\sigma}}^T$ is an additional constraint for a reciprocal medium.

reciprocity may be understood, and we need to carefully consider how integration over the closed antenna surfaces A_1 and A_2 is to be performed.

We imagine that each antenna is an arbitrary metallic structure connected to a transmission line, as illustrated in Figure 9.29. Just as in Section 9.3, we divide $A = A_1 + A_2$ into subsurfaces such that $A = A_1' + A_1''' + A_2' + A_2''' + A''$. The surfaces A_1' and A_2' intercept all of the power flowing in the transmission line leading to antennas 1 and 2, respectively, and A_1''' and A_2''' are the corresponding outer surfaces of the perfectly conducting transmission lines. Surface A'' is the boundary at infinity surrounding both antennas, and these surfaces are pictured in Figure 9.29. Note the similarity between Figures 9.29 and 9.3.1. Because $\overline{E}_A \times \overline{H}_B - \overline{E}_B \times \overline{H}_A$ vanishes at infinity from (9.6.12) and is also parallel to

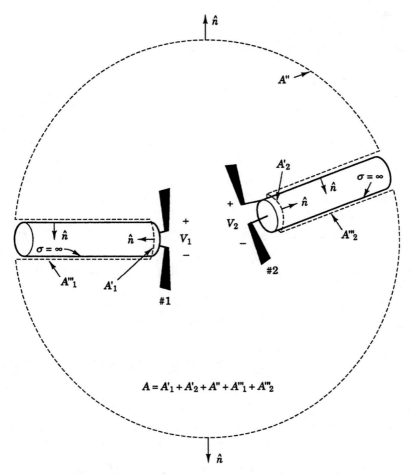

$$A = A_1' + A_2' + A'' + A_1''' + A_2'''$$

Figure 9.29   Closed surface area for reciprocity calculation.

the conducting surface of the transmission lines, only the A_1' and A_2' surfaces contribute to (9.6.15):

$$\int_{A_1'+A_2'} (\overline{E}_A \times \overline{H}_B - \overline{E}_B \times \overline{H}_A) \cdot \hat{n}\, da_1 = 0 \qquad (9.6.16)$$

The *TEM* electric and magnetic fields inside the transmission lines 1 and 2 are necessarily transverse to the axial ($\hat{z}$) direction of power flow. From Section 5.3 we recall that these transverse electric fields are uniquely determined by

$$\nabla \times \overline{E}_T^{A,B} = 0$$
$$\nabla \cdot \overline{E}_T^{A,B} = 0$$
$$\hat{n} \times \overline{E}_T^{A,B} = 0 \qquad (9.6.17)$$
$$\nabla \cdot \overline{E}_T^{A,B} = \sigma^{A,B}$$

while the transverse magnetic fields obey

$$\nabla \times \overline{H}_T^{A,B} = 0$$
$$\nabla \cdot \overline{H}_T^{A,B} = 0$$
$$\hat{n} \times \overline{H}_T^{A,B} = \overline{J}^{A,B} = \hat{z}\, J^{A,B} \qquad (9.6.18)$$
$$\nabla \cdot \overline{H}_T^{A,B} = 0$$

Therefore, the electric field $\overline{E}_A$ is parallel to $\overline{E}_B$ and $\overline{H}_A$ is parallel to $\overline{H}_B$. All these electric and magnetic fields lie in the x–y plane so that $\overline{E}_A \times \overline{H}_B$ and $\overline{E}_B \times \overline{H}_A$ are $\hat{z}$-directed power fluxes.

We evaluate the quantity $\int_{A_1'} (\overline{E}_A \times \overline{H}_B) \cdot \hat{n}\, da_1$ by recalling that $\hat{n}\, da_1 = \hat{z}\, da_1 = d\overline{\ell}_1 \times d\overline{s}_1$, where $d\overline{\ell}_1$ is integrated from the inner to outer conductor to *TEM* line 1 and $d\overline{s}_1$ is a closed contour encircling the inner conductor of this line. Therefore,

$$\int_{A_1'} (\overline{E}_A \times \overline{H}_B) \cdot \hat{n}\, da_1 = \int_{A_1'} (\overline{E}_A \times \overline{H}_B) \cdot (d\overline{\ell}_1 \times d\overline{s}_1)$$

$$= \int_{A_1'} (\overline{E}_A \cdot d\overline{\ell}_1)(\overline{H}_B \cdot d\overline{s}_1) - (\overline{E}_A \cdot d\overline{s}_1)(\overline{H}_B \cdot d\overline{\ell}_1) \qquad (9.6.19)$$

where the vector identity $(\overline{A} \times \overline{B}) \cdot (\overline{C} \times \overline{D}) = (\overline{A} \cdot \overline{C})(\overline{B} \cdot \overline{D}) - (\overline{A} \cdot \overline{D})(\overline{B} \cdot \overline{C})$ has been used. But the second term in (9.6.19) is zero since the integral of $\overline{E}_A$ around any closed contour in the x–y plane is zero, so

$$\int_{A_1'} (\overline{E}_A \times \overline{H}_B) \cdot \hat{n}\, da_1 = \int_1 (\overline{E}_A \cdot d\overline{\ell}_1) \oint_1 (\overline{H}_B \cdot d\overline{s}_1) \equiv -V_1^A I_1^B \qquad (9.6.20)$$

The voltage on transmission line 1 created by electric field $\overline{E}_A$ is V_1^A, defined as $-\int_1 \overline{E}_A \cdot d\ell_1$ from (5.3.10), and the current I_1^B is just $\oint_1 \overline{H}_B \cdot d\overline{s}_1$ from (5.3.14). Each of the four terms in (9.6.16) can be evaluated similarly, yielding the circuit equivalent of the reciprocity theorem

$$V_1^A I_1^B - V_1^B I_1^A + V_2^A I_2^B - V_2^B I_2^A = 0 \qquad (9.6.21)$$

In experiment A, antenna 2 is open-circuited so that $I_2^A = 0$, and in experiment B, antenna 1 is open-circuited so that $I_1^B = 0$. Equation (9.6.21) then reduces to

$$\left.\frac{V_1^B}{I_2^B}\right|_{I_1^B=0} = \left.\frac{V_2^A}{I_1^A}\right|_{I_2^A=0} \qquad (9.6.22)$$

The left side of (9.6.22) is the definition of Z_{12} in (9.5.2) while the right side is Z_{21} from (9.5.1). Therefore, we have just demonstrated that $Z_{12} = Z_{21}$ for two interacting antennas in a reciprocal medium. Adding other antennas to the system and then rerunning Experiments A and B might change the measured values of Z_{12} and Z_{21} but would not change the fact that these mutual impedances are equal; nothing in the above discussion of the reciprocity theorem prevents the inclusion of additional antennas within the volume V. Therefore, the reciprocity theorem can be generalized to a system of N interacting antennas where $Z_{ij} = Z_{ji}$ for any two individual antennas i and j. The relationship between the terminal voltages and terminal currents for each of the N antennas may thus be represented by an $N \times N$ impedance matrix $\overline{\overline{Z}}$, where

$$\overline{\overline{Z}} = \overline{\overline{Z}}^T \qquad (9.6.23)$$

for reciprocal media. This is the most compact form of the reciprocity theorem for circuits.

9.7 LONG-WIRE ANTENNAS

Many practical antennas are constructed from wires that are long compared to a wavelength, for which the propagation delay differences between signals originating from different parts of the wire cannot be neglected and the current distribution on the wire may be uncertain. Although the exact current distribution on a radiating wire is difficult to determine, a reasonable approximation can often be made by treating the wire as the inner conductor of a *TEM* coaxial line with outer diameter much larger than that of the wire, as suggested in Figure 9.30. Since the fields very close to the wire must satisfy boundary conditions at the wire surface, they will resemble *TEM* coaxial line fields, provided the antenna wire is thin relative to its length. Because the electromagnetic fields near the wire decay as $1/\rho$, where ρ is the distance from the wire axis, most of the energy propagates within a few wire diameters of the antenna and the character of wave propagation is determined

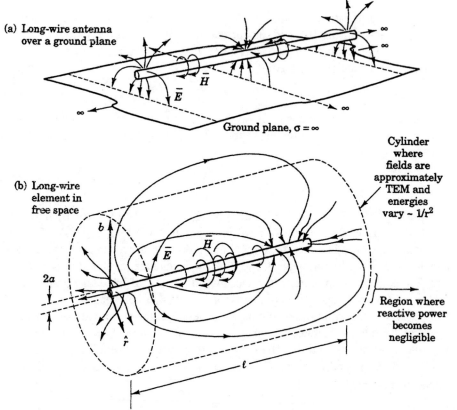

(a) Long-wire antenna
    over a ground plane

Ground plane, $\sigma = \infty$

(b) Long-wire
    element in
    free space

Cylinder
where
fields are
approximately
TEM and
energies
vary ~ $1/r^2$

Region where
reactive power
becomes
negligible

Figure 9.30 *TEM*-like fields near a long-wire antenna.

by the electric and magnetic fields near the wire. Since the current at the ends of the wire must go to zero, a standing wave pattern is established on the wire with wavelength $\lambda = 1/f\sqrt{\mu\epsilon}$.

Typical instantaneous antenna current distribution for long wire antennas are suggested in Figure 9.31(a). The 1.1λ center-fed antenna illustrated in Figure 9.31(b) shows a typical standing wave pattern for the current and further shows how currents at the antenna terminals can have magnitudes significantly smaller than the peak currents flowing in the antenna. Figure 9.31(c) illustrates how an off-center feed can produce peak currents that are different on the two sides of the antenna structure. When the driving point currents are small relative to the peak currents, radiation effects become more important and this simple standing wave model is less accurate.

As an example, consider a center-driven antenna of length d. Assume the feed lines are at $z = 0^+$ and $z = 0^-$. Then the current on this antenna is nominally

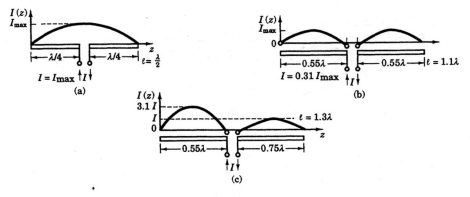

Figure 9.31   Current distributions on long-wire antennas.

$$I(z') = \hat{z}\, I \sin k \left(\frac{d}{2} - |z'| \right) \tag{9.7.1}$$

with driving point current $I(z'=0) = I \sin kd/2$.

Substituting this expression for current into the farfield vector potential of a $\hat{z}$-directed wire of length d given by (9.2.20) and integrating yields

$$\overline{A}_{\text{ff}}(\overline{r}) = \hat{z}\, \frac{\mu_0}{4\pi r}\, e^{-jkr} \int_{-d/2}^{d/2} I(z')\, e^{jkz'\cos\theta}\, dz'$$

$$\cong \hat{z}\, \frac{\mu_0 I}{2\pi kr}\, e^{-jkr} \left[\frac{\cos\left(\frac{kd}{2}\cos\theta\right) - \cos\left(\frac{kd}{2}\right)}{\sin^2\theta} \right] \tag{9.7.2}$$

where θ is the usual spherical coordinate angle and (9.7.2) reduces to (9.2.26) when $d = \lambda/2$. Using (9.2.7), the farfield electric field is just

$$\overline{E}_{\text{ff}}(\overline{r}) \cong \hat{\theta}\, \frac{j\eta_0 I}{2\pi r}\, e^{-jkr} \left[\frac{\cos\left(\frac{kd}{2}\cos\theta\right) - \cos\left(\frac{kd}{2}\right)}{\sin\theta} \right] \tag{9.7.3}$$

The radiation pattern, proportional to the square of (9.7.3), is illustrated in Figure 9.32 for $d = \lambda/2$, $d = 3\lambda/2$, and $d = 2\lambda$.

The standing-wave current approximation (9.7.1) fails for wire antennas that are too many wavelengths long because radiation losses along the wire diminish the currents that reach the ends. In this case, the current distribution is not only nonuniform along the wire, but the waves near the feed points take on a less stationary character and become more like traveling waves. Since the amount of radiation depends on the current distribution and vice versa, some procedure for finding a self-consistent solution is necessary to find the exact current $I(z')$.

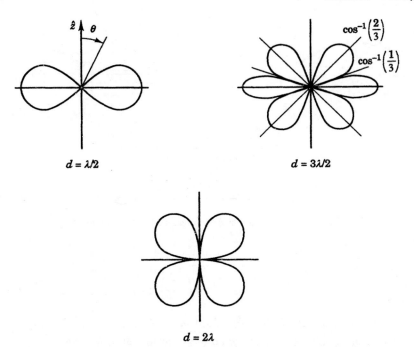

Figure 9.32   Radiation patterns for wire antennas of varying lengths.

By modeling the current distributions on long wires as *TEM*-like standing waves of spatial frequency k, we are now prepared to analyze several commonly used multiple-wire antennas. Perhaps the simplest such antenna is the *folded dipole*, illustrated in Figure 9.33(a). It consists of a wire of length λ, which is bent into a long, thin rectangle with sides approximately $\lambda/2$ in length and is fed from the middle of the long sides. Each end of the antenna can be modeled as a short-circuited *TEM* line one-quarter wavelength long. The currents flowing on these parallel wires can be considered as a superposition of two modes—the differential and common modes. The *differential mode* refers to the *TEM* mode for which the currents flow on the two wires with equal magnitude but in opposite direction. *Common mode* currents flow with equal magnitudes in the **same** direction in both wires. The differential mode currents do not radiate well because their 180° out-of-phase, equal magnitude currents produce fields that essentially cancel for an observer far from the wires. In contrast, the in-phase common mode currents in the wires radiate constructively. By short-circuiting the ends of the folded dipole, the current is a maximum at the ends; one-quarter wavelength back at the feed, the *TEM* current is basically negligible. Therefore the current I driving the center-fed folded dipole excites only the common mode, so the currents in the parallel wires

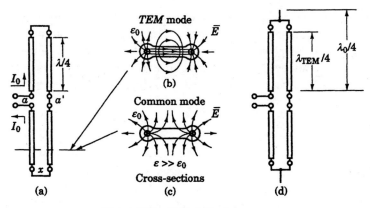

Figure 9.33   Folded-dipole antennas.

are approximately equal and flow in the same direction. These common mode currents must be zero at the ends of the antenna, and therefore the total current flowing in the $\hat{z}$ direction is $2I \cos kz'$. The total power radiated by this structure is therefore four times that of an ordinary half-wave dipole driven with the same terminal current I because the radiated power is proportional to $|I|^2$. The radiation resistance of the folded dipole also quadruples to $\sim 4 \times 73 \ \Omega = 292 \ \Omega$. However, the farfield radiation pattern is the same for both folded and half-wave dipoles. Had the ends of the folded-dipole antenna been open-circuited instead, the impedance looking into the center of the dipole would have been a short circuit and differential-mode (*TEM*) currents would dominate. This antenna would be essentially useless, however, because the fields radiated by the two closely spaced but oppositely directed and equal magnitude currents would approximately cancel. The radiation resistance for this antenna would be correspondingly small.

Folded dipoles are often constructed of twin-lead *TEM* lines, a cross-section of which is suggested in Figure 9.33(b). Because the metal wires are bonded in plastic, the differential mode electric field trapped within the plastic has a wavelength that is typically 20 percent less than it would be in free space ($\lambda \propto 1/\sqrt{\mu\epsilon}$), whereas the common-mode field largely escapes the plastic and hence has an intermediate wavelength. Because the wavelengths of the two modes are different, a "pig-tail" is often added at the ends of the structure to adjust the length of the common-mode wires so that both modes are resonant at the same frequency, as in Figure 9.33(d).

Another very important type of antenna involves arrays of long wire elements, some of which are free floating *parasitic elements* indirectly excited by radiation from the *driven elements* connected directly to the antenna feed lines. The currents that flow in the parasitic elements can be greatly enhanced near their resonant frequencies, which occur when the links are approximate multiples of a half wavelength. The degree to which these currents are enhanced depends on the Q of these isolated wire resonators. The Q of an isolated straight wire $\lambda/2$ long can be estimated by assuming a standard sinusoidal current distribution and then calculating the total radiated power and the total energy stored in the *TEM* region of the wires, which typically dominates the total energy. The total Q is then evaluated using (8.3.3). Values near 10 are typical, with higher quality factors obtainable with thinner wires. In Example 9.7.2, the Q of a resonant wire is calculated using the procedure just outlined.

The phase between the current induced on the resonant parasitic element and the exciting electric field can be adjusted by raising or lowering the resonant frequency of the element (i.e., by changing its length). Elements slightly longer than $\lambda/2$ have a lower resonant frequency, which means that they are driven **above** resonance when $f = c/\lambda$. These elements thus lag in phase with respect to the driven elements. Conversely, elements slightly shorter than $\lambda/2$ exhibit a phase lead compared with the driven elements. A high-Q parasitic element that radiates very little power will have a phase difference of approximately $\pm90°$, depending on its length.[17]

If a resonant parasitic element is placed parallel to and some distance away from a driven element, its scattered fields can be made to interfere either constructively or destructively with those of the driven element, depending on the separation distance between the two and the propagation direction of interest. Parasitic elements that enhance radiation in the direction opposite to that from which they receive radiation are called *reflector elements*, while parasitic elements that enhance radiation in the same forward direction it is emitted are called *director elements*.

High-gain *Yagi antennas*, consisting of arrays of reflectors and directors are very practical, easily constructed antennas, often used for TV reception. The exact analysis of Yagi antennas is difficult, but it is observed experimentally that closely spaced parasitic elements spaced less than $\lambda/4$ apart can yield gains of 20 dB and higher.

Another type of wire antenna was developed by Kraus in 1946, the *helical*

---

17. We also observe this phase shift about resonance in a simple parallel RLC circuit with impedance

$$Z = \frac{1}{R + j\omega L + 1/j\omega C}$$

At the resonant frequency $\omega_0 = 1/\sqrt{LC}$, the impedance is just $Z = 1/R$ with no imaginary part ($0°$ phase). If the circuit is driven slightly off-resonance at $\omega_0 + \Delta\omega$ where $|\Delta\omega| \ll \omega_0$,

$$Z \cong \frac{1}{R}\left(1 - \frac{2j\Delta\omega L}{R}\right)$$

and the phase is $\tan^{-1}(-2\Delta\omega L/R)$, which is positive when $\Delta\omega < 0$ and negative when $\Delta\omega > 0$. Note that as $R \to 0$ so that the resonator has a large Q, the off-resonance phase approaches $\pm90°$ with a very narrow transition between the two phase extremes.

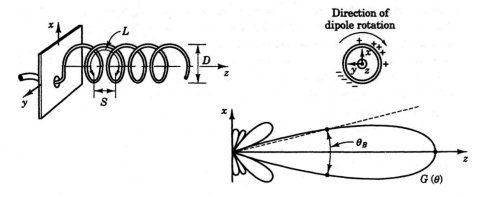

Figure 9.34   Helical antennas.

antenna illustrated in Figure 9.34. The helical antenna produces a sharp beam of circularly polarized radiation along its axis, and has nearly uniform resistive input impedance over a bandwidth that is a large fraction of an *octave* (an octave is a factor of two in frequency). It is also easy to build. As a first approximation, a helical antenna may be understood in terms of a single current wave of constant amplitude that propagates along the conductor wire in *TEM* fashion. If the helix has N turns, each of length L and separation distance S, then the gain may be found by multiplying the farfield element patterns of a single helix loop of length L by the array factor of an N-element array of points separated by S, with adjacent points having a phase difference of $\psi = -kL$. (The traveling wave at the $n + 1^{st}$ point is shifted by e^{-jkL} with respect to the nth point.) Because the array factor dominates the array pattern if N is large, the gain for this helix is approximately

$$G(\theta) \simeq \left| \sum_{i=0}^{N-1} \left(e^{-jkL + jkS\cos\theta} \right)^i \right|^2 \propto \frac{\sin^2 Nu}{\sin^2 u}$$

by analogy with (9.4.16). Here, $u = k(S\cos\theta - L)/2$ where S and L are defined in Figure 9.34. Such antennas normally operate in an *endfire* configuration, producing maximum gain along the helix axis ($\theta = 0$); $k(L - S)/2 = n\pi$ or, equivalently, $L - S = n\lambda$ for this endfire helix. These relations can also be expressed in terms of D, the diameter of the helix, since $L = \sqrt{(\pi D)^2 + S^2}$. Because the maximum gain of an array increases as N^2, helices with more turns are increasingly directive. Since the currents flowing on a helix interact with each other, the pure traveling wave model is only approximate. If $L - S = \lambda$, then the instantaneous current distribution perceived by a distant observer on the helix axis $\hat{z}$ is equivalent to that of a linear dipole lying in the x–y plane with dipole axis at some angle ϕ. This dipole rotates as the waves propagate outward, one rotation per frequency cycle. The helix of Figure 9.34 and its radiation are left-hand circularly polarized.

Although Yagi and helical antennas typically have bandwidths that are fractions of an octave, they may be modified to increase the bandwidth to one or even several octaves. One common frequency-independent antenna has a shape specified only in terms of angles. Because there is no length scale for this type of antenna, it radiates well at all wavelengths simply by using different parts of its self-similar structure. The *planar log-spiral antenna* is an example of such a frequency-independent antenna and is illustrated in Figure 9.35(a). The equation

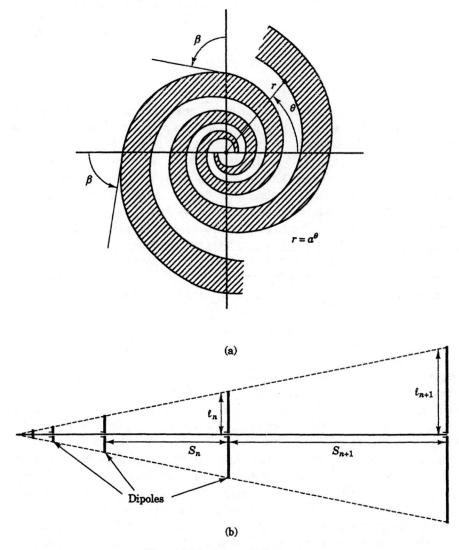

(a)

(b)

Figure 9.35 Log-periodic antennas.

for a logarithmic spiral is $r = b^\theta$, where r and θ are cylindrical coordinates and b is a constant. The behavior of the antenna is determined by the angle β between a line tangent to the spiral and the radial line from the origin. If β is constant, the spiral is self-similar and completely frequency-independent if infinitely large. For a logarithmic spiral, $\tan \beta = r/(dr/d\theta) = 1/\ln b$. If the finite two-armed spiral of Figure 9.35(a) is excited at its center, and if most of the radiation is emitted before the currents progress to the perimeter, then the diameter of the antenna is not critical and the spiral can be extremely broad-band.

Another popular frequency-independent antenna is the *log-periodic dipole array* illustrated in Figure 9.35(b). Its structure is made self-similar by increasing the dipole lengths d along the array such that the ratio of any two adjacent dipole lengths and the ratio of the spaces s between consecutive pairs of elements is the same constant K: $K = d_{n+1}/d_n = s_{n+1}/s_n$, where d_n and s_n are defined in Figure 9.35(b). Because the array is fed from the small end and radiates most efficiently for those few dipole lengths near resonance, the radiating region of the array shifts toward the longer dipoles as the frequency decreases. There is again no length scale in an infinite array of this type, so the infinite log-periodic dipole array has a completely flat frequency response. A finite-sized log-periodic array can be used at any wavelength where the radiating region of the array is well-removed from its low-frequency and high-frequency ends. In a similar fashion, helical antennas can also be made log-periodic and will then yield multioctave bandwidths. Such helices are generally fed from their small forward ends. Yagi antennas are also often made periodic to broaden their bandwidths.

Example 9.7.1

Design an endfire helical antenna for use in the FM band at 100 MHz, where a gain of 16 dB is desired along the helix axis. What is the approximate bandwidth?

Solution: The gain is proportional to $\sin^2 Nu / \sin^2 u$ when $u = k(S - L)/2 = \pi(S - L)/\lambda$ along the helix axis ($\theta = 0$) so $G_{\max} \propto N^2$. If the helix is to have a gain of at least 16 dB relative to a single loop antenna, then

$$16 \geq 10 \log_{10} N^2$$

and $N = 7$ is the minimum number of turns necessary. If we choose $S = \lambda/4 = 0.75$ m to eliminate grating lobes, we see that

$$(\pi D)^2 + S^2 = (\pi D)^2 + \frac{\lambda^2}{16} = L^2$$

But $L - S = n\lambda$; for $n = 1$, we find

$$L = S + \lambda = \frac{5\lambda}{4} = 3.75 \text{ m}$$

$$D = \sqrt{\frac{3}{2}\frac{\lambda}{\pi}} = 1.17 \text{ m}$$

If we now wish to vary λ, we must keep u within the approximate region $[-\pi - \pi/2N, -\pi + \pi/2N]$. When $S - L = -\lambda$, $u = -\pi$. Alternatively, we find that

$$\frac{2N}{2N+1} < \frac{\lambda}{\lambda_0} < \frac{2N}{2N-1}$$

places a limit on the percentage we can alter λ from its original value $\lambda_0 = 3$ m. For $N = 7$, λ can vary only about 7 percent; as N becomes larger, the bandwidth becomes smaller.

Example 9.7.2

Just as a plucked piano string can excite resonances in neighboring strings, so can isolated wires in space be excited near their resonances and reradiate (scatter) the incident energy. This reradiated energy may be adjusted to enhance or diminish antenna gain in various directions. What is the Q of a half-wave dipole antenna of radius a and length $\ell = \lambda/2$?

Solution: If there were no radiation or dissipation losses, then energy in the standing wave could resonate indefinitely and $Q \to \infty$. But since the half-wave dipole radiates efficiently but not perfectly, we may estimate its Q using (8.3.3):

$$Q = \frac{\omega_0 \mathsf{w}_T}{P_d}$$

At resonance, the total stored energy w_T is simply related to $I(z)$

$$\mathsf{w}_T = \langle \mathsf{w}_e \rangle + \langle \mathsf{w}_m \rangle = 2\langle \mathsf{w}_m \rangle = 2 \int_V \frac{1}{4} \mu_0 |\overline{H}|^2 \, dv$$

$$\cong 2 \int_{-\ell/2}^{\ell/2} dz \int_a^{\ell/2} dr \int_0^{2\pi} r d\phi \frac{1}{4} \mu_0 \left| \frac{I(z)}{2\pi r} \right|^2$$

$$\cong \frac{\mu_0 \ell |I|^2}{8\pi} \ln\left(\frac{\ell}{2a}\right)$$

where the integration is only over the region for which the magnetic field is approximately inversely proportional to radius r and where most of the energy resides. The current distribution on the dipole is given by (9.2.25): $I(z) \simeq I \cos kz$.

The power radiated is $\frac{1}{2} R_{\text{rad}} |I|^2$ where $R_{\text{rad}} \cong 73$ Ω for a half-wave dipole. The resonant frequency ω_0 is $\omega_0 = kc = 2\pi c/\lambda = \pi c/\ell$. Therefore,

$$Q \simeq \frac{(\pi c/\ell)(\mu_0 \ell |I|^2/8\pi) \ln(\ell/2a)}{R_{\text{rad}} |I|^2/2} = \frac{\eta_0}{4 R_{\text{rad}}} \ln\left(\frac{\ell}{2a}\right)$$

If the wire is 1 mm in diameter and 10 m long, then $Q \simeq 11.9$. Note that approximating the range of the energy integral as $r = a$ to $r = \ell/2$ was not very critical because the maximum radius appeared only logarithmically. The formula suggests that Q might approach unity as the wire diameter approaches one-third the antenna length, but then the field approximations and the upper bound of the integral become more critical.

This result is useful in understanding how thin wires can efficiently scatter energy near resonances and how a small group of resonant wires orbiting Earth has been used as an experimental reflecting belt for centimeter wavelength communications. The scattering cross-section of an isolated wire can be found from its equivalent circuit, viewed from terminals at the center of the wire. This circuit is a series resonator with $R_{eq} \cong 73\ \Omega$, L_{eq}, and C_{eq}. We know $\omega_0 = 1/\sqrt{L_{eq}C_{eq}}$ and $Q = \omega_0 L_{eq}/R_{eq}$ because the equivalent circuit is short-circuited at its terminals rather than being connected to an external system. The scattering cross-section of such an isolated wire may thus be estimated as a function of frequency near isolated resonances by using the equivalent circuit to determine the dipole currents and reradiated power. The equivalent circuit of the excited dipole is a Thevenin voltage source of $\sim (\lambda/2)E_{inc}\cos\theta$ V in series with R_{eq}, L_{eq}, and C_{eq}; this gives I and therefore the scattered fields.

9.8 APERTURE ANTENNAS AND DIFFRACTION

If an antenna is to have modest gain, then a simple array of a few radiating elements· usually suffices, but a high-gain antenna requires using a large aperture. Such large structures can be most easily characterized by the electromagnetic fields they excite in a convenient aperture plane; the transmitting and receiving properties of the antenna can then be deduced from those fields. In this section, we first consider a simple, approximate way in which to find the diffracted fields radiated by an aperture and then discuss some of the more common types of aperture antennas.

An *aperture antenna* may be created by cutting a hole in a perfectly conducting sheet and allowing a uniform plane wave to be incident on one side of this sheet, as shown in Figure 9.36. The plane wave will pass through the aperture and will also induce currents on the sheet, which then reradiate, leading to a *diffracted field* on the other side of the aperture. Because we cannot find the induced currents without knowing the tangential magnetic fields on both sides of the aperture plane and we cannot find the diffracted magnetic field without knowing the induced currents,

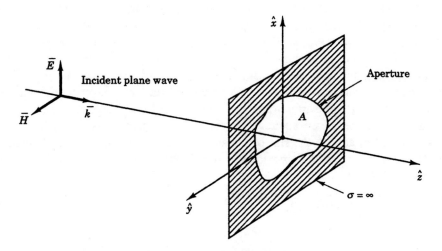

Figure 9.36   Aperture antenna.

the diffracted field must in general be found self-consistently. However, in keeping with previous methods of solving antenna problems, we use an approximate method for finding these diffracted fields instead.

Consider a uniform current sheet at $z = 0$ on which the surface current $\overline{J}_s = -\hat{x} J_s$ (A/m) flows. This current will generate outwardly propagating uniform plane waves with electric and magnetic fields given by

$$\left.\begin{array}{l} \overline{E} = \hat{x} E_0 e^{-jkz} \\[2mm] \overline{H} = \hat{y} (E_0/\eta_0) e^{-jkz} \end{array}\right\} \quad \text{for } z > 0 \qquad (9.8.1)$$

and

$$\left.\begin{array}{l} \overline{E} = \hat{x} E_0 e^{jkz} \\[2mm] \overline{H} = -\hat{y} (E_0/\eta_0) e^{jkz} \end{array}\right\} \quad \text{for } z < 0 \qquad (9.8.2)$$

where the tangential electric field is continuous at $z = 0$. The tangential magnetic field satisfies the boundary condition

$$\hat{z} \times \left[\overline{H}(z = 0^+) - \overline{H}(z = 0^-) \right] = \overline{J}_s \qquad (9.8.3)$$

where $\overline{H}(z = 0^{\pm}) = \pm\hat{y} E_0/\eta_0$ from (9.8.1) and (9.8.2), and thus $E_0 = \eta_0 J_s/2$. Given a uniform plane wave, we can therefore replace it by the equivalent current sheet that could generate the wave, and vice versa. (See also the discussion in Section 4.5.)

If we now return to the discussion of aperture antennas, we note that the conducting screen blocks most of the incident plane wave so that only that portion of the wave that passes through the hole in the screen is transmitted. We can therefore represent the plane wave segment by a current sheet of magnitude $J_s = 2E_0/\eta_0$, where E_0 is the amplitude of the incident wave and the current sheet is no longer infinite, but instead has the same extent as the aperture. This representation is **not** exact, as we have not imposed the boundary condition that the tangential electric field vanish on the conducting sheet, and additional currents will be induced at the edges of the aperture to ensure that all boundary conditions are met. The correction currents are negligible beyond approximately a wavelength from the aperture edge, so these currents flow in an area of roughly λP square meters, where P is the perimeter of the aperture. By contrast, the current sheet that spans the aperture has an area A the size of the aperture, so the current sheet replacement of the aperture is a valid approximation only if the aperture is many wavelengths across ($\lambda P \ll A$).

A general plane wave that propagates in the $\hat{z}$ direction and has nonuniform amplitude $E_0(x, y)$ can be expressed as

$$\overline{E} = \hat{t} E_0(x, y) e^{-jkz} \qquad (9.8.4)$$

where $\hat{t}$ is a transverse unit vector that is perpendicular to $\hat{z}$ ($\hat{x}$, $\hat{y}$, or $(\hat{x} - \hat{y})/\sqrt{2}$

are all possible choices for this unit vector). The electric field must satisfy $\nabla \cdot \overline{E} = 0$ and is incident from the left on a large aperture of area A cut in a perfectly conducting screen. We can approximate the diffracted fields to the right of the aperture by replacing the screen with the surface current distribution

$$\overline{J}_s(x', y') = -\hat{\imath} \frac{2E_0(x', y')}{\eta_0} \quad (\text{A/m}) \tag{9.8.5}$$

over the area A. The vector potential for this current sheet is then given by (9.2.1)

$$\overline{A}(\overline{r}) = -\hat{\imath} \frac{\mu_0}{4\pi} \int_A \frac{2E_0(x', y')}{\eta_0} \frac{e^{-jk|\overline{r}-\overline{r}'|}}{|\overline{r}-\overline{r}'|} da' \tag{9.8.6}$$

where the surface integral includes only the area of the aperture. If we are observing the diffracted fields far from the aperture in the region $r \gg \lambda/2\pi$, $r \gg r'$, and $r \gg 8r'^2/\lambda$, the usual Fraunhofer farfield approximation (9.2.4) can be made and (9.8.6) becomes

$$\overline{A}_{\text{ff}}(\overline{r}) \cong -\hat{\imath} \frac{\mu_0}{4\pi r} e^{-jkr} \int_A \frac{2E_0(x', y')}{\eta_0} e^{jk\hat{r}\cdot\overline{r}'} da' \tag{9.8.7}$$

which may be compared to (9.2.5). The farfield electric field is then computed from (9.2.7)

$$\overline{E}_{\text{ff}}(\overline{r}) \cong j\omega\hat{r} \times \left(\hat{r} \times (-\hat{\imath}) \frac{\mu_0}{4\pi r} e^{-jkr} \int_A \frac{2E_0(x', y')}{\eta_0} e^{jk\hat{r}\cdot\overline{r}'} da' \right)$$

$$\cong \hat{\theta} \, \eta_0 \frac{jk}{4\pi r} e^{-jkr} \cos\theta \int_A \frac{2E_0}{\eta_0}(x', y') e^{jk\hat{r}\cdot\overline{r}'} da' \tag{9.8.8}$$

where $\hat{r} \times (\hat{r} \times \hat{\imath}) \simeq -\hat{\theta} \cos\theta$ for directions close to the $\hat{z}$-axis.

Equation (9.8.8) is just the farfield electric field for a continuous distribution of $-\hat{\imath}$-directed Hertzian dipoles, where the factor outside the integral is the far field of $-\hat{\imath}$-directed dipole with unit magnitude Id. The integral over the surface current $J_s(x', y') = 2E_0(x', y')/\eta_0$ in the aperture is just the array factor for the continuous array of dipoles.[18] If we define $\overline{k} \equiv k\hat{r}$, then for $\theta \cong 0$, we can rewrite (9.8.8) as

$$\overline{E}_{\text{ff}}(z > 0) \cong \hat{\imath} \frac{j}{\lambda r} e^{-jkr} \int_A dx'dy' E_0(x', y') e^{jk_x x' + jk_y y'} \tag{9.8.9}$$

where $\overline{k} \cdot \overline{r}' = k_x x' + k_y y'$. From (9.8.9) we now see that the farfield electric field ($r \gg \lambda/2\pi$, $r \gg |\overline{r}'|$, $r \gg 8r'^2/\lambda$) is just proportional to the *two-dimensional*

---

18. A more exact expression for the farfield electric field is found by replacing $\cos\theta$ in (9.8.8) by $(1 + \cos\theta)/2$, giving the Huygen equation. Since both expressions are identical near $\theta = 0$, we confine our attention to the region $\theta \cong 0$.

Fourier transform of the incident field over the aperture![19]   A small aperture thus
has a farfield radiation pattern spanning a wider solid angle than does a large aperture, and radiation from a large aperture can therefore be more directed, yielding
higher gain.

Consider the general case of a rectangular aperture of dimensions L_x and
L_y that is excited by an $\hat{x}$-directed plane wave normally incident from $z < 0$ and
having a uniform electric field strength E_0. The farfield electric field is then

$$\overline{E}_{ff}(z > 0) = \hat{x} \, \frac{j}{\lambda r} \, e^{-jkr} \int_{-L_x/2}^{L_x/2} dx' \int_{-L_y/2}^{L_y/2} dy' E_0 \, e^{jk_x x' + jk_y y'}$$

$$= \hat{x} \, \frac{j}{\lambda r} \, e^{-jkr} L_x L_y E_0 \, \frac{\sin(k_x L_x/2)}{k_x L_x/2} \, \frac{\sin(k_y L_y/2)}{k_y L_y/2} \qquad (9.8.10)$$

from (9.8.9). This radiation pattern is illustrated in Figure 9.37, where we note
that (9.8.10) has a maximum at $\theta = 0$ ($k_x = k_y = 0$) with nulls at $k_x L_x/2 = m\pi$
and $k_y L_y/2 = n\pi$ (m, n integers). The angles at which these nulls occur can be
approximated when θ is small by defining α_x and α_y, representing angular offsets
from the $\hat{z}$ direction in the $\hat{x}$ and $\hat{y}$ directions, respectively. We see that $k_x =
k \sin\theta \cos\phi \simeq k\alpha_x$ and $k_y = k \sin\theta \sin\phi \simeq k\alpha_y$ when θ is small, so that $\alpha_{x,\text{null}} =
2m\pi/kL_x = m\lambda/L_x$ and $\alpha_{y,\text{null}} = n\lambda/L_y$. The full width half maximum (FWHM)
of the main diffracted beam is approximately λ/L_x radians in the x–z plane and
λ/L_y radians in the y–z plane, so the beam size is large for a small aperture and
small for a large aperture.

The total power radiated from this aperture could be found by integrating
the farfield radiation pattern over all angles, but this is difficult because (9.8.10)
is valid only near $\theta \cong 0$. Instead, the total radiated power is found by integrating
the incident electric field over the aperture area. By conservation of energy, this
power eventually finds its way into the far field. Thus,

$$P = \int_A \frac{|E_0(y')|^2}{2\eta_0} \, da' = \frac{|E_0|^2 A}{2\eta_0} \qquad (9.8.11)$$

---

19. Recall the Fourier transform relations:

$$x(t) = \int_{-\infty}^{\infty} X(\omega) \, e^{j\omega t} \, \frac{d\omega}{2\pi}$$

$$X(\omega) = \int_{-\infty}^{\infty} x(t) \, e^{-j\omega t} \, dt$$

In the diffraction formulation, the complementary coordinates (x', y') and (k'_x, k'_y) are related just as t
and ω are above.

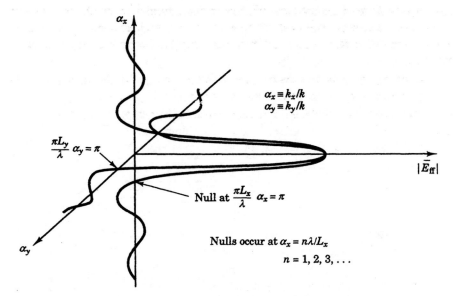

Figure 9.37 Radiation for a rectangular aperture.

where the latter equality holds for a uniform incident electric field. The gain of this aperture is therefore

$$G(k_x, k_y) = \frac{|\overline{E}_{\text{ff}}(\overline{r})|^2/2\eta_0}{P/4\pi r^2}$$

$$= \frac{4\pi A}{\lambda^2} \frac{\sin^2(k_x L_x/2)}{(k_x L_x/2)^2} \frac{\sin^2(k_y L_y/2)}{(k_y L_y/2)^2} \qquad (9.8.12)$$

$$= \frac{4\pi}{\lambda^2} A_{\text{eff}}$$

which has a maximum value of $4\pi A/\lambda^2$ at $k_x = k_y = 0$ ($\theta = 0$). For a uniformly illuminated rectangular aperture, $A_{\text{eff}} = A$ along the $\hat{x}$-axis; $A_{\text{eff}} \leq A$ in general. The ratio A_{eff}/A is called the *aperture efficiency* η_A and is degraded from unity as either the excitation magnitude or phase becomes nonuniform over the aperture.

So far, we have been exclusively considering the Fraunhofer farfield region, where the observer is both physically far ($r \gg r'$) and electrically far ($r \gg \lambda/2\pi$, $r \gg 8r'^2/\lambda$) from the aperture. We now discuss diffraction in the *Fresnel region*, for which $r \gg \lambda/2\pi$, but the observer is not necessarily far from the aperture. We again approximate $|\overline{r} - \overline{r}'|$ in (9.8.6) using a Taylor series:

$$|\overline{r} - \overline{r}'| = \sqrt{r^2 - 2\overline{r} \cdot \overline{r}' + r'^2}$$

$$\cong r - \frac{\overline{r} \cdot \overline{r}'}{r} + \frac{r'^2}{2r} \qquad (9.8.13)$$

This enables us to quantify the boundary of the Fraunhofer region: $r'^2/2r$ can be neglected when the phase $kr'^2/2r$ does not change appreciably over an aperture of nominal radius R, or $kR^2/2r \ll \pi/8$. (The maximum phase error of $\pi/8$ is somewhat arbitrary.) The observer is thus in the Fraunhofer region when $r \gg 8R^2/\lambda = 2D^2/\lambda$ and in the Fresnel region when $r < 2D^2/\lambda$; $D = 2R$ is the aperture diameter and $r \gg \lambda/2\pi$ holds for both regions.

To illustrate the Fresnel region, we consider diffraction of a $\hat{z}$-propagating, $\hat{x}$-directed uniform plane wave from a uniformly illuminated straight-edge conducting sheet in the half-plane $z = 0$, $x < 0$ as shown in Figure 9.38(a). If we observe the diffracted fields close to the $\hat{z}$-axis, $|\bar{r} - \bar{r}'|$ may be approximated as

$$|\bar{r} - \bar{r}'| = \sqrt{(x - x')^2 + (y - y')^2 + z^2}$$
$$\cong z + \frac{(x - x')^2}{2z} + \frac{(y - y')^2}{2z} \qquad (9.8.14)$$

and the vector potential in the Fresnel zone is thus

$$\bar{A}(\bar{r}) \cong -\hat{x}\frac{\mu_0}{4\pi z}e^{-jkz}\int_A dx'dy'\frac{2E_0}{\eta_0}e^{-jk(x-x')^2/2z}e^{-jk(y-y')^2/2z} \qquad (9.8.15)$$

after (9.8.6) and (9.8.14) are combined. The Fresnel electric field is then computed using (9.2.7), which only requires that $r \gg \lambda/2\pi$:

$$\bar{E}(\bar{r}) \cong \hat{x}\frac{jE_0}{\lambda z}e^{-jkz}\int_0^\infty dx'\int_{-\infty}^\infty dy'\,e^{-jk(x-x')^2/2z}e^{-jk(y-y')^2/2z} \qquad (9.8.16)$$

If we now replace the variables x', y' with u', v' such that

$$u' = \sqrt{\frac{k}{\pi z}}(x - x')$$
$$v' = \sqrt{\frac{k}{\pi z}}(y - y') \qquad (9.8.17)$$

then

$$\bar{E}(\bar{r}) \cong \hat{x}\frac{jE_0}{2}e^{-jkz}\left(\int_{-\infty}^\infty dv'\,e^{-j\frac{\pi}{2}v'^2}\right)\left(\int_{-\infty}^{x\sqrt{k/\pi z}}du'\,e^{-j\frac{\pi}{2}u'^2}\right) \qquad (9.8.18)$$

The definite integral in (9.8.18) is easily evaluated by recalling that $\int_{-\infty}^\infty dx\,e^{-\alpha x^2} = \sqrt{\pi/\alpha}$ for complex values of α, so that $\int_{-\infty}^\infty dv'\,e^{-j\frac{\pi}{2}v'^2} = \sqrt{2/j} = \sqrt{2}e^{-j\pi/4}$. The second, indefinite integral is called the Fresnel integral and must be evaluated numerically. A plot of the magnitude of the diffracted electric field as a function of x is given in Figure 9.38(b). Note that for large negative values of x, (9.8.18) is attenuated, indicating that little of the incident field penetrates behind the conducting screen. On the other hand, for large positive values of x, the indefinite integral in (9.8.18) approaches $\sqrt{2/j}$ and the diffracted field is then just $\bar{E}(x \to \infty) \cong \hat{x}E_0e^{-jkz}$. Far from the straight-edge, the incident wave remains unaffected. We

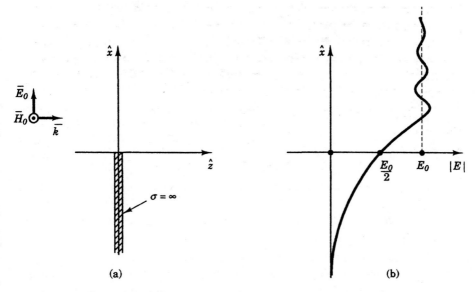

Figure 9.38   Straight-edged aperture; illustration of Fresnel zone.

can also compute the electric field at $x = 0$ since $\int_{-\infty}^{0} du' \, e^{-j\frac{\pi}{2}u'^2} = \frac{1}{2}\sqrt{2/j}$ and thus $\overline{E}(x=0) \cong \hat{x}\,(E_0/2)\,e^{-jkz}$, which is just one-half the incident field.[20]

---

20.   Although a discussion of Fresnel integrals is beyond the scope of this text, we can find approximate solutions for various ranges of x. If we integrate the Fresnel integral by parts, we find

$$\int_{-\infty}^{X} e^{-j\frac{\pi}{2}u'^2} du' = \left. \frac{j}{\pi u'} e^{-j\frac{\pi}{2}u'^2}\right|_{-\infty}^{X} + \int_{-\infty}^{X} \frac{j}{\pi u'^2} e^{-j\frac{\pi}{2}u'^2} du'$$

$$= \frac{j}{\pi X} e^{-j\frac{\pi}{2}X^2}\left(1 + \frac{j}{\pi X^2} + \ldots\right) \tag{1}$$

where $X \equiv x\sqrt{k/\pi z}$ and we neglect all but the leading term in the parentheses if $|X|$ is large. This integral vanishes as $x \to -\infty$, so

$$\overline{E}(x \ll 0) \cong -\hat{x}\,\frac{E_0}{x}\sqrt{\frac{z}{2\pi k}}\,e^{-j(kx^2/2z + \pi/4)}$$

For $x \gg 0$, (1) does not have the correct limiting behavior, since we have already shown that $\overline{E}(x \to \infty) = \hat{x}\,E_0\,e^{-jkz}$. We therefore rewrite the Fresnel integral as

$$\int_{-\infty}^{X} e^{-j\frac{\pi}{2}u'^2} du' = \int_{-\infty}^{\infty} e^{-j\frac{\pi}{2}u'^2} du' - \int_{X}^{\infty} e^{-j\frac{\pi}{2}u'^2} du'$$

$$= \sqrt{\frac{2}{j}} - \frac{j}{x}\sqrt{\frac{z}{\pi k}}\,e^{-jkx^2/2z}$$

where the second integral is evaluated using (1). The electric field for large positive x is then

$$\overline{E}(x \gg 0) \cong \hat{x}\,E_0\,e^{-jkz}\left(1 - \frac{j}{x}\sqrt{\frac{z}{2\pi k}}\,e^{-j(kx^2/2z - \pi/4)}\right)$$

consistent with Figure 9.38(b).

Using the preceding discussion of diffraction from apertures, we may now proceed to analyze some common aperture antennas. The most common reflector antenna is the *circular parabolic dish* with surface given by $z = (x^2 + y^2)/4f$ where the *focal length* f is the distance between the origin and the *focal point*, which lies at $x = y = 0$, $z = f$. A radiating source or *antenna feed* placed at this focal point to illuminate the reflector gives rise to a uniform plane wave propagating in the $+\hat{z}$ direction, as shown in Figure 9.39, because the radiation path lengths from the focal point to aperture plane via the parabolic reflector are the same for each ray, and each emergent ray from the aperture plane is $\hat{z}$-directed. (A parabola is the only surface with these properties). The aperture plane passes through the edge of the dish and has an area $A = \pi R^2$. A typical aperture excitation function is shown in Figure 9.39 together with a typical gain pattern, proportional to the square of the Fourier transform of the excitation function in the Fraunhofer region. *Parabolic cylinder antennas* are sometimes used, but require a *line feed* at the *focal line* of the parabola.

The design of such antennas involves tradeoffs between the aperture efficiency η_A and the sidelobe level, which includes *spillover*, the fraction of energy that leaves the antenna feed without impacting the reflector. Because of the Fourier relationship between aperture excitation and farfield radiated fields (and therefore gain), efforts to decrease the sidelobes and reduce spillover by tapering the aperture excitation smoothly and strongly toward the edge of the reflector unfortunately result in broader antenna beams and reduced values of antenna efficiency η_A. Figure 9.39 illustrates a typical compromise where the sidelobes are ~ 20 dB below peak gain, the beam width is $\sim 1.3\lambda/D$ ($D = 2R$), and the aperture efficiency is ~ 60 percent. Typical ratios of f/D are $\sim 0.3 - 0.6$, with the more shallow dishes (larger f/D ratios) exhibiting better polarization purity in off-axis directions and better imaging capability if more than one feed is employed side by side.

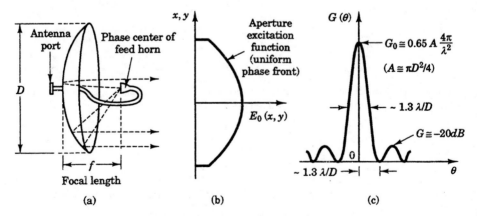

Figure 9.39   Typical parabolic antennas.

Two common variations of the parabolic antenna are the off-axis (offset) parabola and the Cassegrain configurations. The *off-axis parabolic antenna*, illustrated in Figure 9.40(a), is used when it is important to avoid sidelobes caused by aperture excitation energy blocked or scattered by the antenna feed structure. The structure is similar to a normal parabolic antenna except that the feed illuminates only one sector of the reflector such that no reflected energy impinges on the feed assembly; the unilluminated portions of the reflector can be eliminated. However, polarization purity is generally not as good for the offset parabola as for the parabolic dish antenna.

A *Cassegrain antenna* replaces the prime-focus feed with a hyperbolic reflector located between the focal point and the main reflector. This subreflector permits the feed to be located near the reflector surface, as illustrated in Figure 9.40(b). The virtues of this configuration are that (1) the effective f/D ratio is generally larger, permitting the use of more adjacent feeds, (2) the electronics can be much larger and more accessible without physically blocking the aperture plane, and (3) Cassegrain dishes can be deeper than parabolic dishes, with feeds located closer to the primary reflector surface because the effective f/D ratio of the feeds is larger. More compact structures with lower antenna backlobes and far sidelobes can thus be achieved compared to the standard parabolic dish.

Also useful is the *spherical reflector antenna*, which can be scanned (the

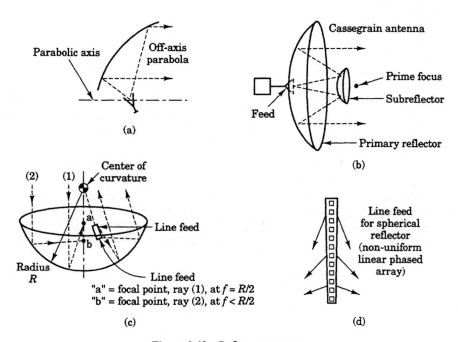

Figure 9.40   Reflector antennas.

main beam can be moved) over large angles by steering only the feed. The surface of a spherical reflector of radius R centered at $(0, 0, R)$ is given by $x^2 + y^2 + (z - R)^2 = R^2$ or $z - z^2/2R = (1/2R)(x^2 + y^2)$, which resembles a parabolic dish for small values of z if $f = R/2$. If a point feed at this paraxial focal point is allowed to illuminate only this small region of the sphere, then the emerging wave is approximately uniform and planar. If the point feed at $f = R/2$ is then rotated about the center of the sphere, the emerging plane wave can be scanning, using a limited portion of the reflector for each angle. Because spherical reflectors naturally focus radiation along a line rather than a point, the radiation source is typically a line feed, often a linear phased array designed to illuminate the central part of the sphere from a distance $R/2$, and the rim from closer distances, as illustrated in Figure 9.40(c). The antenna can be steered by physically rotating the line feed about the center of the sphere. The virtues of spherical reflector antennas are (1) steering of very large motionless reflectors (such as the 1000 foot reflector at Arecibo, Puerto Rico) is possible by moving only the feed, (2) many feeds can be used simultaneously and independently, and (3) very rapid scanning can be achieved because only a light-weight feed must be physically moved instead of the large reflector.

The most common scanning feed reflector is a torus, which has a circular surface in the plane through which the antenna is to be scanned and a parabolic surface in the orthogonal plane. Such a *torus antenna* has been proposed for use in several satellites. The torus resembles the antenna of Figure 9.40(a) and is scanned by rotating the feed axis around the axis shown in the figure.

Example 9.8.1

A communication link consists of a half-wave dipole located parallel to and $\lambda/4$ m in front of a vertical perfectly conducting sheet. This combination transmits signals to a uniformly illuminated circular parabola antenna 1 m in diameter located 10 km away. Describe the transmitter antenna pattern in the plane perpendicular to the half-wave dipole and determine approximately what transmitter power is required to produce a received signal greater than 0.1 nW.

Solution: The image of the half-wave dipole in the reflecting plane will have oppositely directed, equal-magnitude current located $\lambda/2$ m away from the true dipole. As a result, the signals will add coherently in the horizontal direction perpendicular to the sheet and cancel in the vertical direction. Although computation of the antenna gain would require integrating the radiated power over all directions, we approximate the gain as roughly twice that of a half-wave dipole or ~ 3 in the direction of the parabola. The received power is simply the transmitted power multiplied by $G_{tr} A_{rec}/4\pi r^2$ where $G_{tr} \simeq 3$ and A_{rec} is the effective aperture of the parabolic dish and is approximately equal to the physical aperture πR^2 since the dish is uniformly illuminated. Therefore,

$$P_{tr} \simeq \frac{4\pi r^2 P_{rec}}{G_{tr} A_{rec}} = \frac{4\pi (10^4 \text{ m})^2 \times 10^{-10} \text{ W}}{3\pi (0.5 \text{ m})^2} = 53 \text{ mW}$$

Example 9.8.2

Find the farfield electric field along the $\hat{z}$-axis for a uniformly illuminated circular aperture of radius R, where $z \gg \lambda/2\pi$ but R is not necessarily small. Show that as the aperture becomes infinite, (i.e., the screen disappears), the diffracted field becomes the incident field. If the aperture is finite and we observe the far field ($z \gg 8R^2/\lambda$), show that the axial gain is just $4\pi A/\lambda^2$ where $A = \pi R^2$. Sketch the intensity of $\overline{E}$ as a function of z.

Solution: If $z \gg \lambda/2\pi$ but z is not necessarily larger than R, we cannot use the Fraunhofer farfield approximation. Instead, we use the Fresnel approximation with (9.8.6) to find the farfield vector potential. If the incident electric field is given by $\hat{x} E_0 e^{-jkz}$, then

$$\overline{A}(z > 0) \cong -\hat{x} \frac{\mu_0}{4\pi} \int_A \frac{2E_0}{\eta_0} \frac{e^{-jk|\vec{r}-\vec{r}'|}}{|\vec{r}-\vec{r}'|} \, da'$$

$$\cong -\hat{x} \frac{\mu_0 E_0}{2\pi \eta_0} \int_0^R dr' \int_0^{2\pi} r' d\phi' E_0 \frac{e^{-jk\sqrt{r'^2+z^2}}}{\sqrt{r'^2+z^2}} \tag{1}$$

as illustrated in Figure 9.41(a), where $da' = r'dr'd\phi'$. If we make the substitution $r' = z \tan\alpha$, we may rewrite (1) as

$$\overline{A}(z > 0) \cong -\hat{x} \frac{\mu_0 E_0 z}{\eta_0} \int_0^{\tan^{-1}(R/z)} e^{-jkz\sec\alpha} \sec\alpha \tan\alpha \, d\alpha$$

$$\cong \hat{x} \frac{E_0}{j\omega} e^{-jkz\sec\alpha} \Big|_0^{\tan^{-1}(R/z)}$$

$$\cong \hat{x} \frac{jE_0}{\omega} \left(e^{-jkz} - e^{-jk\sqrt{z^2+R^2}} \right) \tag{2}$$

Equation (2) is valid for all z such that $z \gg \lambda/2\pi$, and we can easily compute the electric field from (9.2.7) when $z \gg \lambda/2\pi$ and $\hat{r} = \hat{z}$:

$$\overline{E}(z > 0) = j\omega\hat{z} \times (\hat{z} \times \overline{A}(z > 0))$$

$$= \hat{x} E_0 \left(e^{-jkz} - e^{-jk\sqrt{z^2+R^2}} \right) \tag{3}$$

In the far field, $z \gg R$ and a Taylor expansion of $\sqrt{z^2 + R^2}$ yields

$$\sqrt{z^2 + R^2} \cong z + \frac{R^2}{2z} \tag{4}$$

Therefore,

$$\overline{E}_{\text{ff}}(z \gg R) \cong \hat{x} E_0 e^{-jkz} \left(1 - e^{-jkR^2/2z} \right)$$

$$\cong \hat{x} \frac{jkR^2 E_0}{2z} e^{-jkz} \tag{5}$$

where $e^{-jkR^2/2z} \cong 1 - jkR^2/2z$. The gain of this aperture along the $\hat{z}$-axis is then

$$G(\theta = 0) = \frac{|E_{\text{ff}}|^2/2\eta_0}{P/4\pi z^2} = \frac{\pi k^2 R^4}{A} = \frac{4\pi A}{\lambda^2} \tag{6}$$

where (9.8.11) and (5) have been used. The axial gain of any uniformly illuminated aperture

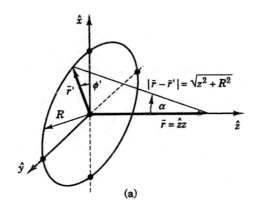

(a)

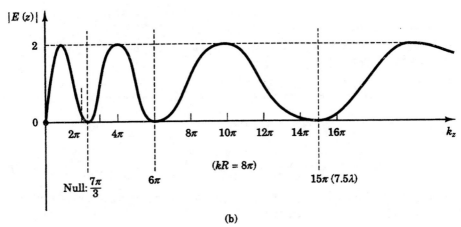

(b)

Figure 9.41   Circular aperture in Example 9.8.2.

is $4\pi A/\lambda^2$, as was also illustrated for a rectangular aperture in (9.8.12). The intensity of the electric field along the $\hat{z}$-axis is found by multiplying (3) by its complex conjugate

$$I = |\overline{E}|^2 = |E_0|^2 \left[2 - 2\cos k(z - \sqrt{z^2 + R^2}) \right]$$

$$= 4|E_0|^2 \sin^2 \frac{k}{2}(z - \sqrt{z^2 + R^2}) \tag{7}$$

and (6) is plotted as a function of kz in Figure 9.41(b) ($kz \gg 1$) for $kR = 8\pi$. Note that the intensity fluctuates markedly with z in the Fresnel region. The last on-axis null occurs sufficiently far away that $z_{null} \gg R$; we approximate $\sqrt{z_{null}^2 + R^2}$ as $z_{null} + R^2/2z_{null}$. Therefore, $\frac{k}{2}(z_{null} - z_{null} - R^2/2z_{null}) = -n\pi$ and $z_{null} \cong R^2/2n\lambda$, which is maximum when $n = 1$. The distance to the last null is thus $R^2/2\lambda$, which is one-sixteenth the distance to the Fraunhofer region ($\sim 8R^2/\lambda$). Example 9.8.4 describes a more intuitive way to understand the origin and placement of these on-axis nulls.

Example 9.8.3

What is the far field for the double-slit aperture illustrated in Figure 9.42? Each slit has width w and length ℓ, and the slits are separated by a distance a. What happens if there are N slits instead of just two?

Solution: The farfield electric field of the two-slit configuration is given by (9.8.9), where the x' coordinate is integrated from $-(a+\ell)/2$ to $-(a-\ell)/2$ for one slit and then from $(a-\ell)/2$ to $(a+\ell)/2$ for the other, and the y' coordinate ranges over $-w/2 \leq y' \leq w/2$. For a uniformly illuminated aperture,

$$
\overline{E}_{\text{ff}}(z > 0) \cong \hat{x}\,\frac{jE_0}{\lambda r}\,e^{-jkr}\left[\int_{-(a+\ell)/2}^{-(a-\ell)/2} dx'\,e^{jk_x x'} + \int_{(a-\ell)/2}^{(a+\ell)/2} dx'\,e^{jk_x x'}\right] \times
$$

$$
\left[\int_{-w/2}^{w/2} dy'\,e^{jk_y y'}\right]
$$

$$
= \left[\hat{x}\,\frac{jkE_0}{2\pi r}\,e^{-jkr}\ell w\,\frac{\sin(k_x\ell/2)}{(k_x\ell/2)}\frac{\sin(k_y w/2)}{(k_y w/2)}\right] \times \left[2\cos(k_x a/2)\right] \qquad (1)
$$

But we note that this result could also be derived by convolving a single-slit aperture of length ℓ and width w with two unit impulses located on the $\hat{x}$-axis at $x' = \pm a/2$. Because the radiation pattern is the Fourier transform of the incident field, and convolution becomes multiplication in the transform domain, we see that (1) is the product of the rectangular aperture transform given by (9.8.10) and the transform of the unit impulses at $x' = \pm a/2$. The latter Fourier transform is

$$
\int_{-\infty}^{\infty}\int_{-\infty}^{\infty} dx'dy'\left[\delta(x' - a/2) + \delta(x' + a/2)\right]\delta(y')\,e^{jk_x x' + jk_y y'}
$$

$$
= e^{jk_x a/2} + e^{-jk_x a/2} = 2\cos(k_x a/2)
$$

which is the second factor in the rectangular brackets of (1). From this result, we can easily derive the transform for N slits, which is the convolution of the single rectangular slit with

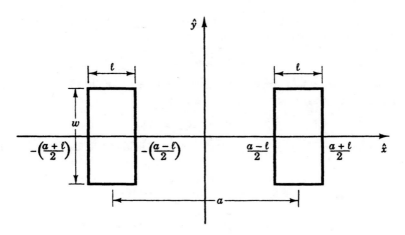

$$
-\left(\frac{a+\ell}{2}\right) \qquad -\left(\frac{a-\ell}{2}\right) \qquad \frac{a-\ell}{2} \qquad \frac{a+\ell}{2} \qquad \hat{x}
$$

Figure 9.42   Double-slit aperture in Example 9.8.3.

N unit impulses spaced a distance a apart. The Fourier transform of this pulse train is

$$\int_{-\infty}^{\infty} \int_{-\infty}^{\infty} dx'dy' \sum_{i=0}^{N-1} \delta(x' - ia) \, \delta(y') \, e^{jk_x x' + jk_y y'}$$

$$= \sum_{i=0}^{N-1} e^{jk_x ia} = e^{jk_x (N-1)a/2} \, \frac{\sin(k_x N a/2)}{\sin(k_x a/2)} \tag{2}$$

Therefore, the farfield electric field for the N slits is given by the product of the rectangular aperture transform [the first term in brackets in (1)] and the pulse train transform (2). Notice that the Fourier transform of the pulse train is just the same as the array factor for N equally spaced in-phase ($\psi = 0$) dipoles given in (9.4.15).

Example 9.8.4

On-axis diffraction by a circular aperture, solved exactly in Example 9.8.2, can be understood geometrically using the construction of *Fresnel zones*. We consider a uniform $\hat{x}$-polarized plane wave normally incident upon the aperture and observe the diffracted field at $z > 0$. The aperture is divided into a set of concentric annular regions such that the nth ring has outer radius R_n and inner radius R_{n-1}, illustrated in Figure 9.43(a). The radius R_n is chosen so that the distance from the outer edge of the nth annular zone to z is n half-wavelengths greater than the distance from the origin to z along the $\hat{z}$-axis, as shown in Figure 9.43(b). The diffracted field can thus be expressed as a sum of fields from each of the annular zones. Find R_n and show that the area of each zone is constant for $R_n \ll z, z_0$. If the aperture contains an even number of zones, show that the diffracted field is zero, but is a constant (maximum) for an odd number of zones. If we manufacture a zone plate by making the odd-numbered zones opaque, what happens to the diffracted electric field intensity?

Solution: Using Figure 9.43(b), we find the outer radius R_n of the nth annular Fresnel zone from

$$\sqrt{z^2 + R_n^2} - z = \frac{n\lambda}{2} \tag{1}$$

If $R_n \ll z$, then the first term of (1) can be replaced by its Taylor expansion $\sqrt{z^2 + R_n^2} \cong z + R_n^2/2z$, leading to an expression for R_n

$$R_n \cong \sqrt{n\lambda z} \tag{2}$$

The radius of each annular Fresnel zone is thus directly related to the observation point of the field. The area of the nth annular Fresnel zone is

$$A_n = \pi R_n^2 - \pi R_{n-1}^2 \cong \pi\lambda z \tag{3}$$

which is independent of n (for $R_n \ll z$). The electric field at z is then found from the vector potential (9.8.6)

$$\overline{A}(z > 0) \cong -\hat{x} \, \frac{\mu_0}{2\pi \eta_0 z} \int_0^{R_n} dr' \int_0^{2\pi} r' d\phi' \, E_0 \, e^{-jk\sqrt{z^2 + r'^2}} \tag{4}$$

where $|\overline{r} - \overline{r}'| \cong \sqrt{z^2 + r'^2}$ in the phase of (4) as may be seen from Figure 9.43(b), but $|\overline{r} - \overline{r}'| \cong z$ in the denominator of (4). The area integral in (4) may be replaced by a

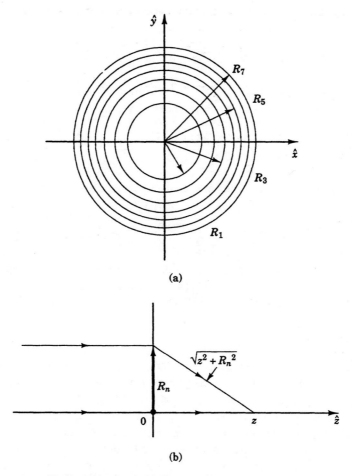

Figure 9.43 On-axis diffraction by a circular aperture in Example 9.8.4.

summation over each of the annular Fresnel zones comprising the aperture

$$\int_0^{R_N} dr' \int_0^{2\pi} r' d\phi' \rightarrow \sum_{n=1}^{N} A_n$$

and thus (4) becomes

$$\overline{A}(z > 0) \cong -\hat{x}\, \frac{\mu_0 E_0}{2\pi \eta_0 z} \sum_{n=1}^{N} A_n\, e^{-jk\sqrt{z^2 + R_n^2}}$$

$$\cong -\hat{x}\, \frac{\mu_0 E_0 (\pi \lambda z)}{2\pi \eta_0 z}\, e^{-jkz} \sum_{n=1}^{N} e^{-jkn\lambda/2} \tag{5}$$

where A_n is given by (3) and $\sqrt{z^2 + R_n^2}$ is evaluated using (1). The electric field at z is then found using (9.2.7) when $z \gg \lambda/2\pi$ and $\hat{r} = \hat{z}$

$$\overline{E}(z>0) \cong -j\omega\overline{A}(z>0) = \hat{x}\, j\pi E_0\, e^{-jkz} \sum_{n=1}^{N} e^{-jn\pi} \tag{6}$$

because $k\lambda = 2\pi$. If N is even, $\sum_{n=1}^{N} e^{-jn\pi} = 0$ and the diffracted field vanishes, but if N is odd, $\sum_{n=1}^{N} e^{-jn\pi} = -1$ and $|E_N(z)| = \pi|E_0|$ is a constant maximum independent of N.

For an aperture of radius R, the diffracted field will have nulls for those values of z for which $R = R_N$ (N even) or $z = R_N^2/N\lambda$ (N even) for $z \gg R_N$ from (2). Therefore, the most distant null is located at $z = R^2/2\lambda$ in agreement with Example 9.8.2. If we make all of the odd-numbered Fresnel zones opaque, then the magnitude of the diffracted field is

$$|E_N(z)| \propto \left| \pi E_0\, e^{-jkz} \sum_{\substack{n=2 \\ (n\,\text{even})}}^{N} e^{-jn\pi} \right| = \begin{cases} \pi N|E_0|/2 & N\ \text{even} \\ \pi(N-1)|E_0|/2 & N\ \text{odd} \end{cases} \tag{6}$$

if the aperture radius is R_N. The field intensity can thus be increased by a factor of $\sim N^2$ by this construction. Flat Fresnel zone plate lenses can be constructed in this way, even for x-rays, which are easier to block than to reflect.

9.9 SUMMARY

Antennas couple electromagnetic waves traveling in guides to waves traveling in space. Because most antennas couple to an infinite number of plane waves in space, our traditional methods for solving boundary value problems are of limited utility, as finding the antenna current distribution $\overline{J}$ satisfying all boundary conditions is generally very difficult. Except for a few simple cases, we are often reduced to guessing the current distribution on the antenna using approximate physical arguments, and then computing the resulting vector potential field $\overline{A}$, where $\overline{B} = \nabla \times \overline{A}$. We may then find the fields $\overline{E}$ and $\overline{H}$ from $\overline{A}$.

Although antennas can produce intense near fields within a fraction of a wavelength of the antenna structure, the radiated far fields are usually of greater interest. These fields result from the *superposition integral* for the farfield vector potential

$$\overline{A}_{\text{ff}}(\overline{r}) = \frac{\mu_0}{4\pi r} e^{-jkr} \int_{V'} \overline{J}(\overline{r}')\, e^{jk\hat{r}\cdot\overline{r}'}\, dv'$$

A *Hertzian dipole* is the simplest radiating element, consisting of a current I of infinitesimal length d (such that the dipole moment Id is finite). The electric and magnetic farfields for this basic antenna are:

$$\overline{E}_{\text{ff}}(\overline{r}) = \hat{\theta}\, \frac{j\eta_0 kId}{4\pi r} e^{-jkr} \sin\theta$$

$$\overline{H}_{\text{ff}}(\overline{r}) = \hat{\phi}\, \frac{jkId}{4\pi r} e^{-jkr} \sin\theta$$

If the dipole is short but not infinitesimal, d is replaced by the *effective dipole length* d_{eff} where $d_{eff} \approx d/2$ for a realistic *short dipole*. The *farfield Poynting power* for a Hertzian dipole is

$$\langle \overline{S}_{ff}(\overline{r}) \rangle = \frac{1}{2} \, \mathrm{Re} \left\{ \overline{E}_{ff}(\overline{r}) \times \overline{H}_{ff}^{*}(\overline{r}) \right\} = \hat{r} \, \frac{\eta_0}{2} \left| \frac{k I d_{eff}}{4\pi r} \right|^2 \sin^2 \theta$$

so it radiates $\eta_0 |kId|^2/12\pi$ W total power.

The directionality with which an antenna radiates power can be described by its *gain* $G(\theta, \phi)$, defined as the ratio of the farfield power radiated in a particular direction to the power that would be radiated by an isotropic source

$$G(\theta, \phi) = \frac{\langle \overline{S}_{ff}(\overline{r}) \cdot \hat{r} \rangle}{P/4\pi r^2}$$

where P is the total power available to the antenna at its input. The *directivity* of the antenna $D(\theta, \phi)$ is equal to $G(\theta, \phi)/\eta_{rad}$, where the *antenna efficiency* factor η_{rad} is the discrepancy between the power available at the antenna terminals and the power actually radiated. Both the directivity and gain of a lossless, matched Hertzian dipole are equal to $(3/2) \sin^2 \theta$, independent of wavelength.

The linearity of Maxwell's equations implies that antennas can be represented by Thevenin or Norton equivalent circuits with input impedance Z. The resistive part of the antenna impedance results both from dissipative losses and *radiation resistance*, where the latter corresponds to power radiated into space; for a short dipole,

$$R_{rad} = \frac{2P_{rad}}{|I|^2} = \frac{\eta_0}{6\pi} (kd_{eff})^2$$

The reactive part of the antenna impedance is dominated by stored electric energy in the near fields.

Antenna arrays composed of identical elements can be analyzed by superposing the fields from each element while being careful to account for the phase difference $e^{j\overline{k} \cdot \Delta \overline{r}}$ between elements. A variety of techniques are used to analyze these arrays, including *phasor addition*, the *visible window method*, and the inclusion of *image sources* if necessary. The array gain is the product of an *element factor*, describing the details of the individual array element, and an *array factor*, describing the orientation and excitation of each element in the array. In a *reciprocal medium* (which excludes magnetized plasmas and ferrites), the directionality of a receiving antenna is identical to that of the same antenna while transmitting. We characterize antenna reception by its *effective area* A_{eff}:

$$A_{eff}(\theta, \phi) = \frac{\lambda^2}{4\pi} G(\theta, \phi)$$

The current distribution on long wire antennas is close to that of a *TEM* line, with approximately sinusoidal variation along the length of the wire and current nulls at the wire ends. *Multiple-wire antennas* can be analyzed by superposing the fields of the individual wire elements and making allowances for the propagation delays between wires. Because of these phase lags the contributions from different elements often interfere, producing radiation patterns with multiple lobes.

The directional properties of radiation emanating from finite apertures can also be computed by integrating the contributions associated with the plane wave strength at each point within the aperture. *Huygen's equation* for the far field is

$$\overline{E}_{\text{ff}}(\overline{r}) \cong \hat{\theta} \, \frac{j}{2\lambda r} \, e^{-jkr}(1 + \cos\theta) \int_A E_0(x', y') \, e^{jk\hat{r}\cdot\overline{r}'} \, da'$$

Because this farfield *diffraction field* $\overline{E}_{\text{ff}}$ is the two-dimensional Fourier transform of the aperture field $\overline{E}_0$, aperture antennas are sometimes used in signal processing.

9.10 PROBLEMS

9.2.1 An antenna has a gain $G(\theta, \phi)$ proportional to $\sin^3\theta \cos^2\phi$. What is the directivity of this antenna? If the maximum gain is measured to be $8/\pi$, what is the efficiency η_{rad} of this antenna?

9.3.1 The losses in a 1-cm Hertzian dipole can be modeled by a 10^{-5}-Ω resistor.

    (a) At which frequency is the dipole driven if the resistive and radiative losses in the antenna are to be comparable?

    (b) What is the antenna efficiency in this case? Assume that the reactive fields of the antenna can be modeled by a 0.01-pF capacitor.

9.3.2 Show that the stored electric energy dominates the stored magnetic energy in the vicinity of a Hertzian dipole. What is this excess stored energy $\Delta\langle W \rangle = \langle W_e \rangle - \langle W_m \rangle$? Show that the reverse is true for a current-loop antenna and find $\Delta\langle W \rangle$ for this antenna (see Example 9.3.1).

9.4.1 Each dipole in a 100-dipole array is excited identically. The length of the array is L, and the dipoles are spaced equally and are oriented in the $\hat{z}$ direction.

    (a) Sketch $G(\phi)$ in the x–y plane for the case where the dipoles lie on the $\hat{x}$-axis and: (i) $L = 99\lambda/2$, (ii) $L = 99\lambda$.

    (b) Quantitatively determine the location of the first nulls in each of these two cases.

9.4.2 A dipole array consists of two parallel half-wavelength dipoles horizontally positioned above and on either side of a flat metal roof sloped at $45°$. The dipoles are fed $180°$ out of phase with each other at a wavelength of $\sqrt{2}$ meters, are separated by 1 m, and are each 0.5 m from the roof and parallel to it (so that a straight line could be passed through both dipoles and the apex of the roof).

    (a) Sketch the arrangement and excitations of the original two dipoles together with any images resulting from the roof.

    (b) Sketch the farfield radiation pattern produced in the x–y plane perpendicular to the dipoles.

    (c) If each dipole is excited with a peak current of 1 A, what is the maximum radiated power (W/m^2) at a distance of 10 km?

(d) Explain why this arrangement and excitation of dipoles is inefficient for transmitting power horizontally. Can you design a physical arrangement and excitation scheme using these two dipoles and the same roof so that the signals radiated by the two dipoles and their images add coherently only in the horizontal direction? Assume each dipole still carries a peak current of 1 A and that you can choose the phase difference between the two dipoles arbitrarily. Some thought may be required here, but simple answers exist.

9.4.3 (a) Sketch the antenna gain $G(\phi)$ in the x–y plane for the following two-dipole arrays, where the dipoles are oriented with their axes in the $\hat{z}$ direction and lie on the $\hat{x}$-axis. The dipoles are separated by D and have dipole moments of Id at $x = 0$ and aId at $x = D$. Quantitatively compute the angles of any nulls for (i) $D = \lambda$, $a = -1$, (ii) $D = 3\lambda$, $a = 1$, (iii) $D = \lambda/4$, $a = j$, and (iv) $D = \lambda/4$, $a = -j/2$.

(b) Sketch the gain pattern $G(\phi)$ in the x–y plane for a four-dipole array, with the $\hat{z}$-oriented dipoles at the corners of a square with edge length λ. All dipoles are excited equally and in phase.

(c) Sketch the gain $G(\theta)$ for array of part (b) in the x–z plane.

9.4.4 A rancher has a one-sided flat metal roof sloped $30°$ from horizontal, above which he wishes to place a horizontally polarized short-dipole antenna operating at wavelength λ. How far above the roof vertically should he place the dipole to maximize transmission toward the horizon in the direction the roof slopes?

9.4.5 (a) A radio station is located on the west coast, 15 km west of a city. The transmitting antenna tower may be modeled as a vertical Hertzian dipole antenna of dipole moment $I_0\ell$. To maintain the FCC standard of 25 mV/m field strength in the city, how much power must be provided to the station?

(b) The radio station can now afford to erect another antenna tower that is driven with equal-magnitude dipole moment. Relative to the first antenna tower, at what distance d should the second tower be placed and with what phase difference ψ should it be fed, so there is a null in the radiation pattern in the direction of the ocean ($\phi = 180°$) and no "dead" spots in the urban reception area.

(c) With the configuration of the two antenna towers as in part (b), how much power must be provided to maintain the same urban field strength as in part (a)?

(d) Down the coast, another radio station with two identical antenna towers faces a different problem. This radio station must service two cities, sending most of the power over the land areas and little power in the direction of the ocean ($\phi = 180°$), with maximum radiation in the direction of the two cities ($\phi = \pm 60°$), and no nulls, or "dead" spots, in the reception area ($|\phi| \leq 90°$). Determine the spacing d of these two antenna towers and the relative phase difference ψ satisfying these requirements.

9.4.6 All dipole antennas in this problem are pointing in the $\hat{z}$ direction. For each configuration listed below, sketch the radiation patterns in the x–y plane. Indicate locations of all maxima and minima on your sketch. Use pattern multiplication whenever possible.

(a) Two equally excited dipoles $\lambda/2$ apart on the $\hat{y}$-axis.

(b) Same as part (a), but excited $180°$ out of phase.

(c) Same as part (b), but λ apart.

(d) Four equal amplitude dipoles $\lambda/2$ apart on the $\hat{x}$-axis. The left two dipoles are excited 180° out of phase with the right two.

(e) Three dipoles $\lambda/2$ apart on the $\hat{y}$ axis, excited with I_0, $3I_0$, and $2I_0$, respectively.

9.5.1 A wrist-watch manufacturer believes he can place a 10-mW transmitter inside a watch together with a receiver of 10^{-12} W sensitivity. They operate in the 400-MHz land-mobile band.

(a) If the antenna is a wire fastened inside a shirt to form a $\sim \lambda/2$ dipole antenna of gain $\sim 3/2 \sin^2 \theta$ (ignore the strong effects of the person for purposes of this problem), then what is the maximum distance two people can be separated and still communicate? Assume both antennas are perfectly matched to their respective circuits.

(b) Repeat part (a) for the case where ground reflections are present. Assume both antennas are located above a planar ground plane ($\sigma \cong \infty$) at an optimum height h. Find the minimum value of h that maximizes power transfer for two people separated 100 m for horizontal polarization, and then for vertical polarization.

9.5.2 A bird scatters incident electromagnetic radiation isotropically, i.e., uniformly into 4π steradians. A certain uniformly illuminated radar antenna has an effective area of $1000\ \text{m}^2$ and transmits 1-MW pulses at 15 GHz. The same antenna is shared by the transmitter and receiver, and the radar scans the sky at a range of R meters.

(a) What is the maximum reflected power received by this radar as a function of range R, antenna effective area A_{eff}, and target cross-section σ_t [σ_t is that area (m^2) that would intercept and reradiate isotropically power consistent with the power received at the radar station]. For example, a target that scatters 1 W isotropically when placed in a $10\ \text{W/m}^2$ beam has $\sigma_t = 0.1\ \text{m}^2$.

(b) If the receiver has a state-of-the-art maser amplifier, it can detect 10^{-18} W. In this case, if the bird were 100 km away (line-of-sight), then how small could its cross-section σ_t (m^2) be and still be detectable?

9.5.3 A perfectly conducting short-dipole antenna is transmitting at 1 MHz to a receiver employing a small two-turn loop antenna of diameter $D \ll \lambda/2\pi$. The short dipole is oriented for maximum reception by the loop.

(a) What is the magnetic field strength $|\overline{H}|$ (A/m) in the vicinity of the receiver in the far field at distance r in the x–y plane if 30 W are transmitted?

(b) In what direction(s) should the loop be oriented to maximize the receiver power? (List all directions if there are more than one.)

(c) Assume the answer to part (a) is H_0. Then what is the maximum open-circuit voltage V_L for this loop antenna?

9.5.4 Consider an elementary ground-to-air communication system. An airplane is flying parallel to the ground at an altitude d; the path of the airplane is directly over a transmitting antenna. Both receiving and transmitting antennas are short dipoles that remain parallel during the flight and perpendicular to the ground. Assume throughout that the altitude $d \gg \lambda$.

(a) Let the total time-averaged power radiated by the ground antenna be P. Determine the power received by the receiving antenna on the airplane, assuming the receiving antenna is matched to its ground transmitter. Express your answer in terms of θ, d, λ, and P where θ is the angle from zenith to the airplane as seen from the ground antenna. For what value of θ is the received power maximized?

(b) Assume 10 kW is radiated, the receiver sensitivity is 10^{-10} W (i.e., the minimum detectable signal is 10^{-10} W), and the wavelength is 10 cm. How far away from the transmitter will the signal be detectable in the airplane? (Hint: For $r \gg d$, $d/\cos\theta = r \simeq x$ and $\sin\theta \simeq 1$.)

9.5.5 The migratory patterns of caribou in Alaska are to be monitored by means of small transmitters attached to the animals. Periodically, the transmitter radiates a coded pulse train with a total radiated power of 10 W at 300 MHz. Assume the transmitting antenna is an isotropic radiator (with gain 1), but the receiving antenna is an array with 30-dB gain. The receiver is matched to the antenna, which has an impedance of 50 Ω and can detect the coded signals if the voltage across the receiver terminals is greater than 1 μV. Assuming the path is line-of-sight, at what distance can the caribou be tracked? Can they be tracked from a geosynchronous satellite at 36,000 km altitude?

9.7.1 An antenna consists of a wire bent into a square, each side of which is one-half wavelength long. The antenna is fed from the middle of one side.

(a) Sketch a typical instantaneous current distribution on the antenna.

(b) Sketch the antenna gain $G(\phi)$ in the plane of the square (the x–y plane).

(c) Sketch the antenna gain for $\hat{x}$-polarization in the y–z plane intersecting the center of the square.

(d) Repeat part (c) for the orthogonal polarization.

9.8.1 A human eye has an iris opening of ~ 2 mm diameter.

(a) At the wavelength of yellow light (580 nm), what is the resolution of the human eye (in degrees)? For simplicity, assume the opening is square, 2 mm on a side, and the resolution is the angle between the visual axis and the first null.

(b) Sketch and dimension the diffraction pattern of a square human iris along the $\hat{x}$-axis.

9.8.2 A uniformly illuminated square aperture antenna 1 m on a side has an electric field strength in the aperture of 10 V/m at 30 GHz.

(a) What is the total power radiated by this aperture? (Hint: Calculate the power in the aperture directly, not by using the radiation pattern.)

(b) Sketch the resulting antenna pattern and give a quantitative value for its half-power width.

(c) What is the farfield electric field strength on-axis at a distance 10 km from the aperture?

(d) If a 10-cm square in the center of the 1-m uniformly illuminated aperture were obstructed by a subreflector or an absorber, then what would be the answer to part (c)? Show by integration that the on-axis gain would be reduced by approximately 2 percent. Since only 1 percent of the power in the aperture is obstructed, where does the other 1 percent of the power go?

9.8.3 A certain radar antenna can be modeled as a uniformly illuminated rectangular aperture 10 cm $\times$ 1 m. At 30 GHz and 10 km distance, what is the farfield radiated intensity $P(\alpha_x, \alpha_y)$ (W/m^2) for $|\phi_{x/y}| \ll 1$, where α_x and α_y are as defined in Figure 9.37? Sketch and dimension your answer for $P(\alpha_x)$ along the $\hat{x}$-axis, where $\alpha_y \equiv 0$.

10

ACOUSTICS

10.1 INTRODUCTION

Many of the techniques developed in this text for understanding electromagnetic wave phenomena can also be used to characterize linear waves in nonelectromagnetic systems. One of the most important examples of this parallelism is the remarkable similarity between electromagnetic and acoustic wave behavior, discussed in this chapter. The introduction of key concepts in acoustics follows the same logical sequence used for electromagnetic waves in Chapters 1–9, covering the wave equation, plane waves, power, waves at interfaces, waves in ducts (acoustic transmission lines and waveguides), acoustic resonators, and antennas (speakers). Because we shall make strong analogies between the two types of wave behavior, this chapter may also be used to review electromagnetic phenomena.

Although acoustic structures behave quite similarly to their electromagnetic counterparts, significant differences do exist and we should be wary of blindly applying electromagnetic formulas to problems in acoustics. One of the principal differences is that acoustic waves are inherently nonlinear. We shall be considering only small-amplitude linearized acoustic waves in this section except for a brief discussion of the full nonlinear acoustic wave equation in Section 10.3. Nonlinear effects that can arise include wave mixing and multiplication, shock wave formation, amplification and so on, and are beyond the scope of this text. The other major difference is that acoustic waves have longitudinal field components (gas velocities are usually parallel to the direction of wave propagation), whereas electromagnetic waves are typically characterized by transverse fields. Some discrepancies between acoustic and electromagnetic behavior arise from this difference.

10.2 LINEARIZED ACOUSTIC WAVE EQUATION

We observe *acoustic waves* in solids as well as in liquid or gaseous fluids. Local pressure gradients produce local accelerations of the fluid particles, and the resulting velocity gradients lead to local density variations and therefore to new pressure gradients. This coupling between pressure, velocity and density results in acoustic waves, described by a simple, linearized model of fluid flow in this section. Section 10.3 describes the nonideal, nonlinear fluid waves that actually propagate, but this more rigorous treatment of acoustic waves is unnecessary to understand the subsequent sections of this chapter. All acoustic waves are thus described by the following acoustic field quantities:

$$P(\bar{r}, t): \ fluid \ pressure(N/m^2)$$

$$\rho(\bar{r}, t): \ fluid \ mass \ density(kg/m^3)$$

$$\overline{U}(\bar{r}, t): \ fluid \ velocity(m/s)$$

Note that these fluid fields characterize macroscopic, continuum fluid properties, and so each "infinitesimal" fluid element must contain a sufficient number of particles that average fields can be meaningfully defined. This theory is thus not appropriate for describing microscopic fluid behavior, such as Brownian motion.

In order to derive the acoustic wave equation, we must apply both conservation of mass and conservation of linear momentum to the fluid and link the resulting expressions by an equation of state for the specific fluid medium. Both conservation relations involve taking the time derivative of a volume integral that is deforming in time. The *Liebnitz identity* for any scalar field quantity $q(\bar{r}, t)$ greatly facilitates this process:

$$\frac{d}{dt} \int_{V(t)} q(\bar{r}, t) \, dv = \int_{V(t)} \frac{\partial q(\bar{r}, t)}{\partial t} \, dv + \oint_{A(t)} q(\bar{r}, t) \overline{U}_A(\bar{r}, t) \cdot \hat{n} \, da$$

$$= \int_{V(t)} \left[\frac{\partial q(\bar{r}, t)}{\partial t} + \nabla \cdot \left(q(\bar{r}, t) \overline{U}_A(\bar{r}, t) \right) \right] dv \tag{10.2.1}$$

The second equality results from Gauss's divergence theorem (4.1.1); $V(t)$ and $A(t)$ are the volume and closed surface area, respectively, of a time-varying volume, and $\overline{U}_A(\bar{r}, t)$ is velocity of the surface $A(t)$. This identity simply says that the rate of change of a quantity $q(\bar{r}, t)$ contained in a moving volume $V(t)$ is the sum of the intrinsic rate of change of $q(\bar{r}, t)$ and the rate at which the moving boundaries are encompassing $q(\bar{r}, t)$.

Conservation of mass states that

$$\frac{d}{dt} \int_{V(t)} \rho(\bar{r}, t) \, dv = 0 \tag{10.2.2}$$

where $V(t)$ is any volume of fixed identity in the fluid that deforms and moves with fluid velocity $\overline{U}(\overline{r}, t) = \overline{U}_A(\overline{r}, t)$.[1] If we substitute $\rho(\overline{r}, t)$ for $q(\overline{r}, t)$ in (10.2.1) and use (10.2.2), we find that

$$\frac{\partial \rho}{\partial t} + \nabla \cdot (\rho \overline{U}) = 0 \tag{10.2.3}$$

because $V(t)$ is an arbitrary volume and thus the integrand of (10.2.1) must vanish.

Conservation of linear momentum in the fluid may be expressed by Newton's force law. In the $\hat{x}$ direction,

$$\frac{d}{dt} \int_{V(t)} \rho(\overline{r}, t) U_x(\overline{r}, t) \, dv = \int_{V(t)} F_x(\overline{r}, t) \, dv \tag{10.2.4}$$

which is the continuum generalization of the force expression $d(m\overline{u})/dt = \overline{f}$ for a single particle having mass m and velocity $\overline{u}$. The total force density acting on the fluid is $\overline{F}(\overline{r}, t)$ N/m^3 so both sides of (10.2.4) have dimensions of newtons;[2] F_x and U_x are the x components of $\overline{F}$ and $\overline{U}$, respectively. We assume here that the force density arises exclusively from pressure gradients in the fluid, and we therefore ignore viscous, gravitational, electromagnetic, and external forces.

To understand how $\overline{F}$ is related to the fluid pressure P, consider the small rectangular fluid volume element of dimensions Δx, Δy, Δz illustrated in Figure 10.1. The force density F_x is the total force f_x per unit volume $\Delta x \Delta y \Delta z$ where f_x is proportional to the pressure differential on the surfaces at x and $x + \Delta x$:

$$
\begin{aligned}
F_x &= \lim_{\Delta x, \Delta y, \Delta z \to 0} \frac{f_x}{\Delta x \Delta y \Delta z} \\
&= \lim_{\Delta x, \Delta y, \Delta z \to 0} \frac{P(x, y, x)\Delta y \Delta z - P(x + \Delta x, y, z)\Delta y \Delta z}{\Delta x \Delta y \Delta z} \\
&= \lim_{\Delta x \to 0} \frac{P(x, y, z) - P(x + \Delta x, y, z)}{\Delta x} \\
&= -\frac{\partial P}{\partial x}
\end{aligned} \tag{10.2.5}
$$

The negative sign in (10.2.5) occurs because the force vector points from regions of high pressure to regions of low pressure; (10.2.5) generalizes to $\overline{F} = -\nabla P$ in three

---

1. A volume of fixed identity may be understood by the following thought experiment. If at $t = 0$, we dye blue all the particles contained within volume V_0, then the volume of fixed identity $V(t)$ at a later time t is that volume which contains all the original blue particles (and no other particles). Even if $V(t)$ no longer resembles V_0 in shape or size because the particles have moved, it must necessarily contain the same mass because both volumes contain the same blue particles, so $\int_{V_0} \rho(\overline{r}, t_0) \, dv = \int_{V(t)} \rho(\overline{r}, t) \, dv = $ constant. Taking the total time derivative of both sides of this equation leads to (10.2.2).

2. One newton (N) is roughly equivalent to the force of gravity acting on one-quarter pound (actually 0.225 lbs) of butter and will accelerate one kilogram at one m/s^2.

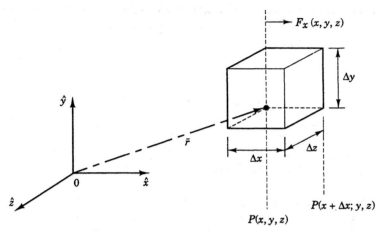

Figure 10.1   Incremental volume of a fluid.

dimensions. If we now let $q(\bar{r}, t)$ represent the $\hat{x}$-directed momentum density $\rho(\bar{r}, t)U_x(\bar{r}, t)$ in (10.2.1) and use (10.2.4)–(10.2.5), we find that

$$\frac{\partial(\rho U_x)}{\partial t} + \nabla \cdot (\rho U_x \bar{U}) = -\frac{\partial P}{\partial x} \qquad (10.2.6)$$

where the differential form of conservation of momentum holds because the volume $V(t)$ in (10.2.1) is arbitrary. The three-dimensional generalization of the differential form of Newton's force law for a fluid (10.2.6) is just

$$\frac{\partial(\rho \bar{U})}{\partial t} + \nabla \cdot (\rho \bar{U}\,\bar{U}) = -\nabla P \qquad (10.2.7)$$

The two conservation equations (10.2.3) and (10.2.7) thus serve as the basis from which a simple wave equation may be derived.[3]

Because we wish to consider only a linearized model of acoustic wave propagation, we now represent each of the three field quantities ρ, P, and $\bar{U}$ as an average, constant field plus a small fluctuation

$$\rho = \rho_0 + \rho_1(\bar{r}, t)$$

$$P = P_0 + p(\bar{r}, t) \qquad (10.2.8)$$

$$\bar{U} = \bar{U}_0 + \bar{u}(\bar{r}, t)$$

where the density fluctuation ρ_1 has a subscript to differentiate it from the total density ρ. Because ρ_0, P_0, and $\bar{U}_0$ are constants, all spatial and temporal derivatives of these quantities are zero. We assume the fluctuations in pressure and density are small: $|\rho_1| \ll |\rho_0|$ and $|p| \ll |P_0|$. On average, the fluid is also assumed

---

3. The quantity $\rho \bar{U}\,\bar{U}$ is a second-rank tensor (matrix), which means that $\nabla \cdot (\rho \bar{U}\,\bar{U})$ is a first-rank tensor (vector). Note that $\nabla \cdot (\rho \bar{U}\,\bar{U}) = (\nabla \rho \cdot \bar{U})\bar{U} + \rho(\bar{U} \cdot \nabla)\bar{U} + \rho \bar{U}(\nabla \cdot \bar{U})$. This term disappears in the simple, linearized equations that follow.

to be at rest: $\overline{U}_0 = 0$. If we substitute (10.2.8) into (10.2.3), we find the conservation of mass equation becomes

$$\frac{\partial \rho_1}{\partial t} + \rho_0 \nabla \cdot \overline{u} = 0 \qquad (10.2.9)$$

where the nonlinear term $\nabla \cdot (\rho_1 \overline{u})$ is neglected because it is the product of two (small) first-order fields and is thus second-order. Similarly, the force equation (10.2.7) is linearized as

$$\rho_0 \frac{\partial \overline{u}}{\partial t} = -\nabla p \qquad (10.2.10)$$

where the second-order terms $\partial(\rho_1 \overline{u})/\partial t$ and $\nabla \cdot (\rho_0 \overline{u}\,\overline{u})$ and the third-order term $\nabla \cdot (\rho_1 \overline{u}\,\overline{u})$ are all ignored.

Because (10.2.9) and (10.2.10) are two equations in the three variables ρ_1, $\overline{u}$, and p (ρ_0 and P_0 are constants), we need a constitutive relation linking two of these variables. If the acoustic frequency is sufficiently high that the heat flux in a gaseous fluid is zero (the gas does not have time to exchange heat with its container), then pressure and density are related "adiabatically" by the *isentropic acoustic constitutive relation*[4]

$$\frac{\partial \rho}{\partial P} = \frac{\rho}{\gamma P} \qquad (10.2.11)$$

where the adiabatic exponent γ is the (dimensionless) ratio of the specific heat at constant volume to the specific heat at constant pressure in the gas. An ideal monatomic gas has $\gamma = 5/3$; typical values for real gases are ~ 1–2. The linearized version of this constitutive relation is thus:

$$\rho_1 = \left(\frac{\rho_0}{\gamma P_0} \right) p \qquad (10.2.12)$$

We can now express the linearized conservation equations (10.2.9) and (10.2.10) in terms of the two coupled variables p and $\overline{u}$

$$\nabla p = -\rho_0 \frac{\partial \overline{u}}{\partial t} \qquad (10.2.13)$$

$$\nabla \cdot \overline{u} = -\frac{1}{\gamma P_0} \frac{\partial p}{\partial t} \qquad (10.2.14)$$

where the spatial derivatives have been written on the left and the temporal derivatives on the right, analogous to Faraday's and Ampere's laws (1.2.1)–(1.2.2). Equations (10.2.13) and (10.2.14) are known as the *acoustic differential equations*. We eliminate the incremental gas velocity $\overline{u}$ by taking the divergence of (10.2.13) and

---

4. A gas is defined to be isentropic if the particles comprising it cannot exchange energy; entropy therefore does not change with time.

the time derivative of (10.2.14), and in so doing obtain an *acoustic wave equation* for the incremental gas pressure p:

$$\nabla^2 p - \frac{\rho_0}{\gamma P_0} \frac{\partial^2 p}{\partial t^2} = 0 \qquad (10.2.15)$$

Likewise, we obtain a wave equation for $\bar{u}$ by taking the gradient of (10.2.14) and the time derivative of (10.2.13)

$$\nabla(\nabla \cdot \bar{u}) - \frac{\rho_0}{\gamma P_0} \frac{\partial^2 \bar{u}}{\partial t^2} = 0 \qquad (10.2.16)$$

where both wave equations are similar in form to (1.3.5). The most general solution to (10.2.15) is

$$p(\bar{r}, t) = p(\omega t - \bar{k} \cdot \bar{r})$$

which, when substituted into (10.2.15), yields the *acoustic dispersion relation*

$$k^2 = \omega^2 \frac{\rho_0}{\gamma P_0} \qquad (10.2.17)$$

analogous to (1.3.10). From (10.2.13) we see that $\partial \bar{u}/\partial t$ points in the direction of ∇p and so acoustic waves are *longitudinal waves*; $\bar{k}$ is parallel to ∇p and therefore to $\bar{u}$. In contrast, electromagnetic waves have vector components perpendicular to the direction of wave propagation and are therefore transverse waves.

The phase and group velocities of this acoustic wave are

$$v_g = \left(\frac{\partial k}{\partial \omega} \right)^{-1} = v_p = \frac{\omega}{k} = \sqrt{\frac{\gamma P_0}{\rho_0}} \equiv c_s \qquad (10.2.18)$$

where c_s is the *velocity of sound*, analogous to the speed of light given by (1.3.14) for electromagnetic waves in free space. For a sound wave propagating adiabatically in air at $0°$ C, the acoustic parameters are $\gamma = 1.4$, $\rho_0 = 1.29$ kg/m^3, and $P_0 = 1.01 \times 10^5$ N/m^2. The speed of sound in air is typically close to 330 m/s.

For a liquid or solid, the acoustic constitutive relation (10.2.11) must be replaced with

$$\frac{\partial \rho}{\partial P} = \frac{\rho}{K} \qquad (10.2.19)$$

which is linearized as $\rho_1 = (\rho_0/K)p$. The *bulk modulus* K of the medium has units of pressure (N/m^2). The coefficient $1/\gamma P_0$ in (10.2.14) must then be replaced by $1/K$ and $\rho_0/\gamma P_0 \to \rho_0/K$ in the wave equations (10.2.15)–(10.2.16). The speed of sound in a liquid or solid is thus $c_s = \sqrt{K/\rho_0}$ m/s. Typical velocities are 900–2000 m/s in liquids (~ 1500 m/s in water) and 1500–13,000 m/s in solids (~ 5900 m/s in steel). Although gases are dominated by longitudinal wave behavior, solids and liquids may also support transverse *shear waves* caused by nonnegligible viscous forces.

To examine the behavior of one-dimensional uniform plane acoustic waves in a stationary gas ($\overline{U}_0 = 0$), we define the phasor fields

$$p(z, t) = \text{Re}\{\underline{p}(z)\, e^{j\omega t}\}$$

$$\overline{u}(z, t) = \hat{z}\,\text{Re}\{\underline{u}(z)\, e^{j\omega t}\} \tag{10.2.20}$$

where the phasor underbars will henceforth be dropped. Note again that $\overline{u} = \hat{z}\,u$ is parallel to the direction of ∇p so that (10.2.20) describes a longitudinal wave. Substituting (10.2.20) into the linearized equations of momentum and mass conservation (10.2.13)–(10.2.14) yields

$$\nabla p = \frac{dp(z)}{dz} = -j\omega\rho_0 u(z)$$

$$\nabla \cdot \overline{u} = \frac{du(z)}{dz} = -j\omega\left(\frac{1}{\gamma P_0}\right) p(z) \tag{10.2.21}$$

which has the uniform plane wave solutions

$$p(z) = p\, e^{-jkz}$$

$$u(z) = u\, e^{-jkz} = \frac{k}{\omega\rho_0}\, p\, e^{-jkz} = \frac{1}{\rho_0 c_s}\, p\, e^{-jkz} \tag{10.2.22}$$

where (10.2.18) has been used. The density $\rho_1(z)$ is easily found from either constitutive relation and we note that the field quantities ρ_1, p, and $\overline{u}$ are all in phase.

The ratio of sound pressure $p(z)$ to fluid velocity $u(z)$ is the *characteristic acoustic impedance* η_s of the fluid:

$$\eta_s = \frac{p(z)}{u(z)} = \rho_0 c_s = \begin{cases} \sqrt{\gamma\rho_0 P_0} & \text{gases} \\ \sqrt{\rho_0 K} & \text{liquids, solids} \end{cases} \tag{10.2.23}$$

For air at room temperature, $\eta_s \cong 425$ N·s/m³. Notice that acoustic impedance and electrical impedance do not have the same units.

Equations (10.2.13) and (10.2.14) also serve as the starting point for the derivation of an acoustic power conservation law similar to the Poynting theorem. The acoustic power flux $p(\overline{r}, t)\,\overline{u}(\overline{r}, t)$ has units of W/m², just as does the Poynting vector $\overline{S}(\overline{r}, t) = \overline{E}(\overline{r}, t) \times \overline{H}(\overline{r}, t)$, so we begin by taking the divergence of $p\,\overline{u}$:

$$\nabla \cdot (p\,\overline{u}) = \overline{u} \cdot \nabla p + p\nabla \cdot \overline{u}$$

$$= \overline{u} \cdot \left(-\rho_0 \frac{\partial \overline{u}}{\partial t}\right) - \frac{1}{\gamma P_0} p \frac{\partial p}{\partial t} \tag{10.2.24}$$

$$= -\frac{\partial}{\partial t}\left(\frac{1}{2}\rho_0\overline{u} \cdot \overline{u} + \frac{1}{2\gamma P_0} p^2\right)$$

where the linear acoustic differential equations (10.2.13) and (10.2.14) have been used to rewrite ∇p and $\nabla \cdot \overline{u}$. Equation (10.2.24) is the differential form of lossless acoustic power conservation and should be compared to its electromagnetic analog (1.6.5). If we integrate both sides of (10.2.24) over an arbitrary volume V and apply Gauss's divergence theorem (4.1.1), we obtain the *acoustic power conservation law*:

$$\oint_A p\bar{u} \cdot \hat{n}\, da = -\frac{d}{dt} \int_V \left(\frac{1}{2}\rho_0 |\bar{u}|^2 + \frac{1}{2\gamma P_0}p^2\right) dv \qquad (10.2.25)$$

where A is the closed surface surrounding V, and $\hat{n}$ is an outwardly directed unit vector normal to this surface. We interpret $p\bar{u} \cdot \hat{n}$ as the component of *acoustic power flux* (W/m^2) normal to the surface A leaving the volume V, while $W_k = \frac{1}{2}\rho_0|\bar{u}|^2$ is the *acoustic kinetic energy density* (J/m^3) and $W_p = p^2/2\gamma P_0$ is the *acoustic potential energy density* (J/m^3) of the wave. If the fluid is a liquid with a relation between p and ρ given by (10.2.19), $\gamma P_0 \rightarrow K$ and $W_p = p^2/2K$.[5] Thus, the right side of (10.2.25) is the rate of decrease of the total energy stored in V, equal to the total instantaneous acoustic power flowing out of the closed surface A. No losses have been included in (10.2.25).

If the fields p and $\bar{u}$ are time-harmonic, meaning that $p(\bar{r}, t) = \text{Re}\{p(\bar{r})e^{j\omega t}\}$ and $\bar{u}(\bar{r}, t) = \text{Re}\{\bar{u}(\bar{r})e^{j\omega t}\}$, then an *acoustic intensity* $\bar{I}(\bar{r})$ may be defined, where

$$\bar{I}(\bar{r}) = p(\bar{r})\,\bar{u}^*(\bar{r}) \quad (\text{W/m}^2) \qquad (10.2.26)$$

The complex acoustic intensity phasor $\bar{I}(\bar{r})$ is analogous to the complex Poynting vector $\bar{S}(\bar{r}) = \bar{E}(\bar{r}) \times \bar{H}^*(\bar{r})$ and an equivalent *acoustic conservation of complex power* equation may be constructed for lossless systems by taking the divergence of $\frac{1}{2}\bar{I}(\bar{r})$:

$$
\begin{aligned}
\frac{1}{2}\nabla \cdot (p\,\bar{u}^*) &= \frac{1}{2}\bar{u}^* \cdot \nabla p + \frac{1}{2}p(\nabla \cdot \bar{u}^*) \\
&= \frac{1}{2}\bar{u}^* \cdot (-j\omega\rho_0\bar{u}) + \frac{1}{2}p\left(\frac{j\omega}{\gamma P_0}p^*\right) \\
&= -2j\omega\left(\frac{1}{4}\rho_0|\bar{u}|^2 - \frac{1}{4\gamma P_0}|p|^2\right) \\
&= -2j\omega\left(\langle W_k\rangle - \langle W_p\rangle\right)
\end{aligned}
\qquad (10.2.27)
$$

---

5. If the fluid particles through which the wave propagates are modeled by masses connected by springs, then the kinetic and potential energies of a single particle are given by $w_k = \frac{1}{2}m|\bar{u}|^2$ (J) and $w_p = \frac{1}{2}K_0|\bar{x}|^2 = |\bar{f}|^2/2K_0$ (J) from elementary mechanics, where $\bar{u}$ is the particle velocity, m is the particle mass, K_0 is the spring constant, $\bar{x}$ is the particle displacement, and $\bar{f} = -K_0\bar{x}$ is the restoring force of the spring. By making the replacements $m \rightarrow \rho$, $K_0 \rightarrow K$, and $|\bar{f}|^2 \rightarrow p^2$, we recover the acoustic energy densities W_k and W_p (J/m^3). This superficial analogy between particles in a fluid and particles connected by a spring lattice also helps give meaning to the bulk modulus K; K is the "stiffness" of the fluid.

where the complex form of the acoustic equations (10.2.21) is used. The time-averaged kinetic and potential energy densities are

$$\langle W_k \rangle = \frac{1}{4} \rho_0 |\bar{u}|^2$$

$$\langle W_p \rangle = \frac{1}{4\gamma P_0} |p|^2 \tag{10.2.28}$$

This lossless acoustic analog to the complex Poynting theorem states that the divergence of complex acoustic power is equal to the reactive acoustic energy density. Since p and $\bar{u}$ are in phase, $\frac{1}{2} p \bar{u}^*$ is a purely real quantity and therefore the imaginary terms in (10.2.28) must vanish; $\langle W_k \rangle = \langle W_p \rangle$. Just as the time-averaged electric and magnetic densities were equal for a uniform plane electromagnetic wave, so are the time-averaged kinetic and potential energy densities equal for a uniform plane acoustic wave. The time-averaged acoustic power intensity is given by

$$\langle \bar{I}(\bar{r}) \rangle = \frac{1}{2} \text{Re}\{\bar{I}(\bar{r})\} = \frac{1}{2} \text{Re}\{p \bar{u}^*\} \tag{10.2.29}$$

and is exactly analogous to the time-averaged Poynting power $\langle \bar{S} \rangle = \frac{1}{2} \text{Re}\{\bar{S}\} = \frac{1}{2} \text{Re}\{\bar{E} \times \bar{H}^*\}$ for an electromagnetic wave. For the $\hat{z}$-directed uniform plane acoustic wave with p and $\hat{z} u$ given by (10.2.22), we find

$$\langle \bar{I}(\bar{r}) \rangle = \hat{z} \frac{p^2}{2\rho_0 c_s} = \hat{z} \frac{p^2}{2\eta_s} = \hat{z} \frac{\eta_s u^2}{2} \tag{10.2.30}$$

The second and third equalities result from the definition of acoustic impedance in (10.2.23).

Example 10.2.1

A 3-kHz, 1-kW acoustic plane wave is confined to propagate within an area $A = 1 \text{ m}^2$ near sea level ($c_s = 330$ m/s). What is the wavelength, peak particle velocity, maximum particle displacement, and average energy density of this wave?

Solution: The wavelength is easily found from the dispersion relation (10.2.17):

$$\lambda = \frac{2\pi}{k} = \frac{c_s}{f} = \frac{330 \text{ m/s}}{3000 \text{ s}^{-1}} = 0.11 \text{ m}$$

Since the plane wave has 1 kW total power in one square meter, $|\langle \bar{I}(\bar{r}) \rangle| = 1 \text{ kW/m}^2$. From (10.2.30),

$$|p| = \sqrt{2\eta_s |\langle \bar{I}(\bar{r}) \rangle|}$$

$$= \sqrt{2 \times 425 \text{ N} \cdot \text{s/m}^3 \times 1000 \text{ W/m}^2} = 922 \text{ N/m}^2$$

where $\eta_s = \rho_0 c_s \cong 425 \text{ N} \cdot \text{s/m}^3$, since $\rho_0 = 1.29 \text{ kg/m}^3$ for air near sea level. We can now check that $|p| \ll P_0 = 1.01 \times 10^5 \text{ N/m}^2$, which is necessary for the simple linear acoustic model to be valid. The peak particle velocity $|\overline{u}|$ is

$$|\overline{u}| = |p|/\eta_s = \frac{922 \text{ N/m}^2}{425 \text{ N} \cdot \text{s/m}^3} = 2.17 \text{ m/s}$$

and therefore the maximum particle displacement is approximately

$$|\overline{r}| \simeq |\overline{u}|/\omega = \frac{2.17 \text{ m/s}}{2\pi \cdot 3000 \text{ s}^{-1}} = 115 \ \mu\text{m}$$

The average acoustic energy density of the wave is

$$\langle W_k \rangle + \langle W_p \rangle = 2\langle W_p \rangle = \frac{|p|^2}{2\gamma P_0}$$

$$= \frac{(922 \text{ N/m}^2)^2}{2 \times 1.4 \times 1.01 \times 10^5 \text{ N/m}^2} \simeq 3.01 \text{ J/m}^3$$

since $\langle W_k \rangle = \langle W_p \rangle$ for a uniform plane wave.

10.3* NONLINEAR ACOUSTIC WAVE EQUATION

For completeness, we include the derivation of the full nonlinear acoustic wave equation, as it gives greater physical insight into how acoustic waves propagate in fluids. Conservation of mass and linear momentum are again the basis for wave propagation and are given by (10.2.3) and (10.2.7), respectively

$$\frac{\partial \rho}{\partial t} + \nabla \cdot (\rho \overline{U}) = 0 \tag{10.3.1}$$

$$\frac{\partial (\rho \overline{U})}{\partial t} + \nabla \cdot (\rho \overline{U}\,\overline{U}) = -\nabla P \tag{10.3.2}$$

where ρ, P, and $\overline{U}$ are exact nonlinear fields. We still assume that the force density in the fluid arises only from pressure gradients, and ignore effects of viscosity, gravity, and externally applied forces.

We now change reference coordinates, from the laboratory reference frame in which we watch the fluid move from a stationary vantage point to a fluid frame in which we move with the fluid. In this latter frame the equations of fluid motion become much simpler. The time derivative of a one-dimensional fluid quantity $q(x, t)$ in the fluid coordinate system is given by

$$\frac{Dq(x,t)}{Dt} = \lim_{\Delta t \to 0, \Delta x \to 0} \frac{q(x + \Delta x, t + \Delta t) - q(x, t)}{\Delta t} \tag{10.3.3}$$

where we include the motion of the coordinate in time Δt from x to $x + \Delta x$ as well as the motion of t to $t + \Delta t$ because we are no longer observing the fluid from a fixed point x. The quantity $q(x + \Delta x, t + \Delta t)$ is expanded in a Taylor series as

$$q(x + \Delta x, t + \Delta t) = q(x, t) + \Delta t \frac{\partial q}{\partial t} + \Delta x \frac{\partial q}{\partial x} + \cdots$$

to first order, so (10.3.3) becomes

$$\frac{Dq}{Dt} = \frac{\partial q}{\partial t} + \lim_{\Delta t \to 0, \Delta x \to 0} \frac{\Delta x}{\Delta t} \frac{\partial q}{\partial x} = \frac{\partial q}{\partial t} + U_x \frac{\partial q}{\partial x} \tag{10.3.4}$$

where $U_x \equiv \lim_{\Delta t \to 0, \Delta x \to 0} \Delta x / \Delta t$. In three dimensions, $U_x \partial / \partial x \to \overline{U} \cdot \nabla$, suggesting that (10.3.4) could be written more generally as

$$\frac{Dq(\overline{r}, t)}{Dt} = \frac{\partial q}{\partial t} + \overline{U} \cdot \nabla q = \left(\frac{\partial}{\partial t} + \overline{U} \cdot \nabla \right) q(\overline{r}, t) \tag{10.3.5}$$

The quantity D/Dt is called the *convective derivative* and the fluid frame of reference is often called the Lagrangian coordinate system in contrast to the stationary Eulerian coordinate system. Note that if the fluid is at rest ($\overline{U} = 0$), the convective derivative reduces to the partial time derivative as expected, since in this case the two coordinate systems are identical.

If we express $\nabla \cdot (\rho \overline{U})$ as $\rho(\nabla \cdot \overline{U}) + \overline{U} \cdot \nabla \rho$, then (10.3.1) may be rewritten as

$$\frac{D\rho}{Dt} + \rho \nabla \cdot \overline{U} = 0 \tag{10.3.6}$$

A fluid is *incompressible* if $D\rho / Dt = 0$, which also means that $\nabla \cdot \overline{U} = 0$. Since we have already shown that $\nabla \cdot \overline{U} \neq 0$ for a longitudinal acoustic wave, exemplified by the one-dimensional plane wave (10.2.22), we conclude that incompressible fluids cannot propagate purely longitudinal acoustic waves.

We may also rewrite $\nabla \cdot (\rho \overline{U} \, \overline{U})$ as $\rho \overline{U}(\nabla \cdot \overline{U}) + \overline{U} \cdot \nabla(\rho \overline{U})$ so that (10.3.2) becomes

$$\frac{D(\rho \overline{U})}{Dt} + \rho \overline{U}(\nabla \cdot \overline{U}) = -\nabla P \tag{10.3.7}$$

and further simplify $D(\rho \overline{U}) / Dt$ using the conservation of mass equation (10.3.6), resulting in

$$\rho \frac{D\overline{U}}{Dt} = -\nabla P \tag{10.3.8}$$

which is Newton's law for an observer moving with the fluid. Equations (10.3.6) and (10.3.8) are thus the simplest forms of the nonlinear relations between ρ, $\overline{U}$, and P. If we take a total time derivative of (10.3.6), using (10.3.6) to eliminate $\nabla \cdot \overline{U}$ and (10.3.8) to eliminate $D\overline{U}/Dt$, we find the full *nonlinear acoustic wave equation*:

$$\frac{D^2 \rho}{Dt^2} - \frac{1}{\rho} \left(\frac{D\rho}{Dt} \right)^2 + \frac{1}{\rho}(\nabla \rho \cdot \nabla P) - \nabla^2 P = 0 \tag{10.3.9}$$

Without knowing anything about the relationship between ρ and P, (10.3.9) is as close to a wave equation as we can get. However, if the fluid is isentropic, the adiabatic constitutive relation (10.2.11) may be easily solved for P [6]

$$\frac{P}{P_0} = \left(\frac{\rho}{\rho_0}\right)^{\gamma} \tag{10.3.10}$$

and (10.3.10) is then substituted into (10.3.9) yielding a final wave equation for the gas density ρ:

$$\frac{D^2\rho}{Dt^2} - \frac{1}{\rho}\left(\frac{D\rho}{Dt}\right)^2 + \frac{\gamma(2-\gamma)P_0}{\rho_0^2}\left(\frac{\rho}{\rho_0}\right)^{\gamma-2}(\nabla\rho \cdot \nabla\rho)$$

$$- \frac{\gamma P_0}{\rho_0}\left(\frac{\rho}{\rho_0}\right)^{\gamma-1}\nabla^2\rho = 0 \tag{10.3.11}$$

It is clear that (10.3.11) is highly nonlinear, each term involving products of field quantities. We can linearize (10.3.11) using the fields given by (10.2.8) and keeping only first-order terms. In the following analysis, we do not insist that $\overline{U}_0 = 0$, and the linearized version of D/Dt is thus

$$\left(\frac{D}{Dt}\right)_L = \frac{\partial}{\partial t} + \overline{U}_0 \cdot \nabla \tag{10.3.12}$$

where the subscript L denotes linearization. The second and third terms of (10.3.11) immediately vanish after linearization because both terms involve products of first-order quantities and are thus at least second-order. Because $\nabla^2\rho = \nabla^2\rho_1$ is first-order, only the zeroeth-order part of $(\rho/\rho_0)^{\gamma-1}$ is kept: $(\rho/\rho_0)^{\gamma-1} = (1 + \rho_1/\rho_0)^{\gamma-1} \simeq 1$. The nonlinear wave equation (10.3.11) thus reduces to

$$\left(\frac{D^2}{Dt^2}\right)_L \rho_1 - \frac{\gamma P_0}{\rho_0}\nabla^2\rho_1 = 0 \tag{10.3.13}$$

which is just the usual linearized acoustic wave equation if the gas is stationary ($\overline{U}_0 = 0$), in which case $(D/Dt)_L \to \partial/\partial t$ from (10.3.12). If $\overline{U}_0 \neq 0$, then the general wave equation solutions to (10.3.13) are plane waves of the form

$$\rho_1 = \rho_1(\omega t - \overline{k} \cdot \overline{r})$$

which we can substitute into (10.3.13) to find the dispersion relation

$$\omega = \pm c_s|\overline{k}| + \overline{k} \cdot \overline{U}_0 \tag{10.3.14}$$

where $c_s = \sqrt{\gamma P_0/\rho_0}$ m/s. If $\overline{U}_0 = \hat{z}\,U_0$ and the wave is propagating in the $\hat{z}$ direction, the most general solution for ρ_1 is

$$\rho_1 = \rho_+(z - U_0 t - c_s t) + \rho_-(z - U_0 t + c_s t) \tag{10.3.15}$$

---

6. This expression relating P and ρ through the adiabatic exponent γ now reveals how γ got its name!

Note that the velocity of sound is still c_s measured by an observer traveling with the fluid at $z' = z - U_0 t$. For a stationary observer (in the laboratory frame), the forward wave appears to be traveling at velocity $U_0 + c_s$ while the backward wave travels at $U_0 - c_s$.

10.4 ACOUSTIC WAVE BEHAVIOR AT A PLANAR INTERFACE

Before we can discuss the reflection and transmission of an acoustic wave at a planar interface between two ideal gases, we need to establish boundary conditions for p and $\overline{u}$. We consider the two regions of space illustrated in Figure 10.2(a) characterized by sound velocities c_1, c_2 and mass densities ρ_{01}, ρ_{02}, respectively. If we integrate (10.2.13) along an infinitesimal path AB, which is parallel to ∇p and crosses the boundary as shown in Figure 10.2(a), we find that

$$\int_A^B \nabla p \cdot \hat{n}\, d\ell = dp|_A^B = p_2 - p_1 = -\int_A^B \rho_0 \frac{\partial \overline{u}}{\partial t} \cdot \hat{n}\, d\ell \tag{10.4.1}$$

because $\nabla p = \hat{n}\, (\partial p/\partial \ell)$. If we now let the distance AB approach zero, the right side of (10.4.1) also vanishes because $\rho_0 \partial \overline{u}/\partial t$ is finite.[7] Pressure is therefore continuous across the boundary:

$$p_1 - p_2 = 0 \tag{10.4.2}$$

Equation (10.4.2) is valid only at the interface between two fluids. At the boundary between a rigid wall and a fluid, the pressure is unconstrained.

We find the second boundary condition by integrating (10.2.14) over the pill-box shaped volume V shown in Figure 10.2(b). This volume has two surfaces of area a parallel to the fluid interface, one in each of the two fluids, and these surfaces are separated by the infinitesimal depth δ. Gauss's divergence theorem (4.1.1) allows us to replace the volume integral with an integral over the entire surface area A of the pill-box:

$$\int_V (\nabla \cdot \overline{u})\, dv = \oint_A \overline{u} \cdot \hat{n}\, da = -\int_V \frac{1}{\gamma P_0} \frac{\partial p}{\partial t}\, dv \tag{10.4.3}$$

If we let $\delta \to 0$, then only the upper and lower surfaces of area a survive the integration since $\overline{u}$ is finite. But $\partial p/\partial t$ is finite also, so the volume integral on the right side of (10.4.3) goes to zero because V is proportional to δ. The surface integral of $\oint_A \overline{u} \cdot \hat{n}\, da$ thus becomes $(u_{1n} - u_{2n})a$, so the normal component of $\overline{u}$ must be continuous across the boundary:

$$\hat{n} \cdot (\overline{u}_1 - \overline{u}_2) = 0 \tag{10.4.4}$$

---

7. The speed of the fluid is obviously finite since all particle velocities must be less than the speed of light. Unless the frequency of particle motion approaches infinity, requiring infinite force, the time derivative of $\overline{u}$ is necessarily finite.

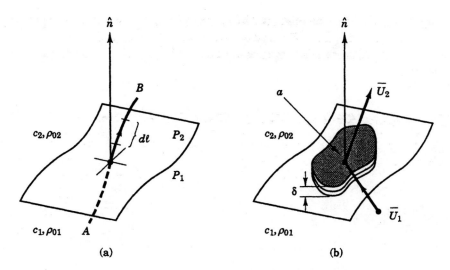

Figure 10.2   Boundary conditions between two ideal fluids.

Therefore, at the boundary between a gas and an ideal rigid wall, $\hat{n} \cdot \overline{u} = 0$. With the two boundary conditions (10.4.2) and (10.4.4), we can now explore the behavior of an acoustic wave at a planar interface.

Consider a uniform plane sound wave incident on an *acoustic planar interface* between two fluids at $z = 0$ with an angle of incidence θ_i, as shown in Figure 10.3. The incident wave propagates in the $\overline{k}_i$ direction, where

$$\overline{k}_i = \hat{x}\, k_{ix} + \hat{z}\, k_{iz} = \hat{x}\, k_i \sin \theta_i + \hat{z}\, k_i \cos \theta_i \tag{10.4.5}$$

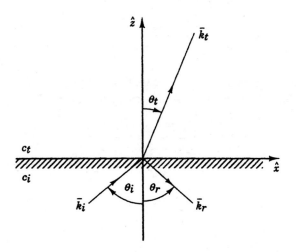

Figure 10.3   Acoustic plane wave incident on a planar interface.

and has a time-harmonic pressure field given by $p = p_i e^{-j\bar{k}_i \cdot \bar{r}}$. The incident sound wave usually generates both reflected and transmitted waves so that (10.4.2) and (10.4.4) are satisfied, with propagation vectors given by

$$\bar{k}_r = \hat{x} k_{rx} - \hat{z} k_{rz} = \hat{x} k_r \sin\theta_r - \hat{z} k_r \cos\theta_r \qquad (10.4.6)$$

$$\bar{k}_t = \hat{x} k_{tx} + \hat{z} k_{tz} = \hat{x} k_t \sin\theta_t + \hat{z} k_t \cos\theta_t \qquad (10.4.7)$$

The negative sign in (10.4.6) is consistent with the propagation direction of the reflected wave.

Both the incident and reflected waves travel in the same medium ($z < 0$), so $k_i = k_r = \omega/c_i$ is the dispersion relation in the incident medium where c_i is the velocity of sound there. Likewise $k_t = \omega/c_t$ for $z > 0$. The pressure waves in the two regions are therefore

$$p(z < 0) = p_i e^{-jk_{ix}x - jk_{iz}z} + \Gamma p_i e^{-jk_{rx}x + jk_{rz}z} \qquad (10.4.8)$$

$$p(z > 0) = T p_i e^{-jk_{tx}x - jk_{tz}z} \qquad (10.4.9)$$

where Γ and T are acoustic reflection and transmission coefficients, respectively. The velocity waves are obtained from the pressure fields using (10.2.21)

$$\bar{u}(z < 0) = -\frac{\nabla p(z < 0)}{j\omega\rho_{0i}} = \left(\frac{\hat{x} k_{ix} + \hat{z} k_{iz}}{\omega\rho_{0i}}\right) p_i e^{-jk_{ix}x - jk_{iz}z}$$

$$+ \left(\frac{\hat{x} k_{rx} - \hat{z} k_{rz}}{\omega\rho_{0i}}\right) \Gamma p_i e^{-jk_{rx}x + jk_{rz}z} \qquad (10.4.10)$$

$$\bar{u}(z > 0) = -\frac{\nabla p(z > 0)}{j\omega\rho_{0t}}$$

$$= \left(\frac{\hat{x} k_{tx} + \hat{z} k_{tz}}{\omega\rho_{0t}}\right) T p_i e^{-jk_{tx}x - jk_{tz}z} \qquad (10.4.11)$$

where ρ_{0i} and ρ_{0t} are the average mass densities in the incident and transmitting acoustic media.

Matching the pressures at the boundary by equating (10.4.8) and (10.4.9) at $z = 0$ yields both the *acoustic phase-matching condition*

$$k_{ix} = k_{rx} = k_{tx} \equiv k_x \qquad (10.4.12)$$

and

$$1 + \Gamma = T \qquad (10.4.13)$$

analogous to (4.3.8) and (4.3.18). The first equality in (10.4.12) and the fact that $|\bar{k}_i| = |\bar{k}_r|$ implies that $\sin\theta_i = \sin\theta_r$ or $\theta_i = \theta_r$, while the second equality in (10.4.12) leads to the *acoustic Snell's law*:

$$k_i \sin\theta_i = k_t \sin\theta_t \quad \text{or} \quad c_t \sin\theta_i = c_i \sin\theta_t \qquad (10.4.14)$$

Equating the normal components of (10.4.10) and (10.4.11) gives

$$\frac{k_{iz}}{\omega \rho_{0i}} - \frac{k_{rz}\Gamma}{\omega \rho_{0i}} = \frac{k_{tz}T}{\omega \rho_{0t}} \tag{10.4.15}$$

If we express (10.4.15) in terms of k_i and k_t and define the characteristic acoustic impedances $\eta_i = c_i \rho_{0i}$ and $\eta_t = c_t \rho_{0t}$ from (10.2.23), then (10.4.15) eventually becomes

$$1 - \Gamma = T/Z_n \tag{10.4.16}$$

where $Z_n = \eta_t \cos \theta_i / \eta_i \cos \theta_t$ is the *normalized acoustic impedance* of the interface. Equation (4.3.25) is the analogous normalized electromagnetic impedance. Combining (10.4.13) and (10.4.16) yields expressions for the reflection and transmission coefficients

$$\Gamma = \frac{Z_n - 1}{Z_n + 1} \tag{10.4.17}$$

$$T = \frac{2Z_n}{Z_n + 1} \tag{10.4.18}$$

which are identical to the electromagnetic versions in (4.3.23) and (4.3.24).

The electromagnetic phenomena of total internal reflection beyond the critical angle and total transmission at Brewster's angle have completely analogous acoustic counterparts because these phenomena are consequences only of phase matching and finite wave velocities. *Total internal reflection* occurs for incident angles greater than

$$\theta_i = \theta_c = \sin^{-1}(c_i/c_t) \tag{10.4.19}$$

where the *acoustic critical angle* is the value of θ_i for which $\theta_t = \pi/2$ ($k_{tz} = 0$). Critical angles exist only when $c_i < c_t$. Incident angles greater than θ_c lead to imaginary values of k_{tz}

$$k_{tz} = \sqrt{k_t^2 - k_{tx}^2} = \sqrt{k_t^2 - k_i^2 \sin^2 \theta_i} = \frac{\omega}{c_t}\sqrt{1 - \left(\frac{c_t}{c_i}\right)^2 \sin^2 \theta_i}$$

$$= -j\frac{\omega}{c_t}\sqrt{\left(\frac{c_t}{c_i}\right)^2 \sin^2 \theta_i - 1} \equiv -j\alpha \tag{10.4.20}$$

where α is real and positive. The transmitted sound pressure is thus

$$p(z > 0) = T p_i\, e^{-jk_x x - \alpha z} \tag{10.4.21}$$

from (10.4.9), which is an evanescent acoustic wave propagating in the $\hat{x}$ direction (along the planar interface) with an amplitude that decays exponentially in the $+\hat{z}$ direction (away from the interface). The fluid velocity $\bar{u}$ is given by (10.4.11):

$$\bar{u}(z > 0) = \left(\frac{\hat{x} k_x - \hat{z} j\alpha}{\omega \rho_{0t}}\right) T p_i\, e^{-jk_x x - \alpha z} \tag{10.4.22}$$

We note that power propagates in the $\hat{x}$ direction parallel and close to the surface of the interface and not in the $\hat{z}$ direction into the transmitting medium because p and u_z are 90° out of phase. This type of wave is called a *surface acoustic wave*.

The incident angle $\theta_i = \theta_B$ for which no reflected wave is observed is called the *acoustic Brewster's angle*. From (10.4.17), $\Gamma = 0$ when $Z_n = 1$ or

$$\eta_t \cos \theta_B = \eta_i \cos \theta_t \qquad (10.4.23)$$

Snell's law (10.4.14) and the condition given by (10.4.23) may be solved uniquely to determine Brewster's angle[8]

$$\theta_B = \tan^{-1} \frac{\eta_t}{\eta_i} \sqrt{\frac{1 - (\eta_i/\eta_t)^2}{1 - (c_t/c_i)^2}} \qquad (10.4.24)$$

In general, a Brewster angle only exists when $\eta_i \leq \eta_t$ and $c_t \leq c_i$, as may be shown by calculating $\sin \theta_B$ and $\cos \theta_B$ from (10.4.15) and (10.4.23). Both trigonometric functions must have magnitudes less than or equal to unity.

We note that there is no distinction between *TE* and *TM* waves in acoustics as there was in electromagnetics because the acoustic plane waves are longitudinal. The real part of $\overline{u}$ is parallel to the direction of wave propagation for both uniform plane waves and evanescent waves. This difference aside, electromagnetic and acoustic waves behave quite similarly at an interface.

Example 10.4.1

The still air immediately above a certain cold lake is 5°C colder than the air more than 10 m above the lake at 27°C. If a child shouts across the lake, what is the critical angle θ_c beyond which the sound is completely reflected downward from the 10-m interface? Assume the air density is inversely proportional to absolute temperature. For a grazing incidence of 0.01 radians, what is the decay rate α (m^{-1}) in the evanescent region above the interface at 1 kHz? Estimate approximately how much louder the sound is at the other side of the 1-km wide lake when this 5°C temperature inversion is present, compared to when

---

8. We find θ_B by squaring both $\eta_t \cos \theta_B = \eta_i \cos \theta_t$ (10.4.23) and $c_t \sin \theta_B = c_i \sin \theta_t$ (Snell's law) and eliminating $\cos \theta_t$:

$$\eta_t^2 \cos^2 \theta_B = \eta_i^2 \cos^2 \theta_t = \eta_i^2 (1 - \sin^2 \theta_t)$$

$$= \eta_i^2 \left(1 - \frac{c_t^2 \sin^2 \theta_B}{c_i^2} \right)$$

Now $\cos^2 \theta_B$ is recast as $1 - \sin^2 \theta_B$ and we find

$$\sin^2 \theta_B = \frac{\eta_t^2 - \eta_i^2}{\eta_t^2 - \eta_i^2 c_t^2/c_i^2}$$

which means that

$$\tan^2 \theta_B = \frac{\sin^2 \theta_B}{1 - \sin^2 \theta_B} = \frac{\eta_t^2 - \eta_i^2}{\eta_i^2 - \eta_i^2 c_t^2/c_i^2} = \frac{\eta_t^2}{\eta_i^2} \left(\frac{1 - (\eta_i/\eta_t)^2}{1 - (c_t/c_i)^2} \right)$$

it is not. In what direction do air particles move acoustically in the evanescent region? In what direction are the pressure wave fronts in this region?

Solution: From (10.4.19), the critical angle is $\theta_c = \sin^{-1}(c_i/c_t)$ where $\begin{Bmatrix} c_i \\ c_t \end{Bmatrix}$ is the speed of sound in the $\begin{Bmatrix} 22°C \\ 27°C \end{Bmatrix}$ air. We recall that $c_s = \sqrt{\gamma P_0/\rho_0}$ where $\gamma = 1.4$ and $P_0 = 1.01 \times 10^5$ N/m^2 for air at sea level. Since only the densities of the two temperature layers of air are different, $c_i/c_t = \sqrt{\rho_t/\rho_i}$. The critical angle is thus $\theta_c = \sin^{-1}\sqrt{\rho_t/\rho_i}$. But we are told that $\rho_0 \propto 1/T$, where T is the absolute (Kelvin) temperature, so

$$\theta_c = \sin^{-1}\sqrt{\frac{T_i}{T_t}} = \sin^{-1}\sqrt{\frac{22 + 273}{27 + 273}} = 82.6°$$

using the conversion $0°C = 273$ K. The critical grazing angle (measured from the horizon) is thus $90° - 82.6° = 7.4°$. In order to find the decay rate of the evanescent wave, we use (10.4.20)

$$\alpha = \frac{\omega}{c_t}\sqrt{\left(\frac{c_t}{c_i}\right)^2 \sin^2\theta_i - 1}$$

where now we actually have to calculate c_t. If we assume air is approximately an ideal gas, we recall that

$$\rho_t = \frac{P_0}{RT_t}$$

where the gas constant $R = 287$ J/(kg·K) for air. Substituting this equation of state into the expression for c_t yields

$$c_t = \sqrt{\frac{\gamma P_0}{\rho_t}} = \sqrt{\gamma RT_t} = \sqrt{1.4 \times 287 \text{ J/(kg · K)} \times 300 \text{ K}} = 347 \text{ m/s}$$

Likewise, we find that $c_i = \sqrt{\gamma RT_i} = 344$ m/s. If the grazing incidence angle is 0.01 radians, then $\theta_i = \pi/2 - 0.01 = 1.561$ radians, and

$$\alpha = \frac{2\pi \cdot 10^3}{347}\sqrt{\left(\frac{347}{344}\right)^2 \sin^2(1.561) - 1} = 2.39 \text{ m}^{-1}$$

which corresponds to an acoustic penetration depth of $1/2.39 = 0.42$ m.

To calculate the amplification present in the child's shout when an inversion layer is present, we imagine the child to be standing at the center of a spherical coordinate system. The shout excites acoustic waves roughly isotropically (since the child is approximately a point source) into the solid angle $0 \le \theta \le \pi/2$ and $-\pi/2 \le \phi \le \pi/2$. The angle θ gives elevation; overhead is at $\theta = 0$ and the horizon is at $\theta = \pi/2$, while ϕ is the azimuthal angle with $\phi = \pi/2$ to the left and $\phi = -\pi/2$ to the right. We assume that no power is transmitted behind the child or below the surface of the lake. Without an inversion layer, the total acoustic power P_{tr} is thus transmitted into π steradians so the acoustic intensity at the opposite edge of the lake is $P_{tr}/\pi r^2$ where $r = 1000$ m is the distance across the lake.

If an inversion layer is present, only waves propagating with grazing angles between $\theta_c = 82.6°$ and $90°$ are transmitted across the lake; the power in all other waves is completely lost (by transmission into the cold air region). However, **all** of the power in the waves traveling with grazing angles $82.6° \le \theta \le 90°$ is carried across the lake since the surfaces of lake and inversion layer act as a waveguide. The power carried by these waves is then

$$P_{\text{rec}} = \left(\frac{P_{\text{tr}}}{\pi r^2} \right) \int_{82.6°}^{90°} d\theta \int_{-90°}^{90°} r^2 \sin\theta d\phi \simeq 0.13 P_{\text{tr}}$$

This received acoustic power emerges approximately uniformly over the area $\pi r h$ where $h = 10$ m is the height of the inversion layer. The acoustic intensity with the inversion layer present is then $P_{\text{rec}}/\pi r h$. Therefore, the ratio of the acoustic intensities with and without the inversion layer is

$$A = \frac{P_{\text{rec}}/\pi r h}{P_{\text{tr}}/\pi r^2} = 0.13 \left(\frac{r}{h} \right) = 13$$

so approximately 13 times as much power is received when the inversion layer is present and the child sounds $10 \log_{10} 13 \simeq 11$ dB louder, explaining why sound carries so well across water. Much greater amplification can result when the inversion layer is slightly dome-shaped, for it then can focus the sound at great distances. Note that $r \gg h$ to ensure that the trapped acoustic energy is approximately uniform over the inversion layer exit.

The particle velocity in the evanescent region can be found from (10.4.22); propagation occurs only in the $\hat{x}$ direction. (Consistent with the spherical coordinate system described above, $\hat{z}$ is the vertical ($\theta = 0$) axis and $\hat{x}$ is the direction of propagation of the surface wave across the lake.) Because the $\hat{x}$- and $\hat{z}$-components of the velocity $\bar{u}$ are 90° out of phase and of unequal magnitude, the particles move in elliptical orbits. With k_{tz} imaginary and k_{tz} real, (10.4.21) suggests that the pressure wave propagates in the $\hat{x}$ direction so that the pressure phase fronts are perpendicular to $\hat{x}$.

10.5 TRANSMISSION LINE MODEL FOR ACOUSTIC WAVES

Acoustic wave propagation in a tube is analogous to wave propagation on transmission lines. Consider the gas-filled tube with rigid side-walls and uniform cross-sectional area A perpendicular to the $\hat{z}$-axis shown in Figure 10.4. The lowest-order acoustic mode that propagates in this tube has uniform pressure in planes transverse to the $\hat{z}$-axis, so p is independent of the transverse coordinates x and y: $p(\bar{r}) = p(z)$. Equations (10.2.13) and (10.2.14) then have the one-dimensional form

$$\frac{\partial p(z)}{\partial z} = -\rho_0 \frac{\partial u(z)}{\partial t} \tag{10.5.1}$$

$$\frac{\partial u(z)}{\partial z} = -\frac{1}{\gamma P_0} \frac{\partial p(z)}{\partial t} \tag{10.5.2}$$

where the velocity has only a $\hat{z}$ component because $\bar{u}$ must point in the direction of ∇p. A $\hat{z}$-directed fluid velocity automatically satisfies $\hat{n} \cdot \bar{u} = 0$ along the rigid side walls of the tube, because the fluid flow is purely tangential there. Further boundary conditions arise from the way in which the tube is driven at one end and terminated at the other.

A = cross-sectional area

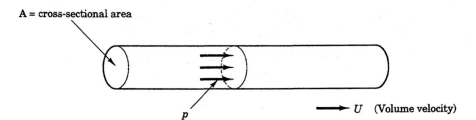

p

U (Volume velocity)

Figure 10.4   An acoustic fluid-filled transmission line.

If we define quantities

$$
\begin{aligned}
P &\equiv pA &&= \text{acoustic energy/length (J/m)} \\
U &\equiv u &&= \text{acoustic velocity (m/s)} \\
M &\equiv \rho_0 A &&= \text{acoustic mass/length (kg/m)} \\
C &\equiv \frac{1}{\gamma P_0 A} &&= \text{acoustic compliance/length (N}^{-1}\text{)}
\end{aligned}
$$

then we may recast (10.5.1)–(10.5.2) as

$$\frac{\partial P}{\partial z} = -M \frac{\partial U}{\partial t} \tag{10.5.3}$$

$$\frac{\partial U}{\partial z} = -C \frac{\partial P}{\partial t} \tag{10.5.4}$$

where (10.5.3) and (10.5.4) are *acoustic telegrapher's equations*, exactly analogous to (5.2.19) and (5.2.20) describing *TEM* transmission lines. If we differentiate both sides of (10.5.3) with respect to z and (10.5.4) with respect to t, we generate the acoustic wave equation

$$\frac{\partial^2 P}{\partial z^2} = MC \frac{\partial^2 P}{\partial t^2} \tag{10.5.5}$$

with a similar expression for U. The velocity of sound in the pipe is $c_s = 1/\sqrt{MC} = \sqrt{\gamma P_0/\rho_0}$ as before. Table 10.1 compares electrical and acoustic transmission lines. Note that time-averaged acoustic power $\frac{1}{2}\operatorname{Re}\{P U^*\}$ has dimensions of watts, in complete analogy with the electromagnetic power $\frac{1}{2}\operatorname{Re}\{V I^*\}$.

The most general time-harmonic solution to the acoustic telegrapher's equations is a superposition of $\pm\hat{z}$-propagating plane waves

$$P(z) = P_+ \left(e^{-jkz} + \Gamma_L e^{+jkz}\right) \tag{10.5.6}$$

$$U(z) = Y_0 P_+ \left(e^{-jkz} - \Gamma_L e^{+jkz}\right) \tag{10.5.7}$$

where the acoustic reflection coefficient Γ_L is the complex ratio of the amplitudes of the backward- and forward-propagating waves. The electromagnetic analogs of (10.5.6) and (10.5.7) are given by (6.1.1) and (6.1.2). These equations serve as the basis for discussions of the gamma plane, Smith chart, ABCD matrices, and so on, in Chapter 6, and all of these tools may be used to describe acoustic transmission

TABLE 10.1 Acoustic and Electromagnetic Counterparts.

Acoustic Line		Electrical Line	
Quantity	Units	Quantity	Units
P	N	V	V
U	m/s	I	A
M	kg/m	L	H/m
C	N^{-1}	C	F/m
$Z_0 = \sqrt{M/C} = \eta_s A$	kg/s	$Z_0 = \sqrt{L/C}$	Ω
$c_s = 1/\sqrt{MC} = \sqrt{\gamma P_0/\rho_0}$	m/s	$c = 1/\sqrt{LC} = 1/\sqrt{\mu\epsilon}$	m/s
$P = \frac{1}{2}\,\mathrm{Re}\{P\,U^*\}$	W	$P = \frac{1}{2}\,\mathrm{Re}\{V\,I^*\}$	W
$\dfrac{dp}{dz} = -j\omega M U$		$\dfrac{dV}{dz} = -j\omega L I$	
$\dfrac{dU}{dz} = -j\omega C P$		$\dfrac{dI}{dz} = -j\omega C V$	

lines. We need only determine how the physical termination of a tube is related to Γ_L.

We first consider a tube that is closed at $z = 0$. From (10.4.4), the normal component of $\bar{u}$ must be continuous at $z = 0$ but the velocity $\bar{u}$ is zero at the stationary tube end: $u = U = 0$ at $z = 0$. From (10.5.7), we see that $\Gamma_L = +1$, which corresponds to an electrical **open circuit**. We knew this anyway because $U = 0 \leftrightarrow I = 0$ at $z = 0$ for the analogous *TEM* line. Pressure and velocity for the standing wave are thus

$$P(z) = 2P_+ \cos kz$$
$$U(z) = -2jY_0 P_+ \sin kz \qquad (10.5.8)$$

and the ratio of $P(z)$ to $U(z)$ is the acoustic impedance $Z(z) = jZ_0 \cot kz$. For a line of length ℓ, the input impedance is $Z_{\text{in}} = Z(z = -\ell) = -jZ_0 \cot k\ell$, and if $k\ell \ll 1$, then $Z_{\text{in}} \cong -jZ_0/k\ell = 1/j\omega C\ell$ where $C\ell$ is the acoustic compliance of the short tube. The acoustic compliance of the short tube is therefore analogous to the capacitance of a short *TEM* line.

In the other limit, we open the tube so that no pressure difference exists across the tube end. The total pressure at $z = 0$ is just P_0 so the pressure fluctuation p is zero there. [We recall from (10.2.8) that $P = P_0 + p(z)$ and we also neglect the acoustic radiation impedance from the tube end.] If $P = pA = 0$ at $z = 0$, then $\Gamma_L = -1$ from (10.5.6), corresponding to an electrical **short circuit**,[9] consistent

---

9. Don't confuse the acoustic and electrical boundary conditions when relating P to V and U to I—the open-ended pipe corresponds to an electrical short circuit while the closed-ended pipe corresponds to an electrical open circuit.

with $P = 0 \leftrightarrow V = 0$. Pressure and velocity in this case are

$$P(z) = -2j P_+ \sin kz$$
$$U(z) = 2Y_0 P_+ \cos kz$$

(10.5.9)

so that $Z(z) = P(z)/U(z) = -jZ_0 \tan k\ell$ and $Z_{in} = jZ_0 \tan k\ell$. If the pipe is short, $Z_{in} \cong jZ_0 k\ell = j\omega M\ell$ where $M\ell$ is the total mass of the fluid in the tube, analogous to the total inductance of a short *TEM* line. If the tube is closed with an acoustic absorbing wire mesh, then the impedance at $z = 0$ may have a resistive component, discussed further in Example 10.5.1.

Example 10.5.1

Assume a closed tube contains a wire mesh at a distance d from the end as shown in Figure 10.5. The mesh has an acoustic resistance given by

$$R = P/U$$

What values of d and R minimize the reflection of incident acoustic waves? Can we build a perfect sound absorber?

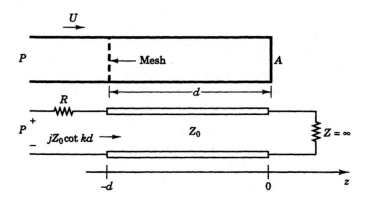

Figure 10.5   Sound absorber in Example 10.5.1.

Solution: Because the tube is closed at $z = 0$, corresponding to infinite load impedance, pressure and velocity are given by (10.5.8), and the impedance $Z(z)$ is

$$Z(z) = \frac{P(z)}{U(z)} = \frac{2P_+ \cos kz}{-2jY_0 P_+ \sin kz} = jZ_0 \cot kz$$

where $Z_0 = 1/Y_0 = \eta_s A$ and A is the cross-sectional area of the tube. The impedance of the tube at $z = -d$ without the wire mesh is then just $Z(z = -d) = -jZ_0 \cot kd$. If we now add the mesh resistance R to the acoustic line impedance at $z = -d$, we find that

$$Z(z = -d) = R - jZ_0 \cot kd \tag{1}$$

As a check, we note that if $R \to 0$, the mesh effectively disappears and (1) becomes the impedance of the acoustic tube. Conversely, as $R \to \infty$, the mesh becomes a rigid wall and $Z(z = -d) \to \infty$ as expected. For no reflection from the mesh to occur, the reactance in (1) must be zero and so the resistance at the mesh must be equal to Z_0; the line is then matched. Therefore, $R = Z_0$ and $kd = \pi/2$ are the two conditions necessary so that $\Gamma = 0$ at $z = -d$. We can then perfectly absorb single-frequency sound with this single (tuned) mesh.

Example 10.5.2

A typical door consists of a thin massive planar sheet blocking the acoustic path. Assume $\rho_0 = 1$ kg/m^3 in the air and $\rho_d = 1000$ kg/m^3 in this door; $c_s = 330$ m/s in air and $c_{sd} = 1050$ m/s in the door. What are the relative characteristic acoustic impedances η_0 and η_d of the air and door? Determine a *TEM* equivalent circuit for a 1-cm thick door shielding normally incident acoustic waves at 100 Hz. What fraction of the 100-Hz incident acoustic power passes through the door and into the air on the other side? Discuss any frequency dependence.

Solution: The door may be modeled by a *TEM* line of length $\ell = 1$ cm and impedance $Z_d = \eta_d A = \rho_d c_{sd} A = 1.05 \times 10^6 A$ N·s/m, where A is the cross-sectional area of the line and is unknown (and irrelevant). The free space on either side of this barrier is represented by *TEM* lines of impedance $Z_0 = \eta_0 A = \rho_0 c_s A = 330 A$ N·s/m, where the right-most line is terminated with the acoustic matched load Z_0 to model the infinite expanse of space to the right of the door. The impedance Z_{in} looking into the door is then given by (6.1.6)

$$Z_{in} = Z(z = -\ell) = Z_d \left(\frac{Z_0 + j Z_d \tan k_d \ell}{Z_d + j Z_0 \tan k_d \ell} \right)$$

Because the acoustic wavelength in the door $\lambda_d = c_{sd}/f = 1.05$ m is much greater than the door thickness $\ell = 1$ cm, $\tan k_d \ell \approx k_d \ell$ where $k_d = 2\pi/\lambda_d$. And because $Z_d \gg Z_0$, the input impedance looking into the door is approximately $Z_{in} \cong Z_0 + j k_d \ell Z_d$. From the expression for Γ given by (10.4.17), the acoustic reflection coefficient at the door is

$$\Gamma(-\ell) = \frac{Z_{in} - Z_0}{Z_{in} + Z_0} = \frac{j k_d \ell Z_d}{2 Z_0 + j k_d \ell Z_d}$$

and the fraction of acoustic power transmitted through the door is

$$f = 1 - |\Gamma|^2 \cong \frac{4 Z_0^2}{4 Z_0^2 + (k_d \ell Z_d)^2}$$

$$= \frac{4 \,(330 A \text{ N} \cdot \text{s/m})^2}{4 \,(330 A \text{ N} \cdot \text{s/m})^2 + (2\pi/1.05 \text{ m}) \,(0.01 \text{ m}) \left(1.05 \times 10^6 \text{ N} \cdot \text{s/m}\right)^2}$$

$$= 0.011$$

which corresponds to a $\log_{10} 0.011 = -19.6$ dB reduction in acoustic power because of the barrier. As the frequency drops below 100 Hz, the fraction of transmitted power approaches unity, and as ω becomes larger (without exceeding the criterion that $\omega \ell / c_{sb} \ll 1$), the transmitted fraction decreases as $1/\omega^2$. Since the dynamic range of hearing generally exceeds 70 dB, the protection of a closed door is often incomplete, especially for low frequencies.

Example 10.5.3

Two acoustic tubes having 1-cm and 2-cm diameters are to be connected to conduct a 1-kHz audio signal. What should be the length and diameter of an acoustic quarter-wave transformer designed to minimize reflections?

Solution: By definition, a quarter-wave transformer has length $\ell = \lambda/4$. For a 1-kHz audio signal, $\lambda = c_s/f = 330/1000 = 33$ cm, so $\ell = 8.25$ cm. The normalized impedance at the load end of the quarter-wave transformer is $Z_n = Z_B/Z_{QW}$ where Z_B is the impedance of the 2-cm pipe and Z_{QW} is the impedance of the transformer. Recalling that a transmission line of length $\lambda/4$ transforms a normalized load impedance into a normalized admittance, we find that the normalized impedance looking into the transformer is Z_{QW}/Z_B which un-normalizes to Z_{QW}^2/Z_B. For the 1-cm line having impedance Z_A to be matched to this load, $Z_A = Z_{QW}^2/Z_B$ so $Z_{QW} = \sqrt{Z_A Z_B}$, exactly as described for an electrical quarter-wave transformer in Section 6.3. The impedance of a tube is directly proportional to the cross-sectional area of the tube ($Z = \eta_s A$), so Z is directly proportional to the square of the tube's diameter. The quarter-wave tube therefore should have a diameter d equal to the geometric mean of the two pipes to be matched: $d = \sqrt{d_A d_B} = \sqrt{2} = 1.41$ cm.

Even if we choose the diameter and length of the quarter-wave transformer to minimize reflections, we will be unable to eliminate scattering from the sharp junctions between the pipes of differing size. While this effect is small for tubes that are close in size to begin with, we could not make a single-step transformer $\lambda/4$ in length linking a 1-cm to a 100-cm diameter pipe without severe reflections. Instead, we could construct an N-step transformer where each segment was 8.25 cm in length and where the ratio of the diameters of tube n to tube $n-1$ did not exceed $\sim\sqrt{2}$. A minimum of 13 tubes with a total length of $13 \times 8.25 = 107.25$ cm would be required for this transformer. Tubes having diameters $\sqrt{2}$, $2\sqrt{2}$, 4, $4\sqrt{2}$, 8, $8\sqrt{2}$, 16, $16\sqrt{2}$, 32, $32\sqrt{2}$, 64, 80 would be one way to link the 1-cm and 100-cm tubes. To reduce the effects of the pipe junctions even further, the pipe could be exponentially tapered, becoming a very broad-band horn. Because the impedance-matching model for tubes breaks down if the cone angle of the horn becomes too large (in our 13-step transformer above, $\theta = \tan^{-1}[(100\text{ cm} - 1\text{ cm})/107.25\text{ cm}] = 42.7°$), most exponential impedance-matching horns are much longer than this, like a French horn or trumpet.

10.6 ACOUSTIC WAVEGUIDES

One of the simplest types of acoustic waveguides is constructed from two parallel rigid walls at $x = 0$ and $x = d$ as shown in Figure 10.6, analogous to the parallel-plate transmission line discussed in Section 7.1. If the fluid between the plates can propagate acoustic disturbances with velocity $c_s = \sqrt{\gamma P_0/\rho_0}$, then the time-harmonic acoustic wave equation for this fluid is given by (10.2.15) with $\partial/\partial t \rightarrow j\omega$:

$$\left(\nabla^2 + \frac{\omega^2}{c_s^2}\right) p = 0 \tag{10.6.1}$$

Because the waveguide walls at $x = 0, d$ are rigid, the pressure p is not constrained there. But the normal component of $\overline{u}$ must be zero at the rigid walls from (10.4.4), so

$$u_x = 0 \qquad \text{at} \quad x = 0, d \tag{10.6.2}$$

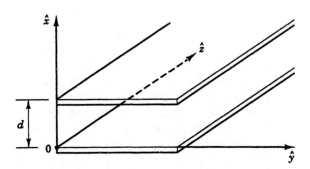

Figure 10.6   Acoustic parallel-plate waveguide.

If we excite $+\hat{z}$-propagating waves in this guide, then the most general solution to (10.6.1) is

$$p(x, z) = A \cos k_x x \, e^{-jk_z z} + B \sin k_x x \, e^{-jk_z z} \qquad (10.6.3)$$

where we restrict ourselves to solutions such that $\partial/\partial y = 0$, as the waveguide does not vary with y, just as we did for electromagnetic parallel-plate waveguides. Substituting (10.6.3) into (10.6.1) yields the acoustic dispersion relation

$$k_x^2 + k_z^2 \equiv k_0^2 = \frac{\omega^2}{c_s^2} \qquad (10.6.4)$$

which is identical to the electromagnetic dispersion relation (7.1.3). In order to determine the constant coefficients A and B, we find $u_x(x, z)$ from (10.2.21) since the acoustic boundary condition (10.6.2) is expressed in terms of the normal fluid velocity:

$$u_x(x, z) = -\frac{1}{j\omega\rho_0} \frac{\partial p}{\partial x} = -\frac{jk_x}{\omega\rho_0} (A \sin k_x x - B \cos k_x x) \, e^{-jk_z z} \qquad (10.6.5)$$

Imposing (10.6.2) on (10.6.5) at $x = 0$ forces $B = 0$, and at $x = d$ we find that $\sin k_x d = 0$ or $k_x d = m\pi$ $(m = 0, 1, 2, \ldots)$. Therefore, the fluid fields for this acoustic waveguide are

$$p(x, z) = p_0 \cos k_x x \, e^{-jk_z z} \qquad (10.6.6)$$

$$\begin{aligned}
\bar{u}(x, z) &= -\frac{\nabla p}{j\omega\rho_0} \\
&= \frac{p_0}{\eta_s} \left(-\hat{x} \frac{jk_x}{k_0} \sin k_x x + \hat{z} \frac{k_z}{k_0} \cos k_x x \right) e^{-jk_z z} \qquad (10.6.7)
\end{aligned}$$

where $A \equiv p_0$ to make the dimensions correct and $\eta_s = c_s \rho_0$. The components of $\bar{k}$ are determined from the following acoustic guidance condition and dispersion relation:

$$k_x = \frac{m\pi}{d}$$

$$k_z = \sqrt{\left(\frac{\omega}{c_s}\right)^2 - \left(\frac{m\pi}{d}\right)^2} \qquad (10.6.8)$$

As in an electromagnetic waveguide, there are various modes of wave propagation between the rigid walls corresponding to $m = 0$, $m = 1$, $m = 2$, and so on. The lowest-order acoustic mode occurs when $m = 0$ ($k_x = 0$, $k_z = \omega/c_s = k_0$), in which case (10.6.6) and (10.6.7) reduce to

$$p(x, z) = p_0 e^{-jk_0 z}$$

$$\overline{u}(x, z) = \hat{z}\,\frac{p_0}{\eta_s} e^{-jk_0 z} \qquad (10.6.9)$$

Because these fields are uniform in planes transverse to the direction of propagation $(+\hat{z})$, the $m = 0$ mode is the mode discussed earlier for acoustic tubes and is analogous to the *TEM* (TM_0) mode in a parallel-plate electromagnetic waveguide.[10]

If the waveguide is excited at frequency ω, where $\omega_{m-1} < \omega < \omega_m$ and $\omega_m = m\pi c_s/d$ is the *acoustic cut-off frequency* of the mth mode, then k_z is imaginary for all modes specified by integers greater than or equal to m and real for the modes specified by integers less than m [see (10.6.8)]. Therefore, modes $0, 1, 2, \ldots, m - 1$ propagate in the waveguide while modes $m, m + 1, m + 2, \ldots$ are cut-off or evanescent, in complete analogy with electromagnetic waveguide propagation. If we choose $\omega < \pi c_s/d$, then only the lowest-order $m = 0$ mode can propagate in this acoustic waveguide.

An acoustic duct, analogous to a rectangular electromagnetic waveguide, is illustrated in Figure 10.7 with rigid walls at $x = 0, a$ and $y = 0, b$. Again, the wave equation (10.6.1) governs the behavior of the fluid in the duct, with boundary conditions on the normal component of $\overline{u}$ such that

$$u_x = 0 \qquad \text{at} \quad x = 0, a$$

$$u_y = 0 \qquad \text{at} \quad y = 0, b \qquad (10.6.10)$$

If $+\hat{z}$-propagating waves are excited in the guide, then the pressure $p(x, y, z)$ is given by

$$p(x, y, z) = p_0 \begin{Bmatrix} \sin k_x x \\ \cos k_x x \end{Bmatrix} \begin{Bmatrix} \sin k_y y \\ \cos k_y y \end{Bmatrix} e^{-jk_z z} \qquad (10.6.11)$$

---

10. Recall that the parallel-plate electromagnetic waveguide has both TE_m and TM_m modes because the vector fields $\overline{E}$ and $\overline{H}$ may have completely transverse $\overline{E}$-field components (TE_m modes) or completely transverse $\overline{H}$-field components (TM_m modes). The *TEM* mode is given by TM_0; TE_0 does not exist. By contrast, p is a scalar field and $\overline{u}$ is uniquely determined from $-\nabla p/j\omega\rho_0$ and (10.6.2) so $\overline{u}$ has only the "polarization" given in (10.6.7).

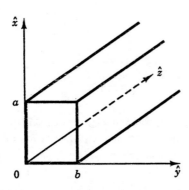

Figure 10.7 Acoustic duct.

where the functions in brackets must be linearly combined to satisfy the boundary conditions in (10.6.10). If we substitute (10.6.11) into (10.6.1), we find the dispersion relation

$$k_x^2 + k_y^2 + k_z^2 \equiv k_0^2 = \frac{\omega^2}{c_s^2} \tag{10.6.12}$$

which is analogous to (7.4.6). Because $u_x \propto \partial p/\partial x$ must be zero at $x = 0, a$ from (10.6.10), we see that $u_x \propto \sin k_x x$ where $k_x a = m\pi$ and $p \propto \cos k_x x$. Likewise, $p \propto \cos k_y y$ since $u_y \propto \partial p/\partial y = 0$ at $y = 0, b$ and thus $u_y \propto \sin k_y y$ with $k_y b = n\pi$. Combining these deductions yields

$$p(x, y, z) = p_0 \cos k_x x \cos k_y y \, e^{-jk_z z} \tag{10.6.13}$$

$$\overline{u}(x, y, z) = \frac{p_0}{\eta_s} \left(-\hat{x} \frac{jk_x}{k_0} \sin k_x x \cos k_y y - \hat{y} \frac{jk_y}{k_0} \cos k_x x \sin k_y y \right.$$
$$\left. + \hat{z} \frac{k_z}{k_0} \cos k_x x \cos k_y y \right) e^{-jk_z z} \tag{10.6.14}$$

with acoustic guidance conditions

$$k_x = \frac{m\pi}{a}$$
$$k_y = \frac{n\pi}{b} \tag{10.6.15}$$

and k_z given by (10.6.12) and (10.6.15):

$$k_z = \sqrt{\left(\frac{\omega}{c_s}\right)^2 - \left(\frac{m\pi}{a}\right)^2 - \left(\frac{n\pi}{b}\right)^2} \tag{10.6.16}$$

Just as in the rectangular electromagnetic waveguide, this acoustic duct can support a variety of modes indexed by the integers m and n. The lowest order mode is given by $m = n = 0$ ($k_x = k_y = 0$, $k_z = k_0$), which has *TEM*-like fields given by (10.6.9). This *TEM*-like acoustic wave has no electromagnetic counterpart; we recall from Section 7.4 that the TE_{10} mode is the lowest order or dominant mode in a rectangular waveguide. The reason that an acoustic *TEM*-like wave can exist in a duct but an electromagnetic *TEM* wave cannot is intimately related to the fact that acoustic waves propagate longitudinally whereas electromagnetic waves are transversely polarized.

Example 10.6.1

If we model a human ear canal as a square tube 3 mm on a side and assume $c_s = 330$ m/s, what are the cut-off frequencies f_{mn} for the four lowest-frequency propagating acoustic modes? Is overmoding a problem for a child with hearing up to 20 kHz? What might be a consequence of hearing when two or more modes propagate?

Solution: The (mn) mode will propagate only when $k_z \geq 0$. From (10.6.16) this constrains the frequency ω_{mn}

$$\omega_{mn} \geq c_s \sqrt{\left(\frac{m\pi}{a}\right)^2 + \left(\frac{n\pi}{b}\right)^2} = \frac{c_s\pi}{a}\sqrt{m^2 + n^2}$$

where $a = b = 3$ mm. Therefore, the cut-off frequencies for the first four propagating modes are 0 Hz for the (00) mode, 55 kHz for the (10) and (01) modes, and 78 kHz for the (11) mode.

As the human ear can hear frequencies between 20 Hz and 20 kHz, the (00) mode will be the only mode that propagates, and overmoding is thus not a problem even for a child with acute hearing. If two or more modes could propagate in the ear, they would create interference patterns, which could even produce nulls at certain frequencies if the two modes were appropriately excited.

10.7 ACOUSTIC RESONATORS

The acoustic resonator shown in Figure 10.8 may be constructed by placing rigid walls at the ends of the acoustic duct illustrated in Figure 10.7. Once again, the fluid in the closed cavity must obey the wave equation (10.6.1) subject to the boundary condition that no normal fluid flow occur at the rigid walls:

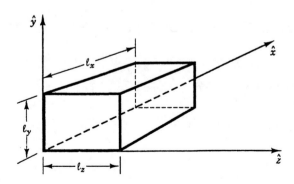

Figure 10.8 Acoustic resonator.

$$\begin{aligned}
u_x &= 0 \quad \text{at} \quad x = 0, a \\
u_y &= 0 \quad \text{at} \quad y = 0, b \\
u_z &= 0 \quad \text{at} \quad z = 0, d
\end{aligned} \tag{10.7.1}$$

The discussion in Section 10.6 may be easily extended to acoustic resonators, yielding the following phasor expressions for pressure and velocity:

$$p(x, y, z) = p_0 \cos k_x x \cos k_y y \cos k_z z$$

$$\bar{u}(x, y, z) = -\frac{jp_0}{\eta_s} \left(\hat{x} \frac{k_x}{k_0} \sin k_x x \cos k_y y \cos k_z z \right.$$

$$\left. + \hat{y} \frac{k_y}{k_0} \cos k_x x \sin k_y y \cos k_z z + \hat{z} \frac{k_z}{k_0} \cos k_x x \cos k_y y \sin k_z z \right)$$

$$\tag{10.7.2}$$

The guidance conditions

$$\begin{aligned}
k_x &= m\pi/a \\
k_y &= n\pi/b \\
k_z &= q\pi/d
\end{aligned} \tag{10.7.3}$$

(m, n, and q integers) are necessary so that the boundary conditions given by (10.7.1) are valid at $x = a$, $y = b$ and $z = d$. If we substitute boundary conditions (10.7.2) into the acoustic wave equation (10.6.1), we find the same dispersion relation (10.6.12) observed for the acoustic duct, which when combined with (10.7.3) yields *acoustic resonant frequencies*

$$\omega_{mnq} = c_s \sqrt{\left(\frac{m\pi}{a}\right)^2 + \left(\frac{n\pi}{b}\right)^2 + \left(\frac{q\pi}{d}\right)^2} \tag{10.7.4}$$

We therefore note that a lossless acoustic resonator, like its electromagnetic counterpart, may only be excited by a discrete set of resonant frequencies that are dependent on the shape of the cavity.

Because p and $\bar{u}$ are 90° out of phase in a lossless resonator, $\text{Re}\{pu^*\} = 0$ and no net power propagates inside the cavity. Instead, energy is stored in kinetic and potential form. The acoustic power conservation law (10.2.25) is expressed as

$$\oint_A p\bar{u} \cdot \hat{n} \, da = -\frac{d}{dt} \int_V (W_k + W_p) \, dv \tag{10.7.5}$$

where V is the volume of the rectangular cavity. The left side of (10.7.5) vanishes since $\hat{n} \cdot \bar{u} = 0$ along the resonator surface A. Therefore, we find that the total acoustic energy $w_k + w_p$ stored in the lossless resonator is a constant in time, where

$$w_k(t) = \int_V W_k(\bar{r}, t) \, dv = \frac{1}{2} \int_V \rho_0 |\bar{u}(\bar{r}, t)|^2 \, dv \tag{10.7.6}$$

$$w_p(t) = \int_V W_p(\bar{r}, t) \, dv = \frac{1}{2} \int_V \frac{1}{\gamma P_0} |p(\bar{r}, t)|^2 \, dv \tag{10.7.7}$$

For the rectangular resonator with $k_x \neq 0$, $k_y \neq 0$ and $k_z \neq 0$, we find that

$$w_k(t) = \frac{\rho_0 p_0^2}{2\eta_s^2} \frac{abd}{8} \sin^2 \omega t \qquad (10.7.8)$$

and

$$w_p(t) = \frac{p_0^2}{2\gamma P_0} \frac{abd}{8} \cos^2 \omega t \qquad (10.7.9)$$

where $\rho_0 p_0^2 / 2\eta_s^2 = p_0^2 / 2\gamma P_0$, so the amplitudes of $w_k(t)$ and $w_p(t)$ are equal. The energy in this acoustic cavity is therefore stored entirely as potential energy at $t = 0$, and one-quarter cycle later ($t = \pi/2\omega$) is stored entirely as kinetic energy. The energy alternates between these two forms indefinitely if there is no loss. The complex lossless power conservation equation is given by (10.2.27) in differential form and may be integrated over the volume of the resonator to give

$$\oint_A \frac{1}{2} p \, \overline{u}^* \cdot \hat{n} \, da = -2j\omega(\langle w_k \rangle - \langle w_p \rangle) \qquad (10.7.10)$$

Again, $\hat{n} \cdot \overline{u} = 0$ at the surface of the resonator and we immediately see that $\langle w_k \rangle = \langle w_p \rangle$.

Until now, we have been assuming that the rectangular cavity is perfectly lossless, and therefore that the total energy stored in the resonator is constant in time. In an acoustic cavity (e.g., a room), the walls have impedance

$$\eta_w = \frac{p}{u_n} = \kappa \rho_0 c_s = \kappa \eta_s \qquad (10.7.11)$$

where the acoustic absorption coefficient κ is typically a large (dimensionless) number ($\kappa \sim 10\text{–}100$). One key loss mechanism in acoustic resonators is therefore the propagation of acoustic waves away from the resonator through its walls. We neglect additional dissipative losses in the carpet, drapery, air, and so on. If the pressure and fluid velocity at the $z = 0$ wall is given by

$$p = p_+ \left(e^{-jkz} + \Gamma e^{jkz} \right)\big|_{z=0} = p_+(1 + \Gamma)$$
$$u = \frac{p_+}{\eta_s} \left(e^{-jkz} - \Gamma e^{jkz} \right)\big|_{z=0} = \frac{p_+}{\eta_s}(1 - \Gamma) \qquad (10.7.12)$$

then

$$\eta_w = \frac{p}{u_z}\bigg|_{z=0} = \eta_s \left(\frac{1 + \Gamma}{1 - \Gamma} \right) = \kappa \eta_s \qquad (10.7.13)$$

and therefore $\Gamma = (\kappa - 1)/(\kappa + 1)$ is the acoustic reflection coefficient at the wall. The power dissipated in the nth mode by the six resonator walls (all having the same value of κ) is

$$P_n = \frac{1}{2}\text{Re}\left\{\oint_A p\bar{u}^* \cdot \hat{n}\, da\right\} = \frac{1}{2}\oint_A \frac{|p|^2}{\eta_w}\, da$$

$$= \frac{1}{2\eta_w}|p_0|^2 2\left(\frac{ab}{4} + \frac{bd}{4} + \frac{ad}{4}\right) = \frac{|p_0|^2}{8\eta_w}A \qquad (10.7.14)$$

where $A = 2(ab + ad + bd)$ is the total surface area of the cavity and (10.7.14) is only valid for *oblique modes* such that $k_x \neq 0$, $k_y \neq 0$, $k_z \neq 0$.[11] We notice that the normal component of $\bar{u}$ is no longer strictly zero at the lossy walls, but use (10.7.11) to express $u_n = p/\eta_w$. We can now find Q_n for this room resonator from (8.3.3)

$$Q_n = \frac{\omega_n \mathsf{W}_n}{P_n} = \frac{\omega_n \rho_0 |p_0|^2 V/16\eta_s^2}{|p_0|^2 A/8\eta_w} = \frac{\kappa}{2}k_n\frac{V}{A} \qquad (10.7.15)$$

where $k_n = \omega_n/c_s$ and W_n is the sum of the kinetic and potential energies (10.7.8) and (10.7.9). In a room with dimensions 20 m $\times$ 15 m $\times$ 10 m and $\kappa \cong 40$ (i.e., $\Gamma = 0.95$), Q at 1 kHz is

$$Q \cong \frac{40}{2}\frac{2\pi \times 10^3 \text{ s}^{-1}}{330 \text{ m/s}}\frac{3000 \text{ m}^3}{1300 \text{ m}^2} \cong 879$$

which is quite substantial. The half-power bandwidth of the resonances near 1 kHz is therefore $\Delta f \cong Q/f_0 = 879/1000 = 0.88$ Hz. The *reverberation time* τ_n is defined as the time it takes for the sound pressure in the room to decay to $1/e$ of its original value. From Section 8.3

$$\tau_n \cong \frac{2Q_n}{\omega_n} \qquad (10.7.16)$$

For this room, the reverberation time for a 1-kHz pulse is $\tau \sim 2 \times 879/2\pi \cdot 10^3 \cong 0.28$ s. We could reduce the reverberation time by putting absorbing material in the walls to reduce κ and therefore reduce $\tau \propto Q \propto \kappa$. The best place to put acoustic absorbers for which loss is proportional to pressure is in the corner of a room where p is a maximum for all modes, as can be seen in (10.7.2). Similarly, speakers with fixed velocity cones transfer the most power when p is maximum. If the acoustic mode density is sufficiently high, then we could place the speakers just about anywhere in the room with little diminution of sound.

Example 10.7.1

If the 3-mm square "human" ear canal described in Example 10.6.1 is 3 cm long, open at one end, and rigidly closed at the other, what are the three lowest resonant frequencies of the ear and the corresponding acoustic modes ? If the four side walls of the canal have

---

11. *Axial modes*, where one or more of the set $\{k_x, k_y, k_z\}$ are zero, store more energy than oblique modes for which the direction of propagation is not parallel to any of the walls. But in large resonators, the number of oblique modes is much greater than the number of axial modes.

an acoustic absorption coefficient $\kappa = 100$, what is Q and the reverberation time τ of the (111) mode?

Solution: If the outer end of the ear canal is open ($p = 0$ at $z = 0$) and the inner end is closed ($u = 0$ at $z = d$) then the pressure and fluid velocities are

$$p(\bar{r}) = p(x, y) \sin k_z z$$

$$\bar{u}(\bar{r}) = \bar{u}(x, y) \cos k_z z$$

where $\sin k_z d = \pm 1$; $k_z d = (q + 1/2)\pi$. The functions $p(x, y)$ and $\bar{u}(x, y)$ are just the usual waveguide solutions with guidance conditions $k_x a = m\pi$ and $k_y b = n\pi$, so the resonant frequencies are

$$f_{mnq} = \frac{c_2}{2} \sqrt{\left(\frac{m}{a}\right)^2 + \left(\frac{n}{a}\right)^2 + \left(\frac{2q + 1}{2d}\right)^2}$$

where $a = b = 3$ mm, $d = 3$ cm, and $c_s = 330$ m/s. The three lowest frequencies are thus 2.75 kHz for the (000) mode, 8.25 kHz for the (001) mode, and 13.75 kHz for the (002) mode.

Because the (111) mode is oblique, we may use (10.7.15) to calculate Q_{111}:

$$Q_{111} = \frac{\kappa V k_{111}}{2A}$$

where

$$k_{111} = \omega_{111}/c_s = \sqrt{(\pi/0.003)^2 + (\pi/0.003)^2 + (3\pi/2 \times 0.03)^2} = 1481 \text{ m}^{-1}$$

which corresponds to $f_{111} = 78$ kHz. Since $A = 4ad$ and $V = a^2 d$, we find

$$Q_{111} = \frac{100 \times 0.003 \text{ m} \times 1481 \text{ m}^{-1}}{8} = 55.5$$

The reverberation time for this mode is $\tau = 2Q/\omega = 2 \times 55.5/330 \times 1481 = 2.3 \times 10^{-4}$ s.

Example 10.7.2

A room has dimensions $20 \times 15 \times 10$ m^3 and therefore a huge number of acoustic resonant frequencies in the audible frequency range 20–20,000 Hz. Find the total number of modes excited below 20 kHz and also the number of resonances within a bandwidth δf about 20 kHz.

Solution: We wish to find the number of modes (mnq) for which

$$f_0 \geq f_{mnq} = \frac{c_s}{2} \sqrt{\left(\frac{m}{a}\right)^2 + \left(\frac{n}{n}\right)^2 + \left(\frac{q}{d}\right)^2} \tag{1}$$

where $f_0 = 20$ kHz and (10.7.4) has been invoked. If there are sufficient modes obeying (1), then the total number of modes excited below f_0 is approximately

$$n_0 \cong \frac{(1/8)(4\pi f_0^3/3)}{(c_s/2a)(c_s/2b)(c_s/2d)} = \frac{4}{3}\pi V \left(\frac{f_0}{c_s}\right)^3 \tag{2}$$

where $4\pi f_0^3/3$ is the volume of a sphere of radius f_0 in frequency space and $(c_s/2a)(c_s/2b)(c_s/2d)$ is the approximate volume of frequency space occupied by one mode. The factor of $1/8$ is necessary because n, m, and q are all positive so only one octant of the frequency sphere is actually used. The volume of the resonator is $V = 20 \times 15 \times 10 = 3000 \ \mathrm{m}^3$ and since there is only one acoustic "polarization" per mode, the extra factor of 2 observed in the electromagnetic case is missing. [Otherwise, this discussion parallels that at the end of Section 8.2 almost exactly; see (8.2.26).] If $c_s = 330$ m/s, then the number of resonances below 1 kHz is

$$n_0(\le 20 \ \text{kHz}) \cong \frac{4}{3}\pi(3000 \ \mathrm{m}^3)\left(\frac{2 \times 10^4 \ \mathrm{s}^{-1}}{330 \ \mathrm{m/s}}\right)^3 = 2.9 \times 10^9$$

If we differentiate (2) with respect to f, we find that

$$\frac{dn}{df} \cong \frac{4\pi V}{c_s}\left(\frac{f_0}{c_s}\right)^2$$

which means that there are approximately $\delta n = 4.2 \times 10^5 \, \delta f$ resonances within a bandwidth δf about $f_0 = 20$ kHz.

10.8 ACOUSTIC RADIATION

From the basic linearized equations of acoustics (10.2.13)–(10.2.14) we can derive the wave equation (10.6.1)

$$(\nabla^2 + k^2)\, p = 0 \tag{10.8.1}$$

where $k = \omega/c_s$ is the dispersion relation in the fluid. We consider an *acoustic monopole source*, which radiates sound uniformly in all directions as shown in Figure 10.9. A spherical balloon with a surface which expands and contracts radially could be such a source. We see that $p(\overline{r})$ must be independent of the coordinates θ and ϕ because of spherical symmetry. In spherical coordinates with $\partial/\partial\theta = \partial/\partial\phi = 0$, (10.8.1) becomes

$$\frac{d^2 p}{dr^2} + \frac{2}{r}\frac{dp}{dr} + k^2 p = 0 \tag{10.8.2}$$

which can also be written as

$$\frac{d^2(rp)}{dr^2} + k^2(rp) = 0 \tag{10.8.3}$$

suggesting the solution $rp \propto e^{\pm jkr}$. For a disturbance traveling in the $+\hat{r}$ direction, the sound pressure is therefore

$$p(r) = \frac{B}{r} e^{-jkr} \tag{10.8.4}$$

and from (10.2.21),

$$\overline{u}(r) = -\frac{\nabla p}{j\omega\rho_0} = \hat{r}\frac{B}{\eta_s r}\left(1 + \frac{1}{jkr}\right)e^{-jkr} \tag{10.8.5}$$

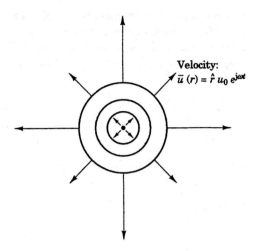

Figure 10.9   Acoustic monopole source.

In the farfield zone, where $r \gg \lambda/2\pi$ or $kr \gg 1$, the second term in (10.8.5) may be neglected, yielding

$$\bar{u}_{\text{ff}}(r) = \hat{r}\,\frac{B}{\eta_s r}\,e^{-jkr} \tag{10.8.6}$$

so that (10.8.4) and (10.8.6) resemble a (uniform) $\hat{r}$-propagating plane wave far from the source, with $|p|/|\bar{u}_{\text{ff}}| = \eta_s$. In the near field, $kr \ll 1$ and

$$\bar{u}_{\text{nf}} = -\hat{r}\,\frac{jB}{\eta_s kr^2}\,e^{-jkr} \tag{10.8.7}$$

which explains why a *velocity microphone* (with electrical output proportional to fluid velocity) held close to one's lips compared to a wavelength will produce a low frequency (bass) boost. Equation (10.8.7) is proportional to $1/k$ or $1/\omega$.

If we consider a small acoustic monopole source of radius a, which moves with a radial velocity u_0 at $r = a$, we may use (10.8.7) to solve for B if $ka \ll 1$; i.e., the radius a is much less than $\lambda/2\pi$. Therefore,

$$u_0 = -\frac{jB}{\eta_s ka^2}\,e^{-jka}$$

so $B \cong j\eta_s ka^2 u_0$ since $e^{-jka} \cong 1$. The fluid pressure and velocity are thus

$$p = \frac{j\eta_s ka^2 u_0}{r}\,e^{-jkr} \tag{10.8.9}$$

$$u = \frac{jka^2 u_0}{r}\left(1 + \frac{1}{jkr}\right)e^{-jkr} \tag{10.8.10}$$

from (10.8.4) and (10.8.5). The time-averaged acoustic power intensity can then be computed from (10.2.29)

$$\langle \overline{I}(\overline{r})\rangle = \frac{1}{2}\,\mathrm{Re}\{p(r)\,\overline{u}^*(r)\} = \hat{r}\,\frac{\eta_s}{2}\left|\frac{ka^2u_0}{r}\right|^2 \quad (\mathrm{W/m^2}) \tag{10.8.11}$$

where we see that the nearfield part of (10.8.10) does not contribute. The total power radiated by this monopole source is found by integrating $\langle \overline{I}(\overline{r})\rangle$ over a sphere of radius r

$$P_T = \int_A \langle \overline{I}(\overline{r})\rangle\, r^2 \sin\theta d\theta d\phi = 2\pi\,\eta_s|ka^2u_0|^2 \quad (\mathrm{W}) \tag{10.8.12}$$

so an *acoustic radiation resistance* R_{rad} may be defined for the acoustic monopole source by analogy with (9.2.13)

$$R_{\mathrm{rad}} = \frac{P_T}{|u_0|^2/2} = 4\pi\,\eta_s(ka^2)^2 \quad (\mathrm{kg/s}) \tag{10.8.13}$$

which has the same dimensions as Z_0 (kg/s), seen in Table 10.1. The acoustic radiation impedance is proportional to a^4/λ^2, so large loudspeakers driving short wavelengths (high frequencies) radiate best if the loudspeaker is a velocity source (analogous to an electromagnetic current source). Conversely, a voltage-like pressure source would radiate best for small values of R_{rad} and speaker size a, but such acoustic pressure sources generally do not exist.

We may create an *acoustic dipole source* by superposing two monopoles of opposite sign separated by a distance d as shown in Figure 10.10. The total pressure is the sum of the fluid pressure fields from each source

$$p = \frac{j\eta_s ka^2u_0}{r_1}e^{-jkr_1} - \frac{j\eta_s ka^2u_0}{r_2}e^{-jkr_2} \tag{10.8.14}$$

where $r_{1,2} \simeq r \pm (d/2)\cos\theta$, as may be seen geometrically in Figure 10.10. As usual, we further approximate the denominators $r_{1,2}$ in (10.8.14) by r because they are not sensitive to the dipole separation distance. The total pressure given by (10.8.14) is thus

$$p \simeq \frac{j\eta_s ka^2u_0}{r}e^{-jkr}\left(e^{j(kd/2)\cos\theta} - e^{-j(kd/2)\cos\theta}\right)$$

$$= -\frac{2\eta_s ka^2u_0}{r}e^{-jkr}\sin\left(\frac{kd}{2}\cos\theta\right) \tag{10.8.15}$$

In the low frequency limit where $kd \ll 1$, we may approximate (10.8.15) as

$$p \simeq -\frac{\eta_s k^2a^2du_0}{r}e^{-jkr}\cos\theta \tag{10.8.16}$$

which differs from the electric field of a Hertzian dipole given by (9.2.9) because (10.8.16) has nulls at the equator ($\theta = \pi/2$) and power maxima at the poles

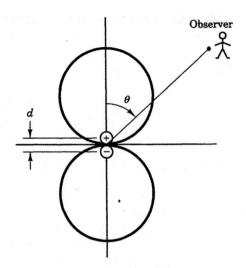

Figure 10.10   Acoustic dipole source.

$(\theta = 0, \pi)$. The acoustic antenna gain of this dipole source is then $3\cos^2\theta$, where the maximum gain $G_{\max} = 3$ is calculated by recognizing that $G_{\max}\cos^2\theta$ integrated over all angles must be 4π for a lossless radiating dipole.

An example of an acoustic dipole source is a low-frequency speaker with its baffle removed, as shown in Figure 10.11(a). With the baffle, the speaker behaves as a monopole source so $p \propto k \propto \omega$ from (10.8.9). The driving source is adjusted so that p is very nearly flat at low frequencies. Because the baffleless looks instead like a dipole source, it exhibits sound pressure $p \propto k^2 \propto \omega^2$ at low frequencies

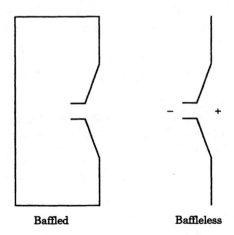

Baffled   Baffleless

Figure 10.11   Speaker with and without baffle.

from (10.8.16). Therefore, its response drops off at a rate of 6 dB/octave at low frequencies relative to the baffle-mounted speaker. The ratio of sound pressures in the two cases is

$$\left| \frac{p_{\text{baffleless}}}{p_{\text{baffled}}} \right| = \left| \frac{p_{\text{dipole}}}{p_{\text{monopole}}} \right| \sim kd \cos \theta \ll 1$$

For this reason, high-fidelity sound systems usually incorporate elaborate baffle systems in their low-frequency speakers as electrical compensation for the speaker deficiencies.

An acoustic source that radiates over a wide angle in the horizontal plane but a relatively narrow angle in the vertical plane can be designed using a vertical array of loudspeakers, each of which acts like a monopole source. Directional microphones can be constructed similarly. Active and passive phased array acoustic systems (sonar) are routinely used for studying undersea phenomena and are analogous to electromagnetic phased arrays. All the array techniques discussed in Chapter 9 may also be used for synthesizing acoustic arrays.

Example 10.8.1

Approximately how far apart can two people stand and communicate by shouting outdoors? Assume that the threshold of hearing is about 30 dB. Neglect atmospheric scattering by microscopic and macroscopic density fluctuations and adsorption by atmospheric humidity. If the shouter stands on a level concrete aircraft runway and yells at 330 Hz, what is the resulting acoustic interference pattern?

Solution: We shall first ignore the effect of the runway on the maximum shouting distance. According to the *Guinness Book of World Records*, the loudest shout is about 112 dB. Assume that the shouter has an oral aperture of 3 cm, and the radiated acoustic power propagates into π steradians as in Example 10.4.1 where the child shouts across a lake. (The mouth is small compared to nominal audio wavelengths so it acts like a monopole source.) The transmitted acoustic intensity of the shouter is then $I_{\text{tr}} = P_{\text{tr}}/\pi(d/2)^2$ and the received acoustic intensity is $I_{\text{rec}} = P_{\text{tr}}/\pi r^2$ where $d = 3$ cm is the diameter of the mouth and r is the distance from the shouter to the hearer. If the shouter can muster 112 dB and the listener can hear 30 dB, then

$$\frac{I_{\text{rec}}}{I_{\text{tr}}} = \left(\frac{d}{2r} \right)^2 \geq 10^{(30-112)/10}$$

so $r = 188$ m.

The effect of a concrete runway is to produce an *acoustic image monopole* source h meters below the concrete where $h = 1.5$ m is an estimate for the distance from the runway to the speaker's mouth. The boundary condition at the concrete is that the normal component of $\bar{u}$ is zero, true of an image monopole in phase with the monopole source. The pressure field for the source and image is then

$$p = \frac{j\eta_s ka^2 u_0}{r} e^{-jk(r+h\cos\theta)} + \frac{j\eta_s ka^2 u_0}{r} e^{-jk(r-h\cos\theta)}$$

$$= \frac{2j\eta_s ka^2 u_0}{r} e^{-jkr} \cos(kh \cos \theta)$$

At $\theta = \pi/2$ (along the runway), we see that the pressure fields interfere constructively, giving a maximum in the radiation pattern. Nulls are observed at $kH \cos\theta_{\text{null}} = \left(n + \frac{1}{2}\right)\pi$, so for $\lambda = 1$ m, there are nulls at $33.6°$, $60°$, and $80.4°$. There are also additional maxima at $\cos\theta_{\text{max}} = n\lambda/2h$ or $\theta_{\text{max}} = 48.2°$, $70.5°$.

A listener 188 m away would be close to the pressure field maximum at $\theta = \pi/2$ and would thus perceive a doubling of the incremental acoustic pressure. The acoustic intensity at the listener's ear would therefore increase by a factor of four, for a net gain of 6 dB, permitting him to stand twice as far away (376 m) and still hear the shouter.

10.9 SUMMARY

The remarkable similarity between electromagnetic and acoustic wave behavior leads to an efficient presentation of acoustic concepts. One key difference between the two is that acoustic wave propagation is inherently nonlinear, forcing us to restrict ourselves to the study of relatively weak waves for which the equations of wave motion can be linearized. The *acoustic waves* that propagate in compressible gases are characterized by a velocity field parallel to the direction of wave propagation, so the second essential difference between acoustic and electromagnetic waves is that acoustic waves are *longitudinal* rather than transverse. Gas waves are thus represented by simpler scalar pressure and velocity fields rather than the vector electromagnetic fields. More complicated acoustic waves can exist in solids and liquids because of nonzero shear viscosities, so the velocity fields in these media can have both longitudinal and transverse components. The longitudinal components generally dominate, however.

The two linearized equations of motion for acoustics arise from *conservation of mass*

$$\frac{\partial \rho_1}{\partial t} + \rho_0 \nabla \cdot \overline{u} = 0$$

and *Newton's force law* $(\overline{f} = m\overline{a})$, also known as *conservation of linear momentum*:

$$\nabla p = -\rho_0 \frac{\partial \overline{u}}{\partial t}$$

The variables p, $\overline{u}$, and ρ_1 represent perturbations to the average *fluid pressure* P_0 ($p \ll P_0$), *velocity* $\overline{U}_0$ ($\overline{U}_0 = 0$ here, as the fluid is at rest overall), and *mass density* ρ_0 ($\rho_1 \ll \rho_0$), respectively. The fluctuation pressure and density are related by the *adiabatic constitutive relations*

$$\rho_1 = \frac{\rho_0}{\gamma P_0} p \qquad \text{gases}$$

$$\rho_1 = \frac{\rho_0}{K} p \qquad \text{solids, liquids}$$

if the wave frequency is sufficiently high that the wave has no time to exchange heat with its container. The *adiabatic exponent* γ is 5/3 for an ideal monatomic gas ($1-2$ for a real gas), and K is the *bulk modulus*. These three equations lead to the *acoustic wave equation* for pressure

$$\nabla^2 p - \frac{\rho_0}{\gamma P_0} \frac{\partial^2 p}{\partial t^2} = 0$$

which has a traveling wave solution $p(\overline{r}, t) = p(\omega t - \overline{k} \cdot \overline{r})$, where the wave number k is related to the frequency ω by the *acoustic dispersion relation* $k = \omega / c_s$ and $c_s = \sqrt{\gamma P_0 / \rho_0}$ is the *velocity of sound* (typically 330 m/s in air). The time-averaged *acoustic intensity* $I = \frac{1}{2} \text{Re}\{p \, u^*\}$ W/m^2 is equal to $p^2 / 2\eta_s = \eta_s u^2 / 2$, where the *characteristic acoustic impedance* $\eta_s = \rho_0 c_s = 425$ N $\cdot$ s/m^3 for air near sea level. An equation for conservation of complex power, analogous to the Poynting theorem, also exists.

Acoustic waves at planar interfaces can be both refracted and reflected, as can their electromagnetic counterparts. The pressure and the normal component of the fluid velocity both must be continuous across an acoustic boundary, leading to an *acoustic phase-matching condition* and an *acoustic Snell's law*: $c_i \sin \theta_t = c_t \sin \theta_i$. *Total internal reflection* occurs when the incident angle θ_i is greater than the *critical angle* $\theta_c = \sin^{-1}(c_i / c_t)$. Beyond the critical angle, *evanescent waves* are produced that propagate parallel to the boundary and decay exponentially away from it; these waves are called *surface acoustic waves*. An *acoustic Brewster's angle* θ_B also exists where reflections are zero: $\eta_t \cos \theta_B = \eta_i \cos \theta_t$.

Acoustic waves in pipes can be related in a simple way to electromagnetic waves on *TEM* transmission lines; line pressure P and velocity U correspond to voltage V and current I, respectively. The characteristic impedance of the acoustic line $\eta_s A$ corresponds to $\sqrt{L/C}$. The general solution to these acoustic transmission lines is a superposition of forward- and backward-moving waves:

$$P(z) = P_+ \left(e^{-jkz} + \Gamma_L e^{+jkz} \right)$$

$$U(z) = Y_0 P_+ \left(e^{-jkz} - \Gamma_L e^{+jkz} \right)$$

Because these equations are directly analogous to the corresponding *TEM* transmission line solutions, we can use the Smith chart, the gamma plane, *ABCD* matrices, and all the other tools of Chapter 6. The complex reflection coefficient Γ_L is equal to $+1$ for a closed tube ($U = 0$) and -1 for an open tube ($P = 0$). Standing waves can be established in tubes terminated by such open or short circuits, and these tubes can also function as *acoustic resonators*. The resonant frequencies of acoustic tubes propagating longitudinal waves are computed in the same manner as their *TEM* line counterparts.

Higher-order modes can also propagate in acoustic waveguides, where the dispersion relation

$$k_x^2 + k_y^2 + k_z^2 \equiv k_0^2 = \frac{\omega^2}{c_s^2}$$

and the *guidance conditions*

$$k_x = \frac{m\pi}{a}$$

$$k_y = \frac{n\pi}{b}$$

imply that acoustic waveguide modes are evanescent above the *cut-off frequency*

$$f_{mn} = \frac{c_s}{2}\sqrt{\left(\frac{m}{a}\right)^2 + \left(\frac{n}{b}\right)^2}$$

Unlike a rectangular waveguide, an acoustic guide can propagate a nontrivial zero frequency (*TEM*-like) mode.

An *acoustic monopole antenna* produces a pressure field

$$p(r) = \frac{j\eta_s k a^2 u_0}{r} e^{-jkr}$$

along with a farfield radial velocity field

$$\bar{u}_{ff}(r) = \hat{r}\, \frac{jka^2 u_0}{r} e^{-jkr}$$

Arrays of acoustic monopoles can be assembled to produce desired radiation patterns, analogous to the patterns available from electromagnetic arrays and aperture antennas. The total power radiated by an acoustic monopole is

$$P_T = 2\pi \eta_s |ka^2 u_0|^2$$

which corresponds to an *acoustic radiation resistance* equal to $4\pi \eta_s (ka^2)^2$ kg/s. These expressions imply that a loudspeaker that acts as a velocity source must produce large amplitude waves to produce significant power at the longest wavelengths. Most loudspeakers are actually dipoles, with monopole-like emission 180° out of phase between the front and back of the speaker membrane. The cancellation this produces at longer wavelengths poses a serious problem, motivating design of clever baffles to absorb or redirect the backward-propagating energy.

10.10 PROBLEMS

10.2.1 A loudspeaker is radiating 10 W of acoustic power approximately as a uniform plane wave at 10 kHz within the 10-cm speaker diameter. The speed of sound c_s is 330 m/s near sea level at 20° C.

(a) What is the wavelength of this plane wave?
(b) What is the peak particle velocity?
(c) What is the maximum particle displacement?
(d) What are the answers to parts (a)–(c) on a mountain top where the pressure P_0 is reduced to $0.6P_0$?

10.2.2 Repeat Problem 10.2.1, parts (a)–(c) for a 10-W wave under water where $\rho_0 = 1000$ kg/m^3 and $c_s = 1500$ m/s.

10.2.3 For the wave of Problem 10.2.1, what is: (i) the time-averaged kinetic energy density (J/m^3)? (ii) the time-averaged potential energy density?

10.3.1 We know that for an ideal gas, $pV = nRT$ where p is the gas pressure, n is the number of moles of the gas, $R = 8.31$ (J/mole-K) is the universal gas constant, and T is the temperature in degrees Kelvin. If $\gamma = 3/2$, what are the maximum and minimum temperatures associated with the wave of Problem 10.2.1?

10.3.2 The loudspeaker of Problem 10.2.1 is radiating in the $+\hat{z}$ direction and traveling at 5 m/s relative to a wind of 2 m/s, all moving in the $+\hat{z}$ direction. Write expressions for $\rho_1(z, t)$ and $\bar{u}(z, t)$ for a stationary observer of this wave located at position z.

10.4.1 What fraction of the acoustic power in a uniform plane wave in air normally incident on a planar interface is reflected by

(a) Air of the same pressure but 10 K warmer?

(b) Air of the same temperature, but of 100 times greater pressure (retained by a massless membrane)?

(c) Water (e.g., a lake)?

10.4.2 For which (if any) of the cases in Problem 10.4.1 is there a Brewster's angle of incidence that permits perfect power transfer? What are the corresponding angles θ_B?

10.4.3 A planar air-air interface between equal-pressure gases having densities of normal air and twice normal air is reflecting uniform plane waves; c_s in the normal air is 340 m/s.

(a) What is the critical angle θ_c?

(b) What fraction of the acoustic power is reflected at $\theta_i = 85°$?

(c) What is the angle of transmission θ_t if $\theta_i = 30°$?

(d) For part (b), what is the decay distance α^{-1} (m) for the evanescent wave if λ_0 in the same medium is 1 m?

10.5.1 Refer to Example 10.5.2.

(a) If the door were twice as thick, what fraction of the power would be reflected at 100 Hz? At 1000 Hz?

(b) If the door were composed of two 1-cm thick slabs with a 1-cm lossless air core between them, what is the acoustic isolation at 100 Hz?

10.5.2 An acoustic tube 1-cm in diameter is fed by a tube of 2-cm diameter. The impedance mismatch is compensated by a closed-end tube of length d_1 m and 2-cm diameter joined to the 2-cm diameter tube at a distance d_2 m from the junction with the 1-cm tube. What are the minimum lengths d_1 and d_2 that can match all 100-Hz acoustic power to the 1-cm tube?

10.5.3 A muffler for an engine producing 100-Hz sine waves consists only of a pipe 4 cm in diameter that broadens abruptly to 8 cm over a length d_1. The wide section is located d_2 meters from the open end of the tail pipe. Assume the impedance of the open end of the tail pipe approximates that of a 12-cm pipe.

(a) What values of d_1 and d_2 minimize the acoustic power radiating from the end of the pipe?

(b) What is the standing wave ratio in the wide pipe for these values of d_1 and d_2?

10.6.1 A rectangular acoustic waveguide is 2×3 cm with $c_s = 330$ m/s and $\rho_0 = 1.29$ kg/m^3.

    (a) What are the four lowest cut-off frequencies and to what modes do they correspond?

    (b) Write expressions for the pressure p and velocity $\bar{u}$ distributions for a 1-W wave at 10 kHz propagating in the (10) mode (the one with the 2nd lowest cutoff frequency).

10.6.2 For the waveguide of Problem 10.6.1,

    (a) What are the waveguide wavelengths for the first three modes at 10 kHz?

    (b) Repeat part (b) of Problem 10.6.1 for a wave of 3 kHz (below cut-off). Normalize the amplitude to unity p_1 at $z = 0$.

    (c) What are the phase and group velocities for the (10) mode at 10 kHz?

    (d) For the wave of part (c), what is the incident angle θ_i for the component plane waves bouncing back and forth between the waveguide walls?

10.6.3 A long square organ pipe resonates at 300 Hz. If we wish to design it so neither (10) nor (01) modes propagate, what is its maximum cross-section?

10.7.1 A ceramic tile bathroom is $1 \times 2 \times 3$ m.

    (a) What are its four lowest resonant frequencies?

    (b) What are the corresponding mode designations [e.g., (211)]?

    (c) What is the average frequency separation between resonances near 200 Hz? Near 800 Hz?

    (d) If the human mouth is modeled as an ideal velocity source, where should one stand to maximize the power input to the (642) mode? Give precise coordinates for one location, and explain briefly.

10.7.2 For the resonator of Problem 10.7.1 oscillating in the (100) mode,

    (a) Sketch the pressure distribution $p(x, y, z)$ when it is maximum. The sketch should apply to a plane that is perpendicular to and bisects the shortest side of the room.

    (b) Repeat part (a) for the (111) mode.

    (c) Repeat part (a) for the velocity distribution of the (100) mode.

    (d) Repeat part (b) for the (111) mode.

10.7.3 The resonator of Problem 10.7.1 has a wall acoustic absorption coefficient of $\kappa = 20$.

    (a) What is the Q for the (100) mode?

    (b) What is the Q for the (111) mode?

    (c) What is the reverberation time for the (100) mode?

10.7.4 A man is singing in a small uncluttered rectangular room with hard wall surfaces that has dimensions $3 \times 4 \times 3$ m.

    (a) Estimate the frequency separation between resonances.

    (b) Estimate the spatial separation between points where excitation of a particular resonance is best.

10.8.1 Master spy James Bond sometimes uses an acoustic antenna 1 m square and uniformly illuminated. When overhearing a conversation centered near 600 Hz, what is the half-power beamwidth of this directional microphone?

10.8.2 An audio conferencing system uses a vertical linear phased array of microphones to discriminate between sound direct from a person and that reflected from the ceiling, floor, or table. The room walls are acoustic absorbers.

   (a) If the array is 2 m tall, what is the half-power acoustic beam width in the vertical direction at 3.3 kHz? Assume $c_s = 330$ m/s.

   (b) What is the approximate antenna pattern in the horizontal plane? Assume the microphones are pressure (not velocity) transducers.

   (c) Assume the same array is remounted horizontally as an endfire "shotgun" microphone, and the separate transducers are phased to respond best to signals coming from the direction in which the microphone points. What is the angle to the first null at 3.3 kHz?

10.8.3* Design a broad-band microphone with a peak response in the $\hat{z}$ direction and a null in the $-\hat{z}$ direction. Use two pressure transducers. Describe any required signal processing steps following the transducers.

10.8.4* Perform an experiment on human auditory direction-finding capability. A subject sits in a chair with eyes closed and points toward the location of keys gently jangled by the experimenter.

   (a) What is the subject's ability to determine the position of the keys in a horizontal plane, including (i) in back of him, and (ii) behind him? Is this consistent with the directionality of a two-element antenna at 10 kHz? At the 20-kHz limit of human hearing? Can you explain any discrepancies?

   (b) Repeat part (a) in the vertical plane bisecting the line between the subject's ears for objects (i) in front of and (ii) behind him.

   (c) Repeat part (a) with a pure single-frequency source like a tuning fork, chime, or digital watch alarm. Discuss any differences between the results of parts (a), (b), and (c) in light of the complex shape of the human ear.

   (d) Can you determine whether the ear possesses any distance measuring capability independent of loudness for the jangling keys? Discuss.

A

A REVIEW OF COMPLEX NUMBERS

A complex number $\underline{A}$ can have both real and imaginary parts and is written in Cartesian form as

$$\underline{A} = A_r + j A_i \tag{A.1}$$

where $j = \sqrt{-1}$ is an imaginary number with unit magnitude; the underbar indicates that $\underline{A}$ is complex. The real and imaginary operators Re{ } and Im{ } extract the real and imaginary parts, respectively, of any complex number; here,

$$\mathrm{Re}\{\underline{A}\} = A_r \tag{A.2a}$$

$$\mathrm{Im}\{\underline{A}\} = A_i \tag{A.2b}$$

We represent $\underline{A}$ in the complex plane that has orthogonal real and imaginary axes, as illustrated in Figure A.1. Typically the positive real axis extends horizontally to the right and the positive imaginary axis extends vertically into the upper half-plane. The point $A_r + j A_i$ corresponds to the Cartesian ordered pair (A_r, A_i) and we plot $\underline{A}$ in Figure A.1, along with the points ± 1 and $\pm j$.

The Cartesian notation in (A.1) is only one way to describe $\underline{A}$, however, as a polar representation is equally valid

$$\underline{A} = A\, e^{j\phi_A} \tag{A.3}$$

where $A = |\underline{A}|$ is the magnitude of $\underline{A}$ and ϕ_A is the phase, defined as the angle between the real axis and $\underline{A}$, as shown in Figure A.1. Note that ϕ_A is not unique,

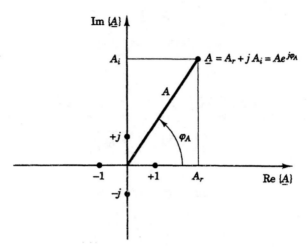

Figure A.1   Representations in the complex plane.

as ϕ_A can be replaced by $\phi_A \pm 2n\pi$, where n is any integer. From this figure, it is clear that

$$A_r = A \cos \phi_A \qquad (A.4a)$$

$$A_i = A \sin \phi_A \qquad (A.4b)$$

or, conversely

$$A = \sqrt{A_r^2 + A_i^2} \qquad (A.5a)$$

$$\phi_A = \tan^{-1}(A_i/A_r) \qquad (A.5b)$$

Equating (A.1) and (A.3) with (A.4) yields De Moivre's theorem

$$e^{j\phi} = \cos \phi + j \sin \phi \qquad (A.6)$$

which can also be proven algebraically by expanding $e^{j\phi}$ in a Taylor series and then grouping the real and imaginary terms:

$$e^{j\phi} = 1 + (j\phi) + \frac{1}{2!}(j\phi)^2 + \frac{1}{3!}(j\phi)^3 + \frac{1}{4!}(j\phi)^4 + \cdots$$

$$= \left[1 - \frac{1}{2!}\phi^2 + \frac{1}{4!}\phi^4 - \cdots \right] + j \left[\phi - \frac{1}{3!}\phi^3 + \frac{1}{5!}\phi^5 - \cdots \right]$$

$$= \cos \phi + j \sin \phi$$

The complex number $+j$ has a polar representation $e^{j(\pi/2 \pm 2n\pi)}$ and $-1 = e^{j(\pi \pm 2n\pi)}$ where n is any integer.

The sum of two complex numbers is found by adding their real and imaginary parts separately

$$\underline{A} + \underline{B} = (A_r + jA_i) + (B_r + jB_i)$$
$$= (A_r + B_r) + j(A_i + B_i) \tag{A.7}$$

which is demonstrated in Figure A.2(a) and is analogous to adding vectors in that the real and imaginary parts are computed separately.

The product of two complex numbers is illustrated in Figure A.2(b)

$$\underline{A}\,\underline{B} = (A_r + jA_i)(B_r + jB_i)$$
$$= (A_rB_r - A_iB_i) + j(A_rB_i + A_iB_r) \tag{A.8a}$$

where we note that $j^2 = -1$. Multiplication in the magnitude-phase representation is simpler:

$$\underline{A}\,\underline{B} = A\,e^{j\phi_A}\,B\,e^{j\phi_B} = AB\,e^{j(\phi_A+\phi_B)} \tag{A.8b}$$

[Using (A.4) and (A.5), one should be able to demonstrate that (A.8a) and (A.8b) are equal!]

A complex number $\underline{A}$ also has a complex conjugate $\underline{A}^*$, which is found by reflecting $\underline{A}$ about the real axis (or by making the substitution $j \rightarrow -j$):

$$\underline{A}^* = (A_r + jA_i)^* = A_r - jA_i \tag{A.9a}$$

$$\underline{A}^* = \left(A\,e^{j\phi_A}\right)^* = A\,e^{-j\phi_A} \tag{A.9b}$$

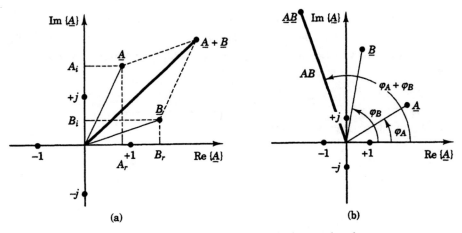

Figure A.2   Addition and multiplication in the complex plane.

From (A.9a), we note that the real and imaginary parts of $\underline{A}$ can be expressed solely in terms of $\underline{A}$ and $\underline{A}^*$

$$A_r = \frac{1}{2}\left(\underline{A} + \underline{A}^*\right) \tag{A.10a}$$

$$A_i = \frac{1}{2j}\left(\underline{A} - \underline{A}^*\right) \tag{A.10b}$$

so that if $\underline{A}$ is real, $\underline{A} = \underline{A}^*$. We also note that $A = \sqrt{\underline{A}\,\underline{A}^*}$.

The division of a complex number by another complex number is performed by rationalizing the denominator by multiplying both numerator and denominator by the complex conjugate of the denominator:

$$\frac{\underline{A}}{\underline{B}} = \frac{A_r + jA_i}{B_r + jB_i} = \left(\frac{A_r + jA_i}{B_r + jB_i}\right)\left(\frac{B_r - jB_i}{B_r - jB_i}\right)$$
$$= \left(\frac{A_r B_r + A_i B_i}{B_r^2 + B_i^2}\right) + j\left(\frac{A_i B_r - A_i B_r}{B_r^2 + B_i^2}\right) \tag{A.11}$$

Of course, $\underline{A}/\underline{B}$ is also equal to $(A/B)\,e^{j(\phi_A - \phi_B)}$ by analogy with (A.8b), and is generally simpler to calculate.

When finding roots or logarithms of complex numbers, it is often useful to use the polar representation (A.3) and use (A.4) to convert back to a Cartesian form if desired. For example,

$$\sqrt{1+j} = \left(\sqrt{2}\,e^{j(\pi/4 \pm 2n\pi)}\right)^{\frac{1}{2}} = 2^{\frac{1}{4}}\,e^{j(\pi/8 \pm n\pi)}$$

which means that

$$\sqrt{1+j} = \begin{cases} 2^{\frac{1}{4}}e^{j\pi/8} = 2^{\frac{1}{4}}\cos(\pi/8) + j2^{\frac{1}{4}}\sin(\pi/8) \\ 2^{\frac{1}{4}}e^{-j7\pi/8} = 2^{\frac{1}{4}}\cos(7\pi/8) - j2^{\frac{1}{4}}\sin(7\pi/8) \end{cases}$$

Note that only two distinct representations of $\sqrt{1+j}$ exist in the complex plane, since there are only two solutions to the equation $z^2 = 1 + j$. (These solutions lie $180°$ apart in the complex plane, so one is the negative of the other.) On the other hand,

$$\ln(1+j) = \ln\left(\sqrt{2}\,e^{j(\pi/4 + 2n\pi)}\right) = \frac{1}{2}\ln 2 + j\left(\frac{\pi}{4} + 2n\pi\right)$$

has an infinite number of solutions.

B

VECTOR IDENTITIES
AND THEOREMS

$$\overline{A} = \hat{x}\,A_x + \hat{y}\,A_y + \hat{z}\,A_z$$

$$\overline{A} + \overline{B} = \hat{x}\,(A_x + B_x) + \hat{y}\,(A_y + B_y) + \hat{z}\,(A_z + B_z)$$

$$\overline{A} \cdot \overline{B} = A_x B_x + A_y B_y + A_z B_z$$

$$\overline{A} \times \overline{B} = \det \begin{vmatrix} \hat{x} & \hat{y} & \hat{z} \\ A_x & A_y & A_z \\ B_x & B_y & B_z \end{vmatrix}$$

$$= \hat{x}\,(A_y B_z - A_z B_y) + \hat{y}\,(A_z B_x - A_x B_z) + \hat{z}\,(A_x B_y - A_y B_x)$$

$$\overline{A} \cdot (\overline{B} \times \overline{C}) = \overline{B} \cdot (\overline{C} \times \overline{A}) = \overline{C} \cdot (\overline{A} \times \overline{B})$$

$$\overline{A} \times (\overline{B} \times \overline{C}) = (\overline{A} \cdot \overline{C})\,\overline{B} - (\overline{A} \cdot \overline{B})\,\overline{C}$$

$$(\overline{A} \times \overline{B}) \cdot (\overline{C} \times \overline{D}) = (\overline{A} \cdot \overline{C})\,(\overline{B} \cdot \overline{D}) - (\overline{A} \cdot \overline{D})\,(\overline{B} \cdot \overline{C})$$

$$\nabla \times \nabla \Psi = 0$$

$$\nabla \cdot (\nabla \times \overline{A}) = 0$$

$$\nabla \times (\nabla \times \overline{A}) = \nabla(\nabla \cdot \overline{A}) - \nabla^2 \overline{A}$$

$$(\nabla \times \overline{A}) \times \overline{A} = (\overline{A} \cdot \nabla)\,\overline{A} - \frac{1}{2} \nabla(\overline{A} \cdot \overline{A})$$

$$\nabla(\Psi \Phi) = \Psi\,\nabla \Phi + \Phi\,\nabla \Psi$$

$$\nabla \cdot (\Psi \overline{A}) = \overline{A} \cdot \nabla \Psi + \Psi \nabla \cdot \overline{A}$$

$$\nabla \times (\Psi \overline{A}) = \nabla \Psi \times \overline{A} + \Psi \nabla \times \overline{A}$$

$$\nabla (\overline{A} \cdot \overline{B}) = (\overline{A} \cdot \nabla) \overline{B} + (\overline{B} \cdot \nabla) \overline{A} + \overline{A} \times (\nabla \times \overline{B}) + \overline{B} \times (\nabla \times \overline{A})$$

$$\nabla \cdot (\overline{A} \times \overline{B}) = \overline{B} \cdot (\nabla \times \overline{A}) - \overline{A} \cdot (\nabla \times \overline{B})$$

$$\nabla \times (\overline{A} \times \overline{B}) = \overline{A} (\nabla \cdot \overline{B}) - \overline{B} (\nabla \cdot \overline{A}) + (\overline{B} \cdot \nabla) \overline{A} - (\overline{A} \cdot \nabla) \overline{B}$$

Gauss's Divergence Theorem:

$$\int_V \nabla \cdot \overline{G} \, dv = \oint_A \overline{G} \cdot \hat{n} \, da$$

Stokes's Theorem:

$$\int_A (\nabla \times \overline{G}) \cdot \hat{n} \, da = \oint_C \overline{G} \cdot d\overline{\ell}$$

EXPLICIT FORMS OF VECTOR OPERATORS

Cartesian (x, y, z):

$$\nabla \Psi = \hat{x} \frac{\partial \Psi}{\partial x} + \hat{y} \frac{\partial \Psi}{\partial y} + \hat{z} \frac{\partial \Psi}{\partial z}$$

$$\nabla \cdot \overline{A} = \frac{\partial A_x}{\partial x} + \frac{\partial A_y}{\partial y} + \frac{\partial A_z}{\partial z}$$

$$\nabla \times \overline{A} = \hat{x} \left(\frac{\partial A_z}{\partial y} - \frac{\partial A_y}{\partial z} \right) + \hat{y} \left(\frac{\partial A_x}{\partial z} - \frac{\partial A_z}{\partial x} \right) + \hat{z} \left(\frac{\partial A_y}{\partial x} - \frac{\partial A_x}{\partial y} \right)$$

$$\nabla^2 \Psi = \frac{\partial^2 \Psi}{\partial x^2} + \frac{\partial^2 \Psi}{\partial y^2} + \frac{\partial^2 \Psi}{\partial z^2}$$

Cylindrical (ρ, ϕ, z):

$$\nabla \Psi = \hat{\rho} \frac{\partial \Psi}{\partial \rho} + \hat{\phi} \frac{1}{\rho} \frac{\partial \Psi}{\partial \phi} + \hat{z} \frac{\partial \Psi}{\partial z}$$

$$\nabla \cdot \overline{A} = \frac{1}{\rho} \frac{\partial (\rho A_\rho)}{\partial \rho} + \frac{1}{\rho} \frac{\partial A_\phi}{\partial \phi} + \frac{\partial A_z}{\partial z}$$

$$\nabla \times \overline{A} = \hat{\rho} \left(\frac{1}{\rho} \frac{\partial A_z}{\partial \phi} - \frac{\partial A_\phi}{\partial z} \right) + \hat{\phi} \left(\frac{\partial A_\rho}{\partial z} - \frac{\partial A_z}{\partial \rho} \right) + \hat{z} \frac{1}{\rho} \left(\frac{\partial (\rho A_\phi)}{\partial \rho} - \frac{\partial A_\rho}{\partial \phi} \right)$$

$$\nabla^2 \Psi = \frac{1}{\rho} \frac{\partial}{\partial \rho} \left(\rho \frac{\partial \Psi}{\partial \rho} \right) + \frac{1}{\rho^2} \frac{\partial^2 \Psi}{\partial \phi^2} + \frac{\partial^2 \Psi}{\partial z^2}$$

Spherical (r, θ, ϕ) :

$$\nabla \Psi = \hat{r}\, \frac{\partial \Psi}{\partial r} + \hat{\theta}\, \frac{1}{r}\, \frac{\partial \Psi}{\partial \theta} + \hat{\phi}\, \frac{1}{r \sin \theta}\, \frac{\partial \Psi}{\partial \phi}$$

$$\nabla \cdot \overline{A} = \frac{1}{r^2}\, \frac{\partial (r^2 A_r)}{\partial r} + \frac{1}{r \sin \theta}\, \frac{\partial (\sin \theta\, A_\theta)}{\partial \theta} + \frac{1}{r \sin \theta}\, \frac{\partial A_\phi}{\partial \phi}$$

$$\nabla \times \overline{A} = \hat{r}\, \frac{1}{r \sin \theta} \left(\frac{\partial (\sin \theta\, A_\phi)}{\partial \theta} - \frac{\partial A_\theta}{\partial \phi} \right) + \hat{\theta} \left(\frac{1}{r \sin \theta}\, \frac{\partial A_r}{\partial \phi} - \frac{1}{r}\, \frac{\partial (r A_\phi)}{\partial r} \right)$$

$$+ \hat{\phi}\, \frac{1}{r} \left(\frac{\partial (r A_\theta)}{\partial r} - \frac{\partial A_r}{\partial \theta} \right)$$

$$\nabla^2 \Psi = \frac{1}{r^2}\, \frac{\partial}{\partial r} \left(r^2 \frac{\partial \Psi}{\partial r} \right) + \frac{1}{r^2 \sin \theta}\, \frac{\partial}{\partial \theta} \left(\sin \theta\, \frac{\partial \Psi}{\partial \theta} \right) + \frac{1}{r^2 \sin^2 \theta}\, \frac{\partial^2 \Psi}{\partial \phi^2}$$

C

LIST OF SYMBOLS

a, A	area	m^2
$\overline{a}$	acceleration	m/s^2
$\overline{a}$	arbitrary vector	
a_n	normal component of $\overline{a}$	
a_t, a_T	transverse components of $\overline{a}$	
$\hat{a}$	unit vector $(\overline{a}/|\overline{a}|)$	
$\overline{a}_i$	displacement vector of ith dipole	m
$\overline{A}$	vector potential	Wb/m
$\overline{B}$	magnetic flux density	Wb/m^2
B	susceptance	Ω^{-1}
c	speed of light	m/s
c_s	speed of sound	m/s
C	capacitance per unit length	F/m
C	acoustic compliance/length	N^{-1}
$C, \mathcal{C}$	capacitance	F
d_{eff}	effective dipole length	m
$d\overline{a} = \hat{n}\,da$	differential area	m^2
$d\overline{\ell}$	differential path length	m
dv	differential volume	m^3
$\overline{D}$	electric displacement field	C/m^2
D	directivity	
e	electron charge	C

540

$\overline{E}$	electric field	V/m
$\overline{\mathcal{E}}$	antenna element factor	
f	frequency	Hz
$\overline{f}$	force	N
$\overline{F}$	force density	N/m^3
F	antenna array factor	
g	gyromagnetic ratio	
G	conductance	Ω^{-1}
G	antenna gain	
$\overline{H}$	magnetic field	A/m
I	current	A
I_N	Norton equivalent current	A
$\overline{\overline{I}}$	identity matrix	
$\overline{I}$	acoustic intensity	W/m^2
j	square root of -1	
$\overline{J}$	electric current density	A/m^2
$\overline{J}_s$	surface current density	A/m
$\overline{k}$	wave vector or wave number	m^{-1}
$\overline{k}'$	Re$\{\overline{k}\}$, propagating part of wave vector	m^{-1}
$\overline{k}''$	$-$Im$\{\overline{k}\}$, attenuating part of wave vector	m^{-1}
K	bulk modulus	N/m^2
L	inductance per unit length	H/m
$\mathcal{L}$	inductance	H
m, M	mass	kg
$\overline{M}$	magnetization	A/m
M	acoustic mass/length	kg/m
n	index of refraction	
$\hat{n}$	unit normal vector	
N	particle number density	m^{-3}
p, P	fluid pressure	N/m^2
p	radiation pattern	
$\overline{P}$	polarization	C/m^2
P	total power	W
P	acoustic energy/length	J/m
P_d	dissipated power	W
q, Q	charge	C
Q	quality factor of a resonator	
(r, θ, ϕ)	spherical coordinate system	
R	resistance	Ω
R	universal gas constant	J/kg$\cdot$K
R_{rad}	radiation resistance	Ω
$\overline{S}$	Poynting vector	W/m^2

t	time	s
$\hat{t}$	unit transverse vector	
T	transmission coefficient	
T	absolute temperature	K
$\bar{u}, \overline{U}$	fluid velocity	m/s
$\bar{v}$	particle velocity	m/s
v_g	group velocity	m/s
v_p	phase velocity	m/s
v, V	voltage	V
V	volume	m^3
V_{Th}	Thevenin equivalent voltage	V
w_i	complex weighting factor of ith dipole	
w_e	stored electric energy	J
w_k	kinetic energy	J
w_m	stored magnetic energy	J
w_p	potential energy	J
W_e	stored electric energy density	J/m^3
W_k	kinetic energy density	J/m^3
W_m	stored magnetic energy density	J/m^3
W_p	potential energy density	J/m^3
(x, y, z)	Cartesian (rectilinear) coordinate system	
X	reactance	Ω
Y	admittance	Ω^{-1}
Y_n	normalized admittance	
Y_0	characteristic admittance of a transmission line	Ω^{-1}
Z	impedance	Ω
Z_n	normalized impedance	
Z_N	Norton impedance	Ω
Z_{Th}	Thevenin impedance	Ω
Z_0	characteristic impedance of a transmission line	Ω
α	imaginary component of the wave vector	m^{-1}
α_0	power attenuation coefficient	m^{-1}
γ	adiabatic exponent	
Γ	reflection coefficient	
δ	skin depth	m
δ	Dirac delta function	m^{-1}, m^{-3}
δ_{nm}	Kronecker delta function	
Δ	penetration depth	m
Δf	half-power frequency bandwidth	Hz
ϵ	permittivity	F/m
ϵ_{eff}	complex dielectric constant	F/m
ϵ_0	permittivity of free space	F/m

η	wave impedance	Ω
η_{rad}	antenna radiation efficiency	
η_s	characteristic acoustic impedance	$\text{N} \cdot \text{s/m}^3$
η_n	normalized impedance	
η_0	characteristic impedance of free space	Ω
θ_B	Brewster's angle	rad
θ_c	critical angle	rad
θ_i	angle of incidence	rad
θ_r	angle of reflection	rad
θ_t	angle of transmission	rad
κ	acoustic absorption coefficient	
μ	permeability	H/m
μ_0	permeability of free space	H/m
ν	collision frequency	s^{-1}
λ	wavelength	m
Λ	magnetic flux	$\text{T} \cdot \text{m}^2$
ρ	electric charge	C/m^3
ρ	fluid mass density	kg/m
ρ_s	surface electric charge density	C/m^2
(ρ, ϕ, z)	cylindrical coordinate system	
σ	conductivity	Ω^{-1}
σ	scattering cross-section	m^2
σ_s	surface charge density	C/m^2
τ	time	s
τ_c	charge relaxation time	s
τ_m	magnetic diffusion time	s
χ_e	electric susceptibility	
Φ	scalar electromagnetic potential	V
ω	frequency	rad/s
ω_c	cyclotron frequency	rad/s
ω_p	plasma frequency	rad/s
Ω	solid angle (steradian)	rad^2
∇	del operator	m^{-1}
∇^2	Laplacian	m^{-2}

D

RATIONALIZED MKS UNITS

Quantity	Name	Symbol	Units
length	meter	m	m
mass	kilogram	kg	kg
time	second	s	s
electric current	ampere	A	A
temperature	kelvin	K	K
frequency	hertz	Hz	s^{-1}
force	newton	N	$kg \cdot m/s^2$
pressure	pascal	Pa	N/m^2
energy	joule	J	$N \cdot m$
power	watt	W	J/s
electric charge	coulomb	C	$A \cdot s$
electric potential	volt	V	W/A
capacitance	farad	F	C/V
electric resistance	ohm	Ω	V/A
conductance	siemens	S	A/V
magnetic flux	weber	Wb	$V \cdot s$
magnetic flux density	tesla	T	Wb/m^2
inductance	henry	H	Wb/A
velocity	—	—	m/s
acceleration	—	—	m/s^2
energy density	—	—	J/m^3
electric field strength	—	—	V/m
electric flux density	—	—	C/m^2
permittivity	—	—	F/m
current density	—	—	A/m^2
magnetic field strength	—	—	A/m
permeability	—	—	H/m

Quantity	Prefix	Symbol
10^{12}	tera	T
10^{9}	giga	G
10^{6}	mega	M
10^{3}	kilo	k
10^{-2}	centi	c
10^{-3}	milli	m
10^{-6}	micro	μ
10^{-9}	nano	n
10^{-12}	pico	p
10^{-15}	femto	f
10^{-18}	atto	a

E

NUMERICAL CONSTANTS

Fundamental Constants

c	velocity of light	2.998×10^8 m/s
ϵ_0	permittivity of free space	8.854×10^{-12} F/m
μ_0	permeability of free space	$4\pi \times 10^{-7}$ H/m
η_0	characteristic impedance of free space	$376.7 \ \Omega$
e	charge of an electron	-1.6008×10^{-19} C
m	mass of an electron	9.1066×10^{-31} kg
e/m	electronic charge/mass ratio	1.7578×10^{11} C/kg
m_p	mass of a proton	1.6725×10^{-27} kg
h	Planck constant	6.624×10^{-34} J $\cdot$ s
k	Boltzmann constant	1.3805×10^{-23} J/K
R	Universal gas constant	8.31 J/mole$\cdot$K
N_0	Avogadro's constant	6.022×10^{23} molec/mole

Electrical Conductivity σ, S/m

Silver	6.14×10^7	Monel	0.24×10^7
Copper	5.80×10^7	Mercury	0.1×10^7
Gold	4.10×10^7	Sea Water	3-5
Aluminum	3.54×10^7	Distilled Water	2×10^{-4}
Tungsten	1.81×10^7	Bakelite	$10^{-8} - 10^{-10}$
Brass	1.57×10^7	Glass	10^{-12}
Nickel	1.28×10^7	Mica	$10^{-11} - 10^{-15}$
Iron (pure)	1.0×10^7	Petroleum	10^{-14}
Steel	$0.5\text{-}1.0 \times 10^7$	Fused Quartz	$< 2 \times 10^{-17}$
Lead	0.48×10^7		

Relative Dielectric Constant ϵ/ϵ_0 at 1 MHz

Styrofoam (25% filler)	1.03	Low-loss glass	4.1
Firwood	1.8-2.0	Nonex glass	4.7
Paper	2.0-3.0	Pyrex glass	5.1
Petroleum	2.1	Muscovite (mica)	5.4
Paraffin	2.1	Mica	5.6-6.0
Teflon	2.1	Magnesium silicate	5.7-6.4
Rubber	2.3-4.0	Porcelain	5.7
Polystyrene	2.55	Aluminum oxide	8.8
Plexiglas	2.6-3.5	Diamond	16.5
Fused quartz	3.78	Distilled Water	81.1
Vycor glass	3.8	Titanium dioxide	100

INDEX

A reference to the principal definition of a term is shown in *italic*; a reference to a problem or example is shown in **bold**.